Topics in Igneous Petrology

Jyotisankar Ray · Gautam Sen · Biswajit Ghosh
Editors

Topics in Igneous Petrology

A Tribute to Professor Mihir K. Bose

Springer

Editors
Jyotisankar Ray
Department of Geology
University of Calcutta
Kolkata, India
jsray65@hotmail.com

Biswajit Ghosh
Department of Geology
University of Calcutta
Kolkata, India
bghosh_geol@hotmail.com

Gautam Sen
Department of Earth Sciences
Florida International University
Miami, FL, USA
seng@fiu.edu

ISBN 978-90-481-9599-2 e-ISBN 978-90-481-9600-5
DOI 10.1007/978-90-481-9600-5
Springer Dordrecht Heidelberg London New York

Cover Illustration: Pillow basalt from Chhitradurga green stone belt, South India. Photograph taken by Biswajit Ghosh

Printed on acid-free paper

Springer is part of Springer Science+Business Media (www.springer.com)

Foreword

"Topics in Igneous Petrology: A Tribute to Professor Mihir K. Bose", as the title appropriately suggests, is a *Festschrift,* ably edited by Professors Jyotisankar Ray, Gautam Sen and Dr. Biswajit Ghosh. It includes 18 scientific papers by 53 authors from around the globe. This *Festschrift* is indeed a welcome tribute, a true "festival of writing", celebrating the life and works of the honoree, the admirable Professor Bose, by his students, colleagues and acquaintances. Unfortunately, Professor Bose passed away on October 1, 2009 before seeing these papers in print. The papers cover a broad range of topics and focus on igneous petrology that is bound to attract attentions of many igneous petrologist-geochemist, and certainly Professor Bose would have found himself drawn to many of these papers with utmost interest.

Professor Bose was born in Calcutta (now Kolkata), India on September 1, 1933 and attended the University of Calcutta for all his academic training, receiving B.Sc., and M.Sc. degrees in Geology. He joined the Department of Geology in Presidency College, Calcutta in 1956 as a Lecturer and taught there throughout his professional career. While teaching, he also enrolled in a doctoral program under the supervision of Professor S. Ray to do research on the alkalic igneous rocks of Koraput in Orissa for which he was awarded a D.Phil. degree in 1965, also from Calcutta University. In the same year he published a paper in the journal *Nature* on the differentiation of alkali basaltic magma, thus earning the attention of the international petrological community, including that of Professor T.F.W. Barth of the University of Oslo in Norway, who invited Dr. Bose to Oslo in 1967 as a post doctoral fellow.

Professor Bose carried out intensive petrological, mineralogical and structural research of a number of plutonic magmatic complexes along the Eastern Ghats Belt in India. Over the course of his distinguished career, starting with the 1965 paper in Nature, he was mostly known in India and abroad as a petrologist of alkalic igneous rocks, contributing many papers on the various nepheline syenite complexes along the Eastern Ghats and also those associated with the Deccan Volcanic Province. Nepheline syenite became widely known among the undergraduate students in Geology in Presidency College, and emerged as the *roche du jour* in geological circles. Professor Bose's fascination for alkalic magmatic rocks led him to investigate them within the vast tholeiitic lavas of the Deccan Traps. This inquiry resulted in a seminal paper in 1980 on alkaline magmatism in the Deccan volcanic province, published in the Journal of the Geological Society of India. He was also a pioneer in India in introducing geochemical modeling and numerical analysis in petrogenetic studies.

Throughout his professional career, Professor Bose received many awards for his innovative research in petrology and mineralogy, beginning with the Indian National Mineral Award in 1972 and the 1976 Bhatnagar Prize in the Earth Sciences in India. He was elected a Fellow of the Indian National Science Academy, and the Indian Academy of Sciences. He received the P.N. Bose Memorial Gold Medal of the Asiatic Society in 2006. Professor Bose was known and respected as a scientist in India and abroad. He was most happy doing science, and teaching and helping his students. On student appeal, later in his teaching career, he wrote a much needed text book on *Igneous Petrology* with relevant Indian context, published by the World Press, Calcutta in 1997. He was an excellent advisor and mentor to his students, and a generous, enthusiastic colleague and collaborator. He will be missed by those who knew him, but his legacy will prevail in his numerous publications and in the writings of the contributors to this *Festschrift.*

Professor Asish R. Basu
Department of Earth and Environmental Sciences
University of Rochester
Rochester, New York 14627, USA

Acknowledgements

We profusely thank all the contributors of this volume for making this special publication a grand success. We are grateful to all reviewers (see next page) who kindly devoted their valuable time to review the submitted papers at a reasonably short time.

We profusely thank Petra D. van Steenbergen and Hermine Vloemans, of Springer for their ever-helping attitude and timely suggestions. We are grateful to Prof. Somdev Bhattacharji of Geology Department, Brooklyn College of the City University of New York, Brooklyn, New York, USA who provided his erudite guidance and thought-provoking ideas during the nucleation-time of this publication.

We are thankful to Sohini Ganguly, Mousumi Banerjee and Abhishek Saha (research-fellows of the Department of Geology, University of Calcutta) for their assistance during final editorial processing of the papers.

Last but not the least; we thank all our colleagues, friends, students and well wishers who helped us in any form or other and provided moral support (and encouragement) in connection with this publication.

Jyotisankar Ray
Gautam Sen
Biswajit Ghosh

The editors are grateful to the following reviewers who kindly provided erudite reviews of the papers submitted for this volume. Their help, support and co-operation are being acknowledged with profound thanks.

Arndt, Nicholas T.: Universite de Rennes, France
Browne, Brandon L.: California State University, Fullerton, California, USA
Bhushan, S.K.: Executive Director, RMM Limited, Hospet, Karnataka, India
Chalapathi Rao, N.V.: Banaras Hindu University, Varanasi, India
Chatterjee, N.: Massachusetts Institute of Technology, Cambridge, USA
Christy, Andrew G.: Australian National University, Canberra, Australia
Cigolini Corrado: University di Torino, Italy
Corfu, Fernando: University of Oslo, Norway
Dixon, Jacqueline: College of Arts and Sciences, University of Miami, Florida, USA
Dostal, Jarda: St Mary's University, Halifax, Nova Scotia, Canada
Fodor, Ronald V.: North Carolina State University, Raleigh, North Carolina, USA
Frey, Frederick A.: Massachusetts Institute of Technology, Cambridge, Massachusetts, USA
Garcia, Mike: Department of Geology and Geophysics, University of Hawaii, USA
Heine, Christian: Drammensveien, Oslo, Norway
Higgins, Michael D.: Université du Québec à Chicoutimi, Québec, Canada
Hollings, Pete: Lakehead University, Canada
Iyer, Sridhar, D.: National Institute of Oceanography, Goa, India
Kaneoka Ichiro: Earthquake Research Institute, University of Tokyo, Japan
Karmalkar Nitin R.: University of Pune, India
Mogessie, Aberra: University of Graz, Austria
Maluski, Henri: USTL, Laboratoire de Geochronologie, Geochimie et Petrologie, Montpellier, France
Mohanty, William K.: Indian Institute of Technology (IIT), Kharagpur, India
Morse, S.A.: University of Massachusetts, Massachusetts, USA
Mukhopadhyay, Ranadhir: National Institute of Oceanography, Goa, India
Newton, Robert: University of California, Los Angeles, USA
Pandey, Kanchan: Indian Institute of Technology (IIT), Bombay, India
Pal, Supratim: Presidency College, Kolkata, India
Pal Tapan: Geological Survey of India, Eastern Region, Kolkata, India
Polat, Ali: University of Windsor, Canada
Putirka, Keith: California State University, Fresno, California, USA
Ramachandra, H.M.: Geological Survey of India, Bangalore, India
Rowe Michael: University of Iowa, Iowa, USA
Sheth, H.C.: Indian Institute of Technology (IIT), Bombay, India
Sinton, John: Department of Geology and Geophysics, University of Hawaii, USA
Steinberger, Bernhard: Norges geologiske undersøkelse, Trondheim, Norway
Tappert, Ralf: University of Adelaide, South Australia

Verma, Surendra P.: Centro de Investigación en Energía, Universidad Nacional Autónoma de México
Wessel, Paul: Department of Geology and Geophysics, University of Hawaii, USA
Whittaker, Jo: University of Sydney, Australia
Yanbin, Wang: Chinese Academy of Geological Sciences, Beijing, China
Zhou, Mei-Fu: University of Hong Kong, Hong Kong

Contents

Ophiolites

Charnockites and Anorthosites

Mineralogy, Mineralization and Earth Dynamics

Contributors

Ajoy K. Baksi
Department of Geology and Geophysics, Louisiana State University,
Baton Rouge, LA 70803, USA
akbaksi@yahoo.com

Pradyot K. Bandyopadhyay
Department of Geology, Presidency College, Kolkata 700073, India
rbanerjee2483@gmail.com

Asish R. Basu
Department of Earth and Environmental Sciences, University of Rochester,
Rochester, NY 14627, USA
abasu@earth.rochester.edu

Somdev Bhattacharji
Department of Geology, Brooklyn College and Graduate Center of the
City University of New York, Brooklyn, NY 11210, USA
somdevlee4@earthlink.net

Elizabeth J. Catlos
Department of Geological Science, Jackson School of Geosciences,
University of Texas at Austin, 1 University Station,
C1100, Austin, TX 78712, USA
and
Department of Geological Engineering,
Middle East Technical University, Ankara, Turkey
e-mail: ejcatlos@gmail.com

Ramananda Chakrabarti
Department of Earth and Environmental Sciences, University of Rochester,
Rochester, NY 14627, USA; Department of Earth and Planetary Sciences,
Harvard University, Cambridge, MA 02138, USA
ramananda@gmail.com; rama@eps.harvard.edu

D. Chandrasekharam
Department of Earth Sciences, Indian Institute of Technology Bombay (IITB), Powai, Mumbai 400076, India
dchandra@iitb.ac.in

Nilanjan Chatterjee
Department of Earth, Atmospheric and Planetary Sciences, Massachusetts Institute of Technology, Room 54-1216, Cambridge, MA 02139, USA
nchat@mit.edu

Rudra Narayan Chatterjee
Department of Earth Sciences, Indian Institute of Technology Bombay (IITB), Powai Mumbai 400076, India
rudra025@gmail.com

Jaroslav Dostal
Department of Geology, Saint Mary's University, Halifax, Nova Scotia B3H 3C3, Canada
jdostal@smu.ca

Chandra S. Dubey
Department of Geology, University of Delhi, Delhi 110007, India
csdubey@gmail.com

Keppie J. Duncan
Instituto de Geologia, Universidad Nacional Autonoma de Mexico, 04510 Mexico D.F., Mexico
duncan@servidor.unam.mx

Robert A. Duncan
College of Oceanic and Atmospheric Sciences, Oregon State University, Corvallis, OR 97331, USA
rduncan@coas.oregonstate.edu

Raymond A. Duraiswami
Department of Geology, University of Pune, Pune 411007, India
raymond_d@rediffmail.com

Fareeduddin
Geological Survey of India, AMSE Wing, Kumaraswamy Layout, Bangalore 560078, India
fareedromani@hotmail.com

N.C. Ghose
Formerly Patna University, G/608, Raheja Residency, Koramangala, 3rd Block, Bangalore 560034, India
ghosenc2008@gmail.com

Biswajit Ghosh
Department of Geology, University of Calcutta, 35, Ballygunge Circular Road, Kolkata 700019, India
bghosh_geol@hotmail.com

Aspen Gillam
Department of Geology, Portland State University, Portland, OR 97207, USA

Trupti Gurav
Department of Earth Sciences, Indian Institute of Technology Bombay (IITB), Powai, Mumbai 400076, India
trupti@iitb.ac.in

Stephen E. Haggerty
Department of Earth and Environmental Science, Florida International University, Miami, FL 33155, USA
haggerty@fiu.edu

Miruts Hagos
Department of Lithospheric Research, Center for Earth Sciences, University of Vienna, Althanstrasse 14, A-1090 Vienna, Austria
miruts2005@yahoo.com

Dhanapati Haldar
Geology Department, Presidency College, Kolkata 700073, India

Kurkura Kabeto
Department of Earth Sciences, Mekelle University, P.O. Box 231, Tigray, Ethiopia
kurkura57@yahoo.com

Crispin Katongo
Department of Lithospheric Research, University of Vienna, Althanstrasse 14, A-1090 Vienna, Austria

E.A. Khalaf
Geology Department, Cairo University, Cairo, Egypt
ezz_khalaf@hotmail.com

M. Khalaf
Nuclear Material Authority (NMA), Kottoymia, Egypt

N. Kochhar
Department of Geology, Panjab University, Chandigarh 160014, India
nareshkochhar2003@yahoo.com

Christian Koeberl
Department of Lithospheric Research, Center for Earth Sciences, University of Vienna, Althanstrasse 14, A-1090 Vienna, Austria
christian.koeberl@univie.ac.at

Friedrich Koller
Department of Lithospheric Research, University of Vienna, Althanstrasse 14, A-1090 Vienna, Austria
friedrich.koller@univie.ac.at

Nicholas W.D. Massey
Geological Survey Branch, British Columbia Ministry of Energy and Mines, Victoria, British Columbia V8W 9N3, Canada
Nick.Massey@gov.bc.ca

Leone Melluso
Dipartimento di Scienze della Terra Università di Napoli Federico II, Via Mezzocannone 8, 80134 Napoli, Italy
melluso@cds.unina.it

J. Brendan Murphy
Department of Earth Sciences, St. Francis Xavier University, Antigonish, Nova Scotia B2G 2W5, Canada
bmurphy@stfx.ca

Yukihisa Nishizono
West Japan Engineering Consultants Inc., Fukuoka 810-0004, Japan
y-nishizono@wjec.co.jp

Theodoros Ntaflos
Department of Lithospheric Research, University of Vienna, Althanstrasse 14, A-1090 Vienna, Austria
theodoros.ntaflos@univie.ac.at

F. Oraby
Nuclear Material Authority (NMA), Kottoymia, Egypt

Frank Ramos
Department of Geological Sciences, New Mexico State University, Las Cruces, NM 88003, USA
framos@nmsu.edu

Jyotiranjan S. Ray
Physical Research Laboratory (PRL), Navrangpura, Ahmedabad 380009, India
jsray@prl.res.in

Jyotisankar Ray
Department of Geology, University of Calcutta, 35 Ballygunge Circular Road, Kolkata 700 019, India
jsray65@hotmail.com

Kaan Sayit
Department of Geological Engineering, Middle East Technical University, Ankara, Turkey
ksayit@metu.edu.tr

Gautam Sen
Vice Provost, Research & Graduate Studies, American University of Sharjah, PO Box 26666, Sharjah, UAE
gsen@aus.edu

Piyali Sengupta
Department of Geology, University of Calcutta, 35 Ballygunge Circular Road, Kolkata 700 019, India; Department of Geology, Presidency College, 86/1, College Street, Kolkata 700073, India
piysen@yahoo.co.in

Sam F. Sethna
Department of Geology, St. Xavier's College, Mumbai 400001, India
samsethna@gmail.com

Hetu C. Sheth
Department of Earth Sciences, Indian Institute of Technology Bombay (IITB), Powai, Mumbai 400076, India
hcsheth@iitb.ac.in

Poovalingam Sivasubramanian
Marine Geochemistry Research Lab,
VO Chidambaram College, SPIC Research Center, Tuticorin 628008, India
spsiva@hotmail.com

Martin J. Streck
Department of Geology, Portland State University, Portland, OR 97207, USA
streckm@pdx.edu

Francis Tembo
School of Mines, Department of Geology, University of Zambia, P.O. Box 32379, Lusaka, Zambia
ftembo@mines.unza.zm

Sadato Ueki
Department of Geophysics, Tohoku University, Sendai 980-8578, Japan
ueki@aob.gp.tohoku.ac.jp

P. Senthil Kumar
National Geophysical Research Institute (NGRI), Council of Scientific and Industrial Research (CSIR), Uppal Road,
Hyderabad 500007, India
senthilngri@yahoo.com

G. Vallinayagam
Department of Geology, Kurukshetra University, Kurukshetra 136119, India
gvallinayagam@rediffmail.com

Akira Yamada
Geodynamics Research Center, Ehime University, Matsuyama 790-8577, Japan
yamada@sci.ehime-u.ac.jp

Dapeng Zhao
Department of Geophysics, Tohoku University, Sendai 980-8578, Japan
zhao@aob.gp.tohoku.ac.jp

Haibo Zou
Department of Earth and Space Sciences, University of California, Los Angeles, CA 90095, USA; Department of Geology and Geography, Auburn University, Auburn, AL 36849, USA
haibo.zou@auburn.edu

Plume and Hotspots

Chapter 1
Upper Triassic Karmutsen Formation of Western Canada and Alaska: A Plume-Generated Oceanic Plateau Formed Along a Mid-Ocean Ridge Nucleated on a Late Paleozoic Active Margin

Jaroslav Dostal, J. Duncan Keppie, J. Brendan Murphy, and Nicholas W.D. Massey

Abstract The 1–6 km thick Upper Triassic Karmutsen basalts extend ca. 2,400 km along the Cordillera from British Columbia to Alaska and are characteristic of the Wrangellia terrane. The basalts were erupted in less than 3.5 million years following rapid >1 km uplift of an extinct Devono-Carboniferous island arc. Eliminating transcurrent dispersal reduces its original distribution to an elliptical area, ca. 1,200 × 700 km^2 in size. The convergent tectonic setting is inferred to be associated with amalgamation of Wrangellia out in the Pacific Ocean with its accretion to Laurentia occurring prior to Middle Jurassic. Geochemical data characterize the basalts as high- and low-Ti within-plate tholeiites derived from two components, an asthenospheric mantle plume at a depth ca. 80–100 km and partial melt of depleted lithospheric mantle enriched during the Paleozoic arc magmatism or of lower crust. Assuming that the ocean floor above the Paleozoic arc lay at a depth of ca. 4.5 km (i.e. 1 km above normal ocean floor) before passage over a plume, and accounting for extrusion of 6 km of basalt, the Karmutsen seamount experienced ca. 4 km of thermal uplift, was ca. 90 km wide and yielded a magma volume of ca. 15,600 km^3. Although the 90 km width is similar to that on Vancouver Island, the predicted

J. Dostal (✉)
Department of Geology, Saint Mary's University, Halifax, Nova Scotia B3H 3C3, Canada
e-mail: jdostal@smu.ca

J.D. Keppie
Instituto de Geologia, Universidad Nacional Autonoma de Mexico, 04510 Mexico D.F., Mexico
e-mail: duncan@servidor.unam.mx

J.B. Murphy
Department of Earth Sciences, St. Francis Xavier University, Antigonish, Nova Scotia B2G 2W5, Canada
e-mail: bmurphy@stfx.ca

N.W.D. Massey
Geological Survey Branch, British Columbia Ministry of Energy and Mines, Victoria, British Columbia V8W 9N3, Canada
e-mail: Nick.Massey@gov.bc.ca

J. Ray et al. (eds.), *Topics in Igneous Petrology*, DOI 10.1007/978-90-481-9600-5_1,

volume is only $^{1}/_{60}$th of the estimated volume. This discrepancy could be explained if the 2,400 km length of the volcanic belt represents a migrating volcanic chain but no diachronism has been detected. Even using the ca. 1,200 × 700 km^2 areal dimension prior to sinistral dispersal yields an unrealistic model plume size. This suggests that the Late Triassic magmatism formed as an oceanic plateau over a triple point or mid-oceanic ridge segment. However, the Karmutsen flood basalts are unusual as they formed over an extinct oceanic arc, which suggests that triple point or ridge may have nucleated on the extinct arc or trench. The modeling also shows that the bulk density of the lithosphere on which the oceanic plateau rests is never less than the asthenospheric density suggesting several possibilities for the plateau's preservation: (1) that the oceanic plateau jammed the trench causing the trench to jump oceanwards: this is consistent with the oceanic arc basement and the unimodal flood basalt magmatism; (2) that Wrangellia formed on the outer margin of Laurentia rather than in the middle of the Pacific Ocean: this is supported by correlation of the Devono-Carboniferous arc basement, but not with the absence of flood basalts on Laurentia; and (3) that the plateau and its arc basement may have been decapitated from the underlying mantle rather than subducted, however this latter possibility is inconsistent with the mild deformation and lack of a basal thrust.

1.1 Introduction

The mantle plume concept is generally regarded as a narrow fixed upwelling, solid-state current originating at the core–mantle or upper–lower mantle boundary that leaves a volcanic chain showing an age progression as the lithospheric plate moves relative to the plume (e.g., Hawaii-Emperor volcanic island chain: Wilson 1973). Although the very existence of mantle plumes has recently been questioned because, among other things they were not detected by seismic data and could not be thermally modeled (e.g., Anderson 2005; and references therein), more recent studies have overcome these difficulties. Thus, an alternate finite-frequency tomographic method for analyzing seismic data has detected roughly circular plumes, 200–800 km in diameter, rising from near the core–mantle boundary and from ca. 670 km depth (Montelli et al. 2004). Also, Farnetani and Samuel (2005) show that thermo-chemical plume models are viable, however, they cannot be applied to ancient plumes and so other criteria are necessary for their identification. In this paper, we examine one of the largest "fossil" plumes in the world, the Triassic Karmutsen Formation (Wrangellia terrane, British Columbia to Alaska), and show that the combination of characteristics, such as the high rate of magma generation, geochemical and thermal modeling, are most compatible with a plume origin. We here present new geochemical data to supplement that already available (Barker et al. 1989; Lassiter et al. 1995; Nixon et al. 2008). Furthermore, the Karmutsen flood basalts are unusual because they rest on an arc basement rather than upon the ocean or continental crust.

1.2 Geological Setting

Upper Triassic flood basalts are characteristic of the Wrangellia terrane (Fig. 1.1) and consist of a sequence bracketed by fossil ages that is ca. 3.5 km thick, subaerial and ca. 231–225 Ma in Alaska, ca. 1 km thick, submarine-subaerial and ca. 236–225 Ma in the Yukon, to ca. 6.3 km thick, mainly submarine and ca. 228–224 Ma on Vancouver Island (Greene et al. 2004, 2005, 2009; Nixon et al. 2008). Wrangellia comprises several large blocks, with those on Vancouver Island and in the Wrangell Mountains being the largest (Fig. 1.1). Barker et al. (1989) and Panuska (1990)

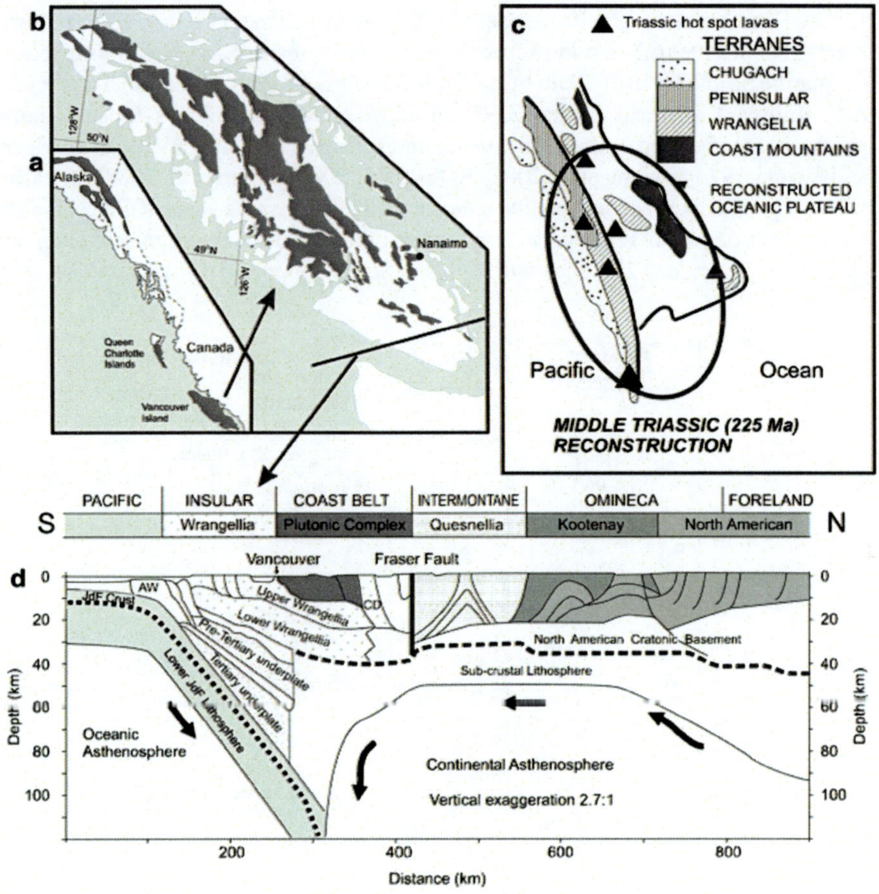

Fig. 1.1 (**a**) Distribution of the Late Triassic tholeiites in the northern Cordillera; (**b**) Geological map of Karmutsen Formation on Vancouver Island (Modified after Nixon et al. 2008); (**c**) Late Triassic plume magmatism after removal of sinistral displacements (Modified after Keppie and Dostal 2001); (**d**) Cross-section across southern British Columbia showing Wrangellia which is capped by the Late Triassic Karmutsen Formation (After Monger and Price 2002). *JdF* Juan de Fuca oceanic plate, *AW* accretion wedge, *CD* Cadwallader terrane

estimated the volume of flood basalts to be 64,000 – >1,000,000 km³, respectively, and paleomagnetic data suggest that they were erupted in the Pacific Ocean at a latitude of 10–17° (Irving and Wynne 1991 and references therein). A Middle Jurassic conglomerate (ca. 180 Ma) in Wrangellia contains pebbles probably derived from the neighbouring Chilliwack terrane, which has Middle–Upper Triassic rocks that have been correlated with the peri-Laurentian Stikine arc (Monger and Struik 2006). These observations indicate that Wrangellia was accreted to the Laurentian margin by the Middle Jurassic, and was subsequently dispersed sinistrally along the orogen (Jones et al. 1977; Monger et al. 1982, 1994). Resurgence of arc volcanism in Wrangellia is recorded in the Early-Middle Jurassic Bonanza arc (DeBari et al. 1999).

The flood basalts of the Karmutsen Formation on Vancouver Island (Fig. 1.2) are mainly submarine, 6.1–6.6 km thick tholeiites and consist of ca. 2,900 m of basal submarine pillow lavas overlain by ca. 600–1,100 m of pillow breccia and aquagene tuff, followed upwards by ca. 2,600 m of massive basalt flows (occasionally pillowed) interbedded with shallow water and subaerial sedimentary rocks (Carlisle and Suzuki 1974; Greene et al. 2004, 2005, 2006, 2009; Nixon et al. 2008). Locally, sills, plugs and dikes accompany the volcanic rocks; however, a sheeted dike complex has not been observed. Fossils constrain the age of the basalts between middle Ladinian (Middle Triassic: ca. 230 Ma) and Carnian-Norian (Upper Triassic: ca. 224 Ma)

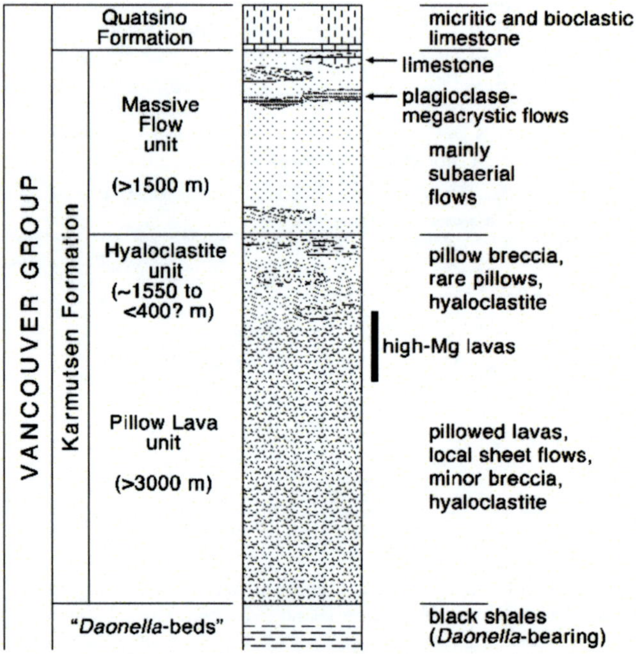

Fig. 1.2 Triassic stratigraphy on Vancouver Island (Modified after Massey 1995a–c; Nixon et al. 2008)

(Carlisle and Suzuki 1974), and suggest that they were erupted rapidly in ca. 2.5–3.5 million years. According to Nixon et al. (2008) the cessation of volcanism was followed by gradual submergence culminating in the deposition of a Late Triassic platform carbonate, known as the Quatsino Formation. Upper Triassic sedimentary rocks interbedded with and overlying the uppermost part of the Karmutsen Formation show a gradually increasing depth of deposition estimated to be no deeper than 60 m (Carlisle and Suzuki 1974; Richards et al. 1991).

Beneath the flood basalts, Middle Triassic rocks consist of <100 m thick, basinal argillite, siltstone and bivalve-bearing limestone (Richards et al. 1991). In Alaska and southwestern Yukon, the basaltic rocks that are underlain by mid- to Upper Permian argillite and radiolarian chert, which indicates deposition below the calcium carbonate compensation depth, i.e. >1 km below the surface of the sea (Richards et al. 1991; Greene et al. 2004). In turn, these rocks rest upon Lower Permian shallow water limestone, sandstone and shale that overlie 2–3 km thick late Pennsylvanian to earliest Permian arc (Skolai Group) volcanic and plutonic rocks consisting of mafic and felsic flows and tuffs (Berg et al. 1972). On Vancouver Island, Mississippian to Permian sedimentary rocks (Buttle Lake Group) unconformably overlie 2,500 m of Devonian-Lower Mississippian arc-related intermediate to felsic volcanic rocks and chert that rest upon mafic volcanic rocks (Sicker Group; Richards et al. 1991; Greene et al. 2005; Nixon et al. 2008) dated at 370–355 Ma (Parrish and McNicoll 1992; Brandon et al. 1986; Sluggett and Mortensen 2003).

Lassiter et al. (1995), Greene et al. (2009) and Nixon et al. (2008) published geochemical data for the Karmutsen Formation in Vancouver Island and Alaska-Yukon. Based on the geochemical parameters, particularly isotopic data, Lassiter et al. (1995) concluded that the Karmutsen flood basalts originated in a mantle plume. In order to augment the geochemical database and provide additional data for thermal modeling, we sampled another section of the Karmutsen Formation on Vancouver Island (Fig. 1.1).

1.3 Petrography

The Karmutsen volcanic rocks in southern Vancouver Island are dark grey feldspar-phyric basalt. They form pillowed flows, pillowed basalt breccias and hyaloclastite breccias that are interbedded with massive flows and sills. Massive flows predominate in the upper part of the formation whereas mainly pillowed flows are found in lower parts. Pillows are usually 1–2 m in size, rounded to irregular in shape and loosely packed with intra-pillow hyaloclastite. The petrography of the Karmutsen volcanic rocks is relatively uniform throughout the sequence (Barker et al. 1989). The basaltic rocks are typically plagioclase- and clinopyroxene-phyric with less than 10% phenocrysts. Plagioclase (labradorite to bytownite) is frequently ragged or forms glomerophenocrysts enclosed in a fine-grained groundmass. Olivine phenocrysts are rare and are typically pseudomorphed by serpentine.

However, pyroxene and plagioclase are mostly fresh. The groundmass contains plagioclase, pigeonite, clinopyroxene, chlorite and Fe–Ti oxides. Many flows are amygdaloidal. Amygdules are usually filled by chlorite, calcite or epidote or rare pumpellyite or prehnite.

The volcanic rocks are associated with mafic subvolcanic sills, dikes and plugs (Massey 1995a). These mafic intrusions are coeval and probably fed the Karmutsen volcanic pile (Massey 1995b). The thickness of the dikes and sills ranges from <1 to 200 m. They consist of fine to coarse-grained, equigranular to plagioclase-phyric gabbro with mineralogy comparable to the Karmutsen lavas. Mafic phenocrysts are generally absent. Locally, the gabbros show compositional layering and/or cumulate textures.

1.4 Analytical Notes

Representative samples were selected from a suite collected by Massey (1995a–c) during the mapping of southern Vancouver Island in the scale of 1:50,000. Major and some trace (Rb, Sr, Ba, Ga, Zr, Nb, Y, Cr, Ni, Sc, V, Cu and Zn) elements were analyzed by X-ray fluorescence. Twenty-three samples were also analyzed by an inductively coupled plasma mass spectrometry for the rare-earth elements (REE), Hf, Ta and Th (Table 1.1). The precision and accuracy are discussed in Dostal et al. (1986, 1994). In general, they are less than 5% for major elements and 2–10% for trace elements.

1.5 Alteration

The Karmutsen volcanic and associated subvolcanic rocks were affected to varying degrees by secondary processes including zeolite facies metamorphism, which modified the chemical composition of these rocks (Massey 1995a–c). For example, several samples yielded elevated LOI values and display secondary mobility of alkali metals indicated by the wide range of K and Rb contents (Table 1.1). The concentrations of most major elements, high-field-strength elements (HFSE, i.e. Ti, Nb, Ta, Zr, Hf), REE and transition elements are thought to reflect the primary magmatic distribution. When these elements including Y and Th are plotted against Zr, which is considered to be a good indicator of the fractionation and is apparently immobile under most metamorphic conditions (e.g., Winchester and Floyd 1977; Dostal et al. 1980), they display distinct correlations (Fig. 1.3). Similarly, these elements display coherent primitive mantle- and chondrite-normalized patterns (e.g., Figs. 1.4 and 1.6). Remobilization during metamorphism is unlikely to produce such a consistent result. The consistency of these trends and their similarities to those of modern volcanic rocks suggest that the distribution patterns of these elements were not significantly modified.

Table 1.1 Chemical analyses of rocks of the Karmutsen Formation

	High-Ti basalt							High-Ti gabbro				Low-Ti basalt		Low-Ti gabbro		
Sample	69	101	46	15	28	6	24	79	84	47	89	70	88	29	32	61
SiO_2 (wt%)	48.11	50.41	48.49	49.57	49.52	49.03	50.25	46.73	48.22	48.85	48.84	47.26	52.22	49.06	48.19	49.52
TiO_2	1.56	1.68	1.70	1.51	1.62	1.44	2.46	0.99	1.33	1.63	1.76	0.90	1.14	1.92	1.71	1.81
Al_2O_3	13.52	13.55	14.77	14.48	14.32	15.68	12.13	15.00	15.62	17.57	16.48	16.81	16.37	14.53	13.95	14.96
Fe_2O_3	3.67	2.93	2.63	0.00	5.90	4.06	9.45	1.06	1.04	1.90	1.55	3.07	3.24	11.29	10.68	3.81
FeO	7.83	8.41	9.05	11.82	6.27	6.62	4.25	7.90	9.08	8.13	9.00	6.93	6.56	2.12	2.62	8.54
MnO	0.16	0.25	0.20	0.20	0.21	0.18	0.24	0.16	0.18	0.16	0.18	0.18	0.20	0.26	0.22	0.19
MgO	8.75	5.97	6.06	7.25	6.08	6.46	5.99	9.10	6.96	5.03	5.31	4.91	3.69	7.03	6.87	6.72
CaO	8.35	11.70	11.39	12.28	12.23	11.35	10.97	12.51	9.68	11.31	11.64	8.72	6.11	6.20	8.58	3.19
Na_2O	3.11	2.37	2.35	2.10	1.80	2.59	2.03	1.41	2.40	2.12	2.02	2.75	4.61	2.82	3.03	4.24
K_2O	0.53	0.28	0.39	0.18	0.16	0.44	0.27	0.28	0.56	0.51	0.32	2.20	1.12	1.15	1.00	1.62
P_2O_5	0.11	0.14	0.13	0.11	0.12	0.11	0.20	0.07	0.12	0.14	0.15	0.26	0.31	0.24	0.19	0.30
CO_2	0.24	0.27	0.10	0.21	2.44	0.00	0.00	0.10	0.26	0.14	0.21	0.82	0.24	0.00	0.00	0.45
LOI	3.11	0.90	1.61	0.58	0.81	1.07	1.06	1.53	2.14	1.88	1.54	3.95	2.77	2.74	2.45	3.21
Total	99.05	98.86	98.87	100.29	101.48	99.03	99.30	96.84	97.59	99.37	99.00	98.76	98.58	99.36	99.49	98.56
Cr (ppm)	375	247	92	352	153	296	56	595	172	84	128	53	41	38	71	43
Ni	179	96	74	110	69	96	61	198	110	80	81	12	6	9	25	13
Sc	35	36	42.9	41.4	39	29	34	0	0	32.4	35.6	28	25	30	38	0
V	319	357	360	351	332	340	417	274	304	282	331	300	287	383	394	352
Cu	97	178	170	177	164	142	258	59	122	165	160	53	57	22	47	22
Rb	12	7	8	0	0	7	6	9	19	11	8	42	16	13	11	24
Ba	116	85	89	39	97	167	140	80	141	291	805	1055	412	577	425	580
Sr	582	294	268	210	218	244	235	236	387	323	256	981	486	341	320	352
Ga	21	23	19	20	20	18	21	18	17	26	22	21	19	23	21	21
Ta	0.45	0.63	0.61	0.56	0.50	0.50	0.89	0.32	0.46	0.55	0.56	0.22	0.25	0.42	0.37	0.60

(continued)

Table 1.1 (continued)

	High-Ti basalt							High-Ti gabbro				Low-Ti basalt		Low-Ti gabbro		
Sample	69	101	46	15	28	6	24	79	84	47	89	70	88	29	32	61
Nb	8.00	9.00	9.00	10.00	8.00	10.00	13.00	0.00	7.00	10.00	10.00	5.00	5.00	8.00	7.00	8.00
Hf	2.46	2.98	2.92	2.51	2.80	2.47	4.23	1.60	2.28	2.84	2.72	1.92	2.99	3.37	2.65	3.87
Zr	94	108	102	88	101	89	151	58	86	101	103	73	108	125	100	149
Y	21	23	22	21	21	19	30	15	18	21	22	21	28	29	25	34
Th	0.48	0.71	0.73	0.60	0.67	0.59	1.16	0.41	0.54	0.71	0.75	0.98	2.04	1.99	1.35	2.22
La	5.63	7.84	7.69	7.51	6.62	6.24	11.17	4.13	5.95	7.47	8.44	9.87	14.37	13.44	11.72	16.54
Ce	14.76	20.08	19.59	17.47	16.74	15.80	27.30	10.60	15.09	19.08	20.90	23.19	32.64	30.66	26.68	37.16
Pr	2.23	2.86	2.77	2.48	2.42	2.25	3.92	1.55	2.19	2.80	3.03	3.21	4.41	4.08	3.57	4.98
Nd	11.21	13.70	13.23	11.73	11.61	10.58	18.65	7.73	10.36	12.98	14.14	14.70	19.59	18.65	16.14	22.47
Sm	3.42	3.85	3.67	3.40	3.36	3.02	5.36	2.39	3.01	3.78	3.97	3.63	4.87	4.82	4.19	5.69
Eu	1.18	1.35	1.37	1.21	1.25	1.11	1.83	0.85	1.01	1.37	1.49	1.07	1.47	1.66	1.44	1.84
Gd	3.93	4.36	4.52	3.95	4.12	3.61	6.15	2.74	3.58	4.47	4.77	3.66	5.27	5.57	4.84	6.32
Tb	0.60	0.69	0.67	0.60	0.61	0.55	0.88	0.44	0.53	0.66	0.73	0.56	0.76	0.84	0.71	1.01
Dy	4.30	4.78	4.58	4.29	4.13	3.75	6.04	3.01	3.79	4.57	4.79	4.11	5.29	5.72	4.92	6.98
Ho	0.80	0.90	0.91	0.88	0.82	0.74	1.17	0.57	0.71	0.82	0.89	0.77	1.04	1.12	0.94	1.32
Er	2.22	2.58	2.63	2.50	2.28	2.05	3.16	1.58	2.00	2.30	2.43	2.25	3.17	3.19	2.67	3.82
Tm	0.32	0.35	0.34	0.31	0.32	0.29	0.45	0.22	0.28	0.32	0.33	0.35	0.45	0.47	0.38	0.56
Yb	1.98	2.34	2.27	2.12	2.01	1.86	2.74	1.39	1.75	1.98	1.99	2.23	2.87	2.89	2.38	3.59
Lu	0.31	0.34	0.33	0.32	0.32	0.29	0.41	0.21	0.27	0.30	0.29	0.35	0.45	0.47	0.36	0.54

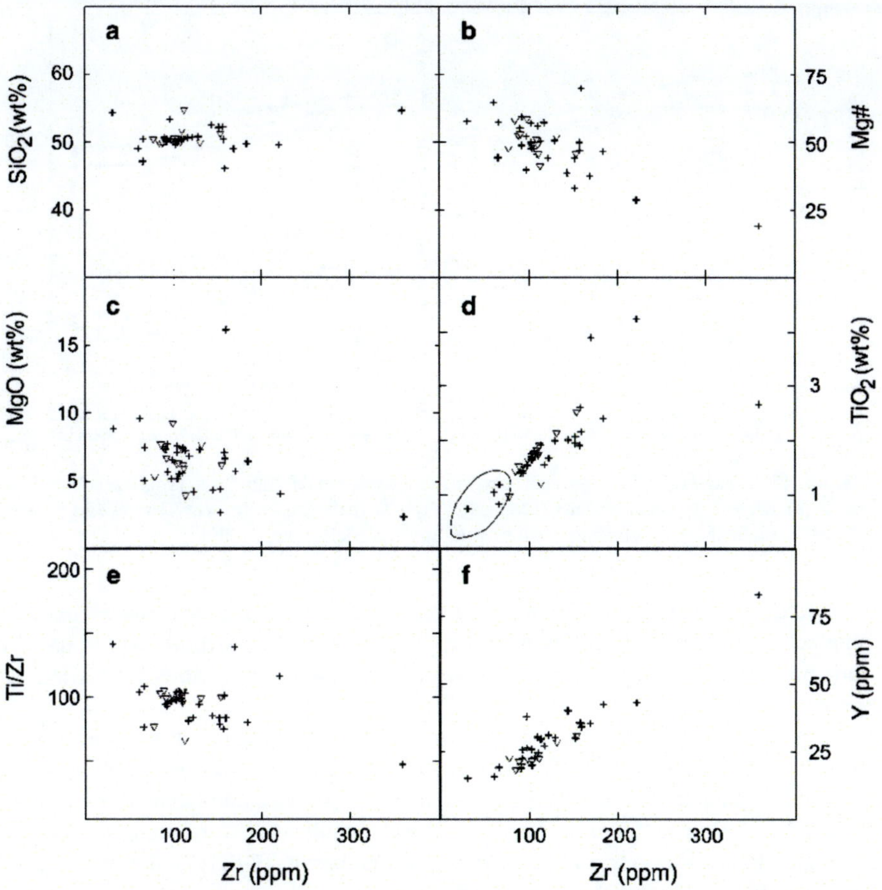

Fig. 1.3 Variations of SiO_2 (%), Mg# (100 × Mg/[Mg + Fe_{tot}]), MgO (wt%), TiO_2 (%), Ti/Zr and Y (ppm) versus Zr (ppm) in the Karmutsen volcanic (*crosses*) and subvolcanic (*triangles*) rocks. The curve in (**d**) outlines the field for island-arc tholeiites (Pearce and Cann 1973; Rollison 1993)

1.6 Geochemistry

The Karmutsen lavas are mafic rocks with <55% SiO_2; most have SiO_2 in a narrow range of 49–52%. The lavas have Mg# (=100 × (Mg/Mg + Fe_{tot})) values typically between 60 and 40 and display tholeiitic Fe- and Ti-enrichment trends with increasing differentiation (Fig. 1.3). The tholeiites include both quartz-normative and olivine-normative types. Associated subvolcanic rocks have a composition corresponding to tholeiitic basalts and are similar to those of the Karmutsen lavas. Compared to the lavas, however, the intrusive rocks display a wider compositional range (Fig. 1.3) with Mg# varying between 67 and 20. Some of the gabbros display

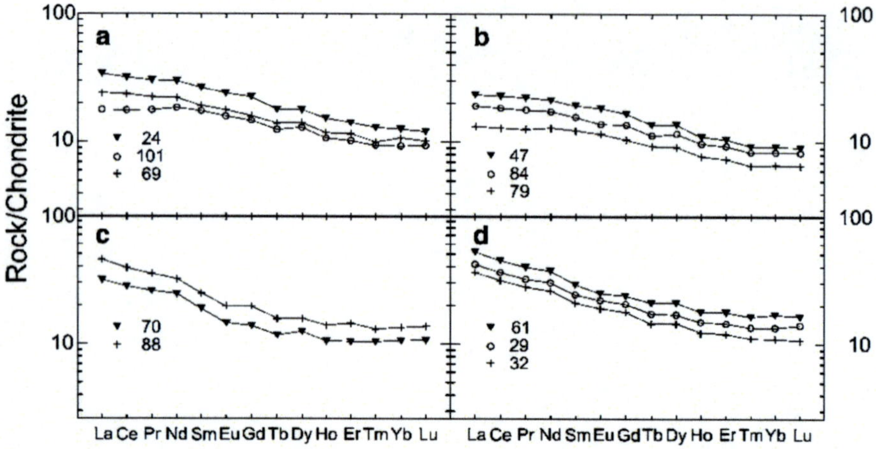

Fig. 1.4 Chondrite-normalized rare earth element abundances in the Karmutsen lavas and subvolcanic rocks. (**a**) High-Ti tholeiitic basalts (lavas); (**b**) high-Ti mafic dikes; (**c**) low-Ti tholeiitic basalts; (**d**) low-Ti mafic dikes (Normalizing values after Sun and McDonough 1989)

cumulate characteristics, such as high Al_2O_3 (>20%). Cumulate phases include plagioclase and ferromagnesian minerals. The compositional similarities of basalts and subvolcanic rocks support field observations that the intrusions are feeders of the volcanic pile.

Compared to island arc tholeiites, the Karmutsen basalts have higher abundances of Ti and other HFSE. Their TiO_2 contents range between 1% and 3% (Fig. 1.3). Compositionally, the Karmutsen lavas resemble mid-ocean ridge basalts (MORB), tholeiites from continental flood basalt (CFB) provinces such as Columbia River, Karoo and Parana, oceanic island basalts (OIB) and oceanic plateau basalts, such as those of the Nauru basin.

These Karmutsen rocks have MgO ranging between 4 and 9 wt% and underwent variable but commonly extensive fractionation. Transition trace elements in the basalts and non-cumulate subvolcanic rocks are lower than those expected for primary mantle melts (Hart and Davis 1978). Even primitive basalts with high Mg# have relatively low contents of transition elements (Table 1.1), indicating that these rocks underwent fractional crystallization. The Karmutsen rocks exhibit decreasing Mg, Ni, Cr and Sc but increasing Ti and V with increasing differentiation at almost constant Si, Al, Sr and Ti/Zr (Fig. 1.3). These trends are consistent with the fractionation of plagioclase, clinopyroxene and olivine and rules out significant titanomagnetite fractionation. Ti/V ratios in the rocks are high (>20) within the range of typical MORB, OIB and continental tholeiites (Shervais 1982).

Like CFB from the Jurassic-Cretaceous flood basalt provinces related to Gondwana (e.g., Karoo, Parana, Madagascar, Antarctica: Hawkesworth et al. 1984, 1999; Dostal et al. 1992; Jourdan et al. 2007) and the Deccan Traps (Lightfoot et al. 1990; Melluso et al. 1995; Peng et al. 1998), the Karmutsen basalts can be further subdivided according to their Ti and HFSE contents into dominant high- and minor low-Ti groups. However, since much of the Ti variation within the Karmutsen lavas and associated subvolcanic rocks is related to the degree of fractional crystallization, the differences are more apparent in incompatible element ratios rather than abundances. Using the criteria of Erlank et al. (1988) and Peate et al. (1992,1999) for CFB of Karoo and Parana, in particular the Ti/Y (ca. 410) ratio, the high-Ti basalts have Ti/Y >410 whereas low-Ti basalts have lower values of this ratio. The differences between the high- and low-Ti tholeiites of Karmutsen are reflected mainly in the distributions of incompatible trace elements. Compared to low-Ti types, the high-Ti basalts have lower Ba/Nb (<20, mostly <10), Th/Nb and La/Nb ratios and P_2O_5 but higher Zr/Y and Ti/V ratios. Like the Karoo and Parana basalts (e.g., Erlank 1984; Piccirillo et al. 1988; Jourdan et al. 2007), the two groups of tholeiites do not show any obvious differences in mineralogy or petrography or field occurrences. The non-cumulate subvolcanic rocks include equivalents of both low- and high-Ti basalts. The high-Ti basalts are the dominant rock type among the samples collected in the southern Vancouver Island area.

Chondrite-normalized REE abundances in the tholeiitic basalts are shown in Fig. 1.4. The tholeiites have REE patterns enriched in light REE (LREE), although some of gabbros and the rocks reported by Barker et al. (1989) and Lassiter et al. (1995) have flat REE patterns. The low Ti-lavas have higher LREE abundances and higher La/Sm and La/Yb ratios compared to high-Ti rocks. The REE patterns of both types (Fig. 1.4) resemble some main basaltic types of the Deccan Traps, Karoo and Parana as well as Hawaiian tholeiites and enriched MORB (E-MORB). Within both basaltic groups of the Karmutsen suite, the REE patterns are approximately parallel, although the total REE abundances are variable. The contents of REE positively correlate with the degree of fractional crystallization, although heterogeneous source and melting conditions also played a role (Fig. 1.5).

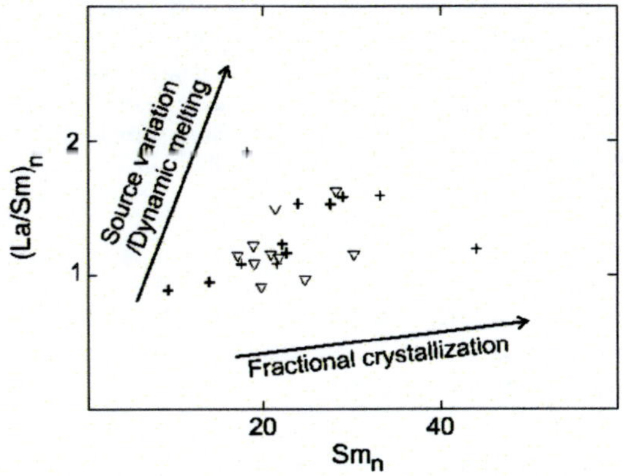

Fig. 1.5 Variation of chondrite-normalized La/Sm versus Sm abundances of the Karmutsen volcanic (*crosses*) and subvolcanic (*triangles*) rocks (Source variation and/or dynamic melting trend and fractional crystallization trend are after Pearce et al. 1995). n – normalized to chondrites

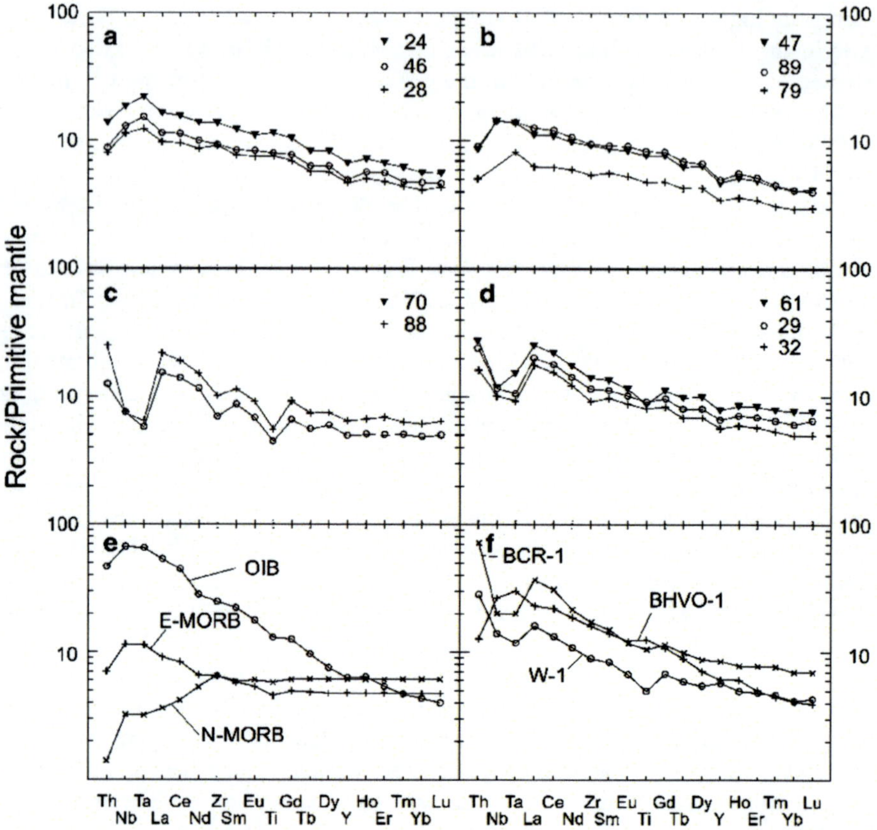

Fig. 1.6 Mantle-normalized incompatible trace element abundances in the Karmutsen lavas and subvolcanic rocks and comparative data for tholeiitic basalts. **(a)** High-Ti lavas (tholeiitic basalts); **(b)** high-Ti mafic dikes; **(c)** low-Ti tholeiitic basalts; **(d)** low-Ti mafic dikes; **(e)** comparative basalts: E-MORB – E-type MORB; N-MORB – N-type MORB; OIB – ocean island basalt (After Sun and McDonough 1989); **(f)** comparative basalts: BCR-1 – Columbia River plateau basalt, Govindaraju (1994); BHVO-1 – tholeiitic basalt from Hawaii, Govindaraju (1994); W-1 – continental tholeiite (diabase), Central Atlantic magmatic province, Centreville, Virginia, USA (After Govindaraju 1994). Normalizing values after Sun and McDonough (1989)

The two types of the tholeiitic rocks also differ in the shapes of their patterns of mantle-normalized abundances of incompatible trace elements (Fig. 1.6). The patterns of the high-Ti basalts peak at Nb–Ta, smoothly decrease to more compatible elements including the heavy REE (HREE) and Y and are depleted in Th relative to Nb. Such a shape is characteristic of OIB, E-MORB and closely resembles Hawaiian tholeiites (Fig. 1.6). Although the patterns are similar to those of the high-Ti lavas in Parana and other CFB provinces, these rocks typically have lower absolute abundances of incompatible trace elements.

The low-Ti tholeiites display a gradual increase from Lu towards large-ion-lithophile elements (LILE) accompanied by a relative depletion in Nb–Ta and Ti with respect to the LILE and LREE. They resemble low-Ti basalts from the other CFB provinces (Fig. 1.6). The patterns mimic basalts from the Bushe Formation of the Deccan Traps (Cox and Hawkesworth 1985; Lightfoot et al. 1990) and low-Ti basalts from Etendeka, Parana and Madagascar (Erlank 1984; Piccirillo et al. 1988; Dostal et al. 1992) and also from Columbia River (Hooper and Hawkesworh 1993). However, they also resemble rocks from subduction related settings, particularly as a Ta–Nb 'trough' is commonly observed on mantle-normalized diagrams for island-arc volcanic rocks (Pearce et al. 1995).

1.7 Petrogenesis

1.7.1 Evolution

The distributions and variability of incompatible elements as well as the available Pb-, Nd- and Sr isotopic data provide information on the processes that gave rise to the Karmutsen tholeiitic basalts and their source regions. Most observed compositional variations of the Karmutsen tholeiites are consistent with low-pressure fractional crystallization of plagioclase, pyroxene and olivine. The dominant plagioclase fractionation is supported by the ubiquitous occurrence of plagioclase phenocrysts in the basalts. The proportions of the separated phases are in agreement with the modal composition of cumulate gabbros from the associated plugs. However, variations of some ratios such as Zr/Nb or Zr/Th cannot be explained by fractional crystallization but require either variations in the degree of melting and/or heterogeneities in the trace element contents of the mantle source (Fig. 1.6). Likewise, the two groups of basalts cannot be related by fractional crystallization involving phenocryst phases.

Several systematic trace element variations in the basalts can reflect the effect of a variable degree of partial melting; model calculations using both garnet and spinel lherzolite source compositions are shown in Fig. 1.7. The graph suggests that the melting of spinel peridotite (mostly low-Ti types) or melting in the transition between spinel and garnet peridotite (most of high-Ti types) that was subsequently followed by fractional crystallization could generate the observed REE distribution in the two groups of the Karmutsen lavas. The spinel–garnet transition zone is typically located between ca. 60 and 80 km, except in mantle plume environments where it occurs at a depth of 80–100 km (McKenzie and O'Nions 1991). However, like fractional crystallization, this partial melting model alone cannot explain the differences between high- and low-Ti tholeiites (Fig. 1.6). The differences between element ratios, such as Zr/Nb, Th/Nb, La/Yb and La/Nb, are too large to be explained by a variable degree of melting and point to a heterogeneous source. This is also supported by the differences in the isotopic ratios (Lassiter et al. 1995).

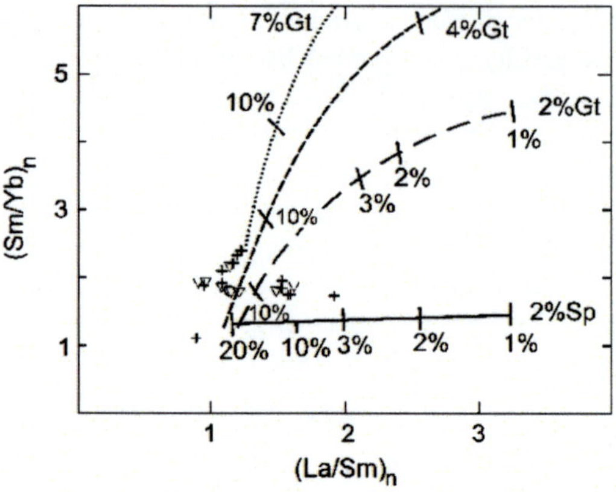

Fig. 1.7 Variation of chondrite-normalized Sm/Yb versus La/Sm ratios for the basaltic lavas (*crosses*) and non-cumulate subvolcanic mafic rocks (*triangles*) of the Karmutsen Formation. The non-modal batch melting curves for garnet and spinel peridotites are shown for comparison (Modified after Jourdan et al. 2007). The *numbers* and *tick marks* on the melting curves refer to percent partial melting. The modal composition of the garnet and spinel peridotite sources and mineral fractions in the liquids are after Jourdan et al. (2007). Source composition: 1.35 ppm La, 0.62 ppm Sm and 0.45 ppm Yb. Partition coefficients after McKenzie and O'Nions (1991). n – normalized to chondrites

1.7.2 Source Regions

The high-Ti basalts of the Karmutsen Formation have mantle-normalized patterns and elemental concentrations similar to those of oceanic island basalts, such as Hawaiian tholeiites and E-type MORB (Fig. 1.6). These patterns are usually thought to be typical of magmas generated by the melting of the asthenospheric upper mantle (e.g., Hooper and Hawkesworh 1993). The asthenospheric mantle source is also consistent with Pb-, Nd- and Sr-isotopic analyses of the Karmutsen lavas (Lassiter et al. 1995; Andrew and Godwin 1989; Andrew et al. 1991; Greene et al. 2009). Lassiter et al. (1995) inferred that the isotopic data are the results of derivation from a plume source with initial ε_{Nd} ~ +6 to +7, $^{87}Sr/^{86}Sr$ ~0.7034 and $^{206}Pb/^{204}Pb$ ~19.0 and support suggestion of Richards et al. (1991) that the high-Ti Karmutsen sequence was related to a mantle plume. A similar conclusion was inferred for the Deccan, Parana and Columbia River flood basalt provinces (e.g., Mahoney 1988; Peate et al. 1990; Hooper and Hawkesworh 1993; Campbell and Griffiths 1990).

However, melting of mantle plume cannot explain some of the geochemical characteristics of the low-Ti basalts of the Karmutsen Formation. The shapes of their incompatible trace element patterns, distinguished by negative Nb, Ta and Ti anomalies associated with high LILE/HFSE and LREE/HFSE ratios are typical of subduction-related rocks. Lassiter et al. (1995) also reported that the basalts with low Nb/Th and Nb/La ratio, which probably correspond to low-Ti types have lower initial ε_{Nd} and $^{206}Pb/^{204}Pb$, but higher $^{87}Sr/^{86}Sr$ ratios, than the dominant basaltic type. These isotope values are comparable to other low-Ti CFB characterized by a Ta-Nb 'trough' (e.g., Pearce 2008).

Many continental volcanic rocks that are not associated with contemporary subduction processes, including CFB, also have similar patterns (e.g., Dupuy and Dostal 1984; Tatsumi and Kogiso 2003; Jourdan et al. 2007; Murphy and Dostal 2007). The high LILE/HFSE component, characterized by a large negative Nb–Ta anomaly has an isotopic and chemical signature commonly associated with crustal rocks and enriched sub-continental lithospheric mantle. This component could be a partial melt from either an enriched mantle lithosphere or the crustal material via crustal contamination (Hooper and Hawkesworh 1993). On the basis of the isotopic data, Lassiter et al. (1995) favoured assimilation of arc material accompanied by fractional crystallization to account for chemical characteristics of the basalts with negative Nb–Ta anomalies. The high-Ti and low-Ti rocks have similar chemical compositions in terms of most major and trace elements (Figs. 1.3 and 1.6) and form continuous variation trends suggesting that they might be derived from the same source region but the low-Ti rocks were modified by crustal contamination. However, there is not enough available data to uniquely distinguish between the crustal contamination and an enriched lithospheric mantle source. This scenario suggests that during ascent, the asthenospheric plume triggered small-scale melting in mantle lithosphere previously enriched by subduction-related processes during the Paleozoic (possibly a mantle wedge of the Paleozoic arc) or in continental crust.

1.7.3 Geochemical Conclusions

The basalts of the Karmutsen Formation and related subvolcanic gabbros include two distinct geochemical groups of basalts, which resemble high-Ti and low-Ti types of Mesozoic continental flood basalt provinces. However, the occurrence of picritic rocks (Nixon et al. 2008) suggests that the sequence closely resembles oceanic plateau rocks (Kerr et al. 2000). The predominant high-Ti tholeiites have convex-shaped mantle-normalized trace element patterns with their peaks at Ta–Nb. These patterns are similar to those of OIB and E-type MORB implying that the high-Ti tholeiites were derived from the asthenospheric upper mantle via a mantle plume. The low-Ti basalts have more irregular mantle-normalized patterns with a Ta–Nb trough. Such patterns, as well as trace element ratios that are different from those commonly observed in mantle plume-derived basalts, suggest that the low-Ti tholeiites were predominantly derived from either lithospheric mantle or from the assimilation of arc material.

The predominance of the high-Ti basalts is consistent with the modeling of McKenzie and Bickle (1988), who concluded that during the upwelling of hot mantle plumes nearly all melting takes place in mantle beneath the mechanical boundary layer (approximately the base of the lithosphere), and that enriched signatures had to be picked up as asthenosphere-derived melts penetrated through the lithosphere. The HREE patterns (Fig. 1.4) indicate that the sources were either: (1) spinel-bearing and were at shallow levels (<60–80 km; McKenzie and O'Nions 1991) in the mantle; or more likely (2) in the transition zone between spinel and garnet field possibly at the depth of 80–100 km (McKenzie and O'Nions 1991).

A deep plume piercing the lithosphere and generating a huge volume (1,000,000 km^3; Panuska 1990) of tholeiite of the Karmutsen Formation in a short time (ca. 4 Ma) indicates an eruption rate of ca. 0.25 km^3/year. Such a rate is greater than that estimated for Hawaii (0.16 km^3/year; Watson and McKenzie 1991), but is comparable to other major oceanic plateau basalt provinces (Bradshaw et al. 1993).

Although the Karmutsen Formation was emplaced on an oceanic arc that probably formed on oceanic crust, it consists of two basaltic types similar to those of CFB provinces emplaced on continental crust. The crustal/arc imprint observed in the low-Ti basalts of CFB provinces is either due to subduction-related processes and sub-lithospheric mantle melting or through crustal contamination. The Karmutsen basalts have relatively flat REE patterns, εNd +4.8 to +7.3, and $(La/Nb)_{pm}$ (pm – normalized to primitive mantle; Sun and McDonough 1989) ca. 1, all of which are similar to oceanic plateau (flood) basalts (Kerr et al. 2000), such as those from the Ontong-Java Plateau (Stoeser 1975) or Nauru basin (Batiza 1981; Saunders 1986; Castillo et al. 1986). However, the Karmutsen basalts have higher contents of incompatible trace elements and include the low-Ti basalts with subduction/crustal contamination-related imprints, which are not commonly found in the oceanic environments. The Karmutsen tholeiites may represent transitional rocks between CFB and oceanic flood basalts.

1.8 Plume Model Test

Estimation of the properties of ancient plumes must be done indirectly because the original connection between plumes, hotspot tracks and seamounts is severed by convergent margin tectonics. However, the uplift history associated with the plume is preserved in the pre-accretionary stratigraphy of seamounts (Murphy et al. 2003). The Devonian–Permian arc history and its effects on the composition and bulk density of the lithosphere into which the Karmutsen volcanics were emplaced are an additional complication to modeling the plume, and so only semi-quantitative models can be developed. Nevertheless, if combined with stratigraphic data, this approach offers hitherto unrecognized perspectives on the plume responsible for the Karmutsen volcanic rocks.

As the Late Devonian–Carboniferous arc represented by the Sicker Group was ensimatic, it was probably developed on typical oceanic lithosphere. If so, then prior to arc activity, its crust would have been about 7 km thick, at a ca. 5.5 km depth below sea level (e.g., Parsons and Schlater 1977) and the bulk density of the oceanic lithosphere would have been ca. 3.27 g/cm^3 (Cloos 1993). Between the development of the arc in the Late Devonian and the onset the Karmutsen magmatism, ca. 5 km of strata were deposited. Although these strata are dominated by basalt, there are significant thicknesses of intermediate to felsic volcanics and clastic rocks (Greene et al. 2005, 2006, 2009). Using standard lithologic densities (e.g., Sclater and Christie 1980), a bulk density of 2.8 g/cm^3 is assigned to these strata, yielding a bulk lithospheric density of 3.25 g/cm^3. Assuming this lithosphere was in isostatic equilibrium prior to the onset of Karmutsen magmatism, this lithospheric bulk density would imply that the top of the crust was about 4.5 km below sea level in the middle Triassic.

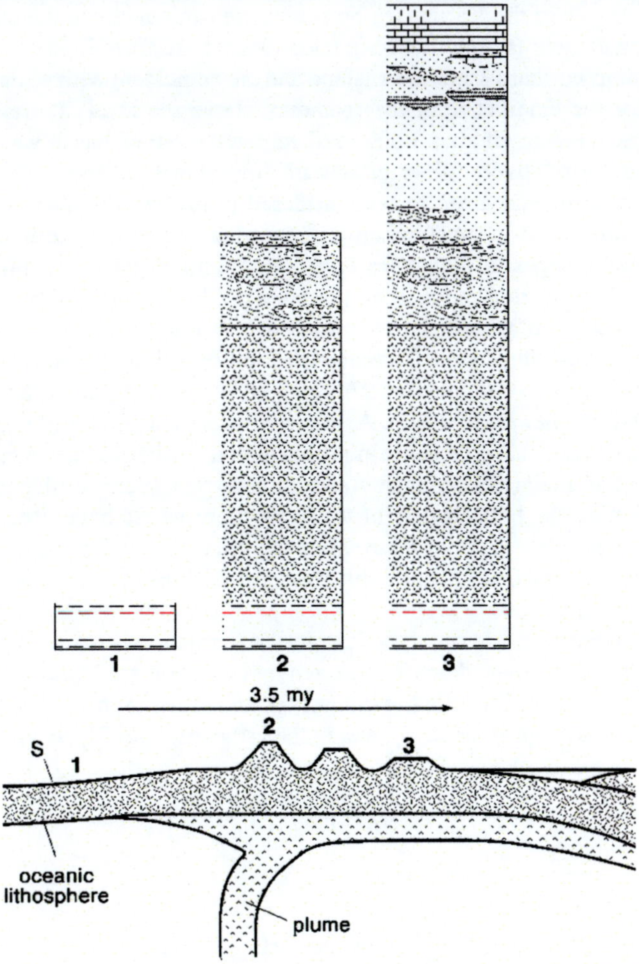

Fig. 1.8 Development of the stratigraphy of the Karmutsen Formation during the passage of a plate over an active plume. Clastic rocks containing Daeonella (1) were deposited before the plate passed over the stagnation point (S). The pillow basalt and subaerial basalts (2) were deposited by the time the plate passed over the plume centre, and the hyaloclastite and carbonate units (3) were deposited during subsidence and atoll development. Symbols are the same as in Fig. 1.2

The shallowing-upward stratigraphy of the Karmutsen Formation culminating in subaerial eruptions is indicative of a Loihi-type setting, i.e. on the upstream side of the plume centre. Thus, the Karmutsen Formation volcanic rocks record development between the stagnation point of the plume (i.e. the point where the radial velocity from the plume is equal and opposite to the mean unperturbed velocity of the asthenosphere, e.g., Richards et al. 1988; Sleep 1987) and the plume head. The rapid subsidence leading to deposition of platformal carbonate is typical of atoll development on the downstream side (Fig. 1.8).

The proximity of the Karmutsen Formation volcanic rocks to the plume head is indicated by the extrusion of high Mg lavas (MgO > 10.0 wt%, Nixon et al. 2008), which occur in northern Vancouver Island and are coincident with sedimentological evidence for the emergence of the seamount above sea level. Therefore the seamount became emergent after 3,000 m of submarine pillow basalt was extruded.

Ignoring the relatively minor effects of lithospheric flexure, three potentially significant components need to be considered to explain the shallowing-upward trend that culminated in subaerial eruptions by the end of deposition of the pillow lava unit: (a) changes in global sea level; (b) the thickness of the pillow basalts; and (c) excess elevation induced by the plume. Global sea level rose throughout most of the Early to Middle Triassic (although estimates vary from about 15–80 m) reaching a peak during the Carnian, followed by a drop of about 30–60 m during the Late Triassic (Vail et al. 1977; Haq et al. 1987; Haq and Schutter 2008; Hallam 1992; Miller et al. 2005). As the sea level change is minor compared to other uncertainties, its effects are not considered further in this analysis. If the lithosphere had maintained isostatic equilibrium, by the end of deposition of the pillow lava unit, the bulk density of the lithosphere would have been 3.24 g/cm^3, corresponding with a depth of 4 km. Thus the thickness of the pillow basalts cannot explain 4 km of the 4.5 km uplift that occurred during the deposition of the pillow basalt unit.

The 4 km excess elevation of the Karmutsen Formation is strong supporting evidence for the plume-related origin indicated by geochemistry. This excess elevation is a combination of uplift from hot lithosphere and asthenosphere. The uplift due to hot material in the lithosphere, U_L, and in the asthenosphere, U_A, is given by (Sleep 1990):

$$U_L = \alpha \Delta T_L L \rho_m / (\rho_m - \rho_w) \tag{1.1}$$

and

$$U_A = \alpha \Delta T_A A \rho_m / (\rho_m - \rho_w) \tag{1.2}$$

where α is the thermal expansion coefficient, ΔT is the excess temperature in the lithosphere or asthenosphere, L is the lithospheric thickness, A is the thickness of the asthenospheric channel of the plume, and ρ_m and ρ_w are the densities of the mantle and water.

Following Sleep (1990), we assume α is $3 \times 10^{-5}\,^\circ C^{-1}$, ρ_m is 3,300 kg/m^3 and that ΔT is 250°C for both lithosphere and asthenosphere. Although the value assigned to ΔT is taken from the theoretical treatment of Sleep (1990), this value is consistent with the presence of high Mg lavas documented by Nixon et al. (2008). The lithospheric thickness (L) is 100 km (e.g., Cloos 1993). Substituting these values into (1.1) yields an U_L of 1.0 km. Since $U_L + U_A = 4$ km, then the asthenospheric component of uplift (U_A) is 3 km. In equation 1.2, therefore, the only unknown is A, and using the above values yields A = 300 km, which is an indication of the thickness of the underplating associated with the plume.

After a further 2 km of subaerial volcanic rock deposition, the bulk lithospheric density would have been 3.23 g/cm^3, which, in the absence of a plume would have yielded been at a depth of 3.6 km. Assuming the plume maintained its vigor during this stage and contributed 4 km of uplift, the seamount would have been 400 m above sea level and therefore its height as it passed over the plume would have been about 4.9 km, measured relative to adjacent oceanic lithosphere unaffected by the plume.

According to Cloos (1993) a seamount 4.9 km high would have a width of about 90 km. A minimum volume of magma produced by the seamount is given by:

$$\int \pi(h) r\,dr = 0.5\pi(h) r^2$$

where h is the height (4.9 km) and r is the radius of the seamount (45 km), yielding a volume of ca. 15,600 km^3.

1.9 Discussion and Conclusions

Although the calculated width and volume are consistent with those of the Karmutsen Formation on Vancouver Island, they are far too small to explain the total ca. 2,400 km length of the belt or the estimated 1,000,000 km^3 volume of magma. Although the area may be somewhat reduced to ca. 1,200 × 700 km^2 by eliminating the sinistral strike-slip offsets (Keppie and Dostal 2001), it is still too large to be explained by a single plume that is stationary relative to the overriding plate. Another possibility is that the magmatism migrated as the overriding plate moved relative to the plume, however, no diachronism in the age of the magmatism has been detected. The size and geochemistry of the Late Triassic magmatism is similar to that in oceanic plateau (Kerr et al. 2000). The synchronicity of the short-lived magmatism along the whole length of the belt could be related to the development of a plume-generated oceanic plateau at a triple junction or mid-oceanic ridge segment that nucleated on or near the old trench associated with the Paleozoic arc (Fig. 1.9). The reconstructed elliptical shape of the oceanic plateau suggests that the relative motion between the overriding plate and the plume was north-south. Movement of the plate perpendicular or at a high angle to the plume track would then have led to termination of magmatism 5–10 million years after inception. Alternatively, melting of a shallow layer of near-solidus, incompatible-element-enriched asthenosphere cannot be discounted (Anderson et al. 1992).

A further complication arises from the model densities, because at no time is the bulk lithospheric density less than that of the asthenosphere. This should result in subduction of the Karmutsen oceanic plateau, as is the case for most modern analogs (Mann and Taira 2004). However, the Karmutsen plateau could have been delaminated with the upper part abducted as the lower part sinks into the mantle (Mann and Taira 2004). This model is unlikely to apply to the

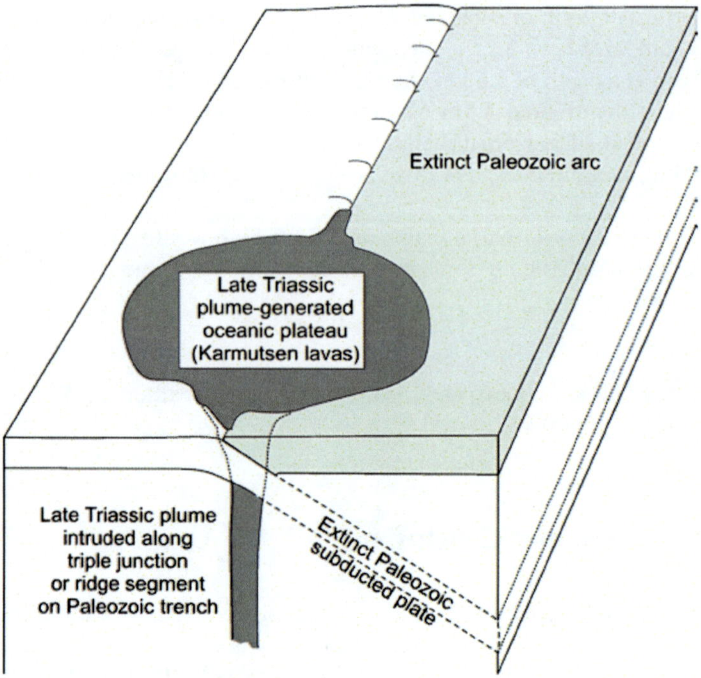

Fig. 1.9 Tectonic cartoon showing superimposition of plume-generated oceanic plateau on extinct Paleozoic arc: the plume(s) are inferred to have nucleated on the extinct arc–trench complex

Karmutsen plateau because it is only mildly deformed and no basal thrust has been encountered. Instead, Wrangellia appears to be bounded by vertical faults related to the sinistral dispersal of Wrangellia (Fig. 1.1) (Monger and Price 2002). The most likely scenario is that Wrangellia was trapped against the Laurentian margin as the subduction zone stepped oceanward. Although Monger and Price (2002) have suggested a potential correlation between the Permo-Carboniferous basement of Wrangellia and in the outer Laurentian margin, the absence of the distinctive Karmutsen Formation in Laurentia suggests they were separated by a considerable, but unknown, distance. Using the terrane tracks generated by Debiche et al. (1987) for points in the Pacific Ocean suggests that Wrangellia lay ca. 4,000–7,000 km west of Laurentia during the Late Triassic (c.f. Keppie and Dostal 2001).

Acknowledgments The continuing support of Natural Sciences and Engineering Research Council of Canada Discovery to JD and JBM and Research Capacity Development grants to JBM is gratefully acknowledged. JDK acknowledges the support of PAPIIT (Programa de Apoyo a Proyectos de Investigación e Innovación Tecnológica) and CONACyt (Consejo Nacional de Ciencia y Tecnología). NWDM thanks the British Columbia Ministry of Energy and Mines Geological Survey Branch for its support. We thank Ali Polat and an anonymous reviewer for helpful reviews that substantially improved the quality of this paper.

References

Anderson DL (2005) Scoring hotspots: the plume and plate paradigms. In: Foulger GL, Natland JH, Presnall DC, Anderson DL (eds) Plates, plumes and paradigms. Geological Society of American Special Paper 388, Boulder, pp 31–54

Anderson DL, Zhang Y-S, Tanimoto T (1992) Plume heads, continental lithosphere, flood basalts and tomography. In: Storey BC et al (eds) Magmatism and the causes of continental break-up. Geological Society of London Special Publication 68, London, pp 99–124

Andrew A, Godwin CJ (1989) Lead- and strontium-isotope geochemistry of the Karmutsen Formation, Vancouver Island, British Columbia. Can J Earth Sci 26:908–919

Andrew A, Armstrong RL, Runkle D (1991) Neodymium–strontium–lead isotopic study of Vancouver Island igneous rocks. Can J Earth Sci 28:1744–1752

Barker F, Southerland Brown A, Budahn JR, Plafker G (1989) Back-arc with frontal-arc component origin of Triassic Karmutsen basalt, British Columbia, Canada. Chem Geol 75:81–102

Batiza R (1981) Trace element characteristics of Leg 61 basalts. Initial Reports of the Deep Sea Drilling Project, 11. U.S. Government Printing Office, Washington, DC, pp 476–478

Berg HC, Jones DL, Richter DH (1972) Gravina-Nutzotin belt-significance of an upper Mesozoic sedimentary and volcanic sequence in southern and southeastern Alaska. U.S. Geological Survey Professional Paper 800-D, Alaska, pp D1–D24

Bradshaw TK, Hawkesworth CJ, Gallagher K (1993) Basaltic volcanism in the Southern Basin and Range: no role for a mantle plume. Earth Planet Sci Lett 116:45–62

Brandon MT, Orchard MJ, Parrish RR, Sutherland-Brown A, Yorath CJ (1986) Fossil ages and isotopic dates from the Paleozoic Sicker Group and associated intrusive rocks, Vancouver Island, Vancouver Island, British Columbia. Current Research, Part A; Geological Survey of Canada, Paper 86-1A, Vancouver, British Columbia, pp 683–696

Campbell IH, Griffiths RW (1990) Implications of mantle plume structure for the evolution of flood basalts. Earth Planet Sci Lett 99:79–93

Carlisle D, Suzuki T (1974) Emergent basalt and submergent carbonate-clastic sequences including the Upper Triassic Dilleri and Welleri zones on Vancouver Island. Can J Earth Sci 11:254–279

Castillo P, Batiza R, Stern J (1986) Petrology and geochemistry of Nauru basin igneous complex: large-volume, off-ridge eruptions of MORB-like basalts during the Cretaceous. Initial Reports of the Deep Sea Drilling Project, 39. U.S. Government Printing Office, Washington, DC, pp 555–576

Cloos M (1993) Lithospheric buoyancy and collisional orogenesis: subduction of ocean plateaus, continental margins, island arcs, spreading ridges, and seamounts. Geol Soc Am Bull 105:715–737

Cox KG, Hawkesworth CJ (1985) Geochemical stratigraphy of the Deccan Traps at Mahabaleshwar, Western Ghats, India, with implications for open system magmatic processes. J Pet 26:355–377

De Bari SM, Anderson RG, Mortensen JK (1999) Correlation among lower to upper crustal components in an island arc: the Jurassic Bonanza arc, Vancouver Island, Canada. Can J Earth Sci 36:1371–1413

Debiche MG, Cox A, Engebretson D (1987) The motion of allochthonous terranes across the North Pacific basin. Geological Society of American Special Paper 207, 49 p

Dostal J, Strong DF, Jamieson RA (1980) Trace element mobility in the mylonite zone within the ophiolite aureole, St. Anthony Complex, Newfoundland. Earth Planet Sci Lett 49:188–192

Dostal J, Baragar WRA, Dupuy C (1986) Petrogenesis of the Natkusiak continental basalts, Victoria Island, N.W.T. Can J Earth Sci 23:622–632

Dostal J, Dupuy C, Nicollet C, Cantagral JM (1992) Geochemistry and petrogenesis of Upper Cretaceous basaltic rocks from southern Malagasy. Chem Geol 97:199–218

Dostal J, Dupuy C, Caby R (1994) Geochemistry of the Neoproterozoic Tilemsi belt of Iforas (Mali, Sahara): a crustal section of an oceanic island arc. Precam Res 65:55–69

Dupuy C, Dostal J (1984) Trace element geochemistry of some continental tholeiites. Earth Planet Sci Lett 67:61–69

Erlank AJ (1984) Petrogenesis of the volcanic rocks of the Karoo Province. Geol Soc S Afr Spec Publ 13:305

Erlank AJ, Duncan AR, Marsh JS, Sweeney RJ, Hawkesworth CJ, Milner SC, Miller R, Rogers NW (1988) A laterally extensive geochemical discontinuity in the subcontinental Gondwana lithosphere. In: Geochemical evolution of the continental crust. Conference proceedings, Pocos de Caldes, Brazil, pp 1–10

Farnetani CG, Samuel H (2005) Beyond the thermal plume paradigm. Geophy Res Lett 32:LO7311. doi:10.1029/2005GL022360

Govindaraju K (1994) 1994 compilation of working values and sample description for 383 geostandards. Geostandards Newslett 18:1–158

Greene AR, Scoates JS, Weis D (2004) Flood basalts of the Wrangellia Terrane, southwest Yukon: implications for the formation of oceanic plateaus; continental crust and Ni-Cu-PGE mineralization. Yukon Exploration and Geology 2004, Yukon Geological Survey, Whitehorse, pp 109–120

Greene AR, Scoates JS, Weis D (2005) Wrangellia Terrane on Vancouver Island, British Columbia: Distribution of flood basalts with implications for potential Ni-Cu-PGE mineralization in southwestern British Columbia. BC Ministry of Energy, Mines and Petroleum Resources, Geological Fieldwork 2004, Paper 2005-1, Victoria, pp 209–220

Greene AR, Scoates JS, Nixon GT, Weis D (2006) Picritic lavas and basal sills in the Karmutsen flood basalt province, Wrangellia, Northern Vancouver Island, British Columbia. BC Ministry of Energy, Mines and Petroleum Resources, Geological Fieldwork 2004, Paper 2006–1, Victoria, pp 39–51

Greene AR, Scoates JS, Weis D, Israel S (2009) Geochemistry of Triassic flood basalts from the Yukon (Canada) segment of the accreted Wrangellia oceanic plateau. Lithos 110:1–19

Hallam A (1992) Phanerozoic Sea-Level. Columbia University Press, New York, 266 p

Haq BU, Schutter SR (2008) A chronology of paleozoic sea-level changes. Science 322(5898): 64–68

Haq B, Hardenbol J, Vail P (1987) Chronology of fluctuating sea levels since the Triassic. Science 235:1156–1167

Hart SR, Davis KE (1978) Nickel partitioning between olivine and silicate melt. Earth Planet Sci Lett 40:203–219

Hawkesworth CJ, Rogers NW, Van Calsteren PWC, Menzies MA (1984) Mantle enrichment processes. Nature 311:331–335

Hawkesworth C, Kelley S, Turner S, Le Roex A, Storey B (1999) Mantle processes during Gondwana break-up and dispersal. J Afr Earth Sci 28:239–261

Hooper PR, Hawkesworh CJ (1993) Isotopic and geochemical constraints on the origin and evolution of the Columbia River basalt. J Pet 34:1203–1246

Irving E, Wynne PJ (1991) Paleomagnetism: review and tectonic implications. In: Gabrielse H, Yorath CJ (eds) Geology of the Cordilleran Orogen in Canada. Geology of Canada, No. 4, Canada, pp 61–86

Jones DL, Silberling NJ, Hillhouse J (1977) Wrangellia: a displaced terrane in northeastern North America. Can J Earth Sci 14:2565–2577

Jourdan F, Bertrand H, Scharer U, Blichert-Toft J, Feraud G, Kampunzu AB (2007) Major and trace element and Sr, Nd, Hf, and Pb isotope compositions of the Karoo Large Igneous Province, Botswana-Zimbabwe: lithosphere vs mantle plume contribution. J Pet 48:1043–1077

Keppie JD, Dostal J (2001) Evaluation of the Baja controversy using paleomagnetic and faunal data, plume magmatism and piercing points. Tectonophysics 339:427–442

Kerr AC, White RV, Saunders AD (2000) LIP reading: recognizing oceanic plateaux in the geological record. J Pet 41:1041–1056

Lassiter JC, De Paolo DJ, Mahoney JJ (1995) Geochemistry of the Wrangellia flood basalt province: implications for the role of continental and oceanic lithosphere in flood basalt genesis. J Pet 36:983–1009

Lightfoot PC, Hawkesworth CJ, Devey CW, Rogers NW, Van Calsteren PWC (1990) Source and differentiation of Deccan Trap lavas: implications of geochemical and mineral chemical variations. J Pet 31:1165–1200

Mahoney JJ (1988) Deccan traps. In: Macdougall JD (ed) Continental flood basalts. Kluwer, Dordrecht, The Netherlands, pp 151–194

Mann P, Taira A (2004) Global significance of the Solomon Islands and Ontong Java Plateau convergent zones. Tectonophysics 389:137–190

Massey NWD (1995a) Geology and mineral resources of the Duncan sheet, Vancouver Island, 92B/13. BC Ministry of Energy, Mines and Petroleum Resources, Paper 1992–4, Victoria

Massey NWD (1995b) Geology and mineral resources of the Alberni-Nanaimo Lakes sheet, Vancouver Island, 92F/1W, 92F/2E and part of 92F/7E. BC Ministry of Energy, Mines and Petroleum Resources, Paper 1992–2, Victoria

Massey NWD (1995c) Geology and mineral resources of the Cowichan Lake sheet, Vancouver Island, 92C/16. BC Ministry of Energy, Mines and Petroleum Resources, Paper 1992–3, Victoria

McKenzie DP, Bickle MJ (1988) The volume and composition of melt generated by extension of the lithosphere. J Pet 29:625–679

McKenzie D, O'Nions RK (1991) Partial melt distributions from inversion of rare earth element concentrations. J Pet 32:1021–1091

Melluso I, Beccaluva L, Brotzu P, Gregnanin A, Gupta AK, Morbidelli L, Traversa G (1995) Constraints on the mantle sources of the Deccan Traps from the petrology and geochemistry of the basalts of Gujarat State (Western India). J Pet 36:1393–1432

Miller KG, Kominz MA, Browning JV, Wright JD, Mountain GS, Katz ME, Sugarman PJ, Cramer BS, Christie-Blick N, Pekar SF (2005) The Phanerozoic record of global sea-level change. Science 310:1293–1298

Monger JWH, Price RA (2002) The Canadian Cordillera: geology and tectonic evolution. Can Soc Econ Geol Rec 27:17–36

Monger JWH, Struik LC (2006) Chilliwack terrane: a slice of Stikinia? A tale of terrane transfer. In: Colpron M, Nelson JL (eds) Paleozoic evolution and metallogeny of pericratonic terranes at the ancient Pacific margin of North America, Canadian and Alaskan Cordillera. Geological Association of Canada Special Paper 45, Canada, pp 351–368

Monger JWH, Price RA, Tempelman-Kluit DJ (1982) Tectonic accretion and the origin of two major metamorphic and plutonic welts in the Canadian Cordillera. Geology 10:70–75

Monger JWH, van der Heyden P, Journeay JM, Evenchick CA, Mahoney JB, (1994) Jurassic-Cretaceous basins along the Canadian Coast Belt: their bearing on pre-mid-Cretaceous sinistral displacements. Geology 22:175–178

Montelli R, Nolet G, Dahlen F, Masters G, Engdahl E, Hung S (2004) Finite-frequency tomography reveals a variety of mantle plumes. Science 303:338–343

Murphy JB, Dostal J (2007) Continental mafic magmatism of different ages in the same terrane: constraints on the evolution of an enriched mantle source. Geology 35:335–338

Murphy JB, Hynes A, Johnston ST, Keppie JD (2003) Reconstructing the ancestral Yellowstone plume from accreted seamounts and its relationship to flat-slab subduction. Tectonophysics 365:185–194

Nixon GT, Larocque J, Pais A, Styan J, Greene AR, Scoates JS (2008) High-Mg lavas in the Karmutsen flood basalts, northern Vancouver Island (NTS 092L): stratigraphic setting and metallogenic significance. Geological Fieldwork 2007. British Columbia Geological Survey Paper 2008-1, Victoria, pp 175–199

Panuska BC (1990) An overlooked world class Triassic flood basalt event. Geol Soc Am Abstr Program 22:A168

Parrish RR, McNicoll VJ (1992) U-Pb age determinations from the southern Vancouver Island area, British Columbia. Radiogenic Age and Isotopic Studies: Report 5; Geological Survey of Canada, Paper 91–2, pp 79–86

Parsons B, Schlater JG (1977) An analysis of the variation of ocean floor bathymetry and heat flow with age. J Geophys Res 32:803–827

Pearce JA (2008) Geochemical fingerprinting of oceanic basalts with applications to ophiolite classification and the search for Archean oceanic crust. Lithos 100:14–48

Pearce JA, Cann JR (1973) Tectonic setting of basic volcanic rocks determined using trace element analyses. Earth Planet Sci Lett 19:290–300

Pearce JA, Baker PE, Harvey PK, Luff IW (1995) Geochemical evidence for subduction fluxes, mantle melting and fractional crystallization beneath the South Sandwich island arc. J Pet 36:1073–1109

Peate DW, Hawkesworth CJ, Mantovani MSM, Shukowsky W (1990) Mantle plumes and flood basalt stratigraphy in the Parana, South America. Geology 18:1223–1226

Peate DW, Hawkesworth CJ, Mantovani MSM (1992) Chemical stratigraphy of the Parana lavas (South America): classification of magma types and their spatial distribution. Bull Volcanol 55:119–139

Peate DW, Hawkesworth CJ, Mantovani MSM, Rogers NW, Turner SP (1999) Petrogenesis and stratigraphy of the high-Ti/Y Urubici magma type in the Paraná flood basalt province and implications for the nature of 'Dupal'-type mantle in the South Atlantic region. J Pet 40:451–473

Peng ZX, Mahoney JJ, Hooper PR, Macdougall JD, Krishnamurthy P (1998) Basalts of the northeastern Deccan Traps, India: isotopic and elemental geochemistry and relation to southwestern Deccan stratigraphy. J Geophys Res 103:29842–29865

Piccirillo EM, Melfi AJ, Comin-Chiaramonti P, Bellieni G, Ernesto M, Marques LS, Nardy AJR, Pacca LG, Roisenberg A, Stolfa D (1988) Continental flood volcanism from the Parana Basin (Brazil). In: Macdougall JD (ed) Continental flood basalts. Kluwer, Dordrecht, pp 195–238

Richards MA, Hager BH, Sleep NH (1988) Dynamically supported geoid highs over hotspots: observation and theory. J Geophys Res 93:7690–7708

Richards MA, Jones DL, Duncan RA, DePaolo DJ (1991) A mantle plume initiation model for the Wrangellia flood basalt and other oceanic plateau. Science 254:263–267

Rollison HR (1993) Using geochemical data evaluation: presentation, interpretation. Prentice Hall, Upper Saddle River, 352 p

Saunders AD (1986) Geochemistry of basalts from the Nauru basin, DSDP Legs 61 and 89: implications for the origin of oceanic flood basalts. Initial Reports of the Deep Sea Drilling Project, 39. U.S. Government Printing Office, Washington, DC, pp 499–517

Sclater JG, Christie PAF (1980) Continental stretching: an explanation of the post-Mid-Cretaceous subsidence of the Central North Sea Basin. J Geophys Res 85:3711–3739

Shervais JW (1982) Ti-V plots and the petrogenesis of modern ophiolitic lavas. Earth Planet Sci Lett 59:101–118

Sleep NH (1987) Lithospheric heating by mantle plumes. Geophys J Res, R Astronom Soc 91:1–12

Sleep NH (1990) Hotspots and mantle plumes: some phenomenology. J Geophys Res 95:6715–6736

Sluggett CL, Mortensen JK (2003) U-Pb age and geochemical constraints on the paleotectonic evolution of the Paleozoic Sicker Group on Saltspring Island, southwestern British Columbia. Geological Association of Canada – Mineralogical Association of Canada, Program with Abstracts, v. 28

Stoeser DB (1975) Igneous rocks from leg 30 of the deep sea drilling project initial reports of the deep sea drilling project, 39. U.S. Government Printing Office, Washington, DC, pp 404–414

Sun SS, McDonough WF (1989) Chemical and isotopic systematics of oceanic basalts: implications for mantle composition and processes. In: Saunders AD, Norry MJ (eds) Magmatism in the ocean basins. Geological Society of London Special Publication 42, London, pp 313–345

Tatsumi Y, Kogiso T (2003) The subduction factory: its role in the evolution of the Earth's crust and mantle. In: Larter RD, Leat ET (eds) Intra-oceanic subduction systems: tectonic and magmatic processes. Geological Society of London Special Publication 219, London, pp 55–80

Watson S, McKenzie D (1991) Melt generation by plumes: a study of Hawaiian volcanism. J Pet 32:501–537

Winchester JA, Floyd PA (1977) Geochemical discrimination of different magma series and their differentiation products using immobile elements. Chem Geol 20:325–343

Vail PR, Mitchum RM, Thompson S (1977) Global cycles of relative changes in sea level. In: Payton CE (ed) Seismic stratigraphy-application to hydrocarbon exploration. Memoir of the American Association of Petroleum Geologists 26, Tulsa, pp 83–97

Wilson JT (1973) Mantle plumes and plate motions. Tectonophysics 19:149–164

Chapter 2
Deccan Traps Flood Basalt Province: An Evaluation of the Thermochemical Plume Model

Gautam Sen and D. Chandrasekharam

Abstract The Deccan volcanic event occurred when India was situated at the present location of Reunion Island in the Indian Ocean. The cause, duration of eruptive pulses of this major volcanic event and its impact on the global climate are controversial. Plume versus non-plume hypothesis for the origin of Deccan melting anomaly is evaluated here based on a review of geochemical and limited geophysical criteria. We know that the most primitive Deccan magmas were picritic in composition that equilibrated in the mantle at about 1,550°C (±25) and 2.5 (±0.3) GPa. These magmas were generated from a large plume. We find that much of the differentiated and contaminated appearing basalts could not have been produced from such picritic magmas, and suggest that such tholeiites are a product of melts derived from the plume, the subcontinental lithospheric peridotites, and from ancient orogenic-type eclogitic blocks embedded within the continental lithosphere.

2.1 Introduction

The Deccan Traps volcanic province is a prominent feature on India's geological map (Fig. 2.1). Bose (1972, 1995) was a pioneer in recognizing the importance of these flood basalts as a volcanic event of global significance and related their origin to global tectonics. This flood basalt province continues to attract global attention today because of its enormous volume and eruption duration that overlapped the Cretaceous–Tertiary boundary (e.g., Mahoney 1988; Duncan and Pyle 1988; Courtillot et al. 1988). Deccan erupted while India was located in the Indian Ocean

G. Sen (✉)
Vice Provost, Research & Graduate Studies, American University of Sharjah,
PO Box 26666, Sharjah, UAE
e-mail: gsen@aus.edu

D. Chandrasekharam
Department of Earth Sciences, Indian Institute of Technology Bombay, (IITB),
Powai, Mumbai 400 076, India
e-mail: dchandra@iitb.ac.in

J. Ray et al. (eds.), *Topics in Igneous Petrology*, DOI 10.1007/978-90-481-9600-5_2,

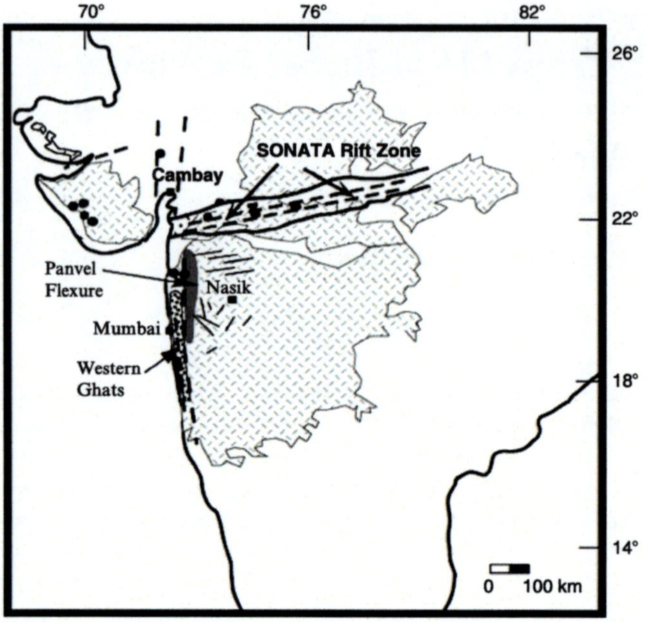

Fig. 2.1 Map of India showing the Deccan Traps and some major intrusive systems related to the Deccan

Fig. 2.2 Location of India where Deccan volcanism actually occurred

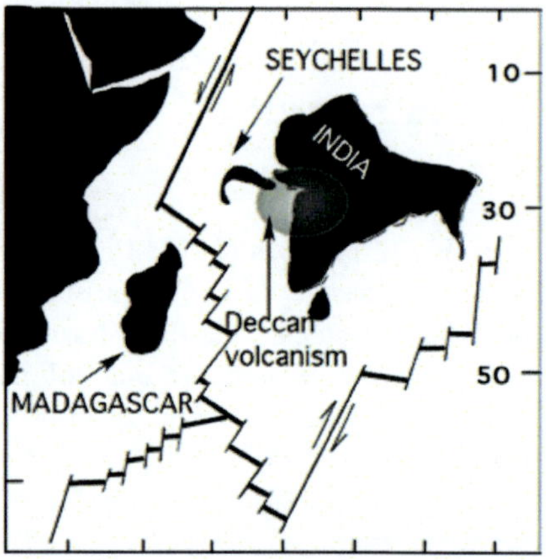

over where Reunion is presently situated (Fig. 2.2; Mahoney 1988). In spite of the past 2 decades of intensive studies of the Traps, our knowledge of its possible impact on global/local climate, oceans, and biota seems limited (Keller et al. 2008; Self et al. 2006, 2008). The 1783–1784 Laki eruption (Iceland) is the closest, albeit being much smaller (Laki – 14 km^3; Deccan – 1.5 million km^3), analog to the Deccan eruption in terms of rate of lava production, basaltic chemistry of lava, and fissure style of eruption (Self et al. 2008). The Laki eruption wiped out over 50% of Iceland's livestock population and led to a famine that killed 25% of the human population. If we extend the Laki eruption to the scale of the Deccan, then its effect could have been colossal at least over India (Self et al. 2008). A few recent studies have suggested that in the Western Ghats, where the lava package is the thickest, there had been short pulses of massive eruptions lasting a few thousand years (Sen et al. 2006; Chenet et al. 2007). If true, such "spikes" would have had significant impact on climate and biota.

The origin of the Deccan Volcanic Province has been much debated. The debate concerns whether the magmas formed by melting a giant mantle plume, normal plate tectonic processes, or impact of a large extraterrestrial bolide (e.g., Mahoney 1988, Richards et al. 1989; Chandrasekharam and Parthasarathy 1978; Chatterjee and Rudra 1992; Sen 1995; Sheth 2005a, b). In the 1960s several authors recognized the difficulties in explaining all volcanic phenomena on earth with the shallow mantle-melting model required in the plate tectonics theory and proposed the concept of hot spots (Wilson 1963). Wilson felt that the Hawaiian island chain, which is located in the middle of a large oceanic plate, could not be explained by shallow mantle process as required by plate tectonics. Wilson's model for the Hawaiian volcanic chain called for generation of magmas from a spatially fixed thermal plume rooted in the deep, non-convecting, mantle. Soon after Wilson, Morgan (1972) presented the concept of mantle plumes. Whereas hot spot *sensu stricto* refers to a hot jet or thermal plume, Morgan's plume would include material transfer from the deep to shallow mantle. In the subsequent years, geochemists and geophysicists have used the term "plume" at random to mean different things; but in general, "plumes are upwellings and downwellings that are maintained either by thermal or chemical buoyancy or by both in concert" (Anderson and Natland 2005).

The plume model as put forth in those early years was quickly gaining recognition once K/Ar dating of the Hawaiian volcanoes showed that the islands get older with distance from the active volcanoes of Kilauea and Mauna Loa on the island of Hawaii (Dalrymple 1969). However, there were others who viewed the plume hypothesis with some skepticism; for example, in a witty article Holden and Vogt (1977) noted "The mantle plume is just the youngest member of a big and colorful family of geological fads and fashions…Since plumes are better hidden from observation than plates, it may take years to prove or disprove their existence."

The Deccan Traps eruption is often used as a spectacular example of mantle plume activity (Fig. 2.2; e.g., Richards et al. 1989; Duncan and Richards 1991). The basic idea is that the Deccan magmas formed by melting a giant "plume head" that rose from the core–mantle boundary. This plume head was anchored to the source by a long tail that created the Chagos-Laccadive volcanic chain. A jump of a ridge that is now the Central Indian Ridge about 40 million years ago broke up the chain, and the original plume tail is now creating volcanism on Reunion Island.

There are several authors who have doubted the plume theory as far as the origin of the Deccan Traps is concerned (e.g., Sheth 2005a). Chandrasekharam and

Parthasarathy (1978) noted that the Deccan basalts predominantly erupted through fissures in the crust and proposed a model of shallow mantle melting in which the preexisting rift zones were reactivated and magma simply poured out of fissures (Chandrasekharam and Parthasarathy 1978). This was a time when the concept of plate tectonics was gaining ground; and so the melting mechanism was considered to be analogous to passive rifting of the lithosphere and magma generation at the global mid-oceanic ridge system. Sheth (2005a, b) has proposed much more revised versions of the shallow melting model that appeals to normal plate tectonics and lithospheric melting processes.

Finally, Chatterjee and Rudra (1992) and Chatterjee et al. (2006) suggested that a large bolide crashed offshore near the "Bombay High" area and melted the lithosphere, generating large volumes of magma that erupted as the Deccan lavas. This hypothesis has not been generally accepted mainly because it is not a testable hypothesis and requires further exploration: for example, the physical evidence is weak for the existence of an offshore impact crater with an onshore crater rim at Panvel flexure. Such an impact would have produced the first batch of magma close to the impact site and thus the oldest lavas should have been found close to Bombay. Instead, the oldest Deccan lavas are found in Pakistan, much further north of Bombay (Mahoney et al. 2002). Also, the impact idea does not explain why there is an age progression from the Deccan through Laccadives to Seychelles to Reunion (Duncan and Richards 1991). Nevertheless, one cannot completely rule out this impact hypothesis until the deeper lavas beneath Bombay area (both off and onshore) are sampled *via* drilling.

In this paper we do not further explore an extraterrestrial cause for the Deccan volcanism and focus the rest of our discussion on internal terrestrial mechanism. Here we re-visit an ongoing debate concerning whether a large plume was at all involved in forming Deccan magmas or such magmas formed due to normal plate tectonic processes. This is a "hot" topic in geology today.

A website (www.mantleplumes.org) and two volumes (Foulger et al. 2005) have been dedicated mainly to disprove the existence of mantle plumes although, in all fairness, the editors have included several papers that defend the plume hypothesis. Those who do not currently subscribe to the plume hypothesis have come up with alternative models that generally imply shallow upper mantle melting and cracks in the lithosphere creating all the volcanism that are usually attributed to plumes. As Campbell (2005) recently noted, even the sharpest critics of the mantle plume hypothesis have failed to make a convincing case for a non-plume origin of the Hawaiian volcanic chain. If lithospheric cracks were the reason then the Molokai fracture zone, which is one of the longest fractures that reaches close to Hawaii, should have caused volcanism, which of course has not happened. It is true that many volcanic regimes randomly attributed to plumes may be explained better with alternative models. In terms of "record keeping", however, the plume hypothesis continues to endure the attacks; and over the past few years the focus has fallen on imaging deep plumes using seismic tomography even though Anderson (2006) has urged caution in using tomography alone to interpret mantle dynamics and proposed additional seismic experiments.

Given the above backdrop, we avoid any genetic implication and use the term "Deccan Melting Anomaly (DMA)" throughout this paper to refer to the Deccan melting source region. Here we review the possible conditions of magma generation and examine the following factors: volume of original (primary) magma produced, the time within which the maximum volume of magma was generated, the source rocks, and the temperature–pressure conditions of magma generation. This evaluation leads us to the conclusion that a complex melting model involving both a relatively Fe-rich plume and lithosphere best explains all features of the Deccan Traps.

2.2 Duration of the Deccan Event and Mean Eruption Rate

An important argument in favor of the plume origin of the Deccan is that too much lava came out in too short a time requiring some special melting mechanism that does not fit the normal plate tectonic schemes (e.g., Sen 1995). The most impressive demonstration of this came through the $^{40}Ar/^{39}Ar$ dating of the 2.7 km thick Western Ghats section by Duncan and Pyle (1988). Since that work, other research groups have dated different parts of the Deccan province but the general picture has remained the same – that is, the bulk of the lavas erupted 65–66 Ma within two magnetic chrons – 29R and 29N; however, one important exception is that Mahoney et al. (2002) obtained a 72 million years. date from some lavas in Pakistan (Fig. 2.3; Mahoney et al. 2002). Another alkalic intrusion of ~69 Ma age was found close to Barmer near the northwestern periphery of the Deccan (Basu et al. 1993). The youngest date of ~61 Ma was obtained by $^{40}Ar/^{39}Ar$ dating of some rhyolitic lava/intrusions in the Bombay area (Lightfoot et al. 1987; Sheth et al. 2001). Figure 2.3 shows a summary of Deccan Trap ages obtained by different methods (slightly modified from Chatterjee and Rudra (1992). It shows that about 90% of the lavas came out within 1 million year during the 29r magnetic chron.

As discussed by Sen (2001), estimates of the original (pre-erosion) volume of the lavas vary significantly. If we accept the volume to be 1.2×10^6 km^3 erupted in 1×10^6 years, then the mean eruption rate could have been 1.2 km^3 per year. Sen et al. (2006) have suggested that the actual duration could have been much shorter – of the order of only 30,000 years, which will mean an average eruption rate of 40 km^3 per year. The Laki eruption of 1783–1784 in Iceland, which produced 14 km^3 of lava, is the only historic eruption that comes close to the Deccan's astounding eruption rate. We should say this with the caveat that much finer age refinement with radioisotope dating of the lavas and a more accurate estimate of the volume of the lavas are needed to have a stronger constraint on the eruption rate of the Deccan. Without such refinement, the mean eruption rate could have varied between 1 and 40 km^3 per year. A paleomagnetic study by Chenet et al. (2007) suggests that there were episodic bursts of lava eruption with intermittent quiescence. The mean eruption rate refers to a cumulative rate that included the quiet periods between eruptions. The picture needs clarity with detailed studies of the individual lava outpourings that produced the

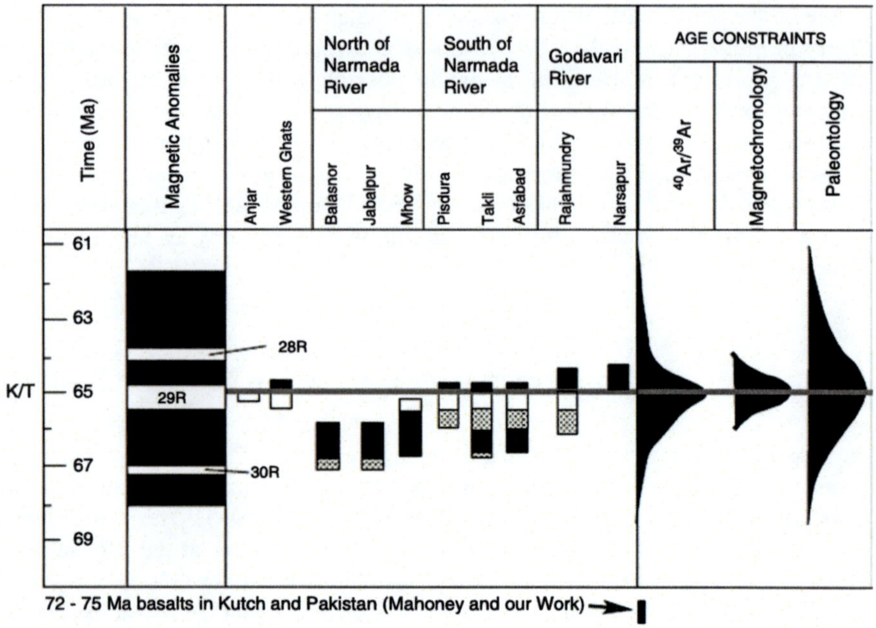

Fig. 2.3 A compilation of ages of the Deccan Traps from various sites and using different age determination techniques. The bulk of the magmatism occurred at 65 ± 1 Ma during the 29R magnetic chron (modified after Chatterjee and Rudra 1992)

geochemically defined formations in the lava province and the intermittent periods when weathering resulted in the formation of "boles" or soil horizons between lava flows.

2.3 Source and Potential Temperature of the Deccan Mantle Anomaly: Message from Geochemistry

2.3.1 General Considerations

The Deccan flood basalt province is predominantly tholeiitic; however, alkalic lavas and intrusions, carbonatite intrusive complexes, and silicic lavas and intrusives also occur. Carbonatites and alkalic lavas, with some of the latter carrying mantle xenoliths, are restricted to the rift zones that occur near the peripheries of the Deccan (Fig. 2.1). Experimental petrology has shown that carbonatites and mafic alkalic melts, such as nephelinites, basanites and alkalic olivine basalts, are generated as near-solidus, low degree melts from carbonated peridotites at pressure of ≥2.8 GPa (e.g., Keshav and Gudfinnsson 2004). Simple volume consideration of the erupted lavas suggests that Deccan tholeiites or their parent magmas were generated in much greater volumes than carbonatites and alkalic magmas.

Presumably, parent magmas of tholeiites were generated at higher potential temperatures and by much greater degrees of melting than the alkalic and carbonatitic magmas (Sen 1995). In the discussion that follows we focus largely on the origin of the tholeiites.

An evaluation of the melting conditions requires an understanding of the composition of the source rock, and conditions of melting, which includes temperature, pressure, and volatiles (CO_2 and H_2O and their fugacities). The first step toward achieving this goal is to "retrieve" the composition of the *primary magma*. Following conventional wisdom we define a primary magma as a natural melt that is in equilibrium with its source rock, which could be an eclogite, garnet pyroxenite, or a peridotite in case of basaltic magmas. Constraining the composition of the primary magma(s) is of first order importance because it sets limits on the thermal conditions reached during magma segregation from the source with which it fully equilibrated at a certain depth or over a depth range. That in turn would allow one to evaluate the excess potential temperature (ΔT_p), defined as potential temperature of the melting anomaly ($T_{p\text{-}MA}$) minus that of Mid-oceanic ridges ($T_{p\text{-}MOR}$). Exactly how much ΔT_p would distinguish a thermal plume (presumably of deep mantle origin) *versus* upper mantle origin of such melting anomalies as the DMA is somewhat debatable but in general, a T_p of >1,450°C would make a strong case for a thermal origin (e.g., Faernetani and Richards 1994; Anderson 2006; Courtier et al. 2007).

Next comes the issue of source rock(s) – peridotite, eclogite, or some other rocks? In the eclogite melting case, the erupted basalts would simply be products of bulk melting (>70%) of a basaltic (eclogite is only a high pressure metamorphosed basalt) source. Also, in this case, the excess temperature required would be very small (<120°C) to generate a large volume of magma (Kogiso et al. 1997; Pertermann and Hirschmann 2003). In the peridotite melting scenario, the primary magmas would necessarily be high temperature picritic magmas that would have lost large amount of olivine by fractional crystallization during their ascent through the crust (e.g., Cox 1980; Sen 1995, 2001). In this scenario, excess temperatures of ≥250°C would be necessary (Sen 1995, 2001). On the other hand, an eclogite/garnet pyroxenite model would not require the significantly excess temperature and could instead favor a shallow mantle origin of the DMA in which the eclogite source could be blobs or blocks derived from old subducted slabs or deep continental crust. Peridotite could be part of the asthenosphere, continental lithosphere, and/or the plume itself.

2.3.2 *Isotopic Composition*

Most geochemists consider helium isotope ratio ($^3He/^4He$) to be an important indicator of the mantle source (e.g., Class and Goldstein 2005). ^3He is primordial whereas ^4He is produced by radioactive decay of the actinide elements U and Th. Mid-ocean ridge basalts (MORB) are generally accepted to result from shallow melting of the asthenosphere, which is also often referred to as "shallow convecting mantle" (which implies that there is a physically and chemically distinct convecting deeper or Lower Mantle). MORB $^3He/^4He$ ratios range between 7 and 9 R/R_A (measured ratio normalized to that of air). Basalts with elevated R/R_A of >10 are considered to be derived from a deeper mantle region with lower time-integrated $(U + Th)/^3He$, and a thermo-chemical plume is usually taken to be responsible for pulling up such mantle materials to shallow levels where they undergo partial melting. An extreme example of a volcano with very high R/R_A (~32) is Loihi, a submarine seamount off Kilauea (Kurz et al. 1983). Basu et al. (1993) determined the R/R_A values of two Deccan alkaline complexes from northern India to be 10–13, which they used as evidence to support a Lower mantle plume origin of Deccan Traps volcanism.

We now look at the Nd, Hf, Pb, and Sr isotopic composition of Deccan basalts in order to assess the nature of the source rock materials (Fig. 2.4). The various isotopic trends exhibited by various Deccan formations diverge away from a "common signature" area, which was suggested by Peng and Mahoney (1995) to be the actual mantle source composition of the Deccan tholeiites. Notably, the "common signature" overlaps the composition of Reunion lavas. A study by Hanyu et al. (2001) showed Reunion source to have a R/R_A of 13, which is consistent with the similarity between Deccan "common signature" and Reunion lavas in terms of He isotope ratio as well.

Many authors have suggested that the cause of isotopic divergence in the Deccan is contamination of the magmas by mostly Archean/Proterozoic crust (e.g., Peng et al. 1994). The problem is that no quantitative modeling scheme that involves the crust could effectively reproduce the contamination signal involving major and trace elements as well as isotope ratios. Chatterjee and Bhattacharji (2008) recently showed that in some cases the isotopic contamination signals would require incorporation of as much as 64% continental crust into the magma generated from a mantle peridotite. The authors therefore noted (as have others before them) that the

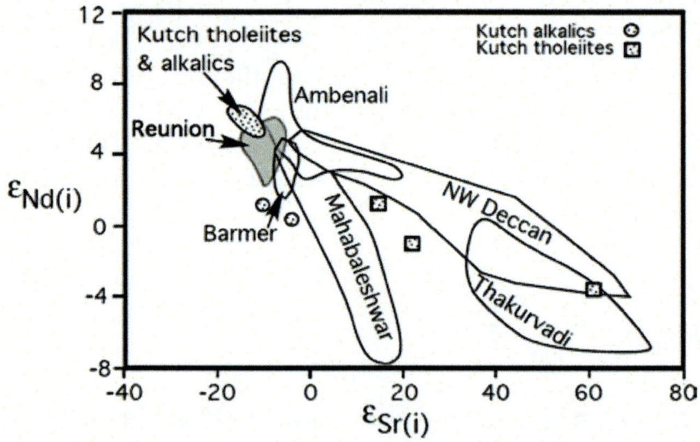

Fig. 2.4 Isotopic variability within some Deccan lava formations (From Sen et al. 2009). The various formations, mantle xenolith-bearing alkalic basalts, and Reunion lavas all seem to converge at a general area

isotopic contamination signal must have been acquired by the magmas prior to entering the shallowest reservoirs where they underwent further mixing and crystallization-differentiation. Sen (1995) deemed these shallow reservoirs to be located no deeper than about 6 km.

In the case of the Deccan, there is a consensus that the isotopic signature of continental crustal contamination is the strongest for the Bushe Formation of the Western Ghats (e.g., Cox and Hawkesworth 1985; Beane and Hooper 1988; Lightfoot and Hawkesworth 1988; Mahoney et al. 2002; Peng et al. 1994; Melluso et al. 2006; Gangopadhyay et al. 2005). For the rest of the formations, the isotopic diversity is perhaps mostly a result of melting the old continental lithosphere and crustal fragments (charnockites, granulites, ecogites) buried within the lithosphere.

The case against bulk crustal contamination at shallow to mid-crustal levels is rooted in the following observations:

1. Geochemists often think of the contamination process in terms of mathematical mixing between a magma and continental crust and do not take into consideration the large amount of heat (enthalpy of fusion) that would be required to accomplish the melting. Glazner (2007) has quantitatively shown that the enthalpy of fusion required to melt such a large chunk of the crust is prohibitively large. Such a large extent of contamination would likely freeze the magma entirely.
2. One may expect to see more silicic and intermediate composition lavas. The fact is that the commonly erupted lavas are tholeiites with 5–7% MgO. Yes, there are some minor rhyolites but andesitic lavas have not been reported anywhere. If we assume that the proportion of erupted lavas are volumetrically representative of the melts generated, then the absence of andesitic melts and presence of tiny volumes of rhyolitic lavas suggest that very little crust was melted, and/or fractional crystallization was not extreme enough to generate the intermediate/silicic melts.
3. Excluding Bushe and other formations that show geochemical evidence of crustal contamination, Deccan tholeiites as a whole follow ordinary liquid line of descent at 2 GPa (Sen 1995). The chemical imprint of fractional crystallization on both major and compatible trace elements significantly supersedes the crustal contamination signal (Figs. 2.6, and 2.7, further discussed in a later section). Bulk assimilation also results in fractional crystallization, for which there is no evidence in the Deccan: for example, there is no correlation between, such fractionation signals as Mg/Fe ratio or TiO_2 content and isotopic signal of contamination.
4. Two recent studies (Sen et al. 2006; Chenet et al. 2007) have suggested that at least some Deccan tholeiites may have erupted at rapid pace, which implies a very short residence time for the magmas in the crust. This would be a problem for models requiring the magmas affected by the continental crust, which would presumably require longer time. This argument is not particularly strong at this point because this line of inquiry is very new and more detailed studies need to be done.

The bottom line is that the spectacular diversity that is exhibited by the isotope ratios is not as dramatic in major element or compatible trace element compositions.

As some authors have said before us, the isotope signal may be partly derived from the melting of complex crustal types already buried in the continental lithosphere and partly from selective contamination by the crust.

2.3.3 *Trace Element Chemistry*

Several authors have modeled rare earth elements in alkalic basalts and tholeiites and suggested that the alkalic magmas were generated from garnet peridotite and the tholeiites, which are predominant in the Deccan province, were mostly produced at a shallower depth within the spinel peridotite field (e.g., Basu et al. 1993; Melluso et al. 2006). Figure 2.5 illustrates this important constraint. It shows a plot of $[Sm]_n$ vs. $[Sm/Yb]_n$ of Deccan tholeiites from across the Deccan (source of data: GEOROC database) and few picrites from northwestern Deccan that were studied by Melluso et al. (2006). We also show the composition of melts formed by equilibrium melting of a spinel versus garnet peridotite that have the same chondritic REE composition

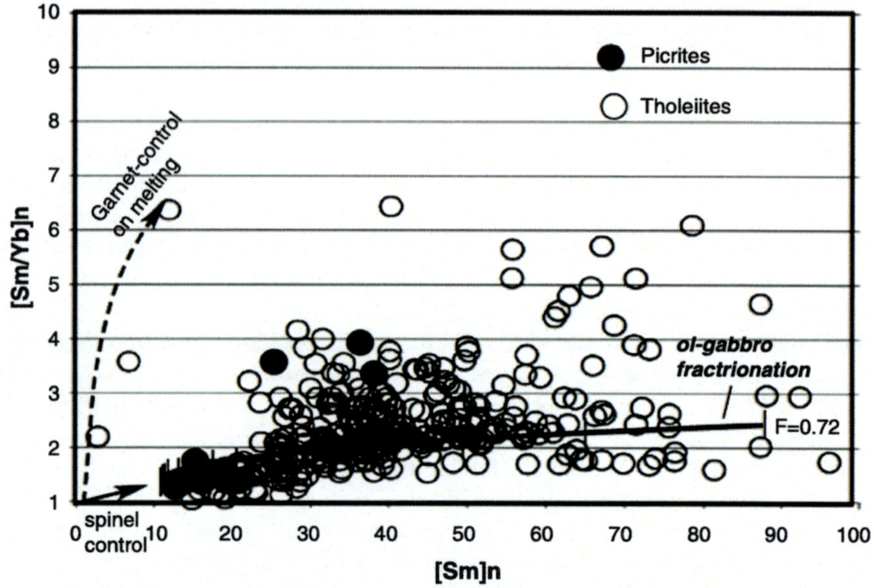

Fig. 2.5 Chondrite-normalized Sm/Yb versus Sm variation in Deccan tholeiites (*circles*, GEOROC database). A few picrites from northwestern Deccan are also shown here (Melluso et al. 2006). Two calculated melting paths in garnet versus spinel peridotite fields from a chondritic source are shown for comparison. The nonmodal equilibrium melting equation was used (additional information is given in Appendix). A hypothetical olivine-gabbro fractionation trend (*bold line with arrow*) is also shown (see Appendix for calculation information). The F reflects melt fraction remaining. Most Deccan tholeiites and the low Sm/Yb picrites appear to form by low pressure melting in the spinel peridotite field (1–2.7 GPa). The range shown by the common Deccan tholeiites can be generated by ~30% olivine-gabbro fractionation

(model calculation data are given in the Appendix). As has long been known, residual garnet selectively retains Yb over Sm and is effective in increasing the Sm/Yb ratio of partial melts. Figure 2.5 shows that Deccan tholeiites, in spite of variable degree of crustal contamination (discussed before), have consistently low Sm/Yb ratio that is best explained by about 30% olivine-gabbro fractionation (see Appendix for calculation) from primary magmas generated from spinel peridotite. We also plotted the high-Ti and low-Ti picrites from Gujarat. The three high-Ti picrites are of alkalic affinity and have higher Sm/Yb ratio, which indicates that they may have had parent magmas that were generated in the garnet peridotite field (Melluso et al. 2006; Sheth and Melluso 2008).

If the source is peridotitic then the minimum pressure at which the residue can have garnet is 2.8 GPa (approximately at 90 km depth); however, if the source is eclogite or garnet pyroxenite then the minimum pressure would be 1.6 GPa (~50 km). As far as REE modeling goes, the eclogite source composition would be expected to vary a great deal in composition because they are basically metamorphosed basalts. The use of this diagram alone, without consideration of other compositional aspects, cannot distinguish between a relatively uniform peridotitic source *versus* a highly variable eclogitic source unless of course garnet stays in the eclogite residue during melting. However, the fact that Deccan tholeiites overwhelmingly exhibit LREE-enriched (chondrite-normalized) character suggest that there was none or extremely minor recycled (previously subducted oceanic crust) eclogite component in its source. The source had to have a predominant LREE-enriched component that could have been old enriched continental lithosphere, old deep crustal metamorphic rocks (e.g., eclogite), or an enriched plume.

2.3.4 *Major Element Composition*

Major elements are critical in obtaining depth and temperature information on primary magmas of a volcanic province, both of which are of relevance in our discussion in the present paper. Picrites have traditionally played an important and sometimes controversial role in all discussions of flood basalt genesis (e.g., Cox 1980; Sen 1988). Here we pay close attention to Deccan Trap picrites from the point of magma generation conditions, which may provide a strong constraint on thermal conditions of the DMA.

Sen (1988) compared the major element composition of Deccan tholeiites with partial melting experiments conducted by Takahashi and Kushiro (1983) and found that tholeiitic picrites with ~10 wt% MgO are almost identical to the experimental melts generated at the solidus from a peridotite at 1 GPa. At 1 GPa, spinel/plagioclase peridotite transition occurs and garnet is not a stable mineral. This is consistent with the trace element discussion of requiring melting in the spinel peridotite field, as discussed by others (e.g., Melluso et al. 2006) and in the previous section.

Sen (1995) calculated potential primary magma compositions for the Ambenali Formation (Table 2.1). Both Sen (1995) and White and McKenzie (1995) applied more complex modeling methods and interpreted the major element variation to

Table 2.1 Calculated Deccan primary magmas and potential temperatures

		Picrite	Adjusted	Sen (1995)	
		Pavagarh	primary	Amb	Amb
Wt%		K'2000	liquid (PMK)	Melt2	Melt1
SiO_2		48.70	*46.40*	48	50.57
TiO_2		1.76	*1.68*	1.3	0.87
Al_2O_3		10.09	*9.61*	13	12.61
FeO*		10.29	*9.91*	11	10.88
MnO		0.18	*0.17*	0.1	0.1
MgO		14.78	*18.96*	15	13.96
CaO		11.76	*11.20*	10.25	9.13
Na_2O		1.49	*1.42*	1.7	1.96
K_2O		0.68	*0.65*	0.12	0.14
Total		99.73	*100.00*	100.47	100.22
Tp-H&G'09	°C	1,453	*1,550*	1,459	1,431
Tp-Put'08–15	°C	1,488	*1,595*	1,519	1,576
Tp-Alb'92	°C	1,435	*1,549*	1,445	1,402
Tp-Lee'09	°C	1,470	*1,490*	1,503	1,474
Mean Tp	°C	1,462	*1,546*	1,482	1,471
Pressure_Lee	GPa	1.90	*2.30*	2.4	1.80

have resulted from accumulation of partial melts of peridotite from about 3 GPa, close to the garnet/spinel transition, to 2 GPa, well within the spinel peridotite field. Based on two different thermometers, Sen (1995) calculated potential temperatures ranging from 1,370°C to 1,460°C. In contrast, White and McKenzie (1995) and Melluso et al. (1995) proposed potential temperatures of 1,470 and 1,430–1,468°C, respectively. Many more detailed studies of Deccan Traps and laboratory melting experiments have taken place since that time and therefore we revisit some aspects of the major elements and evaluate whether the same maximum temperature of 1,470°C for the DMA still holds up.

Figure 2.6 shows a field of Deccan tholeiites, some selected Deccan picrite compositions, global eclogite compositional fields, and parameterized near-solidus compositions from a peridotite over a pressure range of 1–3.5 GPa. Although there are many more picrite analyses to be found in the Deccan literature, a large number of them have been shown to not be melts but melts with excess accumulations of olivine crystals (e.g., Beane and Hooper 1988). We chose only those compositions that have been deemed to be closer in composition to potential melts based on their petrography (aphyric to near-aphyric) and whether or not such compositions exhibit an olivine/melt FeO/MgO partition coefficient close to ~0.3 (e.g., Krishnamurthy et al. 2000; Melluso et al. 2006).

Figure 2.6a, b show that melts produced from a peridotite exhibit a large variation in MgO based on pressure (and also on degree of melting, F, not shown here) but limited variation in Al_2O_3 and SiO_2. Deccan picrites plot either on or close to the mantle melting path shown in Fig. 2.6a; however, the bulk of them consistently plot lower in Fig. 2.6b. In the MgO–Al_2O_3 plot they coincide with melts generated between 1 and 2.5 GPa. These picrites would then seem to be primary or

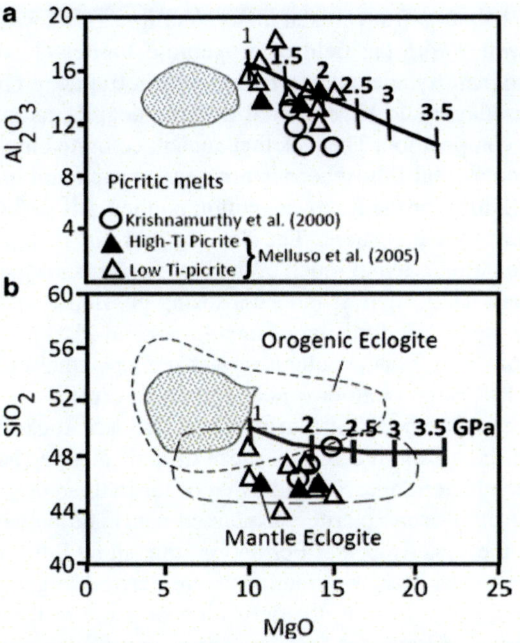

Fig. 2.6 (**a**) Oxide-oxide variation in the Deccan tholeiites (stippled field) and Gujarat picrites. Also shown for comparison is a trend of paramterized near-solidus melt compositions formed from a peridotite source over a pressure range of 1–3.5 GPa. (**b**) The fields of "orogenic eclogites" versus "subducted mantle eclogites" are based on Jacob (2004)

near-primary magmas that could have been generated from a peridotitic source in the DMA between 1 and 2.5 GPa. However, whether there is a possibility of such picrites being generated from mantle (subducted ocean crust) eclogites is discussed next.

Jacob (2004) carried out a global compilation of data on cratonic (presumably deep crustal origin) eclogites and mantle eclogites from kimberlitic intrusions. These fields are shown in Fig. 2.6b. Deccan picrites plot squarely within the mantle eclogite field and therefore, one could perhaps argue for a bulk eclogite melting origin of the picritic magmas. Because subducted oceanic crust (mantle eclogite) is LREE-depleted, one should expect magmas formed by bulk melting of such eclogites to be LREE-depleted; however, the Deccan picrites chosen for this study are all LREE-enriched and do not favor an origin by bulk melting of such eclogite. In fact, Melluso et al. (2006) demonstrated these picrite to have been generated from a garnet lherzolite source. Sheth and Melluso (2008) and others have shown that these picrites are too high in LREE and Ti and could not have generated the more voluminous Deccan tholeiites of the Western Ghats via crystal fractionation.

The tholeiites pose a more complicated problem by virtue of their more chemically evolved character and strong and diverse isotopic contamination. Sen (1995) modeled fractionation paths and showed that all oxide variations, with the exception of SiO_2, can be modeled reasonably well as products of 2 GPa olivine-gabbroic fractionation of primitive magmas. The SiO_2 scatter in the tholeiites is partly due to additional contribution from crustal melts. As Fig. 2.6b shows, common Deccan tholeiites plot well within the field of orogenic eclogites (Jacob 2004). If there were ancient orogenic type eclogites buried within the deep crust/lithosphere of the Indian plate, they could have melted to large degrees and contributed to the average Deccan composition. The idea that ancient eclogite blocks may be embedded in the subcontinental lithosphere is not new as such and has been suggested by Shirey et al. (2001) for the Kaapvaal craton on the basis of 3 Ga Re–Os ages of eclogite xenoliths in kimberlites. The Sr, Pb, and Nd-isotope requirement of strong crustal contamination of much of the Deccan traps may simply reflect such melt contributions. Thus, there is indeed a strong possibility of a mixed origin of the Deccan tholeiites – fractionation of mantle-derived picritic magmas, contribution of melts from deep crustal eclogites, and perhaps smaller amounts of melts contributed by shallow and mid-crustal contaminant crust.

In passing, there is no doubt that peridotite source rocks were involved in generating all of the alkalic magmas because of their olivine (Fo_{89-91}) phenocrysts and the upper mantle peridotite xenoliths that occur in them (e.g., Sen et al. 2009). At least some of the Deccan picrites (explained next) and perhaps other tholeiitic formations, like the vast Ambenali Formation, that show little or no isotopic evidence of ancient crustal melt contributions were derived largely from a peridotite source. In the best possible case, the derivation of some of the common tholeiite melts (e.g., Ambenali Formation) from such picrites would require 25–30% olivine fractionation (Sen 1995). Note that Sheth (2005b) carried out a similar calculation and showed that as much as 35% olivine fractionation would be required.

2.4 Deccan Picrites and Potential Temperature of the DMA: Plume Source Confirmed

Deccan picritic flows erupted early in the lava sequence, particularly in the rift zones (e.g., Krishnamurthy et al. 2000 and refs. therein), where the chance of magma stagnation, contamination, and fractionation would have been much less. They are of interest to us in this particular paper because they also give the highest calculated potential temperatures. Among the various Deccan picrites that are plotted in Fig. 2.6, sample PB-39 from Pawagarh or Pavagadh (Krishnamurthy et al. 2000; composition listed in Table 2.1) is of critical importance in the following discussion because this sample has the most forsteritic olivine ($Fo_{91.5}$). We calculate an olivine-whole rock K_d (FeO/MgO) of 0.22 for this sample, which is not an equilibrium value (0.3 ± .02). In order to calculate the composition of the actual equilibrium melt we added olivine while maintaining a K_d of 0.3 until the computed composition could be in equilibrium with the $Fo_{91.5}$ olivine (procedure as in Sen 1995). Composition of this adjusted primary melt composition (PMK) is also given in Table 2.1. The Pavagadh Traps have been extensively studied by many previous workers (e.g., Hari et al. 1991; Melluso et al. 2006; Sheth and Melluso 2008). As indicated above, the Pavagadh picrites could not be parental to the more voluminous Deccan tholeiites (e.g., Melluso et al. 2006; Sheth and Melluso 2008).

There have been a number of new published thermometers that have come out within the last few years (Putirka 2008; Herzberg and Gazel 2009; Lee et al. 2009). Table 2.1 shows mantle potential temperatures calculated for our Deccan primary melt (PMK, in *italics*) using the newer thermometers and Putirka's method of calculating mantle potential temperature. We also show calculations using Albarède's (1992) thermometer to provide a "baseline" comparison with one of the more established, albeit older, thermometer. Among the calculated potential temperatures, we believe that ~1,550°C (±30) is perhaps the most reasonable. The uncertainty shown above is based on discussions presented by the authors of the various thermometers. The new Deccan primary liquid gives a potential temperature of 70–100°C higher than previously reported temperatures calculated by White and McKenzie (1995), Sen (1995), and Melluso et al. (1995); however, it is almost 100°C lower than the maximum potential temperature ~1640°C calculated by Herzberg and Gazel (2009). It is not clear what Deccan composition was used by Herzberg and Gazel (2009) in their calculation; however, based on our careful examination of the Deccan picrites we consider it likely that they selected a picrite like W-1 listed in Krishnamurthy et al. (2000), which has very high whole-rock MgO (23.78 wt%) and a Fo_{90} olivine phenocryst. The calculated olivine/liquid K_d for this rock is 0.47, which is very different from the commonly accepted K_d value of 0.3 (±0.02). Therefore, this sample has too much excess olivine and is not close to being a liquid composition.

In sum, our newly calculated primary picritic liquid PMK with its ~19% MgO is a reasonable and perhaps the hottest primary magma in the Deccan Traps. A powerful constraint that gives a sense of robustness to this composition is its olivine phenocryst ($Fo_{91.5}$), which in turn constraints the MgO/FeO^{2+} ratio of the primary magma. To be sure, this is by no means the only primary magma as a case has been made for multiple primary magmas in the Deccan in earlier publications (Sen 1995, 2001; Sheth and Melluso 2008; and refs. therein). Nonetheless, the calculated mantle potential temperature of 1,550°C for PMK is important as it provides a strong constraint on the maximum T_p of the DMA.

In a comparative seismic and petrological thermometric study of global volcanic systems, Courtier et al. (2007) determined the potential temperature for mid-oceanic ridges to be 1,350°C (±50). On the other hand, hotspot-influenced ridges and hot spots (their definition) gave T_p of 1,447°C (±16) and 1,484°C (±30). *The newly calculated Deccan Tp of 1,550 puts it at the high end of the hot spots, and therefore makes the best case for the thermal characteristic of the DMA.*

2.5 Fe-Rich Nature of the Deccan Plume?

It is important to recognize here, based on many prior studies of the Deccan, that there are at least two types of strongly mafic primary magmas that are different in Fe, K, and some other elements, which is evident from a comparison between Amb Melt-2 and -1 *versus* the newly calculated primary magma, PMK (Table 2.1). Although not shown here, liquid-line-of descent calculations cannot generate Ambenali-like lavas, which are more common in the Deccan, from the PMK

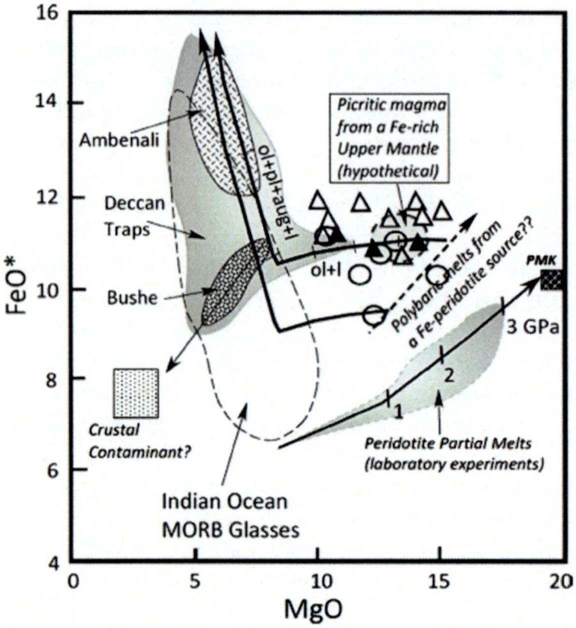

Fig. 2.7 MgO–FeO* relationships in the Deccan (source of data: as in Fig. 2.6) compared with parameterized experimental peridotite partial melt trend over a pressure range of 0–3 GPa. Some equilibrated picrites (see text for discussion) are shown. These require derivation from a Fe-rich source. Two calculated low-pressure (0.2 GPa) fractionation trends are also shown that clearly indicate that Bushe cannot be a result of fractionation of any magma formed by peridotite partial melting. Bushe and other "contaminated" formations are suggested to be products of mixing between peridotite partial melts and eclogitic crustal blocks in the lithosphere (see text for discussion). Indian MORB glass field is shown only as a reference because such glasses have formed by low-pressure fractionation and mixing from relatively Fe-poor asthenospheric partial melts. The calculated primary picrite magma composition for the Pavagadh picrite (PMK) is shown. As discussed in the text, such primary magmas may have produced smaller volumes of magmas at the pre-existing rift zones but did not generate the bulk of the Deccan tholeiites (modified after Sen 2001)

source. Sen (2001) has argued for a Fe-rich peridotitic source for the such lavas based on the MgO–FeO* relationship. We return to this issue here (Fig. 2.7; modified after Sen 2001). In Fig. 2.7 we have plotted Deccan tholeiites (data source: GEOROC Database), selected Deccan picrites (data sources: Krishnamurthy et al. 2000; Melluso et al. 2006), PMK, and MORB glasses from the Indian Ocean (data source: http://www.petdb.org/petdbWeb/index.jsp). Also shown are the fields of the uncontaminated/least contaminated Ambenali Formation basalts and the most contaminated Bushe Formation basalts. We compare them with experimentally derived peridotite melting trend and calculated 0.2 GPa fractionation trends from two hypothetical primary magmas with MgO ~15wt%. This diagram shows the source difference between the PMK Ambenali-like lavas quite well. The "extra Fe" could indicate a Fe-rich peridotitic source as opposed to a more magnesian peridotite that would generate Indian Ocean MORB type melt. Note that in the case of Hawaiian magma source Humayun et al. (2004) also proposed a Fe-rich source peridotite that may have picked up the excess Fe from the Outer Core. In passing, we should mention that Tuff et al. (2005) have suggested that garnet pyroxenite in an otherwise peridotitic plume can also give rise to relatively Fe-rich melts.

We submit that the newly calculated primary magma (PMK) may represent only a small volume of somewhat alkalic or transitional magma that may not have given rise to the more voluminous, Fe-rich, Ambenali-like magmas (compare the computed primary magma composition with Amb Melt 2 and Amb Melt 1 compositions from Sen 1995; and also, Melluso et al. 2006; Sheth and Melluso 2008). Interestingly, our calculated primary magma falls on the experimentally determined magnesian ("normal") peridotite solidus at ~3.5 GPa.

On the basis of the discussion above we conclude that Deccan primary magma generation could have involved multiple sources, but there is no doubt that the primary magmas that yielded the Pavagadh picrites were among the hottest. The highest potential temperature recorded by the new calculated primary magma PMK is about 1,550°C, which can be taken as the maximum temperature within the DMA. We have suggested that the great variation in SiO_2 and Al_2O_3 shown by the more voluminous and evolved Deccan tholeiites was a result of melting of ancient "orogenic" eclogitic pockets buried at various levels within the Deccan lithosphere.

It is interesting to note that Sheth (2005b) presented a model that had some elements very much in common with what we have presented above: his model calls for eclogites of prior subduction origin buried within the lithosphere, and then he attributed the origin of Deccan magmas to large scale melting of these eclogite blocks *via* a lithosphere delamination scheme. While we agree that deeply buried eclogites may have played a role in the production of the more voluminous tholeiites, however, such eclogites could not have had a subduction origin but instead are metamorphosed lower crustal rocks. The biggest difference between Sheth's model and ours is in the process of magma generation – Sheth proposed simple continental extension without the aid of a plume whereas our model requires a large plume as the driving thermal and magma production force.

2.6 Verification of the Picrite Parent Hypothesis: Insights from Geophysics

It is important to consider the constraints imposed on all petrological models from a geophysics perspective. The common Deccan tholeiite basalts with less than 8% MgO have undergone extensive mixing and ol + pl + cpx fractionation within the shallow (6 km) crust, suggesting the presence of extensive network of intrusive complexes at about that depth (Gangopadhyay et al. 2005). However, prior to such fractionation in shallow dikes and sills, the parent magmas could have dropped off as much as 25–35% olivine crystals somewhere deeper in the crust or mantle (Sen 1995; Sheth 2005b). The fraction of olivine could certainly be much less if some of the parental magmas were largely composed of eclogite-derived melts and small amounts of peridotite-derived melts.

Cox (1980) was an early proponent of deriving all flood basalts from picritic primary magma. He argued that ascending picritic melts would stagnate at the Moho because they would be too dense relative to the continental crust. Here these magmas would crystallize significant amount of dunite cumulates and regain the buoyancy that would allow them to rise higher up through feeder fractures.

Based on geochemistry and Cox's conjecture, we have now narrowed the problem to finding thick and laterally extensive dunite cumulate bodies at the Moho and shallower intrusives in the upper crust at ~6 km. The shallow intrusive complexes related to the Deccan have been found along the west coast and more extensively in the Narmada-Son rift (e.g., Sheth et al. 2009). Therefore, their existence is not questionable. One must resort to geophysics to address the bigger question, that is, whether extensive dunite bodies exist at the Moho.

Pandey (2008) made an interesting study of shear wave velocities along a NW–SE transect that cuts across the Deccan – it starts from slightly to the northeast of Mumbai and ends close to the coast, cutting across the thickest part of the Deccan. Figure 2.8 shows smoothed version of one of his two depth – V_s profiles along with the global mean for shields and platforms (Christensen and Mooney 1995, cited by Pandey 2008). The most appealing feature of the depth – V_s profile is that the velocity is significantly greater than the global mean at 0–20 km (shallow to intermediate crust), similar to the global mean in the intermediate-to-deeper

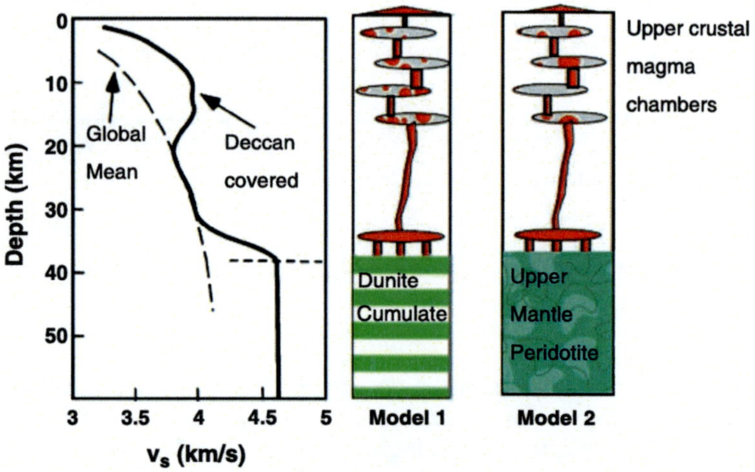

Fig. 2.8 The *left panel* shows a S-wave velocity profile beneath the main Deccan province (Modified from Pandey 2008). The "global mean" represents global mean V_s beneath normal shields and platforms (as quoted by Pandey from Christensen and Mooney 2005). Model 1 is our preferred interpretation of the crustal structure from the velocity profile. Model 2 is our interpretation of Pandey's (2008) inference (see text for further discussion)

crust (20–33 km), and then jumps up to a more mantle-like value, and accordingly, Pandey places the Moho at about 38 km.

We present two interpretative models of Pandey's (2008) velocity profile in Fig. 2.8. Pandey (2008) attributed the large velocity increase in the shallow crust to Deccan Trap intrusive complexes; and he suggested that the material below ~38 km is mantle peridotite. However, we suggest an equally plausible alternative hypothesis: the high velocity materials below ~38 km are dunite cumulates that were the early fractionates from picritic parental magmas of the Deccan lavas. The difference between the two models is that while the "dunite cumulate model" agrees with Cox's (1980) hypothesis, Pandey's model does not allow for large amounts of olivine fractionation and therefore is contrary to it. Moreover, Pandey's model requires the production of great volumes of non-picritic parent magmas. While discussing the elemental geochemistry, we argued for a mixed eclogite–peridotite source for the dominant tholeiites and garnet peridotite source for the picrites. Given the dominance of the tholeiites, and rather restricted occurrence of picrites in rift zones, Pandey's model, i.e., the high velocity material below about 40 km is peridotite, is quite plausible although other materials such as eclogite can also contribute to such high velocity. Unfortunately, seismic velocities are just not good enough to distinguish between these various possibilities.

2.7 A Model

We have considered the magma production volume, rapid eruption, difficulty in explaining the extent of contamination, wide major and trace element diversity of Deccan tholeiites, eclogite melting and the requirement of picritic parent magma (therefore an implied peridotitic source) in some areas. These considerations are best explained with a model that requires the generation of both picritic magma from a peridotite source and the dominant tholeiitic magmas from a mixed source – a mixture of eclogitic blocks and surrounding lithospheric peridotite (dominant). Picritic magmas were produced from the peridotite source at ~1,550°C and >75 km. It is difficult to constrain the generation conditions of the magma from eclogite blocks, but given the extent of thermal anomaly, they would have melted entirely as they were heated (Yaxley 2000). The estimated potential temperature of 1,550°C would have made the DMA significantly hotter than the shallow convecting mantle. This excess temperature, enormous volume of the erupted lavas, short eruption span (the bulk of them erupting in less than 1 million year), age track of Deccan–Laccadives–Maldives-Reunion, and broad isotopic similarity of the "common signature" lavas to those from the Reunion Hot Spot, all favor a plume origin of the Deccan Traps.

We have suggested a dual source, i.e., eclogite or charnockite and a Fe-rich peridotite, as have others (Cordery et al. 1997; Tuff et al. 2005; Sheth 2005b). There is an important difference: the eclogite in our model is a part of the continental lithosphere (also in Sheth 2005b) and not recycled subducted oceanic lithosphere brought up by the plume as suggested by the other authors for other geologic areas (e.g., Campbell 2005). In this sense this model is similar to that presented by Camp and Hanan (2008) for the Columbia River flood basalts of northwestern USA.

We use a four-component isobaric phase diagram (Fig. 2.9; source: Milholland and Presnall 1998) to explain how our dual source melting model "works" even though at least one important component, FeO, is missing from this diagram. We focus on the "eclogite surface" (green) and three invariant points – point "PerM" where a spinel–garnet peridotite would melt, point "ChM" where a charnockitic assemblage quartz + opx + cpx + gt) would melt, and point "M" (refer to Fig. 2.9 and Table 7 in Milholland and Presnall 1998), where a kyanite–quartz eclogite assemblage would melt. The temperature difference between the PerM and point M point is about 170°C whereas that between PerM and ChM is 100°C. Thus, by virtue of having significantly lower melting temperatures, charnockitic and eclogitic blocks (metamorphosed lower crustal blocks), embedded deep in the lithosphere due to ancient orogenic events, will melt to a large degree even though the surrounding peridotitic component of the lithosphere or the plume itself may not melt until the temperature is raised by at least 100°C to reach peridotite solidus (PerM). The hottest zone in the plume would produce relatively picritic magmas, whereas contributions would be produced from melting of old crustal blocks embedded in the lithosphere and some likely by fractionation of picritic magmas.

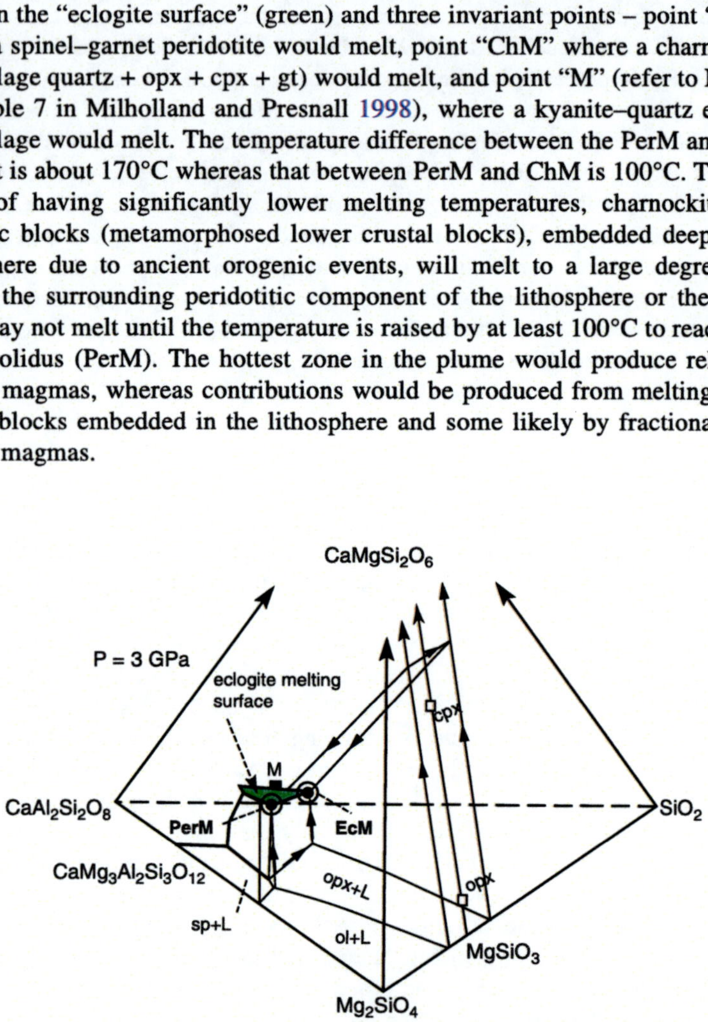

Fig. 2.9 Slightly simplified liquidus diagram CaO–MgO–Al_2O_3–SiO_2 at 3 GPa from Milholland and Presnall (1998). The *colored surface* represents the surface where melts would form from eclogite-like assemblages. "PerM" is the invariant point ol + opx + cpx + gt + sp + liq. "ChM" is the isobaric invariant point opx + cpx + qz + gt + liq. "M" represents kya + qz + cpx + gt + liq. (see text for further discussion)

Sheth (2005b) also suggested the presence of large eclogite bodies in the subcontinental lithosphere; however, instead of a plume, he suggested that these bodies were remnants of ancient subduction; and the Deccan Trap magmas formed when such bodies delaminated from the lithosphere and were heated as the asthenosphere was raised to fill the "void". For reasons stated earlier, we feel that a plume is necessary. Sheth (2007) pointed out that the Deccan lithosphere was not significantly uplifted as would be required by the thermal models. In our model, melting of large eclogite blocks in the continental craton resulting in the lower lithosphere erosion would offset the buoyancy exerted by the plume.

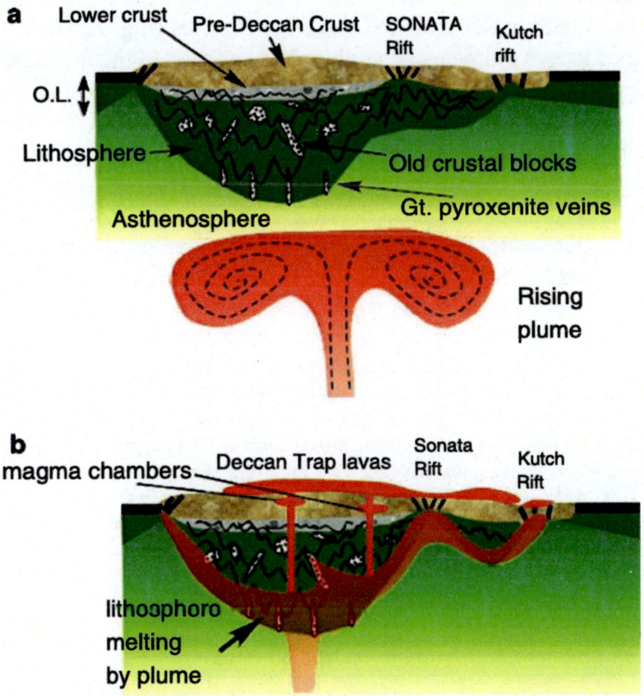

Fig. 2.10 A schematic model for the origin of Deccan Traps volcanism by melting of lithosphere and plume. (**a**) Pre-Deccan lithosphere shown with a thick crust (*brown*) and a thick lithospheric keel. A marginal rift zone, such as the Kutch rift zone, and the interior rift zone (Narmada-Son or SONATA rift zone) are also shown. The lithosphere is assumed to be largely peridotitic or harzburgitic with blocks, lenses, and veins of garnet clinopyroxenite or eclogite. The plume has not impacted the lithosphere in this stage. (**b**) The plume impacts the lithosphere and largely melts the embedded eclogite bodies and some of the surrounding lithosphere. These melts are thought to be the dominant tholeiites. The picrites found in the rift zones near the peripheries of the Deccan Traps are derived from primary magmas such as the PMK and they tap the actual plume materials

In this connection, the investigation by Kumar et al. (2007) shows that the Indian lithosphere to be only ~100 km thick whereas the once contiguous Gondwanaland lithospheres beneath the other southern continents were 180–300 km thick. These authors suggested that the plume that severed the Gondwana continents was responsible for melting the lower lithosphere beneath India although they remarked that the role of the Deccan plume causing such lithospheric erosion could not be ascertained. In our view, the plume that produced Deccan picrites also melted the eclogite bodies embedded in the lower lithosphere and thereby eroded the lower lithosphere.

Figure 2.10 is a model that effectively summarizes our thoughts about Deccan Trap volcanism. Like many authors before us, we suggest that a large plume, rising perhaps from the core/mantle boundary, was the primary cause of this flood basalt event. Following Kumar et al. (2007), it is possible that the pre-Deccan continental lithosphere had an uneven thickness: it was generally 200 km thick beneath cratonic areas and relatively thin (100 km or less) beneath pre-existing rift zones, such as the Kutch and the Son-Narmada (SONATA) rift zones (Fig. 2.10a; e.g., Sen et al. 2009). As the Deccan plume came up it melted and eroded much of the lower lithosphere that had "inclusions" of dispersed blocks of old orogenic eclogite bodies producing contaminated and differentiated appearing tholeiites. Where parts of the plume could rise to shallower levels beneath pre-thinned lithosphere (such as beneath ancient rifts), it produced picritic magmas at ~1,550°C and 2.3 GPa. These picrites produced dunite cumulates near the Moho and gabbroic fractionates in the shallow crust. Our model has components that are similar to hypotheses put forth by "plumists" and "non-plumists" as well and emphasizes a complex magma production history from the plume as well as the ancient continental lithosphere.

Acknowledgments We are pleased to dedicate this review paper to the late Prof. Mihir K. Bose, a leader in Indian petrology and a Deccan Traps enthusiast. We were fortunate to make Professor Bose aware of this volume being prepared in his honor before he passed away. It is unfortunate that neither of us has collaborated with Prof. Bose; however, his papers certainly intrigued us and kept us thinking of the larger issues. We are very grateful to Dr. Hetu Sheth for a thorough review and constructive suggestions that led to some important changes in the original manuscript. We thank Kevin Chau for his help with running Lee et al.'s (2009) program to calculate P, T. We are also thankful to Prof. S. Viswanathan for his critical comments and editing the manuscript.

References

Albarède F (1992) How deep do common basaltic magmas form and differentiate?. J Geophys Res 97:10997–11009

Anderson DL (2006) Speculations on the nature and cause of mantle heterogeneity. Tectonophysics 416:7–22

Anderson DL, Natland J (2005) A brief history of the plume hypothesis and its competitors: concept and controversy. Geol Soc Am Spec Pap 388:119–145

Basu AR, Renne PR, Dasgupta DK, Teichmann F, Poreda RJ (1993) Early and late alkali igneous pulses and a high-3He plume origin for the Deccan flood basalts. Science 261:902–906

Beane JE, Hooper PR (1988) A note on the picrite basalts of the Western Ghats, Deccan Traps, India. In: Subbarao KV (ed) Deccan flood basalts. Memoirs of Geological Society of India 10, Bangalore, pp 117–134

Bose MK (1972) Deccan basalts. Lithos 5:131–145

Bose MK (1995) Deccan basalts, mantle activity and extensional tectonics. Gond Geol Mag Sp 2:267–281

Camp VE, Hanan BB (2008) A Plume-triggered Delamination Origin for the Columbia River Basalt Group. Geosphere 4:480–495

Campbell IH (2005) Large igneous provinces and the mantle plume hypothesis. Elements 1:265–269

Chandrasekharam D, Parthasarathy A (1978) Geochemical and tectonic studies on the coastal and inland Deccan Trap volcanics and a model for the evolution of Deccan Trap volcanism. N Jb Min Abh 132:214–229

Chatterjee N, Bhattacharji S (2008) Trace element variations in Deccan basalts: roles of mantle melting, fractional crystallization and crustal assimilation. J Geol Soc India 71:171–188

Chatterjee S, Guven N, Yoshinobu A, Donofrio R (2006) Shiva Structure: a possible KT boundary impact crater on the western shelf of India. Museum Texas Tech Univ Spec Publ 50:39

Chatterjee S, Rudra DK (1992) KT events in India: impact, rifting, volcanism and dinosaur extinction. Memoir Qld Museum 39:489–532

Chenet A-L, Quidelleur X, Fluteau F, Courtillot V, Bajpai S (2007) ^{40}K-^{40}Ar dating of the Main Deccan large igneous province: Further evidence of KTB age and short duration. Earth Planet Sci Lett 263:1–15

Christensen NI, Mooney WD (1995) Seismic velocity structure and composition of the continental crust: a global view. J Geophys Res 100:9761–9788

Class C, Goldstein SL (2005) Evolution of helium isotopes in the Earth's mantle. Nature 436:1107–1112

Cordery MJ, Davies GF, Campbell IH (1997) Genesis of flood basalts from eclogite-bearing mantle plumes. J Geophys Res 102(20):179–20, 197

Cox KG (1980) A model for flood basalt vulcanism. J Petrol 21:629–650

Cox KG, Hawkesworth CJ (1985) Geochemical Stratigraphy of the Deccan Traps at Mahabaleshwar, Western Ghats, India, with implication for open system processes. J Petrol 26:355–387

Courtier A and 14 others (2007) Comparison of seismic and petrologic thermometers suggest deep thermal anomalies beneath hot spots. Earth Planet Sci Lett 264:308–316

Dalrymple GB (1969) 40Ar/36Ar Analyses of Historic Lava Flows. Earth Planet Sci Lett 6:47–55

Duncan RA, Pyle DG (1988) Rapid eruption of the Deccan flood basalts at the Cretaceous/Tertiary boundary. Nature 333:841–843

Duncan RA, Richards MA (1991) Hotspots, mantle plumes, flood basalts, and true polar wander. Rev Geophys 29:31–50

Faernetani CG, Richards MA (1994) Numerical investigations of the mantle plume initiation model for flood basalt events. J Geophys Res 99:13813–13833

Foulger, G, Natland, JH, Presnall, DC, Anderson, DL (eds) (2005) Plates, plumes, and paradigms. Geological Society of America Special Volume 388, New York, 881 pp

Gangopadhyay A, Sen G, Keshav S (2005) Experimental crystallization of Deccan Basalts at low pressure: effect of contamination on phase equilibrium. A. De Volume. Ind J Earth Sci 75:54–71

Glazner AF (2007) Thermal limitations on incorporation of wall rock into magma. Geology 35:319–322

Hanyu T, Dunai TJ, Davies GR, Kaneoka I, Nohda S, Uto K (2001) Noble gas study of the Reunion hotspot: evidence for distinct less-degassed mantle sources. Earth Planet Sci Lett 193:83–98

Hari KR, Santosh M, Chatterjee AC (1991) Primary silicate-melt inclusions in olivine phenocrysts from the Pavagad igneous suite, Gujarat. J Geol Soc Ind 37:343–350

Herzberg C, Gazel E (2009) Petrological evidence for secular cooling in mantle plumes. Nature 458:619–622

Holden J, Vogt PR (1977) Graphic solutions to problems of plumacy. EOS Trans AGU 56:573–580

Humayun M, Qin L, Norman MD (2004) Geochemical evidence for excess iron in the mantle beneath Hawaii. Science 306:91–94

Jacob DE (2004) Nature and origin of eclogite xenoltihs from kimberlites. Lithos 77:295–316
Keller G, Adatte T, Gardin S, Bartolini A, Bajpai S (2008) Main Deccan volcanism pulse ends near the K-T boundary: evidence from the Krishna–Godavari Basin, SE India. Earth Planet Sci Lett 268:293–311
Keshav S, Gudfinnsson GH (2004) Silica-poor, mafic alkaline lavas from ocean islands and continents: Petrogenetic constraints from major elements. Proc Indian Acad Sci (Earth Planet Sci) 113:723–736
Kogiso T, Tatsumi Y, Nakano S (1997) Trace element transport during dehydration processes in the subducted oceanic crust: 1. Experiments and implications for the origin of ocean island basalts. Earth Planet Sci Lett 148:193–205
Krishnamurthy P, Gopalan K, Macdugall JD (2000) Olivine compositions in picrite basalts and the Deccan volcanic cycle. J Petrol 41:1057–1069
Kumar P, Yuan X, Ravi Kumari M, Kind R, Li X, Chadha RK (2007) The rapid drift of the Indian tectonic plate. Nature 449:894–897
Kurz D, Jenkins WJ, Hart SR, Clague D (1983) Helium isotopic variations in volcanic rocks from Loihi Seamount and the Island of Hawaii. Earth Planet Sci Lett 66:388–406
Lee C-TA, Luffi P, Plank T, Dalton H, Leeman W (2009) Constraints on the depths and temperatures of basaltic magma generation on Earth and other terrestrial planets using new thermobarometers for mafic magmas. Earth Planet Sci Lett 279:20–33
Lightfoot PC, Hawkesworth CJ, Sethna SF (1987) Petrogenesis of rhyolites and trachytes from the Deccan Trap: Sr, Nd and Pb isotope and trace element evidence. Contrib Miner Petrol 95:44–54
Lightfoot P, Hawkesworth C (1988) Origin of Deccan Trap lavas: evidence from combined trace element and Sr-, Nd- and Pb-isotope studies; Earth Planet Sci Lett 91:89–104
Mahoney JJ (1988) Deccan traps. In: Macdougall JD (ed) Continental flood basalts. Kluwer, Dordrecht, pp 151–194
Mahoney JJ, Duncan RA, Khan W, Gnos E, McCormick GR (2002) Cretaceous volcanic rocks of the South Tethyan suture zone, Pakistan: implications for the Réunion hotspot and Deccan Traps. Earth Planet Sci Lett 203:295–310
McKenzie D, Bickle MJ (1988) The volume and composition of melt generated by extension of the lithosphere. J Petrol 29:625–679
Melluso L, Beccaluva L, Brotzu P, Gregnanin A, Gupta AK, Morbidelli L, Traversa G (1995) Constraints on the mantle sources of the Deccan Traps from the petrology and geochemistry of the basalts of Gujarat State (Western India). J Petrol 36:1393–1432
Melluso L, Mahoney JJ, Dallai L (2006) Mantle sources and crustal input in Mg-rich Deccan Trap basalts from Gujarat (India). Lithos 89:259–274
Milholland CS, Presnall DC (1998) Liquidus phase relations in the $CaO–MgO–Al_2O_3–SiO_2$ system at 3.0 GPa: the aluminous pyroxene thermal divide and high-pressure fractionation of picritic and komatiitic magmas. J Petrol 39:3–27
Morgan WJ (1972) Plate motions and deep mantle convection. Geological Society of America Memoir, Boulder
Pandey OP (2008) Deccan Trap volcanic eruption affected the Archaean Dharwar Cratos of southern India: seismic evidences. J Geol Soc India 72:510–514
Peng ZX, Mahoney J, Hooper P, Harris C, Beane J (1994) A role for lower continental crust in flood basalt genesis? Isotopic and incompatible element study of the lower six formations of the western Deccan Traps. Geochim Cosmochim Acta 58(1):267–288
Peng ZX, Mahoney JJ (1995) Drillhole lavas in the Deccan Traps and the evolution of the Reunion plume. Earth Planet Sci Lett 134:169–185
Pertermann M, Hirschmann MM (2003) Anhydrous partial melting experiments on MORB-like eclogite: phase relations, phase compositions and mineral-melt partitioning of major elements at 2–3 GPa. J Petrol 44:2173–2201
Putirka K (2008) Excess temperatures at ocean islands: Implications for mantle layering and convection. Geology 36:283–286

Richards MA, Duncan RA, Courtillot VE (1989) Flood basalts and hotspot tracks: plume heads and tails. Science 246:103–107

Self S, Thordarsson T, Widdowson M, Jay A (2006) Volatile fluxes during flood basalt eruptions and potential effects on global environment: A Deccan perspective. Earth Planet Sci Lett 248:518–532

Self S, Blake S, Sharma K, Widdowson M, Sephton S (2008) Sulfur and chlorine in Late Cretaceous Deccan magmas and eruptive gas release. Science 319:1684–1686

Sen G (1986) Mineralogy and petrogenesis of the Deccan Trap lavas around Mahabaleshwar, India. J Petrol 27:627–663

Sen G (1988) Possible depth of origin of primary Deccan tholeiite magma. In: Subbarao KV (ed) Deccan flood basalts. Memoir of the Geological Society of India 10, Bangalore, pp 35–51

Sen G (1995) A simple petrologic model for the generation of Deccan Trap magmas. Int Geol Rev 37:825–850

Sen G (2001) Generation of Deccan trap basalt. Proc Natl Acad Sci India 110:409–431

Sen G, Borges M, Marsh BD (2006) A case for short duration of Deccan trap eruption. EOS 87(20):197–204

Sen G, Bizimis M, Das R, Paul DK, Biswas S (2009) Deccan plume, lithosphere rifting, and volcanism in Kutch, India. Earth Planet Sci Lett 277:101–111

Sheth H (2005a) From Deccan to Reunion: no trace of a mantle plume. In: Foulger GR, Natland JH, Presnall DC, Anderson DL (eds) Plates, plumes, and paradigm. Geological Society of America Special Papers 388, New York, pp 477–501

Sheth H (2005b) Were the Deccan flood basalts derived in part from ancient oceanic crust within the Indian continental lithosphere? Gondwana Res 8:109–127

Sheth HC, Pande K, Bhutani R (2001) 40Ar-39Ar ages of Bombay trachytes: evidence for a Paleocene phase of Deccan volcanism. Geophys Res Lett 28:3513–3516

Sheth HC, Ray JS, Ray R, Vanderkluysen L, Mahoney JJ, Kumar A, Shukla AD, Das P (2009) Geology and geochemistry of Pachmarhi dykes and sills, Satpura Gondwana Basin, central India: problems of dyke-sill-flow correlations in the Deccan Traps. Contrib Miner Petrol. doi:10.1007/s00410-009-0387-4

Sheth, Melluso L (2008) The Mount Pavagadh volcanic suite, Deccan traps: geochemical stratigraphy and magmatic evolution. J Asian Earth Sci 32:5–21

Shirey SB, Carlson RW, Richardson SH, Menzies A, Gurney JJ, Pearson DG, Harris JW, Wiechart U (2001) Archean emplacement of eclogitic components into the lithospheric mantle during formation of the Kaapvaal Craton. Geophys Res Lett 28:2509–2512

Takahashi E, Kushiro I (1983) Melting of a dry peridotite at high pressures and basalt magma genesis. Am Miner 68:859–879

Tuff J, Takahashi E, Gibson SA (2005) Experimental constraints on the role of garnet pyroxenite in the genesis of high-Fe mantle plume derived melts. J Petrol 46:2023–2058

White R, McKenzie D (1995) Mantle plumes and flood basalts. J Geophys Res 100:17543–17585

Wilson JT (1963) A possible origin of the Hawaiian Islands. Can J Phys 41:863–870

Yaxley GM (2000) Experimental study of the phase and melting relations of homogeneous basalt +peridotite mixtures and implication for the petrogenesis of flood basalts. Contrib Miner Petrol 139:326–338

Chapter 3
A Review of the Radiometric Data Placing the Hawaiian–Emperor Bend at 50 Ma; Placing Constraints on Hypotheses Concerning the Origin of the Hawaiian–Emperor Volcanic Chain

Ajoy K. Baksi

Abstract Recent $^{40}Ar/^{39}Ar$ work reveals that the Bend in the Hawaiian–Emperor Chain (HEB) occurred at 50 Ma, not at 43 Ma, as postulated over 30 years ago. All relevant argon data for seamounts in the Hawaiian–Emperor Chain have been evaluated for statistical validity, as well as for accuracy, based on the freshness of the phases dated. The Pacific Plate motion decreased steadily from ~65 to ~35 Ma. No marked change in the rate of plate motion occurred at the time of the Bend. Hypotheses pertaining to the formation of the Hawaiian–Emperor Chain must (1) be based on radio metric data that have been critically evaluated and (2) explain both the geometry and timing of various features, such as the HEB.

3.1 Introduction

A major theme of plume (hotspot) advocates the progression of ages in the Hawaiian–Emperor Chain (e.g. Grotzinger et al. 2007). Over the last 3 decades the bend in the Chain has been taken to have occurred at ~43 Ma, and thought to denote a major change in motion of the Pacific Plate. Concern has been expressed over this interpretation, since other plates do not show a change in velocity at that time. This rapid change in plate motion from ~345° to 285° at the HEB (in compass terms, measuring clockwise from north) should be linked to a major tectonic event. A recent report (Sharp and Clague 2006) indicates the age of the initiation of the HEB occurred at ~50 Ma, rather than at ~43 Ma. Sharp and Clague (2006) suggested: (a) the earlier (incorrect) ages (Clague and Dalrymple 1975; Dalrymple and Clague 1976), resulted from dating of postshield material at each location; (b) a "jerky" set of changes in Pacific Plate motion occurred from ~65 to ~35 Ma (Sharp and Clague 2006, Fig. 3.2); (c) initiation of subduction in the Izu–Bonin–Marianas Arc (or the Tonga–Kermadec Trench) may have been responsible for the

A.K. Baksi (✉)
Department of Geology and Geophysics, Louisiana State University, Baton Rouge, LA 70803, USA
e-mail: akbaksi@yahoo.com

J. Ray et al. (eds.), *Topics in Igneous Petrology*, DOI 10.1007/978-90-481-9600-5_3,

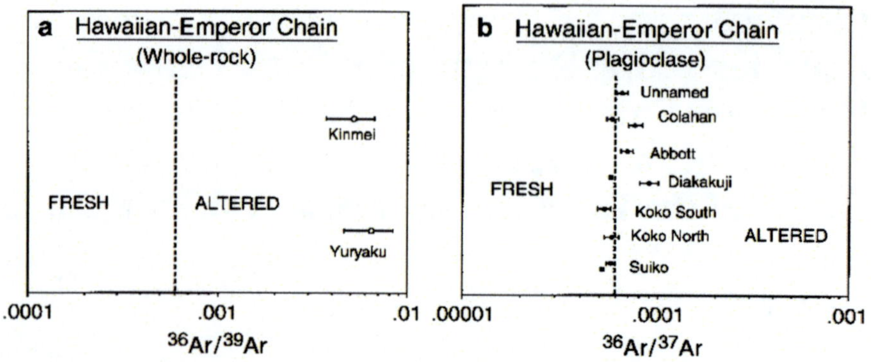

Fig. 3.1 Assessment of the freshness of samples from the Hawaiian–Emperor Chain dated by the $^{40}Ar/^{39}Ar$ stepheating method. The average AI values (plateau sections only) are shown with the associated standard error on the mean, on a *log* scale. (**a**) Whole-rock samples of Dalrymple and Clague (1976). AI ($^{36}Ar/^{39}Ar$) values are considerably > 0.0006, the cutoff value for freshness (see Baksi 2007a). All samples are altered and gave plateau ages that underestimate the time of crystallization. (**b**) Acid leached plagioclase separates of Sharp and Clague (2006). All plagioclase samples show AI values at or below the cutoff ($^{36}Ar/^{37}Ar$ <0.00006 – see Baksi 2007a) for freshness. Samples are unaltered and plateau ages yield good estimates of the time of crystallization. The age of the HEB is 50 Ma (Sharp and Clague 2006)

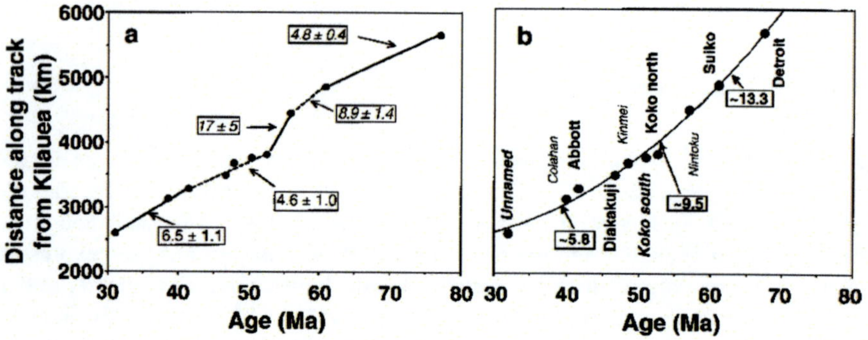

Fig. 3.2 Calculation of the rate of motion of the Pacific Plate from ~80 Ma to ~30 Ma. $^{40}Ar/^{39}Ar$ ages plotted versus the distance of the seamount from Kilauea (following Sharp and Clague 2006). *Values in boxes* denote calculated rates in cm year^{-1}. (**a**) The results of Sharp and Clague (2006) – see their Fig. 2, implies rapid and jerky changes in plate motion. (**b**) Utilizing the most reliable ages as defined herein. A major change in the age of the Detroit Seamount, and the use of higher order polynomials for the fit, leads to smooth changes ("deceleration") in plate motion. See text

HEB. It is shown herein, that (1) the earlier incorrect ages result from dating of altered material (2) a *smooth* change in Pacific Plate motion (rate) occurred from ~65 to ~35 Ma and (3) the hard collision of India with Asia, possibly leading to the creation of subduction in the Izu–Bonin–Marianas Arc, may explain both the timing and nature of the HEB.

3.2 Methodology

Over the past decade, it has been shown that numerous $^{40}Ar/^{39}Ar$ ages for sea floor rocks do not approximate crystallization ages. Plateau/isochron sections of stepheating experiments, must pass the relevant statistical tests for validity (Baksi 1999, 2005, 2007a). The cutoff for acceptance of ages therein was set at >5% probability ($p > 0.05$); Sharp and Clague (2006) suggest a more rigorous test with ages being deemed statistically acceptable only if $p > 0.15$. Altered rocks can yield statistically acceptable plateau/isochron "ages" that generally underestimates the true crystallization age (Baksi 2007a, b) and therefore poses problem. It is critical to quantitatively evaluate the amount of ^{36}Ar in rocks/minerals used for argon dating; levels are low in fresh anhydrous mineral phases, but increase markedly during alteration, resulting from the introduction of water and high amounts of dissolved atmospheric argon. Details of the alteration index (AI) technique to detect chemical weathering, its use in evaluating argon ages, have been set out elsewhere (Baksi 2007a, b).

New $^{40}Ar/^{39}Ar$ dates of carefully selected mineral-separates from rocks in the Hawaiian–Emperor Chain (Sharp and Clague 2006) have led to a revised age for the bend in the Chain. It was argued that the increase in the ages of various seamounts and the new age of the HEB, resulted primarily from the (earlier) dating of post-shield or rejuvenated lavas from individual eruptive centers. Such approach should produce "ages" for the seamounts that are young by 1–2 million years; however, the new ages (Sharp and Clague 2006) are generally 5–7 million years older than the previous ones (e.g. Dalrymple and Clague 1976). The latter utilized large samples (hundreds of milligram) of whole-rock material for step-heating purposes. Careful selection with binocular examination of individual grains was not possible and led to the use of altered material. Utilizing the AI technique (Baksi 2007a) it is seen that the samples of Dalrymple and Clague (1976) were substantially altered (see Fig. 3.1a). Although their plateau/isochron ages pass the statistical tests for robustness, they underestimate the age of crystallization. Subsequent work has shown that leaching ferromagnesian samples with acid (preferably nitric acid), removes alteration products. Acid leached whole-rock samples yield much better estimates of the crystallization age (Baksi and Archibald 1997; Baksi 2007a).

Sharp and Clague (2006), utilized very small proportions (generally a few milligrams) of mineral separates, which had been chemically treated. For feldspars, leaching with hydrofluoric acid prior to dating (Dalrymple and Lanphere 1969) best removes alteration products; use of other acids is less successful (Baksi 2007a). As detailed in the section below, the samples of Sharp and Clague (2006) were generally free of alteration and hence gave reliable crystallization ages. They also utilized the results of whole rock dating of Hawaiian–Emperor Chain seamounts (Keller et al. 1995; Duncan and Keller 2004). Sharp and Clague (2006) concluded that the HEB results from Pacific Plate motion change; they suggested "jerky" changes in Pacific Plate motion from ~65 to ~35 Ma (see Fig. 3.2a). Rather, gradual and smooth changes in plate motion are the norm (see Cande and Kent 1992; Baksi 1993, 1995). Herein, all pertinent ages are evaluated for statistical validity of plateau/isochron ages and also for freshness of material dated by the AI technique. Only those ages with proper statistics ($p > 0.15$), involving fresh whole-rock or mineral separates, have been accepted, as valid estimates of the time of crystallization.

3.3 Critical Evaluation of the Argon (Age) Data Sets

All $^{40}Ar/^{39}Ar$ data, relevant to the age of the HEB are examined critically. Errors in plateau/ isochron ages and initial $^{40}Ar/^{36}Ar$ ratios, are listed at the 2 sigma level. The appropriate measure of scatter in the data, the mean square weighted deviate (MSWD) is designated by F; the corresponding probability value, taking into account the number of steps involved, obtained from Chi Square Tables, is listed as p. Where $F > 1$ and $p > 0.15$, the errors have been multiplied by $(F)^{1/2}$. The methodology and rationale for using ^{36}Ar levels as a quantitative parameter for assessing the alteration of rocks/minerals is elaborated elsewhere; the method works best for whole-rock basalts and plagioclase feldspar, however the method is less refined for potassium feldspar and hornblende (Baksi 2007a, c).

First, I look to the data of Sharp and Clague (2006), whose ages are statistically acceptable, though there are minor errors in calculation of errors and listing of MSWD values for plateau ages – see Table 3.1. Most of the mineral samples are relatively unaltered (see Fig. 3.1b). The ages of Koko (south), Kinmei, Colahan and Unnanmed seamounts are each taken to be ~2 million years older than derived in Table 3.1, ~51, ~48.5, ~40 and ~32 Ma, respectively, as the dated samples belong to the postshield phase of volcanism (Sharp and Clague 2006). In general, the samples used by these authors appear fresh, and yielded accurate estimates of the time of crystallization.

Other ages on the Hawaiian–Emperor Chain are assessed in Table 3.2. Keller et al. (1995) used an age of 27.7 Ma for Fish Canyon Tuff Biotite 3 and Duncan and Keller (2004) used an age of 28.04 Ma for the same standard and a decay constant of 5.53×10^{-10} year^{-1} for ^{40}K. To bring these ages into a line with those for standards used by Sharp and Clague (2006), the correct age for FCT-3 Bio is 28.23 Ma (Baksi 2003; Baksi et al. 1996) and the decay constant needs to be adjusted to 5.543×10^{-10} year^{-1}. Initial $^{40}Ar/^{36}Ar$ ratios on isochron plots measurably lower than the atmospheric value (295.5), denote disturbed samples (Lanphere and Dalrymple 1978) and the corresponding ages are rejected. The major conclusions are: (1) the age of Detroit Seamount is ~67 Ma and not ~77–82 Ma. Three samples yielding the older ages, are based on altered samples, whereas the specimen yielding the age of 67 Ma is fresh. (2) The age of Meiji Seamount is not known accurately. Three ages of 33–38 Ma obtained by Duncan and Keller (2004), were on samples that have suffered alteration, and two other samples did not yield proper (statistically acceptable) plateaus. The (plagioclase) feldspar samples show unusual Ca/K ratios, and result from the inclusion of secondary potassium feldspar (see Dalrymple et al. 1980). The best radiometric estimate for the age of this Seamount is >62 Ma (Dalrymple et al. 1980) and its age is ~72 Ma based on the fossil content of the overlying sediments (Worsley 1973; Sager 2007). (3) The age of Nintoku Sea mount is not well constrained, as all

Table 3.1 Assessment of the $^{40}Ar/^{39}Ar$ ages of Sharp and Clague (2006) on the Hawaiian–Emperor Chain

Seamount	Sample number	Plateau age (Ma)	Statistical evaluation	Alteration index	Validity of age
Suiko	433A-20-1 Plagioclase	61.07 ± 21	F = 0.98(8) p ~ 0.48	0.000052 ± 1	Accepted, sample fresh
Suiko	433B-5-3 Plagioclase	60.64 ± 30	F = 1.33(7) p ~ 0.24	0.000057 ± 3	Accepted, sample fresh
Suiko	433C-31-1 Plagioclase	60.84 ± 48	F = 0.60(5) p ~ 0.66	**0.00011 ± 3**	**Rejected**, sample altered
Koko north	A7-44-5 Plagioclase	52.64 ± 74	F = 0.70(12) p ~ 0.74	0.000058 ± 5	Accepted, sample fresh
Koko south	A7-43-12 Anorthoclase	50.22 ± 14	F = 1.69(4) p ~ 0.17	0.000019 ± 4	Accepted, sample fresh
Koko south	A7-43-23 Anorthoclase	50.40 ± 12	F = 1.14(7) p ~ 0.34	0.000017 ± 3	Accepted, sample fresh
Koko south	A7-43-57 Anorthoclase	49.97 ± 8	F = 1.28(11) p ~ 0.24	0.0000053 ± 4	Accepted, sample fresh
Koko south	A7-43-60 Anorthoclase	49.60 ± 11	F = 2.17(6) **p ~ 0.06**	**0.00008 ± 1**	**Rejected**, p < 0.15, sample altered
Kinmei	A7-51-21 Plagioclase	47.88 ± 25	F = 0.99(10) p ~ 0.45	**0.00014 ± 2**	**Rejected**, sample altered
Diakakuji	A7-55-25 Plagioclase	46.78 ± 16	F = 1.00(6) p ~ 0.42	**0.00028 ± 3**	**Rejected**, sample altered
Diakakuji	A7-55-33 Plagioclase	46.77 ± 12	F = 0.55(8) p ~ 0.80	0.00009 ± 1	Accepted, sample slightly altered
Diakakuji	A7-55-35 Plagioclase	46.55 ± 11	F = 0.31(11) p~0.98	0.000057 ± 1	Accepted, p too high, error underestimated
Abbott	L882NP-D2 Plagioclase	41.49 ± 22	F = 1.03(9) p ~ 0.41	0.000069 ± 5	Accepted, sample slightly altered
Colahan	L882NP-D4-67 Hornblende	38.82 ± 5	F = 0.16(6) p ~ 0.98	**0.00061 ± 3**	**Rejected**, sample altered
Colahan	L882NP-D4- 47 Hornblende	38.66 ± 4	F = 2.39(7) **p ~ 0.03**	**0.00093 ± 2**	**Rejected**, p < 0.15, sample altered
Colahan	L882NP-D4-67 Plagioclase	38.52 ± 9	F = 1.70(9) **p ~ 0.06**	0.000076 ± 6	**Rejected**, p < 0.15,
Colahan	L882NP-D4-47 Plagioclase	38.55 ± 11	F = 2.46(5) **p ~ 0.04**	0.000058 ± 4	**Rejected**, p < 0.15
Unnamed	72-HIG-20 Plagioclase	30.94 ± 8	F = 0.66(12) p ~ 0.78	0.000065 ± 4	Accepted, sample slightly altered

F = MSWD value, number of steps utilized shown in parentheses, p = probability of occurrence. Errors shown at last decimal place. Ages are rejected where p < 0.15. The average alteration index (AI) for the plateau steps is listed with the standard error on the mean. Samples are altered if AI > 0.00006, > 0.00006 and > 0.0002 for plagioclase, anorthoclase and hornblende respectively; the corresponding ages are rejected as valid estimates of the time of crystallization

Table 3.2 Assessment of the $^{40}Ar/^{39}Ar$ ages of Keller et al. (1995) and Duncan and Keller (2004) on the Hawaiian–Emperor Chain

Seamount	Sample number	Plateau age (Ma)	Statistical evaluation	Alteration index	Validity of age
Meiji	6R-2 7-14 Plagioclase	38.25 ± 10(P)	F = 1.72(7) **p ~ 0.11**	**0.00055 ± 7**	**Rejected**, p < 0.15, sample altered
Meiji	6R-2 75-81 Whole-rock	33.52 ± 10(P)	F = 2.34(4) **p ~ 0.07**	**0.0025 ± 5**	**Rejected**, p < 0.15, sample altered
Meiji	6R-3 3-9 Plagioclase	34.10 ± 7(P)	F = 1.28(6) p ~ 0.27	**0.0009 ± 4**	**Rejected**, sample altered
Detroit	10R-6-10 Whole rock	81.5 ± 18(I) IR = 295 ± 8	F = 0.1(5) p ~ 0.96	**0.0046 ± 22**	**Rejected**, sample altered
Detroit	32R-5 58-60 Whole rock	76.25 ± 46(P)	F = 3.58(3) **p ~ 0.03**	**0.0015 ± 5**	**Rejected**, p < 0.15, sample altered
Detroit	35R-4 93-95 Plagioclase	76.8 ± 27(P)	F = 0.36(4) p ~ 0.78	**0.0004 ± 4**	**Rejected**, sample altered
Detroit	38-1 129-131 Whole rock	75.8 ± 9(P)	F = 1.65(5) p ~ 0.16	**0.012 ± 5**	**Rejected**, sample altered
Detroit	59R-2 102-103 Whole rock	78.5 ± 11(P)	F = 1.70(6) **p ~ 0.13**	**0.014 ± 4**	**Rejected**, p < 0.15, sample altered
Detroit	36R-4 102-106 Plagioclase	67.3 ± 16(I) IR = 333 ± 42	F = 0.40(3) p ~ 0.53	0.00003 ± 2	Accepted
Nintoku	5R-2 30-34 Whole rock	55.70 ± 14(P)	F = 0.63(5) p ~ 0.64	**0.015 ± 4**	**Rejected**, sample altered
Nintoku	6R-2 25-28 Plagioclase	55.72 ± 22(P)	F = 2.62(5) **p ~ 0.03**	0.00002 ± 2	**Rejected**, p < 0.15
Nintoku	10R-2 73-75 Whole rock	55.79 ± 10(P)	F = 1.77(8) **p ~ 0.09**	**0.0013 ± 4**	**Rejected**, p < 0.15, sample altered
Nintoku	20R-5 60-65 Whole rock	55.77 ± 18(I) **IR = 265 ± 34**	F = 0.63(7) p ~ 0.95	0.00004 ± 1	**Rejected**, IR low, disturbed sample
Nintoku	39R-1 70-72 Whole rock	56.75 ± 10(I) IR = 300 ± 3	F = 1.11(8) p ~ 0.34	**0.0012 ± 4**	**Rejected**, sample altered
Nintoku	45R-1 123-126 Whole rock	56.85 ± 22(I) **IR = 279 ± 20**	F = 1.73(6) p ~ 0.15	0.0007 ± 3	**Rejected**, IR low, disturbed sample
Koko south	3R-2 116-118 Whole Rock	48.98 ± 66(P)	F = 0.25(4) p ~ 0.86	**0.018 ± 7**	**Rejected**, sample altered
Koko south	10R-1 0-3 Whole Rock	49.1 ± 11(P)	F = 1.01(7) p ~ 0.42	**0.0037± 18**	**Rejected**, sample altered
Koko south	15R-4 37-40 Whole rock	51.1 ± 10(I) **IR = 203 ± 82**	F = 0.62(4) p ~ 0.53	0.0008 ± 5	**Rejected**, IR low, disturbed sample
Koko south	10R-3 70-72 Whole rock	48.78 ± 30(P)	F = 0.53(5) p ~ 0.71	0.0002 ± 1	Accepted
Koko south	24R-3 37-39 Whole rock	49.99 ± 36(P)	F = 0.78(5) p ~ 0.56	**0.0015 ± 4**	**Rejected**, sample altered
Koko south	44R-2 120-125 Whole rock	50.15 ± 22(P) p ~ 0.87	F = 0.24(4) p ~ 0.87	0.00010 ± 3	Accepted

In the age column, P/I denotes plateau/isochron ages, IR = initial $^{40}Ar/^{36}Ar$ ratio. Errors are shown at the last decimal place. Isochron ages are rejected when IR < the atmospheric argon value (295.5). F= MSWD value, number of steps utilized in parentheses, p = probability of occurrence. Ages are rejected when p < 0.15. The average alteration index (AI) for the plateau/isochron steps is listed with the associated standard error on the mean. Samples are altered if AI > 0.0006 (whole-rock) and > 0.00006 (plagioclase); the corresponding ages are rejected as valid estimates of the time of crystallization

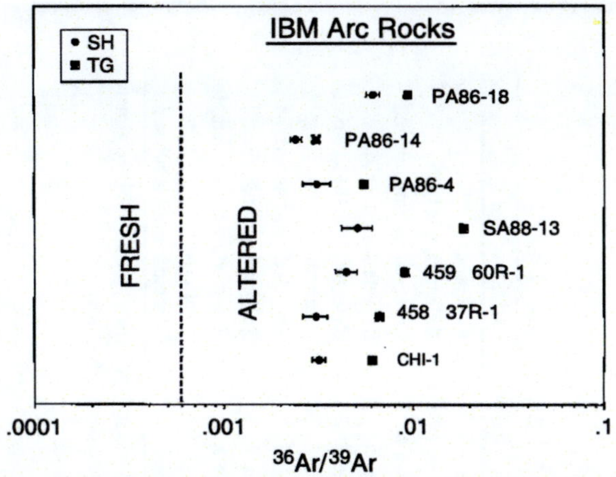

Fig. 3.3 Assessing the freshness of whole-rock samples from the IBM Arc (After Baksi 2007b). *Solid squares* – the AI ($^{36}Ar/^{39}Ar$) for the total gas analyses; *filled circles* – the average value of AI for the isochron sections of Cosca et al. (1998, Table 3.3) plotted on a log scale. SEM errors are shown. All samples are significantly altered ($^{36}Ar/^{39}Ar > 0.0006$) – for details see text

samples dated by Duncan and Keller (2004) were altered and do not yield proper plateaus (see Table 3.2). The best estimate of its age is ~57 Ma (see Table 3.2). (4) The age of Koko (south) is constrained at ~50 Ma based on rock 44R-2.

Next, I evaluate ages from the Izu–Bonin–Mariana (IBM) Arc. The $^{40}Ar/^{39}Ar$ results of Cosca et al. (1998) are examined in detail. An initial AI plot for these rocks (see Fig. 3.3), suggests all samples were altered, unsurprising as the rocks were not acid leached prior to dating. A more detailed examination is attempted in Table 3.3. It has been suggested (Baksi 2007a, b), that rocks/minerals formed by subduction related magmatism should show high (initial) ^{36}Ar levels, and the AI method may then fail to detect subsequent alteration. Geochemists must determine whether high A.I values are due to primary features (Baksi 2007a, b), or due to chemical weathering over geological time. Examination of high quality argon data on very young rocks shows a potential way out of this dilemma, as they are unlikely to have suffered significant alteration. High quality chemical and radiometric argon data for such rocks from South America are available (Singer et al. 2008). The mafic and intermediate rocks show troughs at Ta–Nb in mantle-normalized trace element plots (see Fig. 3.4), denoting derivation by processes driven by the release of water from the subducted slab. As shown in Fig. 3.5, these young rocks show low values of AI (<0.0006). The bulk of the ^{36}Ar released from the dehydrated (subducting) slab, does not enter into the magmatic processes. Large amounts of dissolved atmospheric argon are released as the temperature in the subducted slab rises from ~0°C to ~200°C and do not participate in the onset of magmatism in the overlying mantle wedge. High AI values in convergent zone igneous rocks (such as that in the IBM Arc) result from alteration, and are not magmatic features.

Table 3.3 Assessment of the $^{40}Ar/^{39}Ar$ ages of Cosca et al. (1998) for whole rock samples from the Izu–Bonin–Marianas Arc

Sample location	Sample number	Plateau age (Ma)	Statistical evaluation	Alteration index	Validity of age
Izu–Bonin Arc	786B 02R-1 72-76	45.7 ± 12	F = 0.50(6) p ~ 0.50	0.00025 ± 6	Accepted
Izu–Bonin Arc	786B 02R-1 72-76	46.7 ± 6	F = 1.87(4) **p ~ 0.11**	0.00071 ± 19	**Rejected**, p < 0.15.
Izu–Bonin Arc	786B 21R-2 85-94	54 ± 4	F = 1.50(3) p ~ 0.21	**0.00109 ± 6**	**Rejected**, sample altered
Izu–Bonin Arc	786B 66R-2 128-135	45.7 ± 4	F = 1.48(3) p ~ 0.22	**0.0012 ± 5**	**Rejected**, sample altered
Izu–Bonin Arc	786B 70R-4 27-35	45.3 ± 8	F = 1.98(4) **p ~ 0.09**	**0.0028 ± 3**	**Rejected**, p < 0.15, sample altered
Izu–Bonin Arc	Chi-1	44.8 ± 6	F = 1.68(4) p ~ 0.15	**0.0031 ± 3**	**Rejected**, sample altered
Mariana Trench	458 37R-1	49.2 ± 6	F = 0.86(6) p ~ 0.52	**0.0024 ± 4**	**Rejected**, sample altered
Mariana Trench	459B 60R-2	44.8 ± 14	F = 1.58(7) **p ~ 0.14**	**0.0045 ± 9**	**Rejected**, p < 0.15, sample altered
Palau	PA86-14	51.7 ± 34	F = 1.35(5) p ~ 0.24	**0.0033 ± 9**	**Rejected**, sample altered

F = MSWD value for plateau sections, number of steps utilized in parentheses, p = probability of occurrence. Ages are rejected as valid estimates of the time of crystallization when p < 0.15 or when the AI is > 0.0006. Four (other) samples gave disturbed age spectra

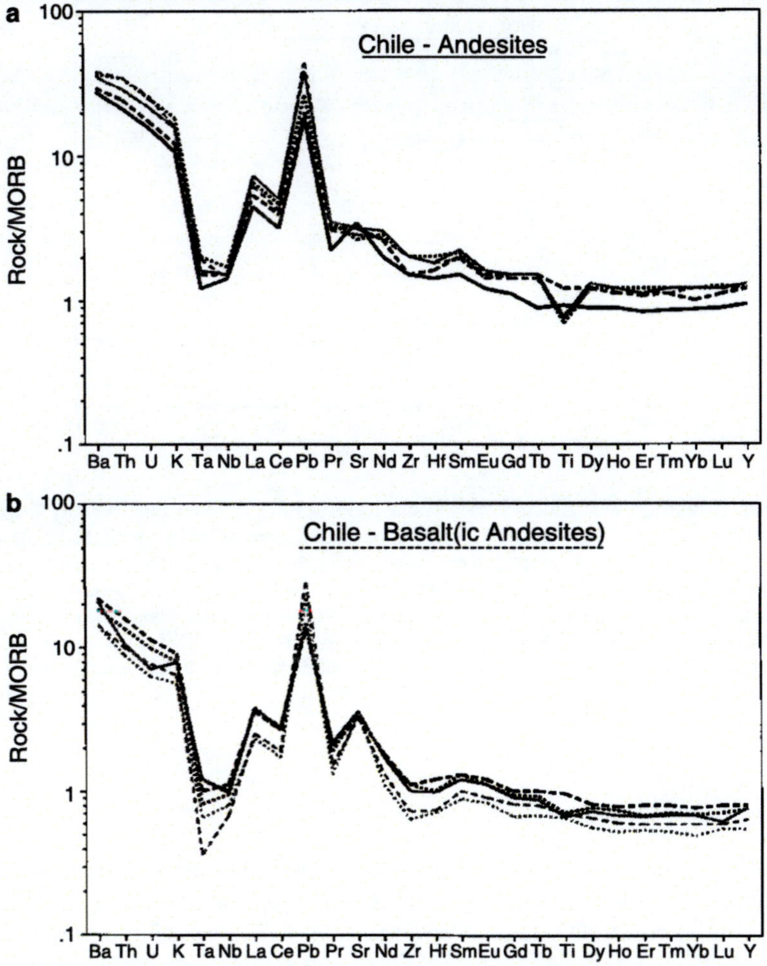

Fig. 3.4 MORB normalized trace element plots for very young (<100 ka) subduction related rocks from Chile (Data taken from Singer et al. 2008); normalizing values from Hofmann (1988) (**a**) Basalts and (**b**) basaltic andesites and andesites. All patterns show marked troughs at Ta–Nb, denoting derivation by magmatism driven by dehydration of the subducted slab

Table 3.3 shows that most of the ages of Cosca et al. (1998) are suspect, being rejected on statistical analysis of the plateaus. Further, the AI values for the best (plateau) steps are well above the cutoff (0.0006), indicating pervasive alteration. A single acceptable age (see Table 3.3) at ~46 Ma, suggests initiation of subduction somewhat earlier at ~50 Ma. The latter is in agreement with the observations of Stern (2004).

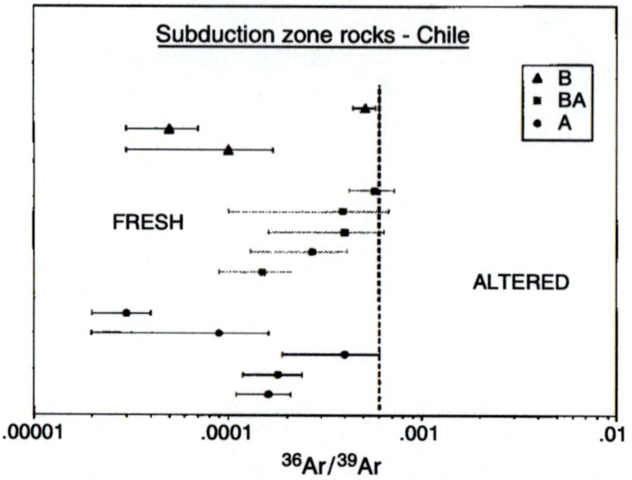

Fig. 3.5 Assessing the freshness of very young rocks (<100 ka) from Chile dated by Singer et al (2008). The average value of the AI of $^{40}Ar/^{39}Ar$ plateau sections are shown with the standard error on the mean plotted on a *log* scale (see Baksi 2007a, b). B = basalt, BA = basaltic andesite, A = andesite. All rocks show values below the cutoff for freshness ($^{36}Ar/^{39}Ar < 0.0006$). These subduction derived igneous rocks (see Fig. 3.4), exhibit low AI values, contrary to the suggestion of Baksi (2007a). High AI values for such rocks, e.g. from the IBM Arc (see Fig. 3.3 and Table 3.3), generally result from chemical weathering (alteration) and are not magmatic features

3.4 Results

The best ages for Seamounts are: Detroit – 67.4 Ma; Suiko – 61 Ma; *Nintoku* – 57 Ma; Koko (north) – 52.6 Ma; *Koko (south)* – 51 Ma; *Kinmei* – 48.5 Ma; Diakakuji 46.7 Ma; Abbott – 41.7 Ma; *Colahan* – 40 Ma; *Unnamed* – 32 Ma. Ages for Seamounts listed in italics are the less satisfactory ones. These ages and the Seamounts' distance from Kilauea, listed in Table 3.1 of Sharp and Clague (2006), are used to derive volcanic migration rates. The results are shown in Fig. 3.2b and yield essentially the same results whether a quadratic or a splined cubic fit is utilized. A steady change in the rate of plate motion is seen, from ~13 cm year^{-1} at ~65 Ma to ~5 cm year^{-1} at ~35 Ma. This is a more realistic scenario than the jerky changes suggested by Sharp and Clague (2006) – see Fig. 3.2a. The new age of the Detroit seamount at ~67 Ma and the smooth "deceleration" seen in Fig. 3.2b, eliminates the need to argue for an *increase* in plate motion from ~5 cm year^{-1} at ~65 Ma, through ~9 to ~17 cm year^{-1} at ~55 Ma (see Sharp and Clague 2006 and Fig. 3.2a herein). Such changes are not supported by data from other sources (e.g. Copley et al. 2010). The smooth change in plate motion over a ~30 million years period reflects a long-term cause for the morphology of the HEB. The rate of Pacific Plate motion decreases much earlier than the abrupt ~60° change in the direction of the Hawaiian–Emperor Chain.

3.5 Discussion

What tectonic event/events underlie the age and morphology of the HEB? Herein, I briefly review a number of hypotheses in this regard. The most critical factor in controlling plate velocity is the "pull mechanism", namely the subduction process. Sharp and Clague (2006) considered a number of geodynamic reasons behind the noted change in plate motion. These include the initiation of subduction in the ~2,000 km IBM arc systems, and the Early Eocene initiation of the ~2,500 km Tonga–Kermadec trench. Critical examination (see above) of the $^{40}Ar/^{39}Ar$ data for magmatism in the IBM arc shows that this event is poorly dated but subduction was probably initiated at ~50 Ma. Whittaker et al. (2007) suggested that plate reconstructions for Australia and Antarctica reveal a major change at 53–50 Ma. They argued that subsequent plate tectonic events, including initiation of the Tonga–Kermadec Trench underlie the origin of the HEB. Tikku and Direen (2008) questioned the style of development and *age* (my italics) of the conjugate Cretaceous Australian–Antarctic passive margins. No hard radiometric data pertains to these events, and as with the case of the hypothesis involving the IBM arc, critical review of relevant radiometric data is a prerequisite for arriving at a cogent explanation of the HEB. In particular, argon age data for sea-floor rocks must be screened to avoid use of data on altered rocks, which often yield incorrect plateau/isochron ages (Baksi 2007a, b).

The hard (continent–continent) collision between India and Asia has been postulated to be the cause of the HEB (e.g. Clague and Dalrymple 1989; Wessel et al. 2006). The timing of this collision has been contentious. Zhu et al. (2005) presented cogent arguments for placing this event at 50 Ma. Island arcs generated on the northern margin of the Tethys Sea would have been accreted to the southern edge of Asia during the period ~65–50 Ma. This should lead to slowing of the Indian Plate, without a marked directional change. A similar situation is noted for the Pacific Plate velocity in this period. At ~50 Ma, the hard collision would lead to an abrupt change in the Indian Plate motion, with strong impediment in the northward movement of the subcontinent itself. This is reflected (this work) in the Pacific Plate velocity, with a sharp change in direction ("escape" – from primarily northwards to westwards), with no abrupt change in the rate (see Fig. 3.2b). Subsequent to 50 Ma, as inertia drove the leading edge of the Indian subcontinent under Asia, the Indian Plate motion decreased smoothly (fell by 50%) in the next ~15 million years (Copley et al. 2010). Figure 3.2b shows a similar decrease in the rate of movement for the Pacific Plate, with little or no directional change. However, there is currently no strong evidence in favor of a steady slowing of the Indian Plate from ~65–50 Ma (see Copley et al. (2010)). Foulger et al. (2004) suggest that there was no change in Pacific Plate motion at 50 Ma. Others (e.g. Winterer 2007) suggest that the continent–continent collision "escape", spread eastward to the Pacific, influencing the morphology of the Hawaiian–Emperor Chain. Richards and Lithgow-Bertelloni (1996) argue that (plate) tectonic ramifications of the India–Asia collision should not be directly transferred to the Pacific realm.

Questions persist about the possible motion of the Hawaiian Hotspot during the time of formation of the Emperor Seamounts (Tarduno et al. 2003; Sager 2007). Paleomagnetic measurements from Emperor Seamounts (Ocean Drilling Program) cores show a progressive offset of paleolatitude from nearly zero, close to the bend, to ~13° near the north end of the Chain (Tarduno et al. 2003). The Hawaiian plume, affected by mantle flow during Late Cretaceous–Early Eocene time, may be the dominant factor in forming the Emperor track and its bend (Tarduno 2009). However, there are large offsets (>5°) between the paleomagnetic and hotspot model predicted apparent polar wander paths (APWP). The latter suggests relatively uniform northward motion with over 30° of polar motion since 80 Ma, whereas paleomagnetic work suggests a polar stillstand from ~80 to ~50 Ma (Sager 2007). The stillstand is in seeming contradiction to the nearly north–south trend of the Emperor Chain from ~75 to ~50 Ma. The explanation may lie in the fact that the hotspot moved rapidly southward while the plate moved primarily east–west (Sager 2007). Whether this explains the smooth change in the "rate" of plate motion noted herein, remains open to question.

Utilizing a hypothesis first proposed by Dana over a hundred years ago, Stuart et al. (2007) look to the propagation of the Hawaiian–Emperor Chain resulting from cooling stress in the Pacific Plate. It requires northwest motion of the Pacific plate with respect to Hawaii. Magma produced by decompression melting at depth close to the tip of the crack tip stress concentration (currently at Hawaii), rises to the surface through a vertical tensile crack. It explains many of the first order features of the Hawaiian–Emperor Chain, but is difficult to quantitatively test, as the conditions of crack propagation, melt production and transport remain uncertain (Stuart et al. 2007).

3.6 Concluding Remarks

For over 2 decades, $^{40}Ar/^{39}Ar$ ages on altered material have been widely quoted, leading to the "standardization of error" (Steffansson 1936). This is well illustrated by the common use of the (incorrect) ~43 Ma age for the HEB (e.g. Grotzinger et al. 2007). It is of critical importance to identify the reasons for earlier incorrect ages. The Italian economist Vifredo Pareto observed "Give me a fruitful error any time, full of seeds, bursting with its own corrections. You can keep your sterile truth for yourself". In the case of the HEB, it leads to a better understanding of the age and morphology of the Hawaiian–Emperor Chain. Critical evaluation of all available argon data, using the techniques set out by Baksi (2007a, b) leads to "revised" ages, in particular for the Detroit Seamount. A steady slowing of the Pacific Plate occurred from ~65 Ma to ~35 Ma. The HEB at 50 Ma exhibits a change in the direction of motion, with little or no attendant change in the speed. Hypotheses seeking to explain the formation of the Hawaiian–Emperor Chain must

(1) Utilize radiometric data that has been critically evaluated, (2) Quantitatively explain the (temporal and morphological) features of the Chain noted herein. Currently, this author favors the hard collision of India with Asia as the underlying

cause of the HEB at ~50 Ma, but recognizes that the hypothesis lacks critical details; in particular relating to the transference of plate tectonic motion from the Indian Ocean to the (western) Pacific.

Acknowledgments I thank Roma Chakravarti and Romila Raman for numerous helpful discussions. Critical comments made during official review of this manuscript, led to considerable revision of a somewhat simplistic overview of the HEB.

References

Baksi AK (1993) A geomagnetic polarity time scale for the period 0–17 Ma, based on $^{40}Ar/^{39}Ar$ plateau ages for selected field reversals. Geophys Res Lett 20:1607–1610

Baksi AK (1995) Fine tuning the radiometrically derived geomagnetic polarity time scale (GPTS) for 0–10 Ma. Geophys Res Lett 22:457–460

Baksi AK (1999) Reevaluation of plate motion models based on hotspot tracks in the Atlantic and Indian Oceans. J Geol 107:13–26

Baksi AK (2003) Critical evaluation of the $^{40}Ar/^{39}Ar$ ages from the Central Atlantic Magmatic Province: iiming, duration and possible migration of magmatic centers. Am Geophys Union Monogr 136:77–90

Baksi AK (2005) Evaluation of radiometric ages pertaining to rocks hypothesized to have been derived by hotspot activity, in and around the Atlantic, Indian and Pacific Oceans. Geol Soc Am Spec Pap 388:55–70

Baksi AK (2007a) A quantitative tool for detecting alteration in rocks and minerals – I; water, chemical weathering and atmospheric argon. Geol Soc Am Spec Pap 430:285–303

Baksi AK (2007b) A quantitative tool for detecting alteration in rocks and minerals – II; application to argon ages related to hotspots. Geol Soc Am Spec Pap 430:305–333

Baksi AK (2007c) Comment on "A $^{40}Ar/^{39}Ar$ and U/Pb isotopic study of the Ilimaussaq complex, South Greenland: implications for the ^{40}K decay constant and the duration of magmatic activity in a peralkaline complex". Chem Geol 244:344–346

Baksi AK, Archibald DA (1997) Mesozoic igneous activity in the Maranhao province, northern Brazil; $^{40}Ar/^{39}Ar$ evidence for separate episodes of basaltic magmatism. Earth Planet Sci Lett 151:139–153

Baksi AK, Archibald DA, Farrar E (1996) Intercalibration of $^{40}Ar/^{39}Ar$ dating standards. Chem Geol 129:307–324

Cande S, Kent DV (1992) A new geomagnetic polarity time scale for the Late Cretaceous and Cenozoic. J Geophys Res 97:13917–13951

Clague DA, Dalrymple GB (1975) Age of the Koko Seamount, Emperor Seamount Chain. Earth Planet Sci Lett 17:411–415

Clague DA, Dalrymple GB (1989) Tectonics, geochronology and the origin of the Hawaiian-Emperor volcanic chain. Geological Society of America DNAG v. N, Boulder, pp 187–217

Copley A, Avouac J-P, Royer J-Y (2010) The India–Asia collision and the Cenozoic slowdown of the Indian plate: implications for the forces driving plate motions. J Geophys Res 115. doi:10.1029/2009JB006634

Cosca MA, Arculus RJ, Pierce JA, Mitchell JG (1998) $^{40}Ar/^{39}Ar$ and K-Ar geochronological constraints for the inception and early evolution of the Izu-Bonin- Mariana Arc system. Island Arc 7:579–595

Dalrymple GB, Clague DA (1976) Age of the Hawaiian-Emperor Bend. Earth Planet Sci Lett 31:313–329

Dalrymple GB, Lanphere MA (1969) Potassium–argon dating. W.H. Freeman, San Francisco, 258 pp

Dalrymple GB, Lanphere MA, Natland JH (1980) K-Ar minimum age for Meiji Guyot, Emperor Seamount chain. Initial Rep Deep Sea Drill Prog 55:677–683

Duncan RA, Keller RA (2004) Radiometric ages for basement rocks from the Emperor Seamounts, ODP Leg 197. Geochem Geophys Geosyst 5 Q08L03. doi:10.1029/2004GC000704

Foulger GR, Anderson DA (2004) The Emperor and Hawaiian volcanic chains; how well do they fit the plume hypothesis? http://www.mantleplumes.org/Hawaii.html

Grotzinger J, Jordan TH, Press F, Siever R (2007) Understanding Earth. W.H. Freeman & Co, San Francisco, p 579

Hofmann AW (1988) Chemical differentiation of the Earth: the relationship between the mantle, continental crust and oceanic crust. Earth Planet Sci Lett 90:297–314

Keller RA, Duncan RA, Fisk MR (1995) Geochemistry and $^{40}Ar/^{39}Ar$ geochronology of basalts from ODP Leg 145 (North Pacific Transect). Proc Ocean Drill Prog Sci Results 145:333–344

Lanphere MA, Dalrymple GB (1978) The use of $^{40}Ar/^{39}Ar$ data in evaluation of disturbed K-Ar systems. U S Geol Survey Open-File Rep 78–701:241–243

Richards MA, Lithgow-Bertelloni C (1996) Plate motion changes, the Hawaiian-Emperor bend and the apparent success and failure of geodynamic models. Earth Planet Sci Lett 137:19–27

Sager WW (2007) Divergence between paleomagnetic and hotspot-model-predicted polar wander for the Pacific plate with implications for hotspot fixity. Geol Soc Am Spec Pap 430:335–357

Sharp WD, Clague DA (2006) 50 Ma initiation of Hawaiian-Emperor Bend records major change in Pacific Plate motion. Science 313:1289–1294

Singer BS, Jicha BR, Harper MA, Naranjo JA, Lara LE, Moreno-Roa H (2008) Eruptive history, geochronology and magmatic evolution of the Puyehue- Cordon Caulle volcanic complex, Chile. Geol Soc Am Bull 120:599–618

Stefansson V (1936) Adventures in error. R.M. McBride and Company, New York, p 299

Stern RJ (2004) Subduction initiation: spontaneous and induced. Earth Planet Sci Lett 226:275–292

Stuart WD, Foulger GR, Barrall M (2007) Propagation of the Hawaiian- Emperor volcano chain by Pacific plate cooling stress. Geol Soc Am Spec Pap 430:497–506

Tarduno JA (2009) On the motion of Hawaii and other mantle plumes. Chem Geol 241:234–247

Tarduno JA, Duncan RA, Scholl DW, Cottrell RD, Steinberger B, Thordarson T, Kerr BC, Neal CR, Frey FA, Tori M, Carvallo C (2003) The Emperor Seamounts; southward motion of the Hawaiian hotspot plume in the Earth's mantle. Science 301:1064–1069

Tikku AA, Direen NC (2008) Comment on "Major Australian–Antarctic plate reorganization at Hawaiian-Emperor Bend time". Science 321:490

Wessel P, Harada Y, Kroenke LW (2006) Towards a self-consistent high resolution absolute plate motion model for the Pacific. Geophys Geochem Geosyst 7, art. no. Q03L12

Whittaker JM, Mueller RD, Leitchenkov G, Stagg H, Sdrolias M, Gaina C, Goncharov A (2007) Major Australian–Antarctic plate reorganization at Hawaiian-Emperor Bend time. Science 318:83–86

Winterer EL (2007) Discussion on "Divergence between paleomagnetic and hotspot- model-predicted polar wander for the Pacific plate with implications for hotspot fixity". Geol Soc Am Spec Pap 430:355–356

Worsley JR (1973) Calcareous nannofossils: leg 19 of the Deep Sea Drilling Project. Initial Rep Deep Sea Drill Prog 19:741–750

Zhu B, Kidd WSF, Rowley DB, Currie BS, Shafique N (2005) Age of initiation of the India–Asia collision in the East-Central Himalaya. J Geol 113:265–285

Chapter 4
Geology, Petrology, and Geochemistry of the Basaltic Rocks of the Axum Area, Northern Ethiopia

Miruts Hagos, Christian Koeberl, Kurkura Kabeto, and Friedrich Koller

Abstract The Axum volcanic rocks constitute an important outcrop of trap basalts at the northern end of the great Ethiopian flood basalt province. They constitute about 5–480-m-thick sequences exposed around Axum and Shire regions covering a total area of about 900 km^2. Even though the Axum volcanic rocks were previously assigned to the low-Ti (LT) tholeiitic basalts of the northwestern Ethiopian volcanic province, our geochemical studies on rocks from this region reveal that these rocks are alkaline to transitional in composition with distinct geochemical characteristics. Their petrologic diversity is very significant, the lavas ranging from basanites to tephrites and trachy-andesites. The variation in the major element (TiO_2 and Fe_2O_3) concentrations, incompatible trace elements, and the rare earth element (REE) concentrations and ratios (e.g., $(La/Yb)n = 5.0–14.0$) indicate a complex petrogenetic evolution of the flood basalts. The Axum volcanic rocks are entirely alkaline to transitional in nature and are classified here into the Lower Sequence (Low-Ti) basalts and the Upper Sequence (High-Ti) basalts. Both these sub-divisions are also reflected in their unique chondrite-normalized REE patterns and multi-element diagrams: the Lower Sequence basalts show enhanced REE fractionation and higher concentrations of the highly incompatible trace elements; whereas the Upper Sequence basalts exhibit less REE fractionation and relatively high concentrations of the moderately incompatible trace elements. The mineralogical and geochemical variations among the basalts of Axum provide clues about the complexity of the northern and northwestern Ethiopian flood basalts and their petrogenetic evolution.

M. Hagos (✉), C. Koeberl*, and F. Koller
Department of Lithospheric Research, University of Vienna, Althanstrasse 14, A-1090 Vienna, Austria
*Current address: Natural History Museum, Burgring 7, A-1010 Vienna, Austria
e-mail: miruts2005@yahoo.com; christian.koeberl@univie.ac.at; friedrich.koller@univie.ac.at

K. Kabeto
Department of Earth Sciences, Mekelle University, P.O. Box 231, Tigray, Ethiopia
e-mail: kurkura57@yahoo.com

J. Ray et al. (eds.), *Topics in Igneous Petrology*, DOI 10.1007/978-90-481-9600-5_4,

4.1 Introduction

Most continental flood basalts are entirely associated with lithospheric extensions resulting from distant and/or local stress fields. The Ethiopian volcanic province is one of the youngest flood basalt provinces on Earth, and is related to the opening of the Red Sea and the Gulf of Aden (Baker et al. 1996; George et al. 1998). These two oceanic rifts are connected to the less developed East African Rift at the Afar triple junction. This triple junction is the center of a large volcanic province that is exposed in Ethiopia, Eritrea, Yemen, and the Republic of Djibouti (Pik et al. 1998).

The study area, Axum, is located in the northern part of the Ethiopian highlands and is mainly covered by Tertiary fissure-basalts and the post-trap, east-west aligned Adwa–Axum phonolite and trachyte plugs. These volcanic rocks are exposed in the north–northwestern part of the Tigray region, northern Ethiopia, and are isolated from the major Ethiopian flood basalts by the deeply cut Tekeze basin and from the Eritrean highland basalts by the Mereb basin (Fig. 4.1). However, little

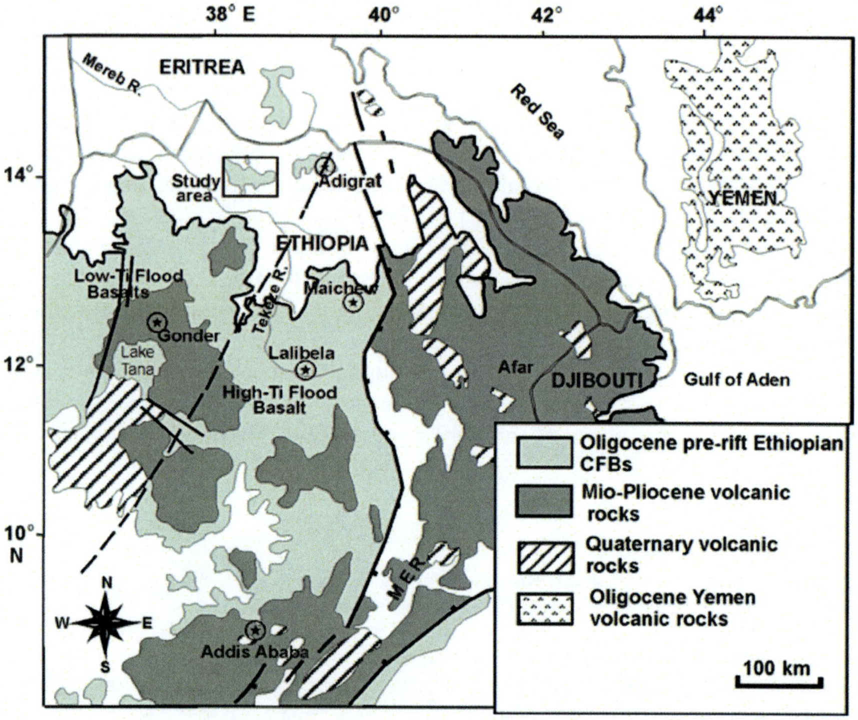

Fig. 4.1 Sketch map of the central, northwestern and northern parts of the Ethiopian flood basalt province and the Afar triple junction with a box showing the location map of the Axum volcanics. Dividing line between high-Ti and low-Ti basalt province is marked by a *dashed line* (Modified from Merla et al. 1979; Pik et al. 1998; Kuster et al. 2005)

attention has been paid to the petrologically important volcanic rocks of the erosional remnants in northern Ethiopia (Axum area).

The southern, central, and northwestern part of the Ethiopian flood basalts have been the focus of numerous studies (e.g., Mohr 1983; George et al. 1998; Pik et al. 1998, 1999; Ebinger and Sleep 1998; Ayalew et al. 1999, 2002; Kieffer et al. 2004). Most of the studies (Alemu 1998; Alene et al. 2000; Asrat et al. 2003, 2004; Tadesse 1996, 1997) in the northern region have focused on the Precambrian basement and Pan-African plutons. The only work dealing with the volcanics of the area was conducted during the regional mapping by the Ethiopian Institute of Geological Survey (1999). According to this study, the volcanic rocks of Axum are classified on the basis of the locality names as the Koyetsa volcanics (stratified flood basalt) and the Adwa trachyte formation (trachytic and phonolitic plugs). The petrologic and geochemical classifications of the volcanics were done on the basis of the rock samples analyzed from the nearby Maichew and northwestern Ethiopian flood basalts. Pik et al. (1998) collected only one sample (i.e., their sample number E210) from the southwestern part of the Axum volcanics and the whole interpretation was made with regard to the single datam that represented about 900 km^2. As a result of incomplete exploration, the petrographic and geochemical characteristics of the volcanic rocks of Axum area remained unsolved. Our aim here is to investigate the petrography, geochemistry, and genesis of the Axum flood basalts, that has remained unknown and unexplored until now.

In this study we have focused on the geochemical (major and trace element) variations between the Axum basalts and the earlier well-defined northwestern tholeiitic basalts, and the possible petrogenetic relationships among them. We have also traced the spatial and temporal geochemical variation of the basaltic rocks of the area with that of the central and northwestern part of the Ethiopian flood basalts.

4.2 Geological Background

The Ethiopian volcanic province covers an area of about 600,000 km^2 with a layer of basaltic and felsic volcanic rocks (Ukstins et al. 2002; Kieffer et al. 2004). An area close to 750,000 km^2 must have existed before plateau uplift and ensuing massive erosion that occurred during Pleistocene time (Mohr 1983) removing about 150,000 km^2. Most Ethiopian flood basalts were extruded on regionally lateritized sandstone horizons (Ukstins et al. 2002) and few on deeply eroded Precambrian basement complexes. Lavas of this province erupted 30 Ma ago, during a short 1–2 Ma period, to form a vast volcanic plateau (Baker et al. 1996; Hofmann et al. 1997). However, rift-triggered Miocene magmatism with an age of 14–5 Ma and pre-rift, early Oligocene (Stewart and Rogers 1996) volcanic rocks are common to the east and south of the main Ethiopian flood basalt, respectively. The flood basalts are partly overlain by younger rhyolitic/trachytic and phonolitic volcanic rocks (Ayalew et al. 1999).

The mineralogical and chemical composition of the flood basalts is relatively uniform (Kieffer et al. 2004). According to these authors, most of the basalts are aphyric to sparsely phyric, and contain phenocrysts of plagioclase and clinopyroxene and sometimes olivine. Inter-bedded with the flood basalts, particularly at upper stratigraphic levels, are felsic lavas and pyroclastic deposits of rhyolitic, or less commonly, trachytic compositions (Ayalew et al. 1999; Kabeto et al. 2004). Despite their compositional similarities, the Tertiary flood basalts of Ethiopia are divided into four types based on their ages (Mohr 1983). From bottom to top these are called the Ashange, Aiba, Alage, and Tarmaber Formations. In contrast to the above classification, the flood basalts are geochemically classified into three distinct magma types: as low-Ti basalts (LT), high-Ti 1 basalts (HT1) and high-Ti 2 basalts (HT2) (Pik et al. 1998, 1999) (Fig. 4.1). These authors recognized a suite of 'low Ti' basalts (LT) characterized by relatively flat rare earth element (REE) patterns and low levels of Ti and incompatible trace elements. These magma groups were defined based on variations of their TiO_2 with MgO contents as well as systematic variations in their immobile trace element contents. The LT rocks are restricted to the northwestern part of the province, as shown in Fig. 4.1. The HT (HT1 and HT2) alkali basalts with higher concentrations of incompatible elements and more fractionated REE patterns are found to the south, southeast and east of the province (Pik et al. 1998). Kieffer et al. (2004), however, disagreed with this simple subdivision and their sampling indicated a slightly wide distribution of the LT rocks. The HT2 basalts are slightly more magnesian than the HT1 basalts and commonly are rich in olivine ± clinopyroxene phenocrysts (Pik et al. 1999). According to these authors, the HT2 basalts have higher concentrations of incompatible elements and show extreme fractionation of the REE.

The regional flood basalt magmatism was succeeded during the Mio-Pliocene by rifting, widespread bimodal alkaline volcanism and crustal uplift (Küster et al. 2005). The development of the Main Ethiopian rift (MER) was accompanied by extensive bimodal volcanic activity that occurred both inside the rift and on its shoulders (Gasparon et al. 1993). The Axum–Adua, Yerer–Tullu Wellel, and Goba–Bonga lineaments (Mohr 1987; Beyth 1991; Abebe et al. 1998) are important E–W trending regional structures, transverse to the MER axis that dissect the rift. The rift-related uplift of the Ethiopian plateau continued until the optimum thinning of the Afar Triangle and led to considerable erosion of the volcanic piles of the Ethiopian highlands. In the northern part of the province (Axum, Adigrat, and central Eritrean highlands) only erosional remnants of flood basalts and erosionally resistant post-Tertiary silicic rocks remain and the subvolcanic phonolites are widely exposed.

4.3 Sampling Strategy and Analytical Techniques

One hundred and twelve (112) rock samples (each weighing 0.75–1 kg) were collected during two field seasons (December 2006 and April 2007) from the entire area of the Axum volcanics but because of weathering and similarity in some of the samples only 30 of them were selected for this study. From these 30 analyzed samples only 19 samples come from the flood basalts and their data are used in this contribution. The remaining 11 samples come from the phonolitic rocks of the Bete-Giorgis and Gobo-Dura ridges.

The basaltic rock samples were first crushed at the University of Vienna in a stainless steel jaw crusher and then powdered in an agate mill in order to avoid contamination by the transition elements (such as Co, Ta, and Nb). Contents of major element oxides, minor and trace element (Nb, Zr, Y, Rb, Sr, Ga, Zn, Cu, Ni, Co, Cr, Sc, V, Ce, Ba, La) abundances were determined using wavelength dispersive X-ray fluorescence (XRF) spectrometry at the Department of Lithospheric Research, University of Vienna, Austria. For this, 1.2 g of powdered rock was mixed with 6 g of lithium tetra-borate to prepare fused pellets used for the analyses of the major element oxides, and 10 g of powdered rock was used to prepare pressed pellets for the analyses of the trace elements.

Contents of the rare earth elements (REE) and other trace elements, namely Sc, Cr, Co, Ni, Zn, As, Se, Br, Rb, Sr, Zr, Sb, Cs, Ba, Hf, Ta, W, Ir, Au, Th, and U, were analyzed by instrumental neutron activation analysis (INAA) at the Department of Lithospheric Research, University of Vienna. For this, samples weighing between about 115 and 130 mg were sealed into polyethylene capsules and, together with the rock standards AC-E (granite), Allende (carbonaceous chondrite), and SDO1 (shale) were irradiated for 8 h at the Triga Mark II reactor of the Atominstitut der Österreichischen Universitäten, Vienna, at a neutron flux of 2×10^{12} n cm^{-2} s^{-1}. Details of the INAA method, including standards, correction procedures, precision, and accuracy, are reported in Koeberl (1993) and Son and Koeberl (2005).

4.4 Rock Type Classification

The Axum volcanics are classified in the field into the fissure-fed continental flood basalts and the phonolite–trachyte plugs. Large parts of the fissure-fed alkali basalts overlie the undifferentiated clastic Mesozoic sedimentary rocks of the region and small parts of the basalts overlie the deeply eroded meta-volcano-sedimentary complexes of the Arabian–Nubian shield (ANS) (Fig. 4.4). The Mio-Pliocene phonolite–trachyte plugs on the other hand are exposed intruding the basalt flows and/or the clastic sedimentary rocks of the area.

During the Tertiary Period, large parts of the present landmass of Ethiopia were flooded by successive lava flows, and a few pyroclastic falls and flows. In the study area, the Mesozoic sedimentary rocks are overlain by Oligocene basaltic lava flows and syn – post trap pyroclastic rocks (Beyth 1972; Zanettin et al. 2006). After a prolonged period of erosion of the uplifted landmasses, the volcanic rocks of the Axum area, range in thickness from a thin veneer (<5 m) at its peripheries to about 500 m in some of the ridges and mountains of the area (e.g., Ferasit and Moror). Based on the field observations and mineral compositions, we subdivide these volcanic rocks broadly into the following three major sequences (Fig. 4.2).

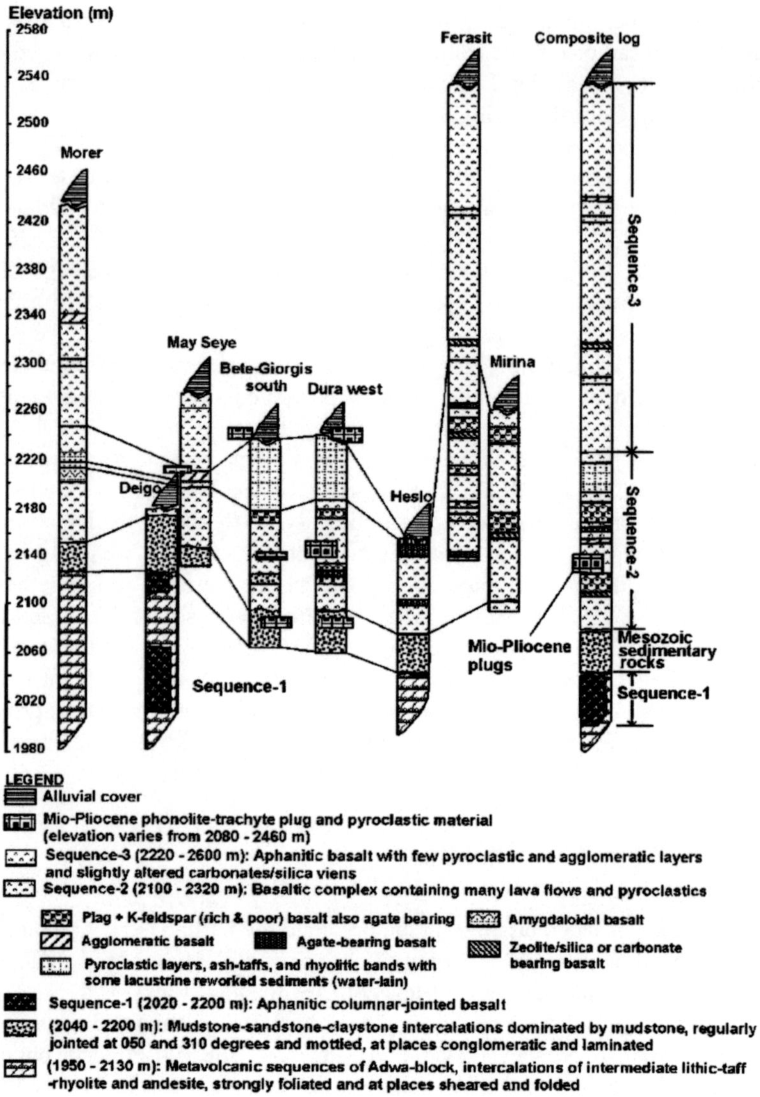

Fig. 4.2 Eight column sections and one composite section (*right*) of the stratigraphy of the Axum area constructed from different traverses; outcrops are logged based on their altitude in the area

4.4.1 Sequence-1

The lowermost samples of the basalts were collected from the southern Axum town or the west and northwest of Deigo village at altitudes of 2,020–2,130 m. This gently inclined (dip not >5°) basaltic outcrop, 5–60 m. thick and unconformably overlying the basement rock, is vesicular, columnar-jointed, and commonly slightly weathered. Unlike the other sequences, which are flat lying, these rocks are linear and

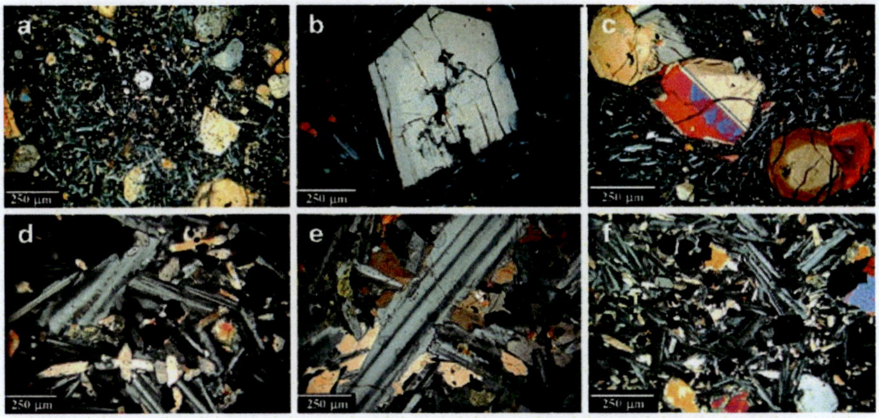

Fig. 4.3 Petrography of the volcanic rocks of the Axum area showing various mineralogical and textural features. All images are taken under crossed polars with a magnification of x125. (**a**) Very fine-grained plagioclase rich (ground mass) alkali basalts with coarse-grained olivine and some clinopyroxenes phenocrysts; the plagioclases are lath-like but with poor flow texture, where as the phenocrysts are rounded to angular and of varying size. (**b**) Large grain of zoned clinopyroxene embedded in fine-grained and altered groundmass of plagioclase. (**c**) Zoned plagioclase and clinopyroxenes imbedded in fine-grained matrices of the same compositions. Thin-sections **a–c** represent the lower basaltic sequence (sequence-1) of the study area. (**d**) Randomly aligned, medium – course-grained plagioclase-laths with few grains of pyroxenes. (**e**) Coarse-grained plagioclase, clinopyroxene and olivine are in association with extremely large plagioclase crystal. (**f**) Medium-grained, plagioclase-laths rich alkali basalt with clinopyroxene and some olivine phenocrysts; dark-colored minerals of Ti oxides and Mg oxides are filling the interstices of other grains. Thin-sections **d–f** represent the mineralogy and texture of alkali basalts of the upper sequences (sequences 2 and 3)

convex downward exposures filling valleys in the underlying basement complexes. Such rocks are laterally discontinuous but well exposed along the big canyons of the southern part of the area.

Petrographically, the basalts are commonly aphyric to phyric/porphyritic, plagioclase-laths with significant phenocrysts (nearly 20–25 vol%). The phenocryst minerals are olivine, clinopyroxene, and plagioclase with a few opaque minerals. Most of the phenocrysts are rounded with some angular grains, and in some thin-sections extremely large and zoned phenocrysts of clinopyroxene are observed (Fig. 4.3a–c). These pyroxenes generally show polygonal outline.

4.4.2 Sequence-2

The second sequence, about 50–140 m thick and commonly exposed at an altitude between 2,080 and 2,240 m, is invariably weathered and intercalated with thin pyroclastic (ash fall) materials and basaltic agglomerate in its top part. The vesiculated and weathered basalts of this sequence also contain some calcite veins in addition to silica and zeolites. Even though it is not more than 140 m thick, this sequence covers a large portion of the mapped area and is emplaced directly over the horizontal,

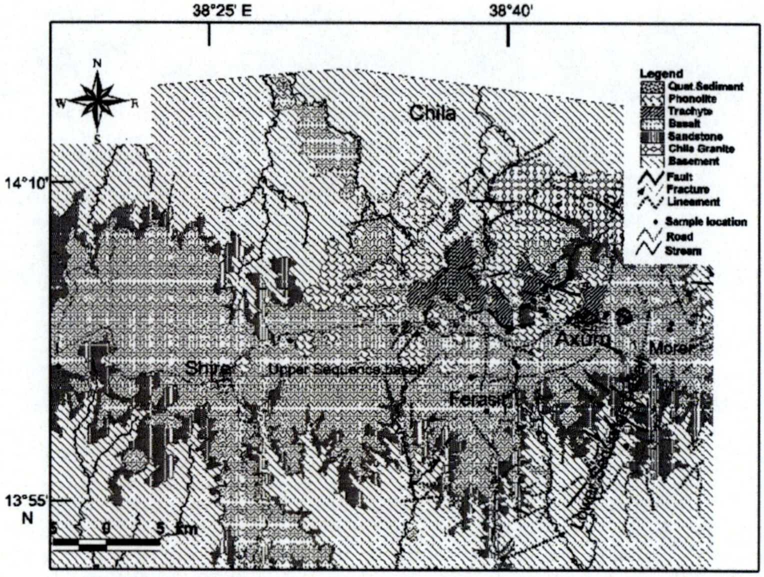

Fig. 4.4 Geological map of the volcanic rocks of Axum area. The 1:50,000 scale topographic maps of Axum, Adwa, Enda-Silasie and Chila were taken as a base map. The spatial variations of rocks, lithologic contacts and the structures were extracted from 1:50,000 scale aerial photographs (Eth94), Landsat ETM+ 2002, and ASTER. Color composite images of Landsat band 7, 5 and 2 are commonly used to differentiate between the clastic Mesozoic sedimentary rocks and the overlaying flood basalts, but geologic structures such as: faults, fractures and lineaments were drawn from aerial photographs. In some areas the high spatial resolution image, the ASTER image was used

clastic Mesozoic sedimentary rocks. This sequence is uniformly made up of coarse-grained, Carlsbad-twinned, plagioclase-phyric, ± K-feldspar bearing (rich and poor) basalts, and grading upwards to agate- and zeolite-bearing amygdaloidal type. Samples of this sequence show significant accumulation of the opaque minerals, but fewer phenocrysts of olivine, clinopyroxene, and other minerals (Fig. 4.3d and e).

4.4.3 Sequence-3

These rocks are essentially aphyric to micro-phenocrystic basalts exposed between 2,200 and 2,600 m with rare ash-fall layers intercalated with them. This sequence is the thickest, 300–400 m thick, and forms the highest peaks of the area. These rocks are extensive in area and form a shield-volcano-type edifice, the Ferasit and Morer ridges. This sequence is relatively fine-grained with few phenocrysts. The phenocryst minerals are clinopyroxene, olivine, and plagioclase without opaque. The groundmass of these basalts consists dominantly of altered plagioclase with the above-mentioned phenocrysts phases, accessory glass, and few zircons (Fig. 4.3f). For a complete petrographic description of the samples, see Table 4.1.

Table 4.1 Detailed petrographic descriptions of the Axum alkali basalts, northern Ethiopia

Sample code	Sample type	Location (in UTM)	Description
AX004	Porphyritic basalt	472099E 1562095N	Needle-like, medium-grained plagioclase-laths and hornblende (>60 vol%) with few anhedral shape micro – mega phenocrysts of olivines, clinopyroxenes and very few opaque minerals. The plagioclases are carlsbad twinned but irregularly aligned. No flow texture.
AX08	Aphanitic basalt	479601E 1564013N	Fine-grained plagioclase (ground-mass) incorporating very few phenocrysts of pyroxenes and some secondary quartz crystals (agate). No olivine crystals were observed.
AX20	Aphanitic basalt	472639E 1551783N	Fine-grained matrix (dominated by plagioclase) and all size (micro – mega) phenocrysts of olivines, clinopyroxenes and some plagioclases. Phenocrysts constitute about 30 vol%. Few opaques are present.
AX21	Columnar basalt	472805E 1551143N	Plagioclase-laths dominated matrix. Euhedral – subhedral shaped, medium-grained phenocrysts of olivine. No opaque minerals.
AX24	Porphyritic basalt	474185E 1550132N	Medium-grained, elongated plagioclase and tabular hornblende (ground-mass) with medium size phenocrysts of olivines and few but very large and perfectly zoned phenocrysts of clinopyroxenes. Opaque minerals are not common.
AX29	Dolerite/ diabase	464013E 1557294N	Coarse-grained plagioclase-laths interfingering with the anhedral shaped, olivine mega-phenocrysts. The plagioclase shows Carlsbad-type twinning but poor flow texture.
AX31	Vesicular basalt	464206E 1557432N	Needle-like, medium-grained plagioclase (ground-mass) with very few pyroxene micro-phenocrysts. The matrix constitute >90 vol% and the plagioclase in the ground mass shows good flow texture.
AX32	Aphanitic basalt	463129E 1560422N	Carlsbad twinned medium-grained plagioclase (ground-mass). No phenocryst minerals found but medium-grained (the same size as the plagioclase) olivines and few opaque minerals are common.
AX34	Aphanitic basalt	473529E 1552248N	Carlsbad twinned, fine-grained plagioclase-laths matrix with varying size (micro–mega) phenocrysts of olivines and pyroxenes. The plagioclases show good flow texture and the phenocrysts are euhedral–anhedral in shape.

(continued)

Table 4.1 (continued)

Sample code	Sample type	Location (in UTM)	Description
AX38	Aphanitic basalt	474033E 1556172N	Fine - medium-grained, elongated – equant plagioclase-laths (50–60 vol%). Micro–mega, Euhedral–subhedral shaped phenocrysts of olivines and zoned and un-zoned phenocrysts of pyroxenes.
AX43d	Columnar basalt	466516E 1564320N	Tabular, medium-grained plagioclase with few olivine phenocrysts. The phenocrysts are subhedral–anhedral in shape. Some opaque minerals are present.
AX44	Basalt	465459E 1563351N	Euhedral–subhedral shaped olivine phenocrysts (about 5–8 vol%) embedded in fine-grained matrix of equant – elongated plagioclase-laths and few olivines.
AX46	Aphanitic basalt	469365E 1562236N	Needle- like plagioclase and hornblende constituting about 90 vol% with very few olivine micro-phenocrysts. No opaques.
AX47b	Aphanitic basalt	474973E 1555646N	Needle-like equant shape, fine-medium-grained plagioclase-laths with medium-size olivine phenocrysts.
AX63	Aphanitic basalt	460147E 15520044N	Extremely large, euhedral–anhedral shaped pyroxene and olivine phenocrysts. The pyroxenes are multiply zoned and some altered/secondary minerals are developed at the rims of the phenocrysts. The ground-mass is still dominated by very fine-grained plagioclase with very few of the above-mentioned minerals.
AX79	Basalt	479293E 1560281N	Needle-like, Medium - course-grained plagioclase with few olivine micro-phenocrysts. No visible pyroxene and opaque minerals.

4.5 Whole Rock Geochemistry

The general characteristics of the Axum volcanic rocks are illustrated in the total alkali-silica (TAS) classification of Le Bas et al. (1986) (Fig. 4.5). In this classification the rocks are basanites, tephrites, trachybasalts, basalts, and basaltic trachy-andesites. All the analyzed rocks have SiO_2 contents ranging from 43 to 55 wt% and also show a wide range in concentrations of other major and minor element oxides (Table 4.2). The Axum volcanics have compositions that plot, with the exception of sample AX12 showing tholeiitic affinity, in the alkaline field and few samples (AX08, AX29, AX42) within the transition field. All the basaltic rocks of the area have low (<49 wt%) SiO_2 contents, zero normative quartz (except AX42, 2.5 wt%), and high normative olivine contents (9–20 wt%) (Tables 4.2 and 4.3). However, slight differences in the alkalinity of the Axum volcanic rocks are observed. All the samples from sequence-1 are highly alkaline (silica-undersaturated), whereas,

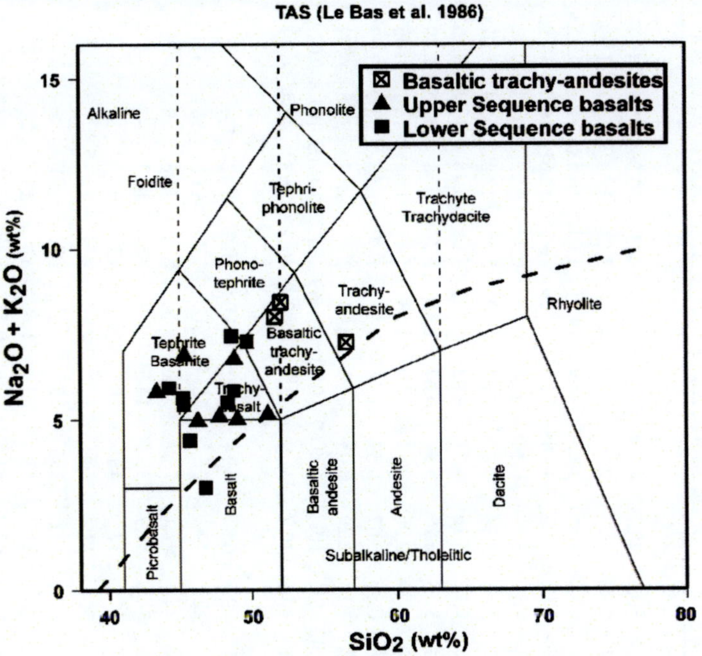

Fig. 4.5 Total alkalis-silica (TAS) diagram comparing the geochemical characteristics of the volcanic rocks of Axum. Oxides are recalculated (on a LOI-free basis) using the SINCLAS program of Verma et al. (2002) for obtaining the TAS names, Mg numbers, and also the normative compositions. The TAS diagram is plotted using the Geochemical Data Toolkit (GCDkit 2.3). The *dashed line* distinguishes alkaline from subalkaline basalts (From Le Bas et al. 1986)

samples from sequences-2 and 3 are a bit stretched, and cover highly – mildly alkaline and transitional fields.

Almost all samples from the sequence-2 and sequence-3 basalts are fractionated mafic rocks with MgO values between 2 and 7.9 wt% and Mg# between 35.3 and 58.3 (Table 4.2). However, sequence-1 basalts (AX12 AX24, AX34, and AX47b) from the Lower Sequence have more primitive compositions with MgO values between 7.7 and 11 wt% and Mg# between 64.1 and 68.8. The distribution of the TiO_2 contents in the basaltic, tephritic, and basanitic rocks of the area is not uniform. Nearly half of the basaltic samples have 2.6–4.4 wt% TiO_2, and the remaining samples from the Lower Sequence have 1.5–2.6 wt% TiO_2. However, the TiO_2 concentrations correlate well with the Fe_2O_3 contents in the rocks. The high-TiO_2 basalts have high Fe_2O_3 content but low MgO and CaO contents (Fig. 4.6). In the binary plot (Harker variation diagram), the distribution of Fe_2O_3 in the basaltic rocks (tephrites, basanites, trachybasalts, and basalts) shows two well defined patterns: the Lower Sequence (sequence-1) is characterized by gently sloping low Fe_2O_3 and the Upper Sequence (sequence-2 and sequence-3) by moderately sloping high Fe_2O_3 patterns (Fig. 4.6c). To simplify our discussion, those samples from

Table 4.2 Major and trace element contents of alkali basalts from the Axum area, northern Ethiopia

	Basalt	TB	Basalt	Basalt	Basalt	Basalt	Basalt	Basalt	Basalt	Basalt	Basalt	Basalt	Basalt	TB	Basalt	Basalt	TB	Basalt	Basalt
	AX004	AX08	AX20	AX21	AX24	AX29	AX31	AX32	AX34	AX38	AX42	AX43d	AX44	AX46	AX47b	AX63	AX79	AX09	AX12
(wt%)																			
SiO_2	43.05	54.37	43.63	44.06	49.20	48.32	48.19	46.82	48.65	44.32	50.68	48.76	47.06	51.60	45.75	44.82	50.96	45.13	45.77
TiO_2	4.39	1.85	2.56	2.95	2.10	3.50	2.12	3.25	2.25	2.44	2.90	2.23	2.85	1.54	2.17	3.07	1.58	3.97	1.99
Al_2O_3	13.88	15.67	15.49	14.85	15.87	15.32	19.22	16.51	15.93	15.58	16.24	16.67	16.38	16.92	14.57	14.32	16.68	15.16	15.53
Fe_2O_3	17.32	9.71	12.94	12.85	10.82	15.39	10.85	14.89	11.12	11.64	13.42	14.57	15.84	12.55	11.68	16.26	12.61	16.58	12.90
MnO	0.21	0.17	0.22	0.20	0.19	0.20	0.21	0.17	0.20	0.20	0.19	0.15	0.14	0.24	0.20	0.24	0.25	0.18	0.17
MgO	7.22	2.01	7.73	7.89	7.68	4.60	3.32	5.22	7.92	6.82	3.63	3.84	3.52	2.45	11.02	5.66	2.64	5.44	10.18
CaO	7.90	5.46	10.59	9.90	9.42	7.08	7.90	6.87	9.35	11.69	7.22	5.16	4.71	4.93	10.72	7.68	5.48	7.07	9.25
Na_2O	4.17	4.01	4.13	4.36	3.96	3.87	5.30	3.86	3.69	4.28	3.35	6.19	5.94	6.87	2.85	5.34	6.38	3.46	2.49
K_2O	1.59	2.97	1.71	1.12	1.93	1.05	2.08	1.12	1.84	1.04	1.74	0.96	0.58	1.50	1.53	1.47	1.54	1.37	0.45
P_2O_5	0.96	0.66	0.69	0.76	0.55	0.53	0.80	0.48	0.53	0.84	0.80	0.67	0.59	1.51	0.57	1.30	1.53	0.60	0.19
LOI	0.72	4.13	1.93	2.81	0.17	1.89	1.41	1.53	0.24	2.57	1.18	2.29	3.08	0.91	0.63	1.05	1.44	2.88	4.02
Total	101.41	101.01	101.62	101.75	101.89	101.75	101.40	100.72	101.72	101.42	101.35	101.49	100.69	101.02	101.69	101.21	101.09	101.83	102.92
Mg#	49.4	35.8	58.27	58.9	64.1	41.1	43.5	46.9	64.2	57.8	40.5	40.7	35.9	33.7	68.8	46.7	35.3	43.4	64.9
(ppm)																			
Sc	13.7	15.4	22.3	18.2	17.7	19.2	6.70	20.6	21.9	24.3	19.5	6.81	6.14	5.42	24.0	10.9	5.42		
V	199	126	254	227	201	207	142	266	212	240	225	79	76	48	242	139	47	384	321
Cr	14.9	17	106	128	273	11.0	9.7	39.3	386	226	23.6	97.1	28.7	9.06	579	8	7.06	39.5	307
Co	44.2	17.9	40.9	33.5	31.9	35.5	20.7	52.0	43.9	40.7	34.6	31.9	38.5	14.3	49.9	32.0	14.3	–	–
Ni	34	14	71	90	143	22	5	52	161	112	27	17	46	10	245	12	8	19	5
Cu	25	17	6	41	66	29	6	30	57	70	36	13	19	2	73	9	4	–	–
Zn	165	137	115	97.8	99.0	140	90.4	165	124	112	171	132	136	155	115	192	150	–	–
Ga	29	22	16	18	18	24	16	24	18	16	22	21	22	25	16	32	25	24	18
As	1.04	0.18	<1.5	1.71	0.45	<1.5	0.45	<1.5	<2.1	<2.3	0.32	<3.2	<1.5	1.36	0.48	1.17	<1.9	–	–
Se	<1.7	<1.64	0.26	2.19	2.19	<2.9	<2.2	3.05	<3.1	<2.4	<2.8	<2.5	<2.3	<2.5	<3.8	<2.8	<2.5	–	–
Br	<0.7	<0.75	<1	1.50	1.37	<3.86	1.91	<3.9	<4.9	<5.2	<3.5	<5.3	<4	<5.1	<3	<5.1	<4.4	–	–

Rb	41.0	37.4	64.7	35.9	36.7	12.3	48.3	19.6	41.7	16.1	51.8	26.7	16.2	25.1	30.0	21.3	29.5	17.7	20.2
Sr	989	559	786	983	796	619	1,163	680	737	927	760	704	1,020	1,776	789	1,207	1,760	614	349
Y	30	44	31	34	28	39	29	36	28	31	45	16	23	61	27	36	62	37	26
Zr	321	389	192	258	197	212	222	208	190	211	269	209	315	637	173	389	621	267	128
Sb	0.10	<0.14	0.09	0.04	0.02	<0.13	0.07	<0.2	0.21	0.10	0.09	<0.1	<0.2	<0.2	<0.1	0.85	0.14	–	–
Nb	34	23	69	84	70	12	77	15	65	73	20	10	20	37	60	40	35	19	5
Cs	0.42	0.34	0.64	0.54	0.26	<0.2	0.89	0.36	0.43	0.74	0.61	0.22	3.18	0.91	0.33	0.48	0.38	–	–
Ba	110	640	523	378	363	221	433	270	431	664	460	178	65.4	176	453	116	150	504	57.7
La	30.4	38.3	46.8	60.8	47.4	15.2	55.5	20	47.4	55.9	31.1	20.9	15.1	42.7	41.1	38.4	41.1	–	–
Ce	72.6	84.2	86.1	110	80.8	38.2	102	47.4	86.5	98.2	72.0	53.6	44.8	108	71.7	90.3	104	47.3	20.2
Nd	46.8	45.2	31.5	45.9	32.5	27	42.3	30.6	34.8	44.3	42.1	38.7	38.5	70.8	35.8	56.9	68.4	–	–
Sm	11.3	9.91	7.29	8.86	6.33	7.27	7.91	8.38	7.62	8.80	10.4	10.3	10.6	17.3	7.24	14.3	16.6	–	–
Eu	4.25	3.79	2.46	3.02	2.24	2.62	2.70	2.85	2.44	2.70	3.27	3.37	3.49	5.73	2.35	4.59	5.66	–	–
Gd	9.0	9.13	6.76	7.69	6.18	9.20	9.07	10.2	9.52	9.81	11.2	9.07	12.4	19.4	8.82	15.1	15.6	–	–
Tb	1.46	1.43	1.07	0.93	0.84	1.16	1.03	1.39	1.03	1.08	1.56	1.22	1.33	2.26	0.86	1.71	2.18	–	–
Tm	0.37	0.68	0.59	0.56	0.47	0.67	0.73	0.75	0.7	0.67	0.82	0.38	0.36	1.08	0.59	0.54	1.08	–	–
Yb	1.80	3.93	2.57	2.69	2.18	2.81	2.63	2.85	2.37	2.76	3.63	1.34	1.37	5.18	2.23	2.03	4.89	–	–
Lu	0.22	0.62	0.37	0.38	0.31	0.43	0.38	0.45	0.32	0.4.0	0.54	0.15	0.17	0.75	0.32	0.24	0.70	–	–
Hf	8.63	10.6	4.16	5.81	4.02	5.23	4.67	6.08	4.49	4.63	7.33	8.27	8.04	17.0	3.52	10.3	16.7	–	–
Ta	2.51	1.33	4.97	5.55	4.53	0.62	5.61	1.13	4.90	4.67	1.32	1.36	1.00	3.12	4.07	2.84	2.86	–	–
W	3.10	<3.42	<4.39	0.41	0.38	<2.4	0.15	<2	<3.6	<4.1	<2.1	2.14	<2.2	<3.6	<1.7	<3.5	<3	–	–
Ir (ppb)	<1.6	<1.6	<1.4	2.99	3.05	<2.4	<1.7	<2.6	<2.9	<2.1	<2.3	<2.2	<1.9	<2	<3.2	<2.3	<2	–	–
Au (ppb)	2.02	0.98	0.8	0.82	0.78	<2.1	1.1	<2.2	<2.3	<0.01	<0.01	<0.01	<2.3	<0.01	<1.8	<2.5	<0.01	–	–
Th	2.53	1.76	5.72	7.14	4.78	0.69	6.58	1.15	4.78	5.79	1.36	1.52	1.07	3.28	3.88	2.92	3.05	2.06	43.4
U	0.67	0.53	1.28	1.92	1.37	<3.86	1.80	<0.8	1.20	1.80	<0.7	<0.6	<0.8	1.12	0.98	<0.9	<0.8	–	–

Table 4.3 CIPW normative composition in wt% for the alkaline basalts of Axum, northern Ethiopia

	Basalt AX004	Basalt AX20	Basalt AX21	Basalt AX24	Basalt AX29	Basalt AX31	Basalt AX32	Basalt AX34	Basalt AX38	Basalt AX42	Basalt AX43D	Basalt AX44	Basalt AX47B	Basalt AX63	Basalt AX09	Basalt AX12
qz	0	0	0	0	0	0	0	0	0	2.47	0	0	0	0	0	0
or	9.47	10.3	6.80	11.3	6.29	12.4	6.76	10.8	6.28	10.4	5.79	3.56	9.04	8.79	8.30	2.72
ab	17.2	6.32	14.2	22.1	33.2	26.4	33.3	22.2	11.6	28.6	42.3	44.3	11.0	23.7	29.7	21.5
an	14.6	18.9	18.1	19.7	21.6	22.7	24.9	21.3	20.7	24.4	15.4	16.9	22.4	10.9	22.3	30.5
ne	9.96	15.8	12.8	6.01	0	10.2	0	4.78	13.8	0	6.02	4.25	7.09	11.9	0.16	0
en	8.81	15.4	14.2	13.2	4.20	4.90	3.08	12.3	17.1	2.67	2.58	1.19	15.9	8.54	4.07	8.38
fs	6.50	8.86	8.00	5.46	4.54	4.70	2.42	4.97	9.81	2.75	2.82	1.59	5.89	7.22	3.90	3.88
di	15.3	24.3	22.2	18.7	8.74	9.59	5.49	17.2	26.9	5.42	5.40	2.78	21.8	15.8	7.97	12.3
Hy	0	0	0	0	10.0	0	3.32	0	0	17.2	0	0	0	0	0	4.55
fo	9.84	8.67	9.53	8.99	3.64	4.26	7.07	9.76	6.60	0	6.00	5.99	14.1	7.22	8.41	13.4
fa	9.17	6.29	6.79	4.69	4.98	5.16	7.02	5.00	4.79	0	8.30	10.1	6.59	7.72	10.2	7.82
ol	19.0	15.0	16.3	13.7	8.62	9.42	14.1	14.8	11.4	0	14.3	16.1	20.7	14.9	18.6	21.2
mt	3.86	2.90	2.92	3.31	3.45	3.37	4.68	3.41	2.63	4.17	5.16	5.07	2.58	5.07	3.76	2.92
il	8.40	4.93	4.78	3.95	6.75	4.06	6.30	4.25	4.74	5.56	4.32	5.62	4.12	5.90	7.73	3.86
ap	2.24	1.62	1.81	1.26	1.25	1.87	1.14	1.22	1.99	1.87	1.58	1.42	1.32	3.05	1.43	0.45

qz quartz, *or* orthoclase, *ab* albite, *an* anorthite, *ne* nepheline, *en* enstatite, *fs* ferrosilite, *di* diopside, *hy* hypersthene, *fo* forsterite, *fa* fayalite, *ol* olivine, *mt* magnetite, *il* ilmenite, *ap* apatite. $Fe_2O_3/Fe_2O_{3T} = 0.15$; $FeO/ Fe_2O_{3T} = 0.85$

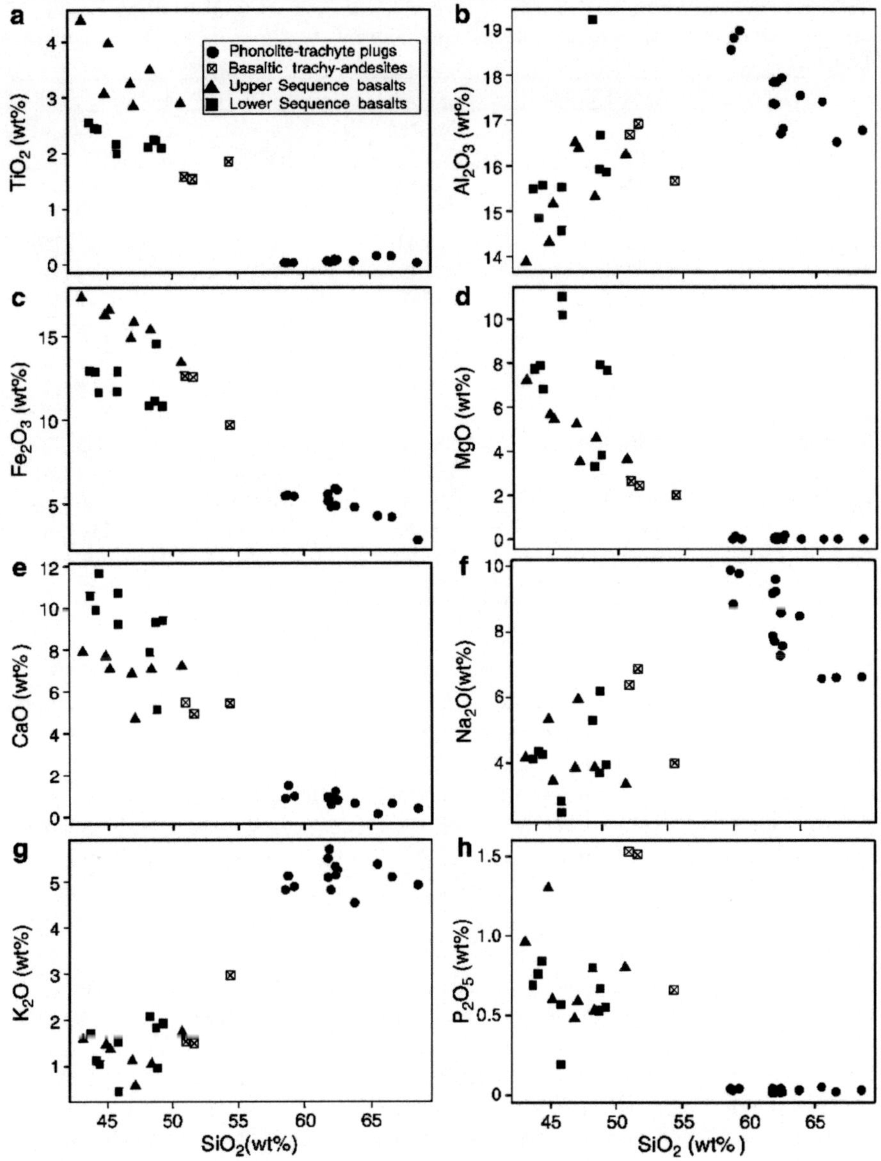

Fig. 4.6 Harker variation diagrams for the volcanic rocks of Axum area. All the major oxides are characterized by their trends as SiO2 ranges from 43 to 67 wt%

sequence-1 are called Lower Sequence basalts and those from sequence-2 and sequence-3 are called Upper Sequence basalts.

The Lower Sequence basalts have the lowest TiO_2 contents (2–2.6 wt%) and are confined to the south and southeastern parts of the mapped area. In contrast,

Table 4.4 Average concentrations and ranges of the contents of selected major elements, trace elements and ratios of the lower sequence and upper sequence basalts of the Axum area, northern Ethiopia

	Lower sequence		Upper sequence	
	Range	Average values	Range	Average values
TiO_2	1.54–2.56	2.26 (9)	2.60–4.40	3.42 (10)
SiO_2	44.1–49.2	46.8 (9)	43.0–48.3	46.2 (10)
Fe_2O_3	10.8–14.6	12.1 (9)	13.4–17.3	15.7 (10)
MgO	3.32–11.0	7.33 (9)	3.53–7.22	5.04 (10)
CaO	5.16–11.7	9.17 (9)	4.71–7.90	6.93 (10)
P_2O_5	0.19–0.84	0.61 (9)	0.48–1.3	0.75 (10)
Y(ppm)	16.0–34.0	27.4 (8)	23–45	34.0 (7)
La(ppm)	21.0–61.0	47.0 (8)	15.0–38.0	24.8 (7)
Ce(ppm)	20.0–118	75.0 (8)	13.0–104	49.6 (7)
Ta(ppm)	1.36–5.61	4.38 (8)	0.62–2.84	1.35 (7)
Nb (ppm)	10.0–84.0	61.7 (8)	12.0–40.0	22.8 (7)
CaO/Al_2O_3	0.31–0.75	0.66 (9)	0.29–0.57	0.40 (10)
Ti/Y	434–853	508 (8)	391–866	630 (7)
Nb/Y	0.63–2.66	2.16 (8)	0.30–1.13	0.70 (7)
Nb/La	0.47–1.49	1.27 (8)	0.64–1.36	0.82 (7)

the Upper Sequence basalts display much higher TiO_2 contents (2.6–4.4 wt%) and are widely distributed in the central, western, and northwestern part of the mapped area. Generally, the Lower Sequence basalts exhibit low Fe_2O_3 (10.8–14.6 wt%, average: 12.1 wt%), low P_2O_5 (0.18–0.84 wt%, average: 0.61 wt%), low Ti/Y (average: 508) and low concentrations of the high field strength trace elements (Zr, Hf, Ga, Tb, Y). However, the Lower Sequence basalts are characterized by significantly higher contents of CaO (5.2–11.7 wt%, average: 9.2 wt%) and MgO (3.3–11 wt %, average: 7.3 wt%), high CaO/Al_2O_3 ratios, and very high contents of Ba, Rb, Th, Nb, Ta, La, and Ce. The Upper Sequence basalts on the other hand display contrasting characteristics, with high contents of Fe_2O_3 (13.4–17.3 wt%, average: 15.7 wt%), P_2O_5 (0.48–1.3 wt%, average: 0.75 wt%), Ti/Y (average: 630), the moderately incompatible trace elements (Sm, Eu, Gd, Zr, Hf, Tb, Y) and low contents of MgO, CaO, low ratios of CaO/Al_2O_3, Nb/Y, Nb/La (Table 4.4), and low concentrations of highly incompatible elements (Ba, Rb, Nb, Ta, La, Ce).

The chondrite-normalized trace element patterns for the mafic rocks of the area show some variation between the various samples (Fig. 4.7a and c). The basaltic rocks show two different REE patterns. The Lower Sequence basalts have highly sloping light REE (LREE) and relatively flat heavy REE (HREE) patterns; in contrast the Upper Sequence basalts have a comparatively flat LREE and moderately sloping HREE pattern. The samples AX21, AX24 and AX31 from the Lower Sequence basalts exhibit very high chondrite-normalized light to heavy REE, (La/Yb)n, ratios of 15.3, 14.7, and 14.3 respectively. The average chondrite-normalized La/Yb ratios (13.1) of the Lower Sequence basalts are higher than the value for Upper Sequence basalts, which is 7.7 (Fig. 4.7a and c). This relatively high (La/Yb)n ratio of the

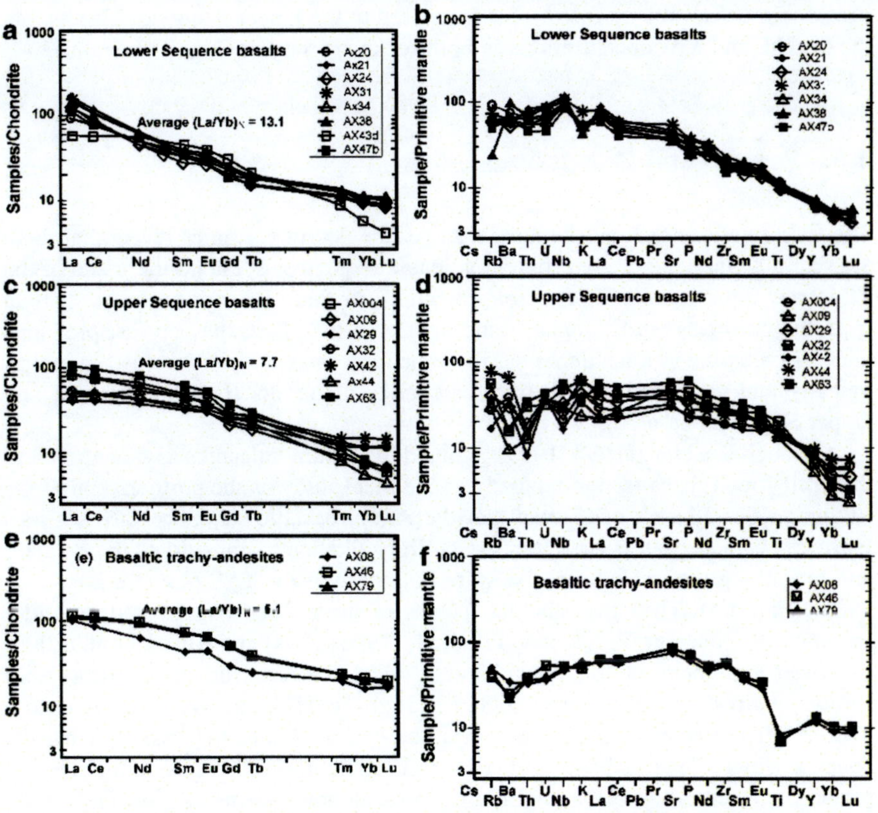

Fig. 4.7 Chondrite-normalized rare earth and primitive mantle-normalized incompatible trace element abundance patterns of the Axum volcanic rocks. Normalization factors after Taylor and McLennan (1985) for the chondrite-normalization, and Sun and McDonough (1989) for the primitive mantle normalization

Lower Sequence basalts shows that basalts of such type are highly fractionated compared with the Upper Sequence basalts. However, both basalts show similar, slightly positive Eu anomalies (average 1.13), and also have similar middle REE (MREE) patterns. There is one sample, AX43d, from the Lower Sequence basalts that has a REE pattern similar to that of the HT basalts.

The geochemical variation of the Lower Sequence and Upper Sequence basalts is highly amplified in the normalized multi-element diagrams. The primitive mantle-normalized incompatible trace element diagrams (Fig. 4.7b and d) show generally coherent behavior with enriched low-field strength elements (Ba, Rb, K, Th) and depleted high-field strength elements (Zr, Eu, Ti) throughout the Lower Sequence suite; however, such behavior in the incompatible trace elements is reversed for the Upper Sequence basalts. In the primitive mantle-normalized multi-element diagram the contents of the elements Th and Nb from the Lower Sequence basalts are enriched by a factor of 3 compared to the Upper Sequence basalts, but the Sr, Nd, and Yb concentrations in both rocks are similar (Fig. 4.7b and d).

4.6 Results and Discussions

The field relations and geochemical data for the Axum volcanics reveal that there are spatially and temporally different volcanic sequences at the northern end of the Ethiopian flood basalts. The Axum volcanics, in this case, appear to be distinct from the northwestern Ethiopian flood basalts in that, first, they consist predominantly of alkaline to transitional basalt sequences; second, they show two mineralogically and geochemically distinct categories of basalts (Lower Sequence and Upper Sequence basalts).

Although Pik et al. (1998, 1999) grouped the Axum volcanics as transitional to tholeiitic low-Ti basalts, our detailed geochemical study on the northern end of the Ethiopian flood basalt confirmed that the Axum basaltic sequences are not tholeiitic in character, but instead they are entirely alkaline to transitional in character and plot along the trachybasalt/basanite boundary in the TAS plot (Le Bas et al. 1986) (Fig. 4.5). The other distinct feature of these flood basalts is their wider range in TiO_2 content. Previously, Pik et al. (1998, 1999) and Kuster et al. (2005) proposed that rocks of the transitional to alkali basalts from the northwestern Ethiopian flood basalts were restricted to high-Ti (HT2 and HT1) suites with a TiO_2 ranging from 2.4 to 5 wt%; however, the Axum alkaline to transitional basalts cover a wider range (1.5–4.4 wt%) in TiO_2 content. They also exhibit a wider range in concentration of the highly incompatible trace elements (Ba, Th, U, Nb, Ta, and La). The concentrations of Ba and Th range from 65 to 664 ppm and from 0.7 to 7.1 ppm, respectively. Based on the field relations, TiO_2 contents and incompatible trace element concentrations, the Axum alkali basalts are grouped into two parts as Lower Sequence alkali basalts (1.5–2.6 wt% TiO_2) and Upper Sequence alkali basalts (2.6–4.4 wt% TiO_2). These LT alkali basalts with their unique geochemical signature are not present in any other part of the Ethiopian flood basalt province, but Zanettin et al. (1999, 2006) and Kieffer et al. (2004) documented the existence of low-Ti alkali basalts in the central Eritrean highlands (Adi Ugri and Asmara basalts) and in the central part of the northwestern Ethiopian volcanic province (Choke and Guguftu shield volcanoes), respectively.

The majority of the basaltic suites of the area (Upper Sequence) are HT type basalts; however, all samples from elongated outcrops in the southern part of the mapped area (Lower Sequence) are LT type basalts (Fig. 4.7). Generally, the Upper Sequence basalts of the area exhibit high TiO_2, Fe_2O_3, P_2O_5, and Y contents and Ti/Y ratios. In contrast, the Lower Sequence basalts exhibit high MgO and CaO contents, and CaO/Al_2O_3, Nb/Y, and Nb/La ratios (Table 4.4). Despite all the major and trace element variations, these two rock suites have similar SiO_2 contents.

The Axum volcanic rocks display a wide range in incompatible trace element abundances that are probably the result of variations in both source composition and melting conditions. The chondrite-normalized REE patterns and the multi-element spider diagram for the basaltic suites of Axum also confirm the existence of these two major groups of alkaline flood basalts. The geochemical characteristics of the Lower Sequence and Upper Sequence basalts of Axum are very different from those of the low-Ti and high-Ti basalts of the north-western Ethiopian flood basalts reported by Pik et al. (1998). According to the authors, the HT basalts were characterized by higher concentrations of incompatible elements and more fractionated REE patterns; however, in our study the opposite is observed.

The Lower Sequence basalts of Axum are characterized by higher concentration of the more incompatible elements (Ba, Rb, K, Nb, Ta, La, and Ce), steeply sloping LREE patterns, and more fractionated REE patterns. These rocks are much more similar with the Choke and Guguftu volcanoes in their trace element characteristics and TiO_2 contents. The Choke and Guguftu volcanoes include high concentrations of incompatible elements, steeply sloping trace element patterns, low TiO_2 contents and positive Nb anomalies (Kieffer et al. 2004). Zanettin et al. (2006) believed that, during the early Miocene (24–22 million year), the whole territory extending from Adi Ugri (Mendefera) and perhaps also from Adwa–Axum to Senafe was involved in basic and acid volcanism, but with a spatial variation in the volumetric ratio between basalt and ignimbrite. Kieffer et al. (2004) also confirmed that the Choke and Guguftu shield volcanoes erupted around 22 Ma ago contemporaneously with the central Eritrean Alkali basalts. Therefore, during the early Miocene period, not only the central Eritrea and Tigray regions were flooded by the alkaline magmatism but also extensive region extending to the central part of the northwestern Ethiopian volcanic province (Choke and Guguftu area) was involved in such magmatism.

Figures 4.6c and d show that, in terms of their trace element contents, the Upper Sequence basalts of Axum resemble Pik et al.'s (1998) HT1 type of flood basalt. The Upper Sequence basalts are characterized by relatively high concentrations of the less incompatible elements (Sm, Zr, Hf, and Tb), nearly flat LREE patterns and moderately sloping HREE patterns. The chondrite-normalized light to heavy REE ratio, $(La/Yb)_n$, of the Upper Sequence basalts of Axum is 7.7, showing the relatively less fractionated nature of the basalts. According to Pik et al. (1998), the ages of HT1 rocks range from about 30 Ma in the north of the province to far younger in the south.

The distinct geochemical signatures of the Lower and Upper Sequence basalts of Axum indicate that the petrogenetic processes that formed these two rock types were different. One of the major element binary plots that can show differences in the petrogenetic processes is the CaO/Al_2O_3 versus SiO_2. At a given SiO_2 content, most of the Lower Sequence basalts are displaced towards higher CaO/Al_2O_3 ratios (Fig. 4.8a). At a given Zr content, the Lower Sequence samples have higher Nb contents than do the Upper Sequence samples (Fig. 4.8b). Kamber and Collerson (2000) have indicated that Nb is more sensitive to variations in degrees of partial melting than Zr, and hence can be used to decipher the influence of variable degrees of melting. The Zr/Nb ratio of the Lower Sequence basalts is very low (2.8–3.1), except for sample AX43d, which has an abnormally high ratio of 21.1; whereas the Zr/Nb ratios of the Upper Sequence basalts are very high (9.4–18.1). A variation of trace element ratios between the Lower and Upper sequence

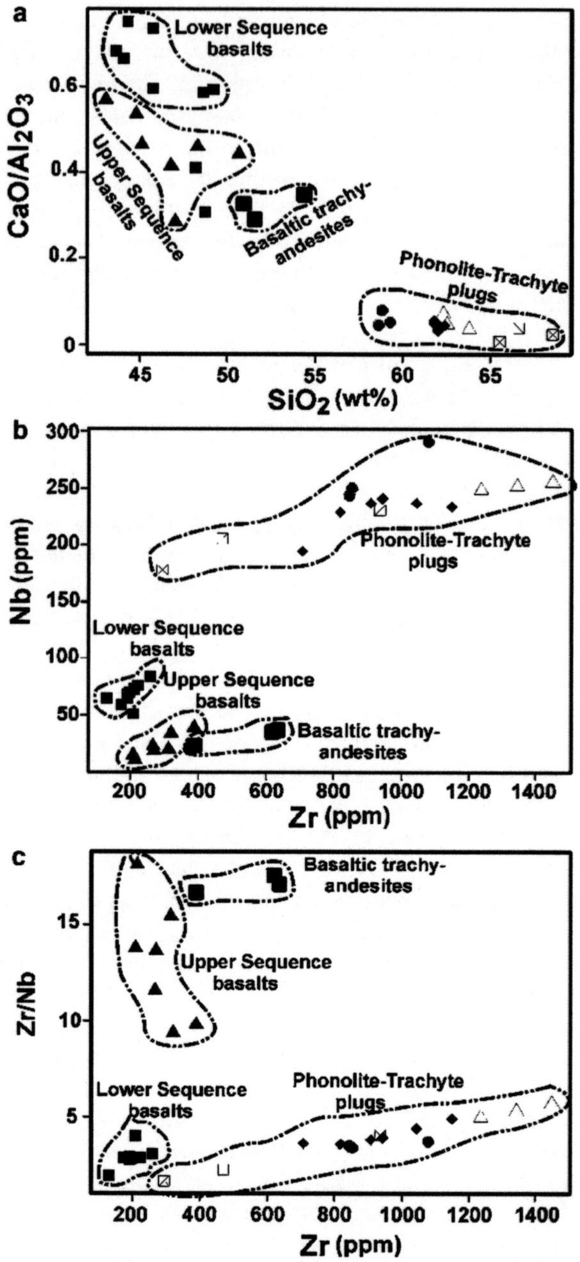

Fig. 4.8 (**a**) Plot of CaO/Al_2O_3 ratio against SiO_2, showing the two apparent evolution trends for the Axum volcanic rocks. All the Upper Sequence basalts plot at lower ratios and most of the Lower Sequence basalts plot at higher ratios. (**b**) Linear correlation between Zr and Nb contents. Samples generally plot along Zr/Nb = 3 and Zr/Nb = 14, which may indicate different petrogenetic processes. (**c**) Zr/Nb versus Zr plot for the Axum volcanic rocks: the Upper Sequence rocks show expanded (7–18) Zr/Nb ratios and the Lower Sequence rocks with narrower and lower (2–5) Zr/Nb ratios

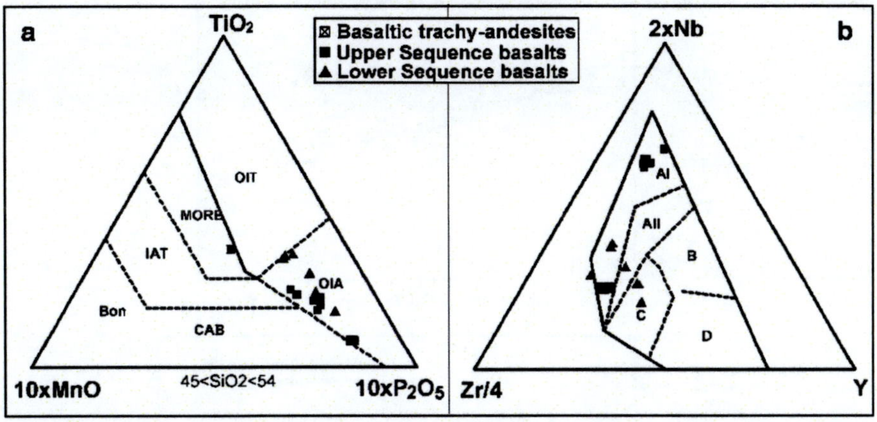

Fig. 4.9 Selected major and trace elements for plotting discrimination diagrams. **(a)** The MnO–TiO_2–P_2O_5 discrimination diagram for basalts and basaltic andesites (45–54 wt% SiO_2) (After Mullen 1983). The fields are MORB – mid-oceanic-ridge basalt; OIT – ocean-island tholeiite; OIA – ocean-island alkali basalt; CAB – island-arc calc-alkaline basalt; IAT – island-arc tholeiite; Bon – boninite. **(b)** The Zr–Nb–Y discrimination diagram for basalts (After Meschede 1986). The fields are defined as follows: AI – within-plate alkali basalts; AII – within-plate alkali basalts and within-plate tholeiites; B – E-type MORB; C – within-plate tholeiites and volcanic-arc basalts, D – N-type MORB and volcanic-arc basalts

samples (e.g., Zr/Nb; Fig. 4.8c) is often cited as strong evidence that fractional crystallization has not been the dominant process in their evolution (Weaver 1977; Wilson et al. 1995).

The MnO–TiO_2–P_2O_5 and Zr–Nb–Y discrimination diagrams for basalts and basaltic andesites indicate that both the Lower and Upper Sequence basalts of Axum have similar source compositions, namely ocean-island alkali basalt (OIA) and/or within-plate alkali basalt (AI and AII), respectively (Fig. 4.9a and b). Although the primitive mantle-normalized incompatible trace element diagram for both the Lower and Upper Sequences shows some deviation from OIB composition, the average of all alkali basalts belongs to the OIB trend except the trace element Sr, which has relatively higher concentrations than those of normal OIB (Fig. 4.10c). The source reservoirs for both the Lower and the Upper Sequence basalts seem to be the same and resemble those of the OIB, but the main petrogenetic change may be most probably in the degree and mechanism of partial melting of the mantle plume. The comparatively small extent and volume of the Lower Sequence basalts in the study area suggests the outpouring of small volume of magma influx during the late-stage of flood basalt volcanism and these basalts are confined to post-volcanic depressions in the basement complex, with clastic sedimentary rock between the volcanics and the basement in some places. The laterally discontinuous but petrologically important rock suites at the base of the Axum volcanic piles are the basis for the classification of Axum basalts into Lower Sequence basalts and Upper Sequence basalts. The Lower Sequence basalts of Axum provide also important clues for further studies regarding the temporal variations of the northwestern Ethiopian flood basalts.

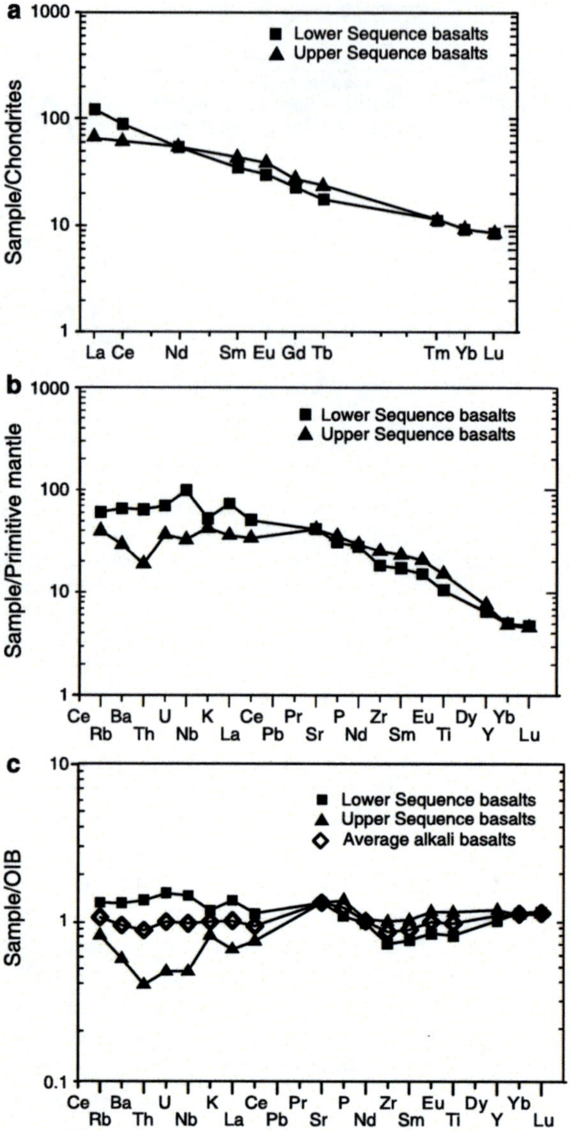

Fig. 4.10 Average plots of the Lower Sequence and Upper Sequence basalts of Axum area. **(a)** Plot of average chondrite-normalized rare earth element abundances (normalizing factors from Taylor and McLennan 1985); **(b)** plot of average primitive mantle-normalized incompatible trace element contents (normalizing factors from Sun and McDonough 1989), and **(c)** plot of ocean island basalt (OIB)-normalized incompatible trace elements (normalizing factors from Sun and McDonough 1989). From the primitive mantle-normalized multi-element plot the Lower Sequence basalts show higher concentrations of the highly incompatible (mobile) trace elements especially Ba, Th and Nb, however the less incompatible trace elements (Zr, Sm, Eu, Ti) are slightly higher in the Upper Sequence basalts. To see further differences between these two suites, the incompatible traces are projected in the OIB normalized plot. The unique character of the chondrite- and primitive mantle-normalized plots is also observed here, but the average of all alkali basalts of Axum agrees well with the composition of OIB, except for the element Sr, which is slightly high for alkali basalts

4.7 Conclusions

Many petrologically different volcanic sequences in the northern Ethiopian plateau have been indistinctly categorized as the low-TiO_2 (LT) transitional to tholeiitic basalt suites of the northwestern flood basalts (Pik et al. 1998, 1999). One of these volcanic sequences is represented by the Axum volcanics, that marks the geochemically complex and lithologically varied part of the northwestern Ethiopian flood basalt province erupted in a narrow time span. The Axum volcanics are highly alkaline to transitional in composition, with distinctive geochemical signatures among the Ethiopian flood basalts.

The Axum volcanic rocks have been subdivided here based on their petrography and geochemistry into two main groups: the Lower Sequence basalts and the Upper Sequence basalts. The Lower Sequence basalts are aphyric - phyric with sparse phenocrysts of clinopyroxene and olivine. They have high MgO and CaO contents, high CaO/Al_2O_3, Nb/Y Nb/La ratios, and high large ion lithophile (LIL) or low field strength incompatible trace element contents. The Upper Sequence basalts, which are plagioclase-phyric, ± K-feldspar bearing (rich and poor), display high TiO_2, Fe_2O_3, P_2O_5 contents and Ti/Y ratio, and relatively high low field strength (LFS) incompatible trace element contents.

The basaltic rocks of Axum area are, therefore, different from the northwestern LT tholeiitic basalts both in respects of mineralogy and geochemistry. The Lower and Upper Sequences of Axum are silica-undersaturated (negative saturation index, SI), alkaline – transitional basalts with basanitic – basaltic–trachyandesitic affinity. The basaltic rocks of Axum are characterized by wide ranges of TiO_2, trace element and REE contents, which helps to obtain information about the mineralogical, geochemical composition and the age temporal variations of the northern Ethiopian volcanic rocks.

Acknowledgments The authors are very grateful to Drs. M. Koch and M. DiBlasi from the Centre for Remote Sensing, Boston University, USA, for their collaboration during the first field session and the local people of Axum for their invaluable help during the field work. We also thank the Tigray Water, Mines and Energy Bureau for their professional and material support during the second fieldwork. We would like to appreciate Axum University staff, especially Drs. K.W-Aregay and W. Fitchanegest, for their invaluable support and discussions during the field session. We thank the reviewers, H.C. Sheth and an anonymous colleague, for their useful suggestions and comments. M.H. is supported by an ÖAD (Austrian Academic Exchange Program) North-South doctoral scholarship at the University of Vienna, Austria.

References

Abebe T, Mazzarini F, Innocenti F, Manetti P (1998) The Yerer – Tullu Wellel volcanotectonic lineament: a transtensional structure in central Ethiopia and the associated magmatic activity. J Afr Earth Sci 26:135–150

Alemu T (1998) Geochemistry of Neoproterozoic granitoids from the Axum Area, northern Ethiopia. J Afr Earth Sci 27:437–460

Alene M, Secco L, Dal Negro A, Sacchi R (2000) Crystal chemistry of clinopyroxene in Neoproterozoic metavolcanic rocks of Tigrai, Northern Ethiopia. Boll Soc Geol Ital 119:581–586

Asrat A, Barbey P, Ludden NJ, Reisberg L, Gleizes G, Ayalew D (2004) Petrology and isotope geochemistry of the Pan-African Negash pluton, northern Ethiopia: mafic-felsic magma interactions during the construction of shallow-level calc-alkaline plutons. J Petrol 45:1147–1179

Asrat A, Gleizes G, Barbey P, Ayalew D (2003) Magma emplacement and mafic–felsic magma hybridisation: structural evidence from the Pan-African Negash pluton, Northern Ethiopia. J Struct Geol 25:1451–1469

Ayalew D, Barbey P, Marty B, Reisberg L, Yirgu G, Pik R (2002) Source, genesis and timing of giant ignimbrite deposits associated with Ethiopian continental flood basalts. Geochim Cosmochim Acta 66:1429–1448

Ayalew D, Yirgu G, Pik R (1999) Geochemical and isotopic (Sr, Nd and Pb) characteristics of volcanic rocks from south-western Ethiopia. J Afr Earth Sci 29:381–391

Baker J, Snee L, Menzies M (1996) A brief Oligocene period of flood volcanism in Yemen: implications for the duration and rate of continental flood volcanism at the Afro-Arabian triple junction. Earth Planet Sci Lett 138:39–55

Beyth M (1972) The geology of central and western Tigray, Ph.D. thesis, Rheinische Friedrich-Wilhelms Universtät, Bonn, 200 p

Beyth M (1991) "Smooth" and "rough" propagation of spreading, southern Red Sea – Afar depression. J Afr Earth Sci 13:157–171

Ebinger CJ, Sleep NH (1998) Cenozoic magmatism throughout East Africa resulting from impact of a single plume. Nature 395:788–791

Ethiopian Institute of Geological Survey (1999) Aksum map sheet ND 37–6 (scale 1:250, 000). Geological Survey of Ethiopia, Addis Ababa

Gasparon M, Innocenti F, Manetti P, Peccerillo A, Tsegaye A (1993) Genesis of the Pliocene to Recent bimodal mafic-felsic volcanism in the Debre Zeyt area, central Ethiopia: volcanological and geochemical constraints. J Afr Earth Sci 17:145–165

George RM, Rogers N, Kelley S (1998) Earliest magmatism in Ethiopia: evidence for two mantle plumes in one flood basalt province. Geology 26:923–926

Hofmann C, Courtillot V, Feraud G, Rochette P, Yirgu G, Ketefo E, Pik R (1997) Timing of the Ethiopian flood basalt event and implications for plume birth and environmental change. Nature 389:838–841

Kabeto K, Sawada Y, Bussert R, Kuster D (2004) Geology and geochemistry of Maychew volcanics, northwestern Ethiopian Plateau. International conference on East African rift system, Addis Ababa. Extended Abstr 1:110–114

Kamber BS, Collerson KD (2000) Zr/Nb systematics of ocean island basalts reassessed—the case for binary mixing. J Petrol 41:1007–1021

Kieffer B, Arndt N, Lapierre H, Bastien F, Bosch D, Pecher A, Yirgu G, Ayalew D, Weis D, Jerram AD, Keller F, Meugniot C (2004) Flood and shield basalts from Ethiopia: magmas from the African superswell. J Petrol 45:793–834

Koeberl C (1993) Instrumental neutron activation analysis of geochemical and cosmochemical samples: a fast and reliable method for small sample analysis. J Radioanal Nucl Chem 168:47–60

Küster D, Dwivedi SB, Kabeto K, Mehary K, Matheis G (2005) Petrogenetic reconnaissance investigation of mafic sills associated with flood basalts, Mekelle basin, northern Ethiopia: implication for Ni-Cu exploration. J Geochem Explor 85:63–79

Le Bas ML, Le Maitre RW, Streckeisen A, Zanettin B (1986) A chemical classification of volcanic rocks based on the total alkali – silica diagram. J Petrol 27:745–750

Meschede M (1986) A method of discriminating between different types of mid-ocean ridge basalt and continental tholeiites with the Nb-Zr-Y diagram. Chem Geol 56:207–218

Merla G, Abbate E, Azzaroli A, Bruni P, Canuti P, Fazzuoli M, Sagri M, Tacconi P (1979) Comments to the geological map of Ethiopia and Somalia. Consiglio Nazionale delle Ricerche (95 p, Firenze)

Mohr P (1983) Ethiopian flood basalt province. Nature 303:577–584

Mohr P (1987) Patterns of faulting in the Ethiopian Rift Valley. Tectonophysics 143:169–179

Mullen ED (1983) $MnO/TiO_2/P_2O_5$: a minor element discriminant for basaltic rocks of oceanic environment and its implications for petrogenesis. Earth Planet Sci Lett 62:53–62

Pik R, Deniel C, Coulon C, Yirgu G, Hoffmann C, Ayalew D (1998) The northwestern Ethiopian Plateau flood basalts: classification and spatial distribution of magma types. J Volcanol Geotherm Res 81:91–111

Pik R, Deniel C, Coulon C, Yirgu G, Marty B (1999) Isotopic and trace element signatures of Ethiopian basalts: evidence for plume-lithospheric interactions. Geochim Cosmochim Acta 63:2263–2279

Son TH, Koeberl C (2005) Chemical variations within fragments of Australasian tektites. Meteoritics Planet Sci 40:805–815

Stewart K, Rogers N (1996) Mantle plume and lithosphere contributions to basalts from Southern Ethiopia. Earth Planet Sci Lett 139:195–211

Sun S, McDonough WF (1989) Chemical and isotopic systematics of oceanic basalts: implications for mantle composition and processes. In: Saunders AD, Norry MJ (eds) Magmatism in the ocean basins. Special Publication of Volcanology. Geological Society of London 42, London, pp 313–345

Tadesse T (1996) Structure across a possible intra-oceanic suture zone in the low-grade Pan-African rocks of northern Ethiopia. J Afr Earth Sci 23:375–381

Tadesse T (1997) The geology of Axum area (ND 37–6). Memoir no. 9. Ethiopian Institute of Geological Survey, Addis Ababa, 184 p

Taylor SR, McLennan SM (1985) The continental crust: its composition and evolution. Blackwell, Oxford, 312 p

Ukstins I, Renne P, Wolfenden E, Baker J, Ayalew D, Menzies M (2002) Matching conjugate volcanic rifted margins: $^{40}Ar/^{39}Ar$ chronostratigraphy of pre- and syn-rift bimodal flood volcanism in Ethiopia and Yemen. Earth Planet Sci Lett 198:289–306

Verma SP, Torres-Alvarado IS, Sotelo-Rodriguez ZT (2002) SINCLAS: standard igneous norm and volcanic rock classification system. Comput Geosci 28:711–715

Weaver SD (1977) The Quaternary caldera volcano Emuruangogolak, Kenya Rift, and the petrology of a bimodal ferrobasalt-pantelleritic- trachyte association. Bull Volcano 40:209–230

Wilson M, Downes H, Cebriá J (1995) Contrasting fractionation trends in coexisting continental alkaline magma series. Cantal, Massif Central, France. J Petrol 36:1729–1753

Zanettin B, Bellieni G, Visentin JE, Haile T (1999) The volcanic rocks of the Eritrean plateau: stratigraphy and evolution. Acta Volcano 11:183–193

Zanettin B, Bellieni G, Visentin JE (2006) New radiometric age of volcanic rocks in the Central Eritrean plateau (from Asmara to Adi Quala): considerations on stratigraphy and correlations. J Afr Earth Sci 45:156–161

Chapter 5
Geological and Geochemical Studies of Kolekole Cinder Cone, Southwest Rift Zone, East Maui, Hawaii

Nilanjan Chatterjee and Somdev Bhattacharji

Abstract The Kolekole cinder cone is located in the southwest rift zone near the triple-junction of the southwest, east and north rifts, and the Haleakala volcano summit area in East Maui, Hawaii. The geochemical and petrological characteristics of the alkaline lavas erupted from the cinder cone provide important information regarding the post-shield evolution of Hawaiian volcanoes. The incompatible element abundances and ratios of the Kolekole lavas are transitional between previously studied post-shield, temporally continuous, Kula and Hana alkaline series. Thus, the incompatible element composition of the alkaline lavas changed gradually during the post-shield stage. Rare earth element ratios such as Tb/Yb indicate that the Kolekole post-shield alkaline lavas originated from garnet-bearing mantle source regions. The Kolekole lavas are predominantly ankaramite containing 25–40% by volume, large (1–5 mm), normal and reverse-zoned olivine and augite phenocrysts. About 74–85% of the olivine and 58–75% of the augite phenocrysts are reversely zoned. The groundmass of the ankaramite is basanitic and similar to alkaline lava elsewhere on East Maui. Barometry indicates that the phenocrysts originated from depths between 15.5 and 48 km, transcending the crust–mantle boundary. The alkaline primary magma probably ponded below solidified, shallow, shield-stage magma chambers and underwent extensive olivine-controlled high pressure fractional crystallization, as indicated by the Fe-rich (Mg# 35–49) composition estimates and major and compatible trace element correlations in the ankaramite groundmass. Large spheroidal bombs (0.6–2.2 m) near the Kolekole vents indicate that the evolved magma eventually erupted by violent volatile-driven explosions.

N. Chatterjee (✉)
Department of Earth, Atmospheric and Planetary Sciences, Massachusetts Institute of Technology, Room 54-1216, Cambridge, MA 02139, USA
e-mail: nchat@mit.edu

S. Bhattacharji
Department of Geology, Brooklyn College and Graduate Center of the City University of New York, Brooklyn, NY, 11210, USA
e-mail: somdevlee4@earthlink.net

J. Ray et al. (eds.), *Topics in Igneous Petrology*, DOI 10.1007/978-90-481-9600-5_5,

5.1 Introduction

Post-shield alkaline lavas constitute about 1% of the lavas on the Hawaiian Islands (Clague 1987). On East Maui, these lavas erupted on top of the voluminous shield-stage tholeiitic lavas from vents along the southwest, east, and north rift zones, and in the Haleakala Crater at the triple junction of the three rifts (Fig. 5.1a; Stearns and Macdonald 1942). A large number of post-shield volcanic vents are located in the southwest and east rift zones, whereas, only a few post-shield vents are located in the north rift zone indicating that the north rift remained inactive as a failed arm of the triple-junction rifts during the post-shield stage (Bhattacharji and Koide 1975). The post-shield alkaline lavas have been subdivided into an older Kula series and a younger Hana series on the basis of stratigraphy (Stearns 1942; Stearns and Macdonald 1942; Macdonald 1978). The Hana series was earlier regarded as a product of rejuvenated volcanism following a period of quiescence

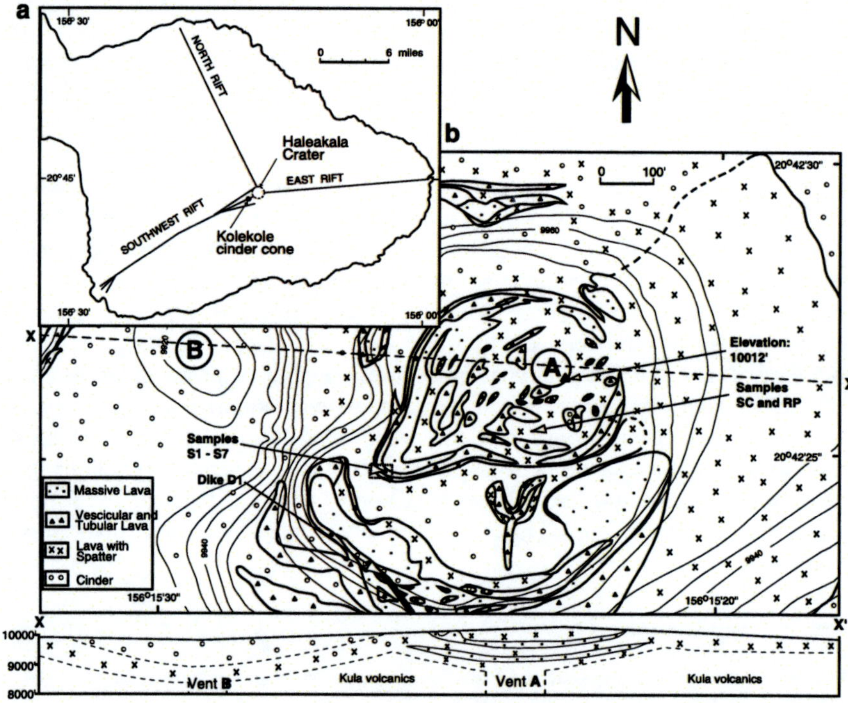

Fig. 5.1 (**a**) Splay branching in southwest rift and the location of Kolekole cinder cone near Haleakala Crater and the junction of the north, east and southwest rifts in East Maui, Hawaii. (**b**) Geological map and schematic cross section along X–X′ of a part of the Kolekole cinder cone showing distribution of ankaramite lava and cinder, and sample locations. 'A' and 'B' are the major vent locations. Contours are in 5-ft intervals. Latitude and longitude are referable to the North American Datum of 1983

after the eruption of the Kula series (Clague and Dalrymple 1987). However, it has recently been shown that the interval between the eruption of the Kula and the Hana series is only 30,000 years (between 150 and 120 ka), and post-shield alkaline volcanism has been continuous on East Maui (Sherrod et al. 2003). The last eruption occurred between AD 1480 and 1600 in the southwest rift zone (U.S. Geological Survey 1999).

The Kolekole cinder cone in the southwest rift zone adjacent to the Haleakala Crater is composed of post-shield alkaline lavas such as ankaramitic picrobasalt and basanite. The Kolekole lavas studied here are not to be confused with the post-shield alkaline lavas of the Kolekole Member of the Waianae Volcanics on Oahu, Hawaii (Presley et al. 1997). Recent K–Ar dating yielded a 128 ± 6 ka age of Kolekole ankaramite from northwest of the Kolekole cinder cone (Sherrod et al. 2003). The studied samples include seven flow samples from a borehole that has been drilled from an elevation above 9960′. All the investigated samples were collected from the southwestern slopes of Kolekole cinder cone. Since the top of the Kula series is dated at 150 ka and the bottom of the Hana at 120 ka (Sherrod et al. 2003), the present samples are probably from the lowermost part of the Hana sequence or the topmost part of the Kula sequence.

As a Hawaiian volcano moves away from the plume axis, there is a change in the thermal structure, melt production rates and the melting regimes under the volcano. The geochemical signatures of the erupted lavas provide clues to these changes. Post-shield volcanoes and cinder cones on shield volcanoes provide an opportunity to compare and contrast post-shield alkaline volcanism from tholeiitic volcanism during the shield building stage. Post-shield cinder cones on shield volcanoes have seldom been studied in detail. Magma reservoir depths under the Kolekole cinder cone have been investigated by Chatterjee et al. (2005). In this study, we examine geological features, mode of eruptions, and geochemical characteristics of Kolekole lavas as an example of Hawaiian post-shield volcanism.

5.2 Previous Work

Stearns (1942) and Stearns and Macdonald (1942) originally studied the geological and geomorphological features of Haleakala volcano and the geology of East Maui. Macdonald and Katsura (1964) later studied the geology and petrology of lavas erupted along the rift zones of East Maui. Macdonald and Powers (1968) first suggested that the Hana magma reservoirs might be deeper than the earliest post-shield Kula alkaline magma reservoirs. However, the chemical compositions of the Hana and Kula lavas (Chen et al. 1990, 1991; West and Leeman 1994; Bergmanis et al. 2000; Sherrod et al. 2003) show considerable overlap and Bergmanis et al. (2000) showed that their equilibrium pressures (2–5 kb) also overlap. Recently, Chatterjee et al. (2005) showed that the phenocrysts in the alkaline lavas of Kolekole originated well below the crust-mantle boundary.

5.3 Geological Setting of Kolekole Cinder Cone

The southwest rift zone, on which the Kolekole cinder cone is situated, is characterized by parallel fissures of varying length, width and depth. This rift zone shows splay branching (Figs. 5.1a and 5.2; Bhattacharji and Koide 1987) partly covered by recent flows at its western end. Detailed field geological study (Figs. 5.1b and 5.2) indicates that Kolekole is an asymmetric volcanic cone with steep (50–55°) western and northwestern slopes near the rim becoming gentler with distance (40–45°). The eastern and southern slopes are also steep near the rim (55–60°), gentler at mid heights (east: 20–25°, south: 30–35°), and very low near the base (~15–20°). Here the flanks of the cinder cone merge with adjacent cinder cones. Much of the northern and northeastern slopes are gentle (10–12°).

Fig. 5.2 Aerial view of Haleakala Crater, Kolekole cinder cone and other cinder cones on the southwest rift zone. The inferred axis and splay branching of southwest rift is shown by *dashed white lines*

Both the eastern and the southern slopes of the Kolekole cinder cone are characterized by composite lava flows agglutinated with spatter, cinder and ash (Fig. 5.1b). The ankaramite lava flows at the crater rims as well as on the slopes are highly vesicular and phenocryst-rich. The lava surface shows alteration to hematite and limonite.

The central crater of Kolekole (site A, Fig. 5.1b) resembles a flattened bowl ponded with ankaramite lava, spatter and pyroclastic ejecta. At this site, as well as in the eastern and southern sides of the crater, ankaramitic (picritic) lava is characterized by a basal zone of sub-columnar to columnar, phenocryst-rich and highly vesicular lava. The vesicles are elongated perpendicular to the columns (Fig. 5.3a). On the northern and steep western slopes, lavas are highly vesicular with spheroidal and oblong vesicles (Fig. 5.3b). Here the base of the lava also shows columnar joints.

In the central vent area, the basal lava zone grades abruptly into highly vesicular aa-type ankaramite lava whose thickness ranges from about 1 (0.6–1 m) to almost 8 m at the northern end. Associated with the aa-type ankaramite lavas are ropy (pahoehoe-type) lavas with tubes of various sizes, as well as pyroclastic ejecta (ash, scoria, lapilli, and volcanic bombs of various sizes). Most of the bombs are

Fig. 5.3 (**a**) Columnar lava at the base of the ponded lava bed showing highly elongated vesicles parallel to the base, in the vent area A. (**b**) Highly vesicular ankaramite lava on the northern slope of Kolekole. Vesicles are spheroidal to oblong and many vesicles coalesce to form ellipsoidal to oblong large vesicles. (**c**) A close-up, megascopic view of unaltered ankaramite lava showing abundant olivine and clinopyroxene phenocrysts. The pen is 15 cm in length. (**d**) Large ellipsoidal bombs (~2.2 m in diameter) and spheroidal bombs (1–1.5 m in diameter) and other pyroclastic materials on the spatter lava and cinder bed north of the main vent A

5–15 cm in size, but they can be as large as 2.2 m. Unaltered ankaramite lava with abundant clinopyroxene and olivine phenocrysts can be commonly seen in hand specimens (Fig. 5.3c). In a vertical section at the main vent site A (Fig. 5.1b, cross-section), pyroclastic materials alternate with several meter thick ankaramite lava, indicating a series of eruptions with alternating lava and pyroclastic materials from the major vent.

A probable site of a secondary vent is a gentle depression at site B (Fig. 5.1b) where the soil is highly weathered to limonite. Surrounding this site, the ground is covered with pyroclastic ejecta, especially large volcanic bombs and lapilli, and broken fragmental lava.

The presence of a large number of spheroidal to oblate shaped bombs (Fig. 5.3d) near the crater areas A and B (Fig. 5.1b), and large dimensions of some of these phenocryst-rich ankaramite bombs (0.6–2.2 m in diameter), indicate the occurrence of volatile-driven explosions and exhalation of gases toward the end phase of eruptions. Such explosive eruptions toward the end phase of volcanism with degassing, threw abundant volcanic bombs and pyroclastic materials on the lava field near the vents A and B (Fig. 5.1b) and its surrounding.

On the Kolekole cinder cone, a near vertical ankaramite dike (D1) is exposed on its southern slope (Fig. 5.1b). This dike and a series of six lava flows (S2 through S7) were sampled from a borehole core on the southwestern slope. In addition, lavas with columnar to sub-columnar joints (SC) and associated tubular, ropy lava (RP) were also sampled (Fig. 5.1a) for chemical analysis. The chemical composition of the lava samples are given in Chatterjee et al. (2005).

5.4 Mineralogy and Petrology

The Kolekole lava contains large (1–5 mm) olivine and clinopyroxene phenocrysts embedded in a dark, vesicular and fine-grained groundmass (Figs. 5.4a, b and 5.5a). Of the eleven samples studied, three are less phenocryst-rich (S6, S4b and RP; groundmass >93%), and the rest are phenocryst-rich ankaramite with 15–18% augite and 11–23% olivine phenocrysts by volume. The olivine to augite ratio of the Kolekole ankaramites is 0.8–2.4 compared to 0.1–2.2 for the Hana and Kula ankaramites from the northeastern coast of Maui (Chen et al. 1990). Plagioclase microphenocrysts (0.5–1 mm) and spinel phenocrysts are rare. The groundmass is composed of augite, plagioclase and Fe-rich oxides.

The olivine and augite phenocrysts show normal (Fe-rich rim), reverse (Mg-rich rim) or cyclic zoning (Fig. 5.5b–d). Reverse zoning dominates: about 74–85% of the olivine and 58–75% of the augite phenocrysts are reversely zoned. The forsterite (Fo) content of phenocrystic olivine is Fo_{65-89} (core: Fo_{76-89}, rim: Fo_{65-83}). Thus, some olivine cores have higher Fo-content than olivine in Hana and Kula lavas from the northeastern coast (Fo_{68-85}, Chen et al. 1990) and olivine from a wide area of Maui (Fo_{68-83}, Fodor et al. 1977), but have similar Fo-content to olivine in the shield-stage Honomanu tholeiitic lavas (Fo_{74-90}, Chen 1993).

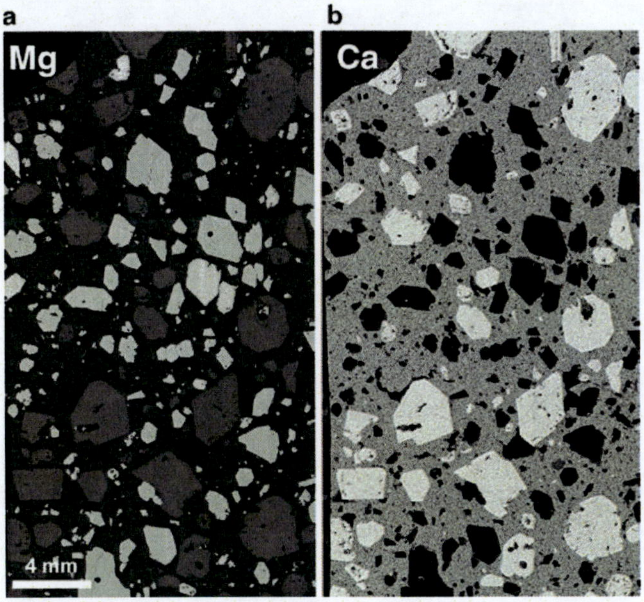

Fig. 5.4 X-ray elemental maps of ankaramite. (**a**) Mg map showing olivine (*light*) and augite (*grey*) phenocrysts. (**b**) Ca map showing augite (*light*) and olivine (*dark*) phenocrysts. The groundmass is Mg-poor and Ca-rich

Compared to the augite in the northeastern coastal samples (Chen et al. 1990), the phenocrystic augite in the Kolekole lava contains higher Al_2O_3 (4.5–12.1 wt%) and TiO_2 (1.0–4.6 wt%); the highest Al_2O_3 and TiO_2 in augite measured by Chen et al. (1990) being 9.6 and 3.5 wt%, respectively. The Mg# (100 × Mg/[Mg + Fe]) of the phenocrystic augite is 66–83 in the cores and 64–82 in the rims. Chen et al.'s (1990) augite data are in a more restricted range (Mg# = 70–75), and augite from a wider area of Maui is of similar composition (Fodor et al. 1975).

Rare plagioclase microphenocrysts show reverse or cyclical zoning. The anorthite (An) content of plagioclase is An_{47-78}, lower than An_{54-87} in Kula lavas (Chen et al. 1990; Keil et al. 1972). The maximum ulvospinel content of rare spinel phenocryst is 64% near the rim, falling within the 54–74% range reported by Fodor et al. (1972). Spinel also occurs as inclusion in olivine. The inclusions closer to the olivine cores are higher in Mg and Al, and lower in Fe than the inclusions closer to the olivine rims.

Chatterjee et al. (2005) showed that the cores of olivine and augite are not in equilibrium with the bulk lava compositions, but the augite rims are in equilibrium with the groundmass compositions estimated through large-area image analysis. Using augite compositions and estimated groundmass compositions, Chatterjee et al. (2005) calculated equilibrium pressures of 4.4–11.2 kb and temperatures of 1,109–1,196°C using the clinopyroxene-liquid equilibrium (Putirka et al. 1996) and composition of liquids saturated with olivine, clinopyroxene and plagioclase (Yang

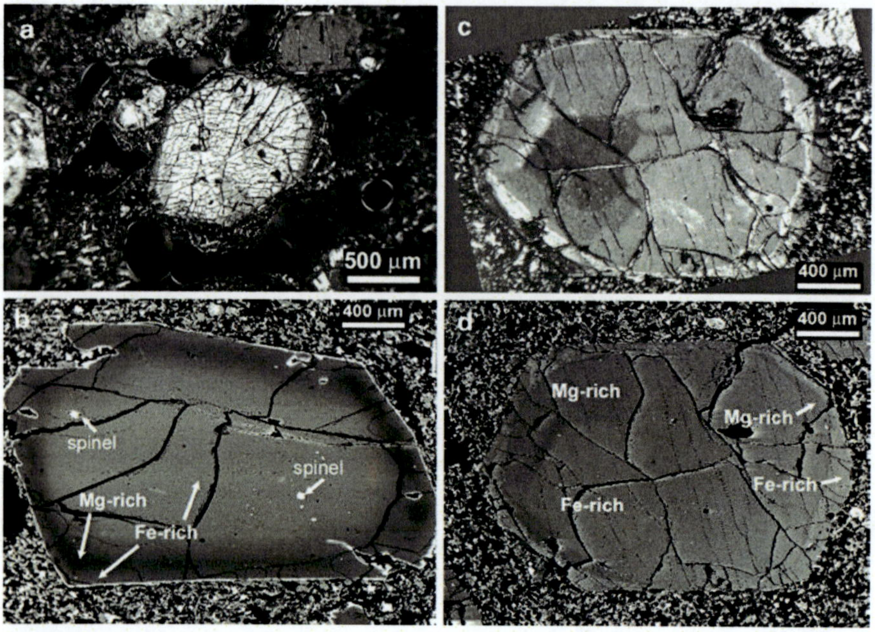

Fig. 5.5 Cross-polarized light (**a**, **c**) and backscattered electron (**b**, **d**) images showing (**a**) large augite phenocryst surrounded by smaller olivine phenocrysts embedded in a vesicular, fine-grained groundmass; (**b**) olivine with cyclic zoning; and (**c**, **d**) augite with cyclic zoning

et al. 1996). The Hana and Kula samples from Chen et al. (1990) yielded similar equilibrium pressures (6.9–10.2 kb) with the same methods. These pressures (4.4–11.2 kb) are higher than those inferred from whole rock data (2–5 kb, West et al. 1988; Bergmanis et al. 2000). Uncertainty in the high-pressure olivine-diopside-plagioclase cotectic (references in Sack et al. 1987), complications due to magma mixing and high accumulated phenocryst-content of many samples in the whole rock data-set may be the reasons for underestimation of pressures.

Pressures calculated with only augite compositions (Nimis 1995) are much lower (0–3.8 kb) than pressures calculated with the methods involving liquid. Although these pressures are underestimates, they provide a measure of the pressure gradients from core to rim of the augite phenocrysts. The cores yield higher pressures than the rims with an average difference of 2.5 kb. The cores thus equilibrated at greater depths than the rims.

5.5 Geochemistry of Kolekole Lavas

The Kolekole samples (Table 5.1) are Mg-rich (Mg# = 50–68) and plot in the alkalic field of Macdonald and Katsura (1964) in a total alkali versus SiO_2 diagram (Fig. 5.6). Four samples (S6, S4b, RP and D1) plot in the basanite field, and the

Table 5.1 Bulk composition of alkaline lavas from the Kolekole cinder cone, East Maui, Hawaii

	D1	S2	S3	S4	S4b	S5	S6	S7	S7b	SC	RP
	Dike	Top flow	flow	flow	flow	flow	flow	Bottom flow	Bottom flow	Sub-columnerlava	Ropy lava
Oxides (wt%)[a]											
SiO_2	41.40	42.58	42.02	42.24	41.74	41.75	42.48	42.27	41.96	42.47	44.88
TiO_2	3.12	2.58	2.73	2.78	3.46	2.65	3.40	2.75	2.65	2.77	3.12
Al_2O_3	14.08	11.31	11.86	11.91	15.24	11.59	15.00	11.96	11.62	12.01	16.23
Fe_2O_3	15.28	15.34	15.27	14.82	15.99	15.53	15.76	15.08	15.31	15.26	13.85
MnO	0.19	0.19	0.19	0.18	0.20	0.19	0.20	0.18	0.19	0.19	0.21
MgO	8.73	13.83	12.45	11.73	6.83	13.00	6.88	12.45	13.23	12.03	6.05
CaO	10.21	11.50	11.85	12.29	10.79	11.88	10.43	12.51	11.85	11.91	9.58
Na_2O	2.71	2.09	2.29	1.90	2.73	1.93	2.58	1.87	1.77	2.11	4.29
K_2O	0.69	0.70	0.50	0.63	0.88	0.40	0.94	0.42	0.38	0.74	1.54
P_2O_5	0.47	0.29	0.30	0.32	0.35	0.31	0.37	0.30	0.28	0.29	0.55
LOI[b]	3.09	−0.09	0.30	1.46	1.69	0.83	1.88	0.61	0.57	0.36	−0.35
Total	99.97	100.32	99.76	100.26	99.90	100.06	99.92	100.40	99.81	100.14	99.95
FeO_t[c]	13.75	13.80	13.74	13.34	14.39	13.97	14.18	13.57	13.78	13.73	12.46
Mg#[d]	57.1	67.8	65.5	64.8	49.9	66.1	50.4	65.8	66.8	64.8	50.4
Trace elements (ppm)[e]											
Rb	11.7	16.2	57.5	13.7	8.0	19.0	21.0	17.4	16.9	16.6	36.9
Ba	549	328	329	375	470	368	492	340	332	410	621
Th	3.15	2.02	2.11	2.14	2.93	2.13	2.88	2.11	2.12	2.21	4.00
U	0.787	0.585	0.613	0.639	0.718	0.617	0.762	0.623	0.584	0.628	1.155
Nb	37.5	24.0	25.6	26.3	37.3	25.1	35.0	25.3	26.7	28.5	51.0
Ta	2.92	1.84	1.96	1.97	2.71	1.92	2.74	1.90	1.91	2.10	3.68
K	5,728	5,811	4,151	5,230	7,305	3,321	7,803	3,487	3,155	6,143	12,784
La	31.5	20.5	21.6	22.4	28.2	21.5	28.3	21.8	21.2	22.5	39.9

(continued)

Table 5.1 (continued)

	D1	S2	S3	S4	S4b	S5	S6	S7	S7b	SC	RP
	Dike	Top flow	flow	flow	flow	flow	flow	Bottom flow	Bottom flow	Sub-columnerlava	Ropy lava
Ce	64.2	46.0	49.2	50.0	56.0	47.8	62.9	49.0	42.2	45.4	85.3
Pb	1.90	1.32	2.34	1.79	1.91	1.15	3.37	1.55	1.32	1.52	2.59
Pr	8.23	5.62	5.99	6.10	7.66	5.80	7.42	6.03	5.83	6.27	10.52
Sr	711	474	508	530	550	483	571	493	475	521	867
Nd	34.4	23.9	25.6	26.0	32.6	24.8	31.4	25.4	25.0	26.4	43.1
Zr	193	129	141	140	164	134	181	137	128	138	242
Hf	4.76	3.61	3.83	3.83	4.56	3.71	4.60	3.80	3.65	3.92	5.52
Sm	7.64	5.60	6.05	6.17	7.25	5.80	7.19	6.04	5.92	6.21	9.22
Eu	2.56	1.88	2.11	2.11	2.42	2.00	2.46	2.10	1.98	2.14	3.04
P	2,051	1,266	1,309	1,397	1,528	1,353	1,615	1,309	1,222	1,266	2,400
Ti	18,702	15,465	16,364	16,664	20,740	15,884	20,380	16,484	15,884	16,604	18,702
Gd	6.81	5.34	5.57	5.80	6.54	5.56	6.82	5.89	5.41	5.80	7.71
Tb	0.979	0.774	0.804	0.830	0.960	0.782	0.981	0.831	0.798	0.838	1.136
Dy	5.20	4.17	4.37	4.49	5.32	4.33	5.30	4.50	4.31	4.52	5.98
Y	25.0	19.5	20.7	21.1	25.3	20.5	26.0	20.7	19.9	20.9	28.2
Ho	0.930	0.716	0.757	0.784	0.939	0.744	0.940	0.761	0.740	0.787	1.012
Er	2.25	1.74	1.83	1.87	2.40	1.83	2.26	1.85	1.80	1.93	2.58
Tm	0.294	0.234	0.244	0.250	0.300	0.246	0.305	0.239	0.229	0.235	0.335
Yb	1.75	1.39	1.42	1.42	1.92	1.40	1.94	1.41	1.40	1.43	2.06
Ni	150	311	296	273	77	311	78	285	296	282	61

Ratios												
K_2O/P_2O_5	1.47	2.41	1.67	1.97	2.51	1.29	2.54	1.40	1.36	2.55		2.80
K/Rb	492	358	72	381	916	174	372	200	186	371		346
Zr/Nb	5.15	5.39	5.50	5.34	4.39	5.35	5.19	5.39	4.78	4.83		4.75
Zr/Y	7.73	6.64	6.80	6.66	6.47	6.56	6.97	6.59	6.42	6.60		8.58
Nb/Y	1.50	1.23	1.24	1.25	1.47	1.23	1.34	1.22	1.34	1.36		1.81
La/Ce	0.49	0.45	0.44	0.45	0.50	0.45	0.45	0.45	0.50	0.50		0.47
La/Eu	12.3	10.9	10.2	10.6	11.7	10.8	11.5	10.4	10.7	10.5		13.1
La/Yb	18.0	14.8	15.2	15.8	14.7	15.3	14.6	15.5	15.2	15.7		19.3
Tb/Yb	0.56	0.56	0.57	0.58	0.50	0.56	0.51	0.59	0.57	0.59		0.55

[a] Determined by ICP-AES
[b] Loss on ignition
[c] Total Fe expressed as FeO
[d] 100 × Mg/(Mg + Fe)
[e] Determined by ICP-MS

phenocryst-rich samples plot in the picro-basalt field of Le Bas et al. (1986). However, the calculated groundmass of the phenocryst-rich samples are Fe-rich (Mg# = 35–49) and plot in the basanite field. Hana samples from previous studies (Chen et al. 1990, 1991; Bergmanis et al. 2000; Sherrod et al. 2003) are also basanites with only a few picro-basalts. Most bulk sample major and minor oxides are negatively correlated with MgO (Fig. 5.7a, b, e–h). Total-FeO shows limited variation (Fig. 5.7c), whereas, Ni is positively correlated with MgO (Fig. 5.7i). Except for CaO and K_2O, the bulk sample major and minor oxide trends coincide with the groundmass trends. For the bulk samples, CaO is negatively correlated with MgO at high MgO values (>12 wt%) but positively correlated at lower MgO values (<12 wt%) (Fig. 5.7d). Hence, bulk compositions of Kolekole lavas with >12 wt% MgO were controlled by olivine, but lavas with <12 wt% MgO show clinopyroxene control. However, CaO of the matrices of all samples are negatively correlated with MgO. Hence, the matrices may have evolved through olivine fractionation.

Although the primitive mantle-normalized incompatible element contents are slightly lower in the Kolekole samples, their patterns are similar to those of previously studied Hana and Kula samples (Fig. 5.8) (Chen et al. 1990, 1991; West and Leeman 1994; Bergmanis et al. 2000; Sherrod et al. 2003). They show prominent Ba and Ta peaks, and Th, K, Pb, Zr and Y troughs. The Ba and Ta peaks and the Th and K troughs are also present in the patterns for the shield stage Honomanu tholeiites and alkali basalts (Chen et al. 1991). However, the Honomanu basalts have a relatively flat profile for the high field strength (HFS) and rare earth (RE) elements and their inter-element ratios are different from those of the Kolekole, Hana and Kula lavas, which have higher contents of highly incompatible elements than even the Honomanu alkali basalt (Fig. 5.8). These differences between the shield stage and post-shield stage basalts may be attributed to a change in the composition of the mantle source with an increasing role of a low melt fraction of a MORB-related depleted mantle component (Chen et al. 1991). The latter authors presented Sr and Pb isotopic evidence to support this hypothesis.

The K_2O/P_2O_5 and K/Rb ratios of the Kolekole samples show a larger scatter compared to previously studied Hana samples (Fig. 5.7j, k) probably due to mobility of these elements during alteration. However, ratios involving HFS and RE elements such as Nb/La, La/Ce and La/Sr are relatively insensitive to hydrous alteration, degree of melting, basaltic fractional crystallization and crystal accumulation and have been used to distinguish among the Honomanu, Kula and Hana lavas (Chen et al. 1991). The Kolekole samples plot within the Hana field in most inter-ratio plots (Fig. 5.9). However, there are significant overlaps in these ratios between Hana and Kula fields. Furthermore, the Zr contents of the Kolekole samples are lower than of other Hana samples (Chen et al. 1990, 1991; Bergmanis et al. 2000). The Kolekole samples are thus similar to Hana, but transitional between Hana and Kula in some respects.

The Tb/Yb ratio has been used to constrain source mineralogy and estimate the degree of melting (Clague and Frey 1982; Chen and Frey 1985). This ratio is expected to remain constant during basaltic fractional crystallization and/or crystal accumulation. The Tb/Yb ratio of the Kolekole samples as well as the Hana samples

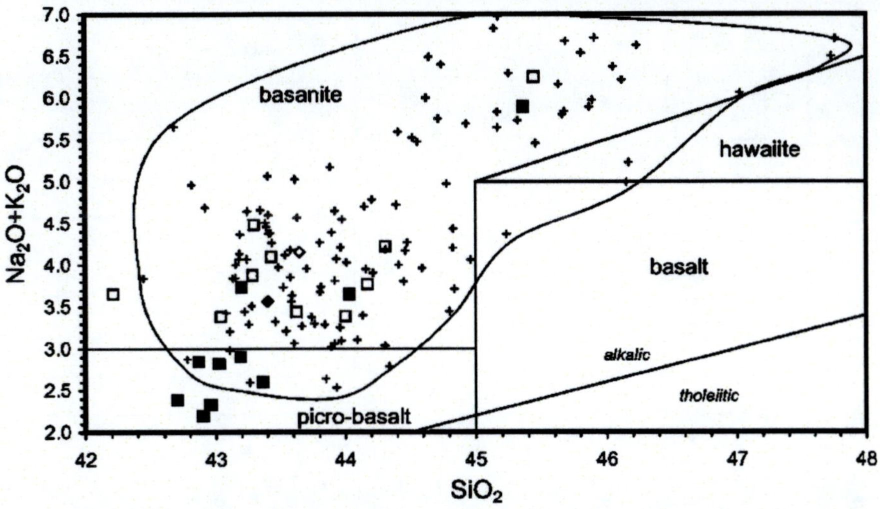

Fig. 5.6 SiO_2 (wt%) versus total alkali (wt%). The alkalic–tholeiitic boundary is from Macdonald and Katsura (1964), and the other boundaries are from Le Bas et al. (1986). The data-field and plusses are for Hana samples from Chen et al. (1990), Bergmanis et al. (2000) and Sherrod et al. (2003). The other symbols are *squares*: flows, *diamond*: dike. *Solid symbols* are bulk compositions, and *open symbols* are calculated groundmass compositions

of Chen et al. (1990) is approximately constant (0.5–0.59) (Fig. 5.7l). However, the higher than chondritic Tb/Yb ratios (0.224 in chondrite; McDonough and Sun 1995) indicate that the Hana lavas may have originated through partial melting of a garnet bearing source rock. For example, a chondritic source containing 5–20% cpx and 5–8% garnet (peridotite modes from Yang et al. 2003) will produce Tb/Yb ratios of 0.5–0.59 by 1–10% modal batch melting.

5.6 Discussion

The alkaline lavas erupted from the Kolekole cinder cone are predominantly phenocryst-rich ankaramite with major element chemistry similar to picro-basalt (Fig. 5.6). However, the groundmass of the ankaramite is basanitic and similar to other Hana and Kula alkaline lavas from the northeastern coast of Maui (Chen et al. 1990, 1991) and southwest rift (Bergmanis et al. 2000; Sherrod et al. 2003). The incompatible element abundances and ratios of the Kolekole lavas are transitional between previously studied post-shield Kula and Hana lavas (Figs. 5.8 and 5.9; also Sherrod et al. 2003). This indicates that although post-shield volcanic activity has been continuous since the end of the shield building stage on East Maui (Sherrod et al. 2003), the incompatible element composition of the lavas have changed gradually. The Tb/Yb ratios of the Kolekole lavas (Fig. 5.7l) indicate a deep,

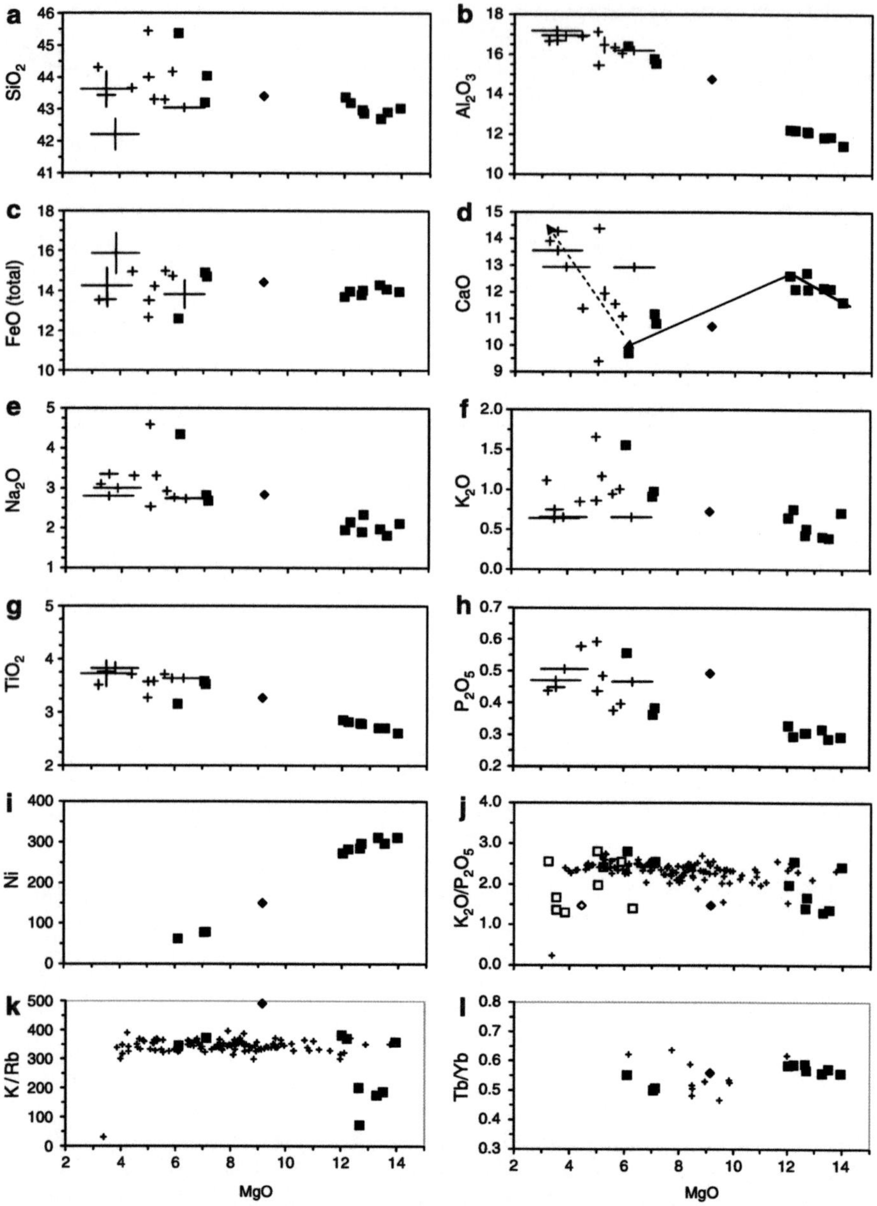

Fig. 5.7 Variation of MgO (wt%) with **(a–h)** other major and minor oxides (wt%), **(i)** Ni (ppm), **(j)** K_2O/P_2O_5, **(k)** K/Rb and **(l)** Tb/Yb. Symbols for bulk compositions in all panels and for groundmass compositions in **(j)** are as in Fig. 4.6. The symbols for groundmass compositions in **(a–h)** are replaced by error bars. The CaO–MgO variations **(d)** show negative correlations for bulk samples with >12 wt% MgO and for all matrices, and a positive correlation for bulk samples with <12 wt% MgO. The errors bars represent extreme compositions of the groundmass calculated using average phenocryst cores and rims

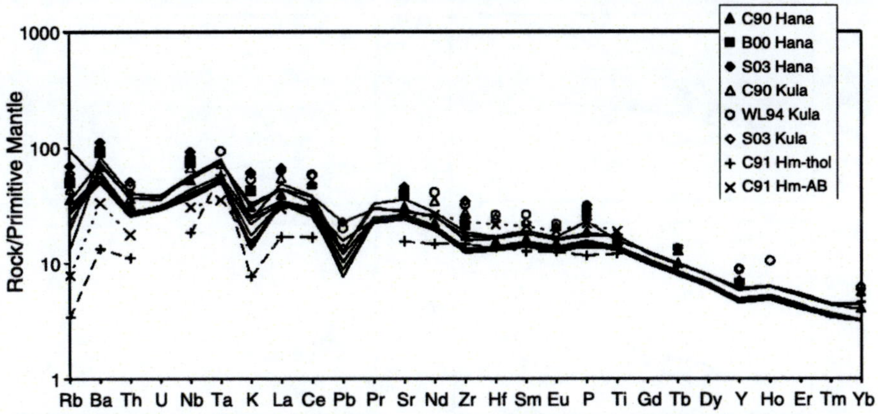

Fig. 5.8 Primitive mantle-normalized (McDonough and Sun 1995) incompatible trace element contents of the Kolekole (*bold lines*) and other Hana (*solid symbols*), Kula (*open symbols*) and Honomanu basalts. Source of data from other studies are C90: Chen et al. (1990); B00: Bergmanis et al. (2000); S03: Sherrod et al. (2003); WL94: West and Leeman (1994); and C91: Chen et al. (1991). Hm-thol (*plus symbol, dashed line*) and Hm-AB (*cross symbol, dotted line*) represent average of five Honomanu tholeiite samples and Honomanu alkali basalt sample with highest alkali content, respectively

garnet-bearing mantle source. Similar conclusions were reached for the rejuvenated-stage lavas of Kauai and Oahu from their high Ce/Yb ratios (Clague and Dalrymple 1988; Clague and Frey 1982).

This study shows that meaningful major element liquid compositions can be estimated even in phenocryst-rich lavas through careful image analysis (Fig. 5.4). These liquid compositions can then be used in thermo-barometric calculations. Barometry indicates that the Kolekole alkaline magma partially crystallized at pressures between 4.4 and 11.2 kb corresponding to depths between 15.5 and 39 km (assuming average crustal and mantle densities of 2.9 and 3.3 g/cm^3). The augite cores crystallized at ~2.5 kb higher pressures than the rims. Hence, crystallization probably began at a depth of about 48 km. The depths of alkaline magma reservoirs below Kolekole in Maui and Mauna Kea in the island of Hawaii (20–45 km, Putirka 1997) are thus similar and extend well below the crust-mantle boundary, which is at about 19 km depth under Maui (Moore 1987). Kilauea shield-stage tholeiitic magma reservoirs are shallower (near summit to 20 km, Putirka 1997). Frey et al. (1990) suggested that a decrease in magma supply during the post-shield stage leads to solidification of the shallow shield-stage magma chambers that serve as a barrier to the later post-shield magma. The alkaline primary magma may have ponded below the solidified shield-stage magma chambers but not at the crust–mantle boundary, as suggested by the upper depth (15.5 km). The relatively low Mg# (35–49) estimates and major and compatible trace element correlations (Fig. 5.7a–i) in the Kolekole ankaramite groundmass indicates that extensive olivine-controlled high pressure fractional crystallization probably occurred in the ponded magma. Furthermore, because of pressure build-up, the alkaline magma probably moved

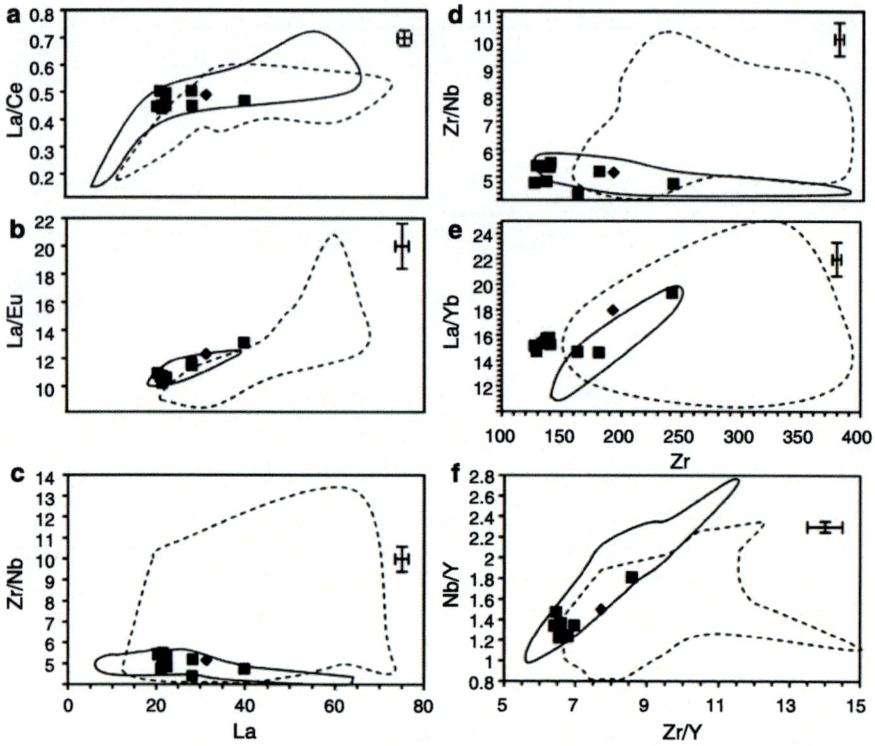

Fig. 5.9 Incompatible ratio variations with (**a–c**) La (ppm), (**d, e**) Zr (ppm), and (F) Zr/Y. Symbols are as in Fig. 4.6. Analytical uncertainties are shown in the upper right corner of each diagram. The data-fields demarcated in *solid lines* (for Hana) and *dashed lines* (for Kula) are from Chen et al. (1990, 1991), Chen and Frey (1985), Bergmanis et al. (2000) and West and Leeman (1994)

upward by brittle fracturing through the upper lithosphere (Putirka 1997) and erupted on the surface by violent volatile-driven explosions as suggested by the occurrence of large spheroidal bombs (0.6–2.2 m) near the Kolekole vent areas (Fig. 5.3d).

5.7 Conclusion

This study demonstrates that the geological, geochemical and petrological characteristics of late stage cinder cones and their eruptive lavas provide valuable information regarding the post-shield evolution of Hawaiian volcanoes. The Kolekole cinder cone located in the southwest rift zone near the triple junction of the southwest, east and north rifts and the Haleakala volcano summit in East Maui exhibits post-shield alkaline volcanism that originated from garnet-bearing source regions. Transitional abundances and ratios of incompatible elements of the Kolekole lavas between Kula and Hana post-shield lavas indicate that the incompatible element composition of the post-shield lavas changed gradually between the temporally continuous Kula and Hana series. The Kolekole lavas are predominantly ankaramite containing 25–40% by volume, large (1–5 mm), normal and reverse-zoned olivine and augite phenocrysts. About 74–85% of the olivine and 58–75% of the augite phenocrysts are reversely zoned. The groundmass of the ankaramite is basanitic and similar to alkaline lava elsewhere on East Maui. Barometry indicates that the phenocrysts originated from depths between 15.5 and 48 km, transcending the crust-mantle boundary. The alkaline primary magma probably ponded below solidified, shallow, shield-stage magma chambers, where extensive olivine-controlled high pressure fractional crystallization occurred, as indicated by Fe-rich (Mg# 35–49) composition estimates, and major and compatible trace element correlations in the ankaramite groundmass. Large spheroidal bombs (0.6–2.2 m) near the Kolekole vents indicate that the evolved magma eventually erupted by violent volatile-driven explosions.

Acknowledgments We are grateful to Professor Fred Frey of Massachusetts Institute of Technology and three other anonymous reviewers for their thoughtful comments that greatly helped in improving this manuscript during peer review. We thank KC Environmental Inc., Makawao, Hawaii for financial support for this research. Field study was carried out by SB with the logistic support from Dr. Charles Fein, Makawao, Hawaii. Remaining part of this study was supported from a Research Foundation Grant of the City University of New York to SB. SB also thanks the Assistant Librarian of Carnegie Institution, Geophysical Laboratory, Washington D.C. for several references on East Maui volcanism made available to him. We thank Late Dr. H. Yoder, Jr. and other reviewers for their valuable comments and suggestions.

References

Bergmanis EC, Sinton JM, Trusdell FA (2000) Rejuvenated volcanism along the southwest rift zone, East Maui, Hawaii. Bull Volcano 62:239–255

Bhattacharji S, Koide H (1975) Mechanistic model for triple junction fracture geometry. Nature 255:21–24

Bhattacharji S, Koide H (1987) Theoretical and experimental studies of mantle upwelling, penetrative magmatism, and development of rifts in continental and oceanic crust. Tectonophysics 143:13–30

Chatterjee N, Bhattacharji S, Fein C (2005) Depth of alkalic magma reservoirs below Kolekole cinder cone, Southwest rift zone, East Maui, Hawaii. J Volcano Geother Res 145:1–22

Chen C-Y (1993) High magnesium primary magmas from Haleakala volcano, east Maui, Hawaii: petrography, nickel and major element constraints. J Volcano Geother Res 55:143–153

Chen C-Y, Frey FA (1985) Trace element and isotopic geochemistry of lavas from Haleakala volcano, East Maui, Hawaii: implications for the origin of Hawaiian basalts. J Geophys Res 90:8743–8768

Chen CY, Frey FA, Garcia MO (1990) Evolution of alkalic lavas at Haleakala volcano, East Maui, Hawaii: major, trace elements and isotopic constraints. Contrib Miner Petrol 105:197–218

Chen C-Y, Frey FA, Garcia MO, Dalrymple GB, Hart SR (1991) The tholeiitic to alkalic basalt transition at Haleakala volcano, Maui, Hawaii. Contrib Miner Petrol 106:183–200

Clague DA (1987) Hawaiian alkaline volcanism. In: Fitton JG, Upton BGJ (eds) Alkaline igneous rocks. Geological Society of London Special Publication 30, London, pp 227–252

Clague DA, Dalrymple GB (1987) The Hawaiian–Emperor volcanic chain. Part 1. Geologic evolution. US Geol Surv Prof Paper 1350:1–54

Clague DA, Dalrymple GB (1988) Age and petrology of alkalic postshield and rejuvenated-stage lava from Kauai, Hawaii. Contrib Miner Petrol 99:202–218

Clague DA, Frey FA (1982) Petrology and trace element geochemistry of the Honolulu Volcanics, Oahu: implications for the oceanic mantle blow Hawaii. J Petrol 23:447–504

Fodor RV, Keil K, Bunch TE (1972) Mineral chemistry of volcanic rocks from Maui, Hawaii: Fe–Ti oxides. Geol Soc Am Abstr Progr 4/7:507

Fodor RV, Keil K, Bunch TE (1975) Contributions to the mineral chemistry of Hawaiian rocks IV. Pyroxenes from the rocks of Haleakala and West Maui volcanoes, Maui, Hawaii. Contrib Miner Petrol 50:173–195

Fodor RV, Keil K, Bunch TE (1977) Contributions to the mineral chemistry of Hawaiian rocks VI. Olivines in rocks from Haleakala and West Maui volcanoes, Maui, Hawaii. Pac Sci 31(3):299–308

Frey FA, Wise WS, Garcia MO, West H, Kwon S-T, Kennedy A (1990) Evolution of Mauna Kea volcano, Hawaii: petrologic and geochemical constraints on postshield volcanism. J Geophys Res 95:1271–1300

Keil K, Fodor RV, Bunch TE (1972) Contributions to the mineral chemistry of Hawaiian rocks, II. Feldspars from rocks from Haleakala and East Maui volcanoes, Maui, Hawaii. Contrib Miner Petrol 37:253–276

Le Bas MJ, Le Maitre RW, Streckeisen A, Zanettin B (1986) A chemical classification of volcanic rocks based on the total alkali-silica diagram. J Petrol 27:745–750

Macdonald GA (1978) Geological maps of the crater section of Haleakala National Park, Maui, Hawaii, U.S. Department of Interior/ US Geol Surv Miscellaneous Investigation Series Map I-1088, Reprinted 1981, 1996

Macdonald GA, Katsura T (1964) Chemical composition of Hawaiian lavas. J Petrol 5:82–133

Macdonald GA, Powers HA (1968) A further contribution to the petrology of Haleakala volcano Hawaii. Geol Soc Am Bull 79:877–888

McDonough WF, Sun S-S (1995) The composition of the earth. Chem Geol 120:223–253

Moore JG (1987) Subsidence of the Hawaiian Ridge. US Geol Surv Prof Paper 1350:85–100

Nimis P (1995) A clinopyroxene geobarometer for basaltic systems based on crystal-structure modeling. Contrib Miner Petrol 121:115–125

Presley TK, Sinton JM, Pringle M (1997) Postshield volcanism and catastrophic mass wasting of the Waianae Volcano, Oahu, Hawaii. Bull Volcano 58:597–616

Putirka K (1997) Magma transport at Hawaii: inferences based on igneous thermobarometry. Geology 25:69–72

Putirka K, Johnson M, Kinzler R, Longhi J, Walker D (1996) Thermobarometry of mafic igneous rocks based on clinopyroxene-liquid equilibria, 0-30 kbar. Contrib Miner Petrol 123:92–108

Sack RO, Walker D, Carmichael ISE (1987) Experimental petrology of alkalic lavas: constraints on cotectics of multiple saturation in natural basic liquids. Contrib Miner Petrol 96:1–23

Sherrod DR, Nishimitsu Y, Tagami T (2003) New K-Ar ages and the geologic evidence against rejuvenated-stage volcanism at Haleakala, East Maui, a postshield-stage volcano of the Hawaiian island chain. Geol Soc Am Bull 115:683–694

Stearns HT (1942) Origin of Hawaiian crater Island of Maui. Geol Soc Am Bull 53:1–14

Stearns HT, Macdonald GA (1942) Geology and ground water resources of the island of Maui. Hawaii Division. of Hydrography Bulletin 7, 344 pp

U.S. Geological Survey (1999) Volcano watch online essay May 7, 2004. http://hvo.wr.usgs.gov/olcanowatch/1999/99_09_09.html

West HB, Leeman WP (1994) The open-system geochemical evolution of alkalic cap lavas from Haleakala Crater, Hawaii, USA. Geochim Cosmochim Acta 58:773–796

West HB, Garcia MO, Frey FA, Kennedy A (1988) Nature and cause of compositional variation among the alkalic cap lavas of Mauna Kea volcano, Hawaii. Contrib Miner Petrol 100:383–397

Yang H-J, Kinzler RJ, Grove TL (1996) Experiments and models of anhydrous, basaltic olivine-plagioclase-augite saturated melts from 0.001 to 10 kbar. Contrib Miner Petrol 124:1–18

Yang H-J, Frey FA, Clague DA (2003) Constraints on the source components of lavas forming the Hawaiian North Arch and the Honolulu Volcanics. J Petrol 44:603–627

Seismic Evidences on Magma Genesis

Chapter 6
New Seismic Evidence for the Origin of Arc and Back-Arc Magmas

Dapeng Zhao, Sadato Ueki, Yukihisa Nishizono, and Akira Yamada

Abstract We used high-resolution seismic tomography to probe into the arc and back-arc magmatism and volcanism of the Japan subduction zone. Prominent zones with low-velocity, high Poisson's ratio, high-attenuation and strong seismic anisotropy are revealed in the crust and uppermost mantle beneath the active volcanoes, and they exist in the central portion of the mantle wedge under the volcanic front and back-arc region, roughly parallel with the subducting oceanic plate. The anomalous zones in the mantle wedge are connected with the subducting slab at a depth of 90–150 km, while they also exhibit along-arc variations. These seismological results indicate that the arc and back-arc magmas are caused by a combination of corner flow (convection) in the mantle wedge and fluids resulting from the dehydration process of the subducting oceanic plate.

6.1 Introduction

The Japan Islands are located in the western Pacific trench – arc – back-arc system and form a typical subduction zone (Fig. 6.1). Many active arc and back-arc volcanoes exist on the Japan Islands, and the arc volcanoes form a clear volcanic front which is parallel with the oceanic trenches in the western Pacific region. After tens of years of multidisciplinary studies since the advent of plate tectonics, geoscientists have achieved a consensus that the formation of these arc and back-arc volcanoes

D. Zhao (✉) and S. Ueki
Department of Geophysics, Tohoku University, Sendai 980-8578, Japan
e-mail: zhao@aob.gp.tohoku.ac.jp

Y. Nishizono
West Japan Engineering Consultants, Inc., Fukuoka 810-0004, Japan
e-mail: y-nishizono@wjec.co.jp

A. Yamada
Geodynamics Research Center, Ehime University, Matsuyama 790-8577, Japan
e-mail: yamada@sci.ehime-u.ac.jp

J. Ray et al. (eds.), *Topics in Igneous Petrology*, DOI 10.1007/978-90-481-9600-5_6,

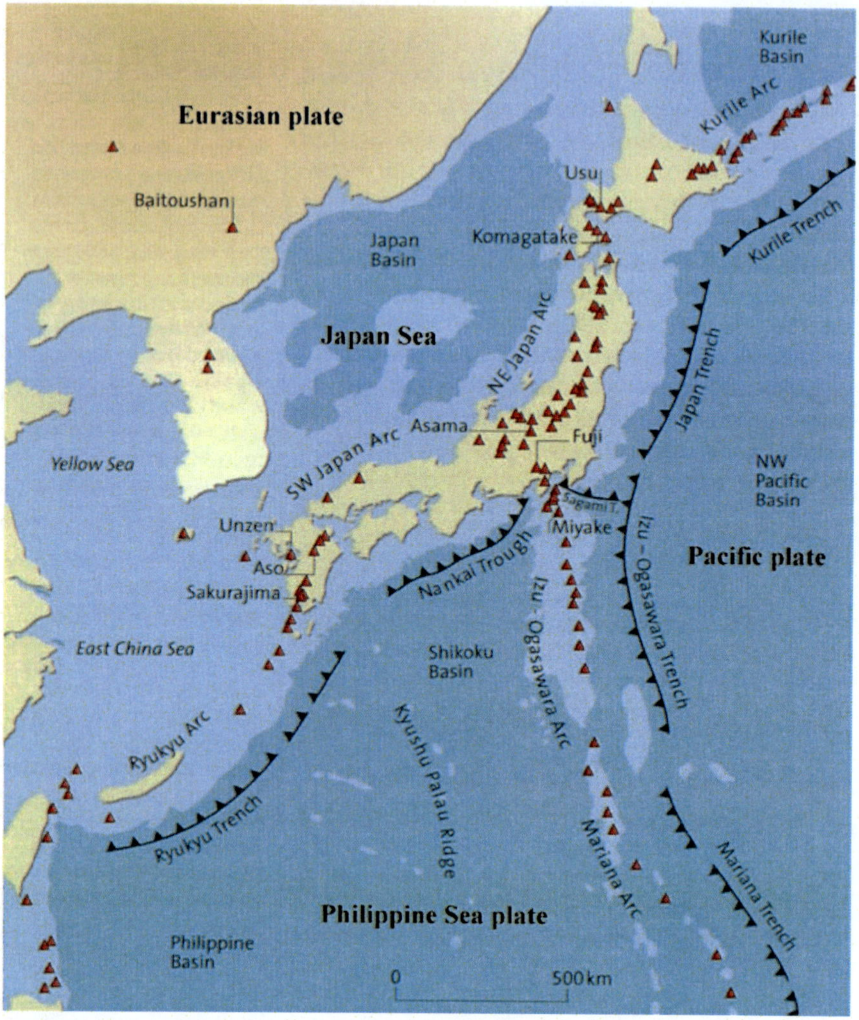

Fig. 6.1 Tectonic background in and around the Japan Islands (After Schmincke 2006). *Red triangles* denote active volcanoes. *Black lines with teeth* symbols show the oceanic trenches where the oceanic plates are subducting into the mantle

is associated with the subduction of the oceanic plates (e.g., Zhao 2001; Stern 2002), though details of the arc and back-arc magmatism and volcanism still need further studies.

The Pacific plate is subducting beneath Hokkaido and northeastern (NE) Japan (Tohoku region) from the Kuril and Japan trenches at a rate of 7–10 km/year. The Philippine Sea plate is subducting beneath the central and southwestern (SW) Japan from the Sagami and Nankai troughs at a rate of 4–5 cm/year. It is generally considered that Hokkaido and NE Japan belong to the Okhotsk (or North American) plate, while SW Japan belongs to the Eurasian plate. Because of the strong interactions of

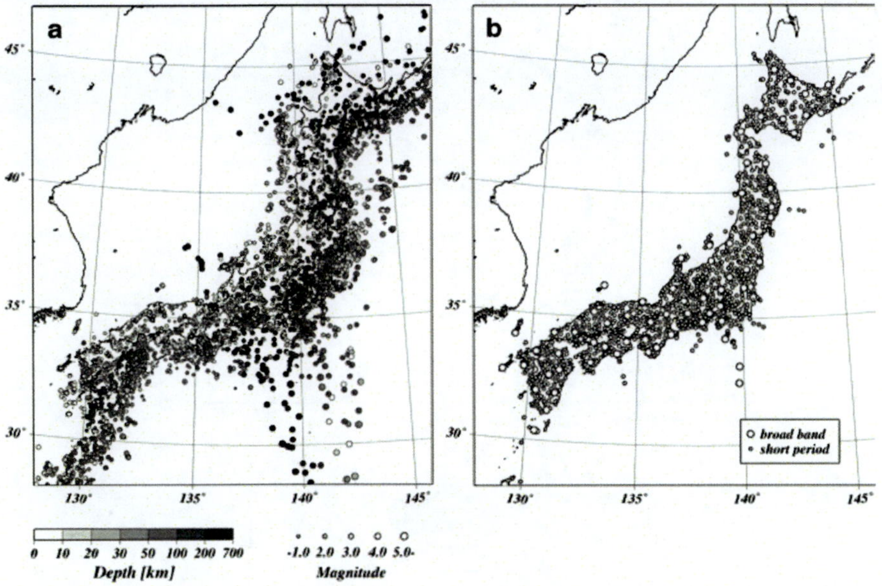

Fig. 6.2 (**a**) Distribution of approximately 11,000 earthquakes occurred in 1 month during August 1–31, 2008. The *size of a dot* denotes the earthquake magnitude. The scales for the earthquake magnitude and focal depth are shown at the *bottom*. (**b**) Short-period (*small circles*) and broad-band (*large circles*) seismic stations on the Japan Islands

the four lithospheric plates in and around the Japan Islands, seismicity is very active in this region. Figure 6.2a shows an example of the seismicity in and around Japan. In just 1 month (August 2008) about 11,000 earthquakes were recorded by the high-quality, dense seismic network on the Japan Islands (Fig. 6.2b). Actually seismicity as shown in Fig. 6.2a occurs every month in Japan, and the pattern of earthquake distribution keeps almost the same.

Because of the high level of seismicity and availability of the dense and high-quality seismic networks operated by the Japanese national universities and research institutions in the past 100 years, a large amount of high-quality seismic data have been accumulated and they are used to determine the three-dimensional (3-D) structure of the crust and upper mantle of the Japan subduction zone. The studies of the 3-D seismic structure have been conducted continuously in the past 30 years, making the Japan Islands the best-studied subduction zone in the world (see the recent reviews by Hasegawa et al. 2005; Zhao 2009; Zhao et al. 2007). These seismological studies have provided important constraints for our understanding of the arc and back-arc magmatism and volcanism.

There are several seismological methods to study the Earth structure, among them seismic tomography (ST) is perhaps the most powerful one. ST is a technique to determine 3-D images of the Earth's interior by combining information from a large number of crisscrossing seismic waves triggered by natural or artificial seismic sources (Zhao 2001). Its principle is similar to that of the medical CT-scan (Fig. 6.3a). Here CT means computerized tomography. In CT-scan, X-rays radiated

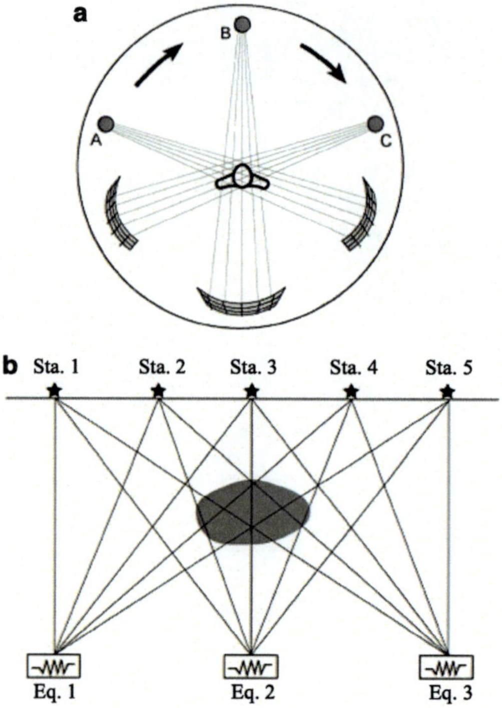

Fig. 6.3 Schematic diagram of Medical CT-scan (**a**) and seismic tomography (**b**) (After Zhao 2009)

from a source can pass through the human body, and the intensity of the X-rays is measured by receivers. The X-ray source is rotated by 360° and so many intensity data of the X-rays can be measured, which can be used to construct a two-dimensional density image (photography) of the human body (Fig. 6.3a). In ST, instead of X-rays, seismic rays are used which are propagating trajectories of seismic waves (Fig. 6.3b). Hence the sources adopted in ST are natural earthquakes or artificial explosions, and the receivers are seismographic stations. The most commonly measured data in ST are travel-time delays of seismic waves, and so ST determines the images of seismic velocity of the Earth's material instead of density. Travel times of the seismic rays which pass through an anomalous body such as a hot magma chamber (or a cold subducting slab) can be delayed (or advanced), while the rays propagating the normal portions of the Earth have no travel-time difference from that calculated for an average one-dimensional velocity model of the Earth's interior. The location, geometry and velocity deviation of the anomalous body can be estimated from the ray paths and travel-time data by using an inverse calculation (called inversion). In CT-scan, both sources and receivers can be controlled and so many data can be measured easily. In contrast, in ST both the sources and receivers are hard to control, and seismic stations are expensive to install and maintain.

Therefore the photos of the Earth's interior cannot be determined as precisely and accurately as the medical CT-scan. Nevertheless, ST with natural earthquake data can determine 3-D velocity tomography of a target area, which is a further refinement of 2-D tomography (e.g., Zollo et al. 1998). Technical details of ST can be found in Zhao (2009).

So far many researchers have applied ST methods to the high-quality data sets recorded by the Japanese seismic networks to study the 3-D seismic velocity structure of the crust and upper mantle of the Japan subduction zone (see a recent review by Zhao 2009). Here we introduce the most recent findings of these seismological studies and discuss their implications for the arc and back-arc magmatism and volcanism.

6.2 Arc Magma Above the Pacific Slab

Figure 6.4 shows vertical cross sections of P and S wave velocity (Vp, Vs) images under Hokkaido from the Kuril Trench to the Japan Sea coast. The two profiles pass through active volcanoes Komagatake, Taisetsuzan, and Meakandake. The subducting Pacific slab is imaged clearly as a high-velocity (high-V) zone with a thickness of about 85 km and seismic velocities 4–6% higher than that of the average mantle. Intermediate-depth earthquakes occur actively in the subducting Pacific slab and they form a clear double seismic zone (Hasegawa et al. 1978). Prominent low-velocity (low-V) zones are visible in the crust and uppermost mantle beneath the active arc volcanoes, and they extend to 150–200 km depth in the central portion of the mantle wedge. The mantle-wedge low-V zone is roughly parallel with the subducting slab, and is close to or connecting with the slab at some depth, e.g., at about 90 km depth in Fig. 6.4b. Many previous studies have shown that the mantle-wedge low-V zones are caused by fluids from the dehydration of the Pacific slab and corner flow (convection) in the mantle wedge (e.g., Zhao et al. 1992, 1997; Hasegawa et al. 1994; Zhao 2001). Low-V zones are also visible in the crust and uppermost mantle under the fore-arc region beneath the Pacific Ocean, which may reflect the shallow dehydration of the Pacific slab (Zhao et al. 2007).

Similar features of seismic images are visible in the crust and upper mantle under the Northeast Japan arc (Fig. 6.5). The double-planed deep seismic zone in the Pacific slab under NE Japan is more clearly visible than that under Hokkaido. Normal crustal earthquakes occur in the upper crust down to about 20 km depths, but some earthquakes occur in the lower crust and uppermost mantle around the Moho discontinuity and they usually exhibit lower frequency than that of the normal crustal earthquakes (Fig. 6.5). It is considered that these low-frequency microearthquakes are associated with the magmatic activity under the active arc volcanoes (Hasegawa and Yamamoto 1994). The low-V zones in the mantle wedge in Fig. 6.5 are wider than those in Fig. 6.4, perhaps because four active volcanoes cluster there which need more magmas to feed.

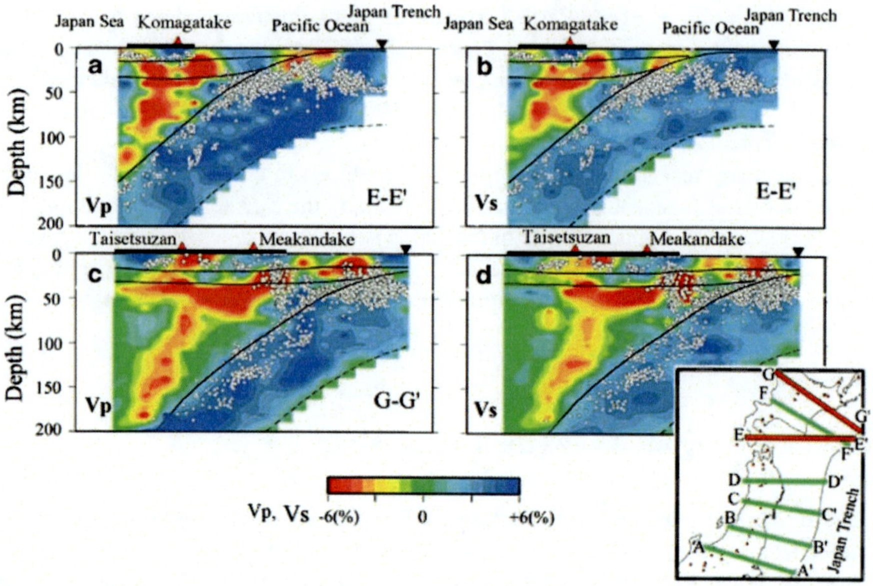

Fig. 6.4 Vertical cross sections of (**a, c**) P- and (**b, d**) S-wave velocity images along profiles E-E′ and G-G′ as shown on the inset map (Wang and Zhao 2005). *Red* and *blue colors* denote low and high velocities, respectively. The velocity perturbation scale (in %) is shown at the bottom. *Red triangles* denote active arc volcanoes. The *reverse triangles* show the location of the oceanic trench. *Small white dots* show earthquakes which occurred within a 15-km width from the profile. The three *curved lines* show the Conrad and Moho discontinuities and the upper boundary of the subducting Pacific slab. The *dashed lines* denote the estimated low boundary of the Pacific slab. No vertical exaggeration for the cross sections in this and following figures. The resolution of the tomographic images is 10–20 km in vertical direction and 25–30 km in the horizontal direction. The uncertainty in the hypocenter locations is 1–4 km under the seismic network on the land area and 3–9 km under the Pacific Ocean

The low-V zones also show high values of Poisson's ratio, indicating that the mantle-wedge low-V zones represent arc magma and fluids (Zhao et al. 2002).

Figure 6.6 shows a N–S vertical cross section of Vp/Vs under NE Japan (Nakajima et al. 2001). It is found that Vp/Vs images show the same pattern as that of Poisson's ratio images (Zhao et al. 1996). High Vp/Vs anomalies exist in the lower crust and uppermost mantle beneath active volcanoes (Naruko, Kurikoma, and Akita-Komagatake), but such a feature is not visible in non-volcanic areas (Fig. 6.6). Low-frequency microearthquakes (red dots in Fig. 6.6) occurred atop the high Vp/Vs zones, indicating the magmatic activity there. The upper crust exhibits low values of Vp/Vs and normal crustal earthquakes occur actively there, suggesting that the upper crust is a brittle seismogenic layer. Many seismic reflectors are detected in the upper crust, which may be sheets of dykes containing fluids from the arc magma (Hasegawa et al. 1994; Horiuchi et al. 1997).

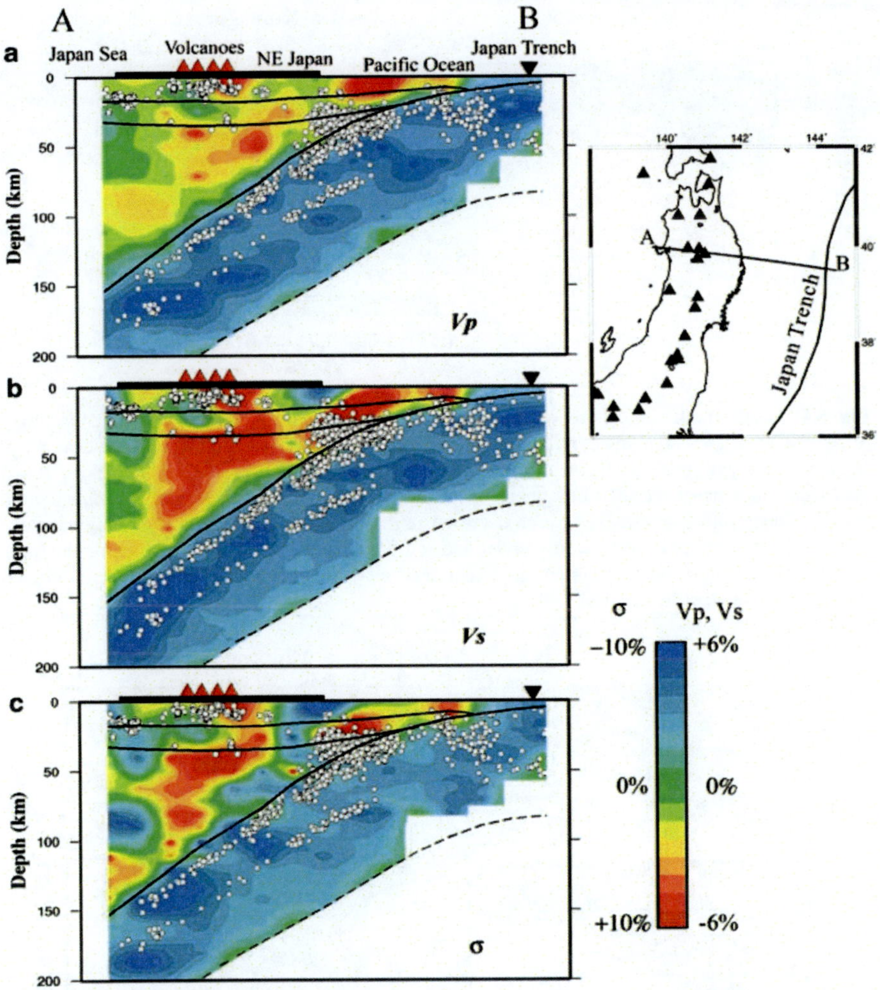

Fig. 6.5 Vertical cross sections of (**a**) P-wave velocity, (**b**) S-wave velocity, and (**c**) Poisson's ratio perturbations along profile AB as shown on the inset map (Zhao et al. 2007). *Red* denotes low velocity and high Poisson's ratio, while *blue* denotes high velocity and low Poisson's ratio. The color scale is shown on the *right*. *Red triangles* denote active arc volcanoes. The *reverse triangles* show the location of the Japan trench. Other labels are the same as those in Fig. 6.4

6.3 Interaction of the Pacific and Philippine Sea Slabs and Arc Magma

Beneath the Kanto district the Philippine Sea (PHS) slab is subducting from the Sagami trough and Suruga trough. The subducting Pacific slab is located beneath the PHS slab and the two slabs are interacting with each other (Ishida 1992). This scenario is clearly imaged by seismic tomography (Fig. 6.7a and b). Subcrustal

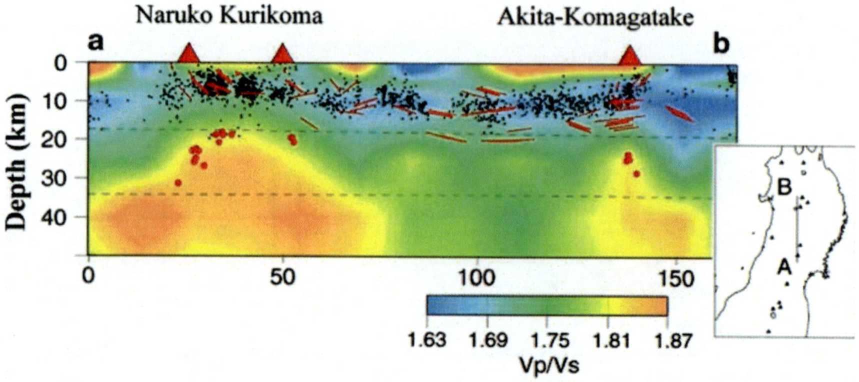

Fig. 6.6 North–south vertical cross section of Vp/Vs along profile AB as shown on the inset map of Northeast Japan (Nakajima et al. 2001). *Red* and *blue colors* denote high and low values of Vp/Vs, respectively; its scale is shown at the *bottom*. *Red triangles* denote active arc volcanoes. *Black dots* show the normal crustal earthquakes occurred within 15 km of the profile. *Red dots* denote low-frequency microearthquakes which occurred in the lower crust or uppermost mantle and are associated with the magmatic activity under active arc volcanoes. *Red short lines* denote the locations of the crustal reflectors which have generated clear seismic reflective waves from crustal earthquakes (see text for details)

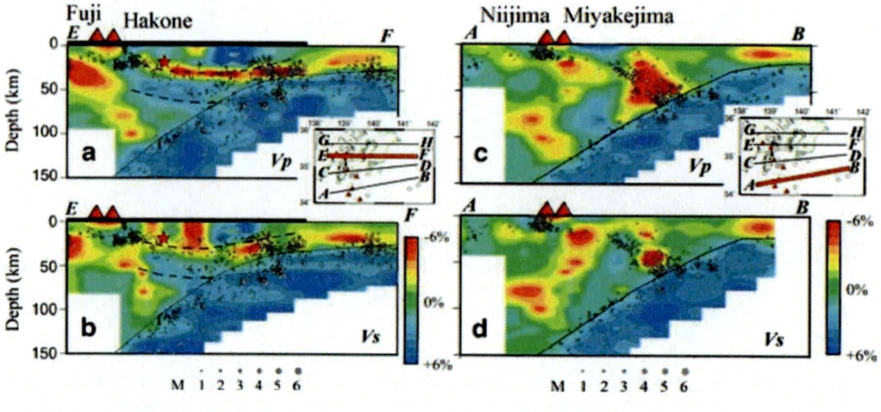

Fig. 6.7 Vertical cross sections of (**a, c**) P- and (**b, d**) S-wave velocity images along profile E–F (**a, b**) and A–B (**c, d**) as shown on the inset map of the Kanto District (Wang and Zhao 2006a). The *solid curved lines* show the upper boundary of the subducting Pacific slab, while the *dashed lines* in (**a, b**) show the estimated upper and lower boundaries of the subducting Philippine Sea slab. The *red star* in (**a, b**) shows the hypocenter of the 1923 Kanto earthquake (M 7.9). Other labels are the same as those in Fig. 6.4

earthquakes are very abundant in the Kanto District because they occur in both the Pacific and PHS slabs. The Pacific slab is old (110–130 Ma) and so it is thick (about 85 km), while the PHS slab is young (20–30 Ma) and so it is thin (about 30 km) (Seno et al. 1993), as clearly visible in the tomographic images (Fig. 6.7). A low-V layer exists atop the PHS slab (Fig. 6.7a and b), which may represent the subducting oceanic crust and slab dehydration. The hypocenter of the 1923 Kanto earthquake (M 7.9) was located at the upper boundary of the PHS slab because it was caused by the underthrusting of the PHS slab beneath the Eurasian plate.

In Hokkaido and NE Japan areas, the subducting Pacific slab is located approximately 100 km beneath the volcanic front (Figs. 6.4 and 6.5). However, under Fuji and Hakone volcanoes in Kanto, the upper boundary of the Pacific slab is about 150 km deep. This westward displacement of the volcanic front is caused by the subduction of the PHS slab under Kanto (Iwamori 2000). However, the mantle-wedge low-V zone is connected with the Pacific slab at about 100 km depth, and it seems that the magma ascending path is deflected toward the west by the PHS slab.

In the Izu–Bonin region south of Kanto, the Pacific plate is subducting beneath the PHS plate. The active volcanoes (Niijima, Miyakejima, etc.) are located about 110–120 km above the Pacific slab. The mantle-wedge low-V zones are connecting with the Pacific slab at that depth, indicating the arc magma under these active volcanoes is related to the dehydration of the Pacific slab (Fig. 6.7c and d). However, the magma upwelling path is not straight but shows a winding image, suggesting that the upwelling magma is deflected by the corner flow in the mantle wedge, just like a mantle plume under a hotspot volcano being deflected by the mantle convection (Zhao 2004, 2009).

6.4 Arc Magma Above the Philippine Sea Slab

Beneath SW Japan except for Kyushu, there are no active volcanoes (Fig. 6.1). Only a few inactive Quaternary volcanoes exist along the coastline of Japan Sea. The reason for the absence of active volcanoes seems that the PHS slab has a very small dipping angle and it is located right beneath the crust under Shikoku. The PHS slab has a complicated geometry in and around Kii Peninsula and it is still unclear if the slab is continuous or disconnected. Thus the mantle wedge above the PHS slab is not well developed to form corner flow which can bring heat from the deeper mantle to generate arc magma.

The PHS slab is subducting beneath Kyushu Island with a dipping angle of about 45°, and several active volcanoes exist in Kyushu (Figs. 6.1, 6.8, and 6.9). A clear volcanic front is also formed. Low-V anomalies are clearly visible in the crust and mantle wedge under the volcanic front in Kyushu (Figs. 6.8 and 6.9), similar to the images under the active volcanoes above the subducting Pacific slab.

Unzen volcano is located about 100 km west of the volcanic front in Kyushu. It is a very active volcano, and its recent eruption occurred in 1991–1995 and killed tens of people. A prominent dipping low-V zone is visible in the crust and mantle wedge under Unzen, which is connecting Unzen at the surface and the PHS slab at about 100 km depth, indicating that the Unzen magma is also related to the dehydration process of the PHS slab at about 100 km depth though the volcano is located at the back-arc side (Fig. 6.8a and b).

Figure 6.8c shows a high-resolution crustal tomography of Unzen volcano (Zhao et al. 2002). A cone-shaped low-V zone is visible in the crust under Unzen, which

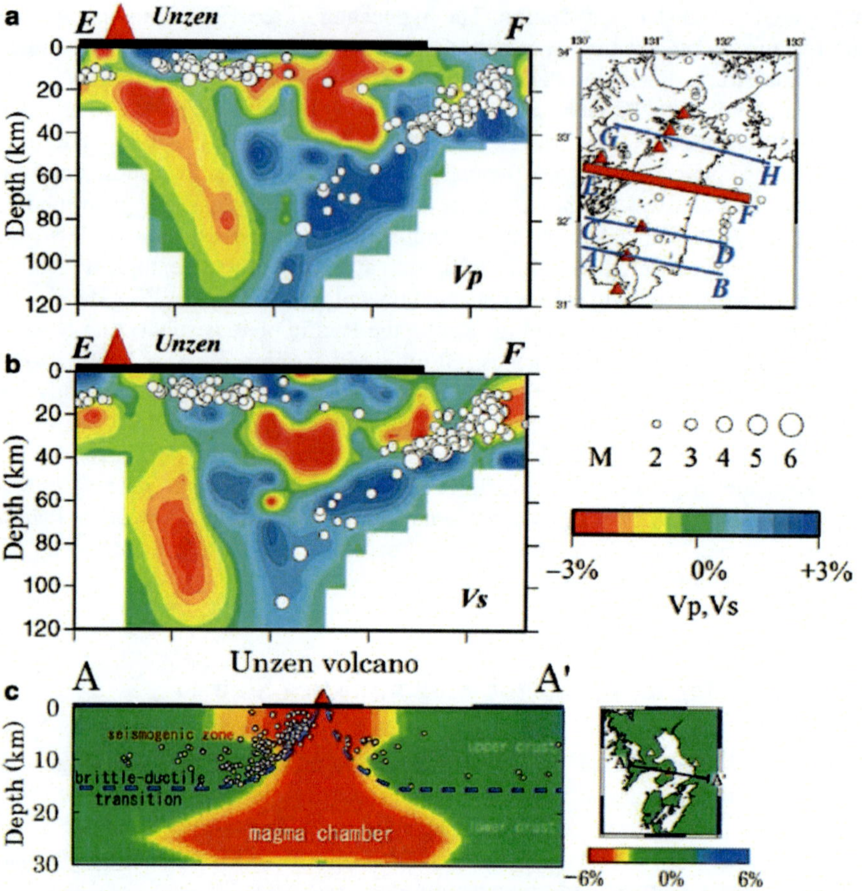

Fig. 6.8 (**a, b**) Vertical cross sections of P- and S-wave velocity images along profile E–F as shown on the inset map of Kyushu Island (Wang and Zhao 2006b). *Red* and *blue colors* denote low and high velocities, respectively. *White dots* denote earthquakes occurred within 20 km of the profile. The scales of earthquake magnitude and velocity perturbations (in %) are shown on the *right*. *Red triangles* denote active arc volcanoes. (**c**) East-west vertical cross section of P-wave tomography of the crust under Unzen volcano in Kyushu (Zhao et al. 2002). The *dashed lines* show the cut-off depth of crustal earthquakes (*white dots*), which may represent the brittle–ductile boundary in the crust. On the inset map the *red triangle* shows Unzen volcano, *white triangles* denote other Quaternary volcanoes

may represent high-temperature anomalies containing magma reservoir. The cutoff-depth of crustal seismicity shallows toward the crater of the volcano, which is in good agreement with the upper boundary of the low-V zone (Fig. 6.8c). These features indicate thinning of the brittle seismogenic layer beneath the volcano. Previous studies suggested that there are large lateral variations in temperature of the crust and in the cutoff-depth of crustal seismicity in the volcanic areas (e.g., Ito 1993; Zhao et al. 2002). Figure 6.8c shows a nice example of these features.

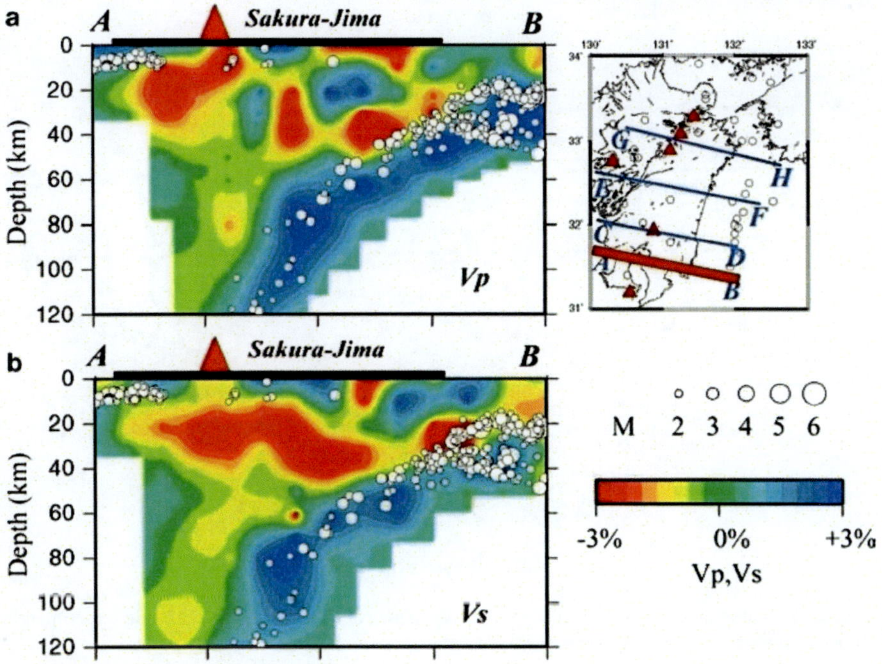

Fig. 6.9 The same as Fig. 6.8a, b but along profile A–B as shown on the inset map of Kyushu Island (Wang and Zhao 2006b)

As compared with the subduction zones above the Pacific slab, more significant low-V anomalies are visible in the fore-arc areas above the subducting PHS slab (Figs. 6.8 and 6.9). The low-V zones may reflect the subducting oceanic crust atop the PHS slab and fore-arc mantle serpentinization resulting from fluids from the PHS slab dehydration (Xia et al. 2008). The PHS slab is much younger and warmer than the Pacific slab and so its slab dehydration can take place at a shallow depth under the fore-arc area.

6.5 Discussion

A qualitative model is proposed to explain these seismic observations in the Japan subduction zone (Fig. 6.10). Two processes are considered to be the most important in a subduction zone. One is corner flow in the mantle wedge that can bring heat to the mantle wedge from the deeper mantle through convection. The other is the dehydration reactions of the subducting oceanic slab. In the fore-arc region, the temperature is lower, and hence magma cannot be produced. Fluids from the slab dehydration may migrate up to the crust. If the fluids enter an active fault in the crust, pore pressures will increase and fault zone friction will decrease, which may

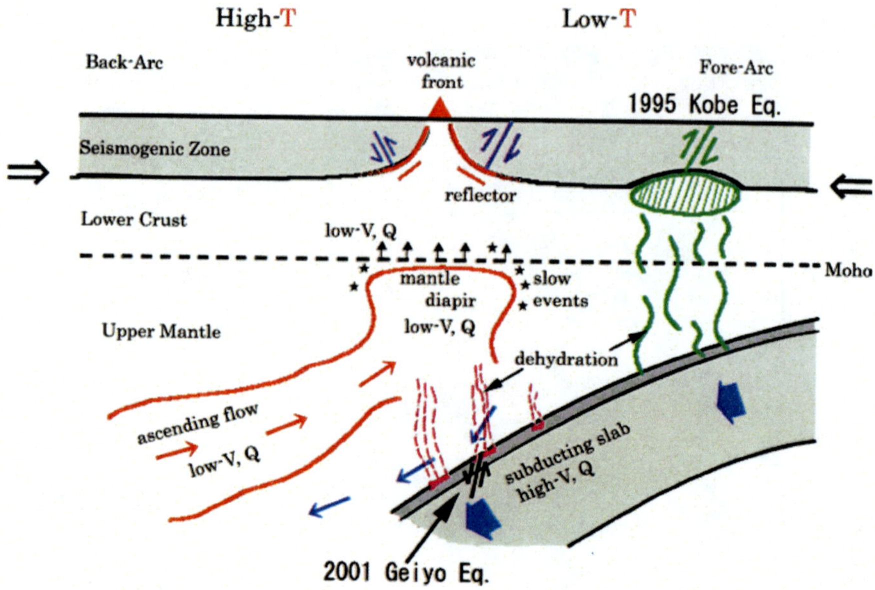

Fig. 6.10 Schematic illustration of across-arc vertical cross section of the crust and upper mantle under a subduction zone region (see text for details)

trigger large crustal earthquakes such as the 1995 Kobe earthquake (M 7.2) (Zhao et al. 1996). Under the volcanic front and back-arc areas, the temperature is high because of the mantle wedge corner flow. Fluids from the slab dehydration play an important role in the generation of arc magma. Migration of magma up to the crust produces arc volcanoes and causes lateral heterogeneities and weakening of the seismogenic upper crust, which can affect the occurrence of large crustal earthquakes. Dehydration reactions of the subducting slab may also trigger a large intraplate earthquake, such as the 2001 Geiyo earthquake (M 6.8) under SW Japan (Zhao et al. 2002).

The low-V zones in the uppermost mantle (Figs. 6.4–6.9) are the manifestation of mantle diapirs associated with the ascending flow of subduction-induced convection in the mantle wedge and dehydration reactions in the subducting slab (Zhao et al. 1992, 1997; Hasegawa et al. 1994, 2005). As mentioned above, magma further rising from the mantle diapirs to the crust causes low-frequency microearthquakes at levels of the lower crust and uppermost mantle, and makes their appearance as S-wave reflectors at mid-crustal levels (Hasegawa et al., 1994). Their upward intrusion raises the temperature and reduces the seismic velocity of crustal materials around them, causing the brittle seismogenic layer above them to become locally thinner and weaker (Fig. 6.8c). Subject to the horizontally compressional stress field in the plate convergence direction, contractive deformations will take place mainly in the low-V areas because of thinning of the brittle seismogenic layer and weakening of the crust and uppermost mantle there due to the higher temperature and the existence of magma- or fluid-filled, thin, inclined reflectors that are incapable of sustaining the applied shear stress. The deformation proceeds partially in small earthquakes but mainly in plastic deformation, causing the crustal shortening, upheaval and mountain building (Zhao et al. 2002; Hasegawa et al. 2005). Large crustal earthquakes cannot occur within the weak low-V zones but in their edge portions where the mechanical strength of materials is stronger than those of the low-V zones (but still weaker than the normal sections of the seismogenic layer). Thus the edge portion of the low-V areas becomes the ideal locations to generate large crustal earthquakes that produce faults reaching to the Earth's surface or blind faults within the brittle upper crust (Fig. 6.10). Examples of such large crustal earthquake are the 2000 Western Tottori earthquake (M 7.3) (see Zhao et al. 2004) and the recent Iwate–Miyagi inland earthquake (M 7.2) occurred on June 14, 2008.

Low-V and high Poisson's ratio (high-PR) zones in the mantle wedge are visible in all the vertical cross sections (Figs. 6.4–6.9). However, in general, the amplitude of the anomalies is larger in the cross sections passing through the active arc volcanoes, while it is smaller in the cross sections without the volcanoes (Fig. 6.6). In addition, low-frequency microearthquakes occurring in the lower crust and uppermost mantle, which are good indicators of magmatic activity (Hasegawa and Yamamoto 1994), are visible only in the cross sections passing through the active volcanoes. These features suggest along-arc variations of the low-V and high-PR zones in the mantle wedge (Tamura et al. 2002; Hasegawa et al. 2005; Zhao et al. 2007). Because the low-V and high-PR zones are generally considered to reflect the arc-magma related hot and wet anomalies caused by the slab dehydration and corner flow in the mantle wedge (Iwamori and Zhao 2000; Stern 2002; Cagnioncle et al. 2007), the along-arc variations of the low-V and high-PR zones indicate the spatial variation in the amount of fluids released from the slab dehydration as well as in the strength of mantle-wedge corner flow along the arc. Causes of the lateral velocity variations can be thermal and/or compositional effects. It was proposed that the origin of the along-arc variation of the velocity anomalies under NE Japan is small-scale convection occurring in the mantle wedge, similar to that under the cooling oceanic lithosphere (Honda et al. 2002; Honda and Saito 2003). Such a small-scale convection can take place because the viscosity in the mantle wedge can be lowered by the water dehydrated from the subducting slab.

The mantle-wedge low-V and high-PR zones extend from the volcanic front to the Japan Sea coast, which is the western limit of the high-resolution tomographic models available now. It is still not very clear whether the mantle-wedge anomalies extend further westward to the Japan Sea. Recent regional and global tomographic models show that the Pacific slab has subducted down to about 600 km depth under the eastern margin of the Asian continent and then the slab becomes stagnant in the mantle transition zone under Eastern China (Zhao 2004; Huang and Zhao 2006). The western edge of the stagnant slab has reached to Beijing, about 2,000 km west of the Japan Trench. A big mantle wedge (BMW) has formed above the stagnant slab, and the BMW exhibits low velocities under the Japan Sea and East China, which is consistent with the existence of active intraplate volcanoes (such as Changbai and Wudalianchi) and a thin lithosphere under Eastern China (e.g., Menzies et al. 2007). It is considered that deep dehydration of the stagnant slab and corner flow in the BMW exist, which induced the upwelling of hot and wet asthenospheric materials and lead to the formation of the intraplate volcanoes and lithospheric thinning in East China (Zhao 2004; Ichiki et al. 2006; Zhao et al. 2009). Because the regional and global tomographic models still have a lower resolution, it is unclear whether the mantle-wedge low-V zones are continuous from the volcanic front on the Japan Islands to East China, or they exist intermittently beneath Japan Islands, Japan Sea, and East China. To resolve this problem, a network of ocean-bottom-seismometer (OBS) stations should be installed in the Japan Sea to determine a high-resolution mantle structure of the BMW.

6.6 Conclusions

Taking advantage of the high-quality of the dense seismograph networks on the Japan Islands and high-level of seismicity in and around Japan, seismologists have determined high-resolution 3-D tomographic images of the Japan subduction zone, which enable us to probe into the origin of arc and back-arc magmatism in this region from a seismological point of view. The tomographic images show prominent low-velocity and high Poisson's ratio zones in the crust and uppermost mantle beneath the active volcanoes. The low-velocity zones are also visible in the central portion of the mantle wedge under the volcanic front and back-arc region, roughly parallel with the subducting oceanic plate. The mantle-wedge low-velocity zone is connected with the subducting slab at a depth of 90–150 km. These seismological results indicate that the arc and back-arc magmas are caused by a combination of corner flow (convection) in the mantle wedge and fluids resulting from dehydration of the subducting oceanic plate. The mantle-wedge low-velocity zones exhibit along-arc variations and they are more prominent in areas with volcanoes on the surface. The along-arc variations may be related to the finer-scale structures and processes in the subducting slab and mantle wedge, and further studies are needed to clarify the problem so as to better understand the arc and back-arc magmatism and volcanism.

Acknowledgments This work was partially supported by grants (Kiban-B 11440134, Kiban-A 17204037) to D. Zhao from Japan Society for the Promotion of Science (JSPS). Two anonymous reviewers provided thoughtful review comments which improved the manuscript.

References

Cagnioncle A, Parmentier E, Elkins-Tanton L (2007) Effect of solid flow above a subducting slab on water distribution and melting at convergent plate boundaries. J Geophys Res 112:B09402

Hasegawa A, Yamamoto A (1994) Deep, low-frequency microearthquakes in or around seismic low-velocity zones beneath active volcanoes in northeastern Japan. Tectonophysics 233:233–252

Hasegawa A, Umino N, Takagi A (1978) Double-planed structure of the deep seismic zone in the northeastern Japan arc. Tectonophysics 47:43–58

Hasegawa A, Horiuchi S, Umino N (1994) Seismic structure of the northeastern Japan convergent margin: a synthesis. J Geophys Res 99:22295–22311

Hasegawa A, Nakajima J, Umino N, Miura S (2005) Deep structure of the northeastern Japan arc and its implications for crustal deformation and shallow seismic activity. Tectonophysics 403:59–75

Honda S, Saito M (2003) Small-scale convection under the back-arc occurring in the low viscosity wedge. Earth Planet Sci Lett 216:703–715

Honda S, Saito M, Nakakuki T (2002) Possible existence of small-scale convection under the back-arc. Geophys Res Lett 29:2043

Horiuchi S, Tsumura N, Hasegawa A (1997) Mapping of a magma reservoir beneath Nikko-Shirane volcano in northern Kanto, Japan, from travel time and seismogram shape anomalies. J Geophys Res 102:18071–18090

Huang J, Zhao D (2006) High-resolution mantle tomography of China and surrounding regions. J Geophys Res 111:B09305

Ichiki M, Baba K, Obayashi M, Utada H (2006) Water content and geotherm in the upper mantle above the stagnant slab: interpretation of electrical conductivity and seismic P-wave velocity models. Phys Earth Planet Inter 155:1–15

Ishida M (1992) Geometry and relative motion of the Philippine sea plate and Pacific plate beneath the Kanto-Tokai District, Japan. J Geophys Res 97:489–513

Ito K (1993) Cutoff depth of seismicity and large earthquakes near active volcanoes in Japan. Tectonophysics 217:11–21

Iwamori H (2000) Deep subduction of H2O and deflection of volcanic chain towards backarc near triple junction duo to lower temperature. Earth Planet Sci Lett 181:41–46

Iwamori H, Zhao D (2000) Melting and seismic structure beneath the northeast Japan arc. Geophys Res Lett 27:425–428

Menzies M, Xu Y, Zhang H, Fan W (2007) Integration of geology, geophysics and geochemistry: a key to understanding the North China Craton. Lithos 96:1–21

Nakajima J, Matsuzawa T, Hasegawa A, Zhao D (2001) Seismic imaging of arc magma and fluids under the central part of northeastern Japan. Tectonophysics 341:1–17

Schmincke H (2006) Volcanism. Springer, Berlin/Heidelberg/New York, 324 pp

Seno T, Stein S, Gripp A (1993) A model for the motion of the Philippine Sea plate consistent with NUVEL-1 and geological data. J Geophys Res 98:17941–17948

Stern R (2002) Subduction zones. Rev Geophys 40:RG000108

Tamura Y, Tatsumi Y, Zhao D, Kido Y, Shukuno H (2002) Hot fingers in the mantle wedge: new insights into magma genesis in subduction zones. Earth Planet Sci Lett 197:105–116

Wang Z, Zhao D (2005) Seismic imaging of the entire arc of Tohoku and Hokkaido in Japan using P-wave, S-wave and sP depth-phase data. Phys Earth Planet Inter 152:144–162

Wang Z, Zhao D (2006a) Suboceanic earthquake location and seismic structure in the Kanto district, central Japan. Earth Planet Sci Lett 241:789–803

Wang Z, Zhao D (2006b) Vp and Vs tomography of Kyushu, Japan: new insight into arc magmatism and forearc seismotectonics. Phys Earth Planet Inter 157:269–285

Xia S, Zhao D, Qiu X (2008) Tomographic evidence for the subducting oceanic crust and forearc mantle serpentinization under Kyushu, Japan. Tectonophysics 449:85–96

Zhao D (2001) Seismological structure of subduction zones and its implications for arc magmatism and dynamics. Phys Earth Planet Inter 127:197–214

Zhao D (2004) Global tomographic images of mantle plumes and subducting slabs: insight into deep Earth dynamics. Phys Earth Planet Inter 146:3–34

Zhao D (2009) Multiscale seismic tomography and mantle dynamics. Gondwana Res 15:297–323

Zhao D, Hasegawa A, Horiuchi S (1992) Tomographic imaging of P and S wave velocity structure beneath northeastern Japan. J Geophys Res 97:19909–19928

Zhao D, Kanamori H, Negishi H (1996) Tomography of the source area of the 1995 Kobe earthquake: evidence for fluids at the hypocenter? Science 274:1891–1894

Zhao D, Xu Y, Wiens D, Dorman L, Hildebrand J, Webb S (1997) Depth extent of the Lau back-arc spreading center and its relation to subduction processes. Science 278:254–257

Zhao D, Mishra OP, Sanda R (2002) Influence of fluids and magma on earthquakes: seismological evidence. Phys Earth Planet Inter 132:249–267
Zhao D, Tani H, Mishra OP (2004) Crustal heterogeneity in the 2000 western Tottori earthquake region: effect of fluids from slab dehydration. Phys Earth Planet Inter 145:161–177
Zhao D, Wang Z, Umino N, Hasegawa A (2007) Tomographic imaging outside a seismic network: application to the northeast Japan arc. Bull Seismol Soc Am 97:1121–1132
Zhao D, Tian Y, Lei J, Liu L, Zheng S (2009) Seismic image and origin of the Changbai intraplate volcano in East Asia: role of big mantle wedge above the stagnant Pacific slab. Phys Earth Planet Inter 173:197–206
Zollo A, Gasparini P, Virieux J et al (1998) An image of Mt. Vesuvius obtained by 2D seismic tomography. J Volcano Geotherm Res 82:161–173

Continental Flood Basalts

Chapter 7
Mineral Compositions in the Deccan Igneous Rocks of India: An Overview

Leone Melluso and Sam F. Sethna

Abstract An overview of the mineral chemical variations observed in the Deccan Trap igneous rocks of western India indicates a very wide and continuous compositional range for olivine (Fo_{92}–Fo_{11}), pyroxene (augite to ferroaugite and pigeonite with rare orthopyroxene in the tholeiitic rocks; diopside to salite with no low-Ca clinopyroxene in slightly more alkaline basalts), spinel (from aluminous to chromiferous and to Ti-magnetite), ilmenite (Mg- to Mn-rich), feldspar (from An_{83} to albite and anorthoclase to K-rich sanidine), and minor phases (biotite, amphibole, pseudobrookite, etc.). These compositions reflect a large variation in temperature and chemical composition of the host rocks (from about 1,200–1,300°C in picritic basalts to 850–800°C in the rhyolites or in the groundmass of basalts). The range of mineral compositions indicates parental magmas with varying degrees of silica saturation (mildly alkaline to tholeiitic), and very similar (low-pressure) crystallization environment for all the tholeiitic magmas throughout the Deccan, regardless of their widely different geochemical and isotopic compositions.

7.1 Introduction

The Deccan Traps (Fig. 7.1) are one of the best known Large Igneous Provinces in the world. They were emplaced roughly 65–67 Ma ago (Chenet et al. 2008 and references therein; Duncan and Pyle 1988; Pande 2002 and references therein) along the western margin and the central part of continental India, a complex structure

L. Melluso (✉)
Dipartimento di Scienze della Terra, Università di Napoli Federico II, Via Mezzocannone 8, 80134 Napoli, Italy
e-mail: melluso@cds.unina.it

S.F. Sethna
Department of Geology, St. Xavier's College, Mumbai 400001, India
and
Tata Colony, Tardeo, Mumbai 400034, India
e-mail: samsethna@gmail.com

J. Ray et al. (eds.), *Topics in Igneous Petrology*, DOI 10.1007/978-90-481-9600-5_7,

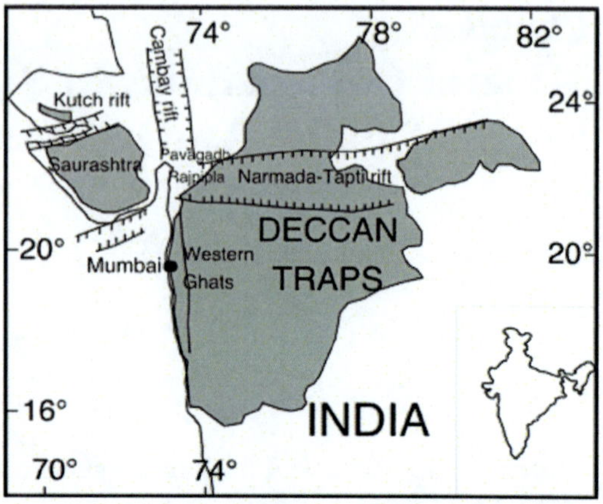

Fig. 7.1 Sketch map of the Deccan Traps

consisting of small basins, crossed rift systems and aborted rift zones (e.g. Krishna et al. 2006). The rocks lie on a Precambrian metamorphic basement mostly made up of rocks of the Dharwar Craton (e.g. Ray et al. 2008). The main outcrops are simple and compound lava flows and dykes, with subordinate red boles, plutonic rocks and pyroclastic rocks (Bose 1972; Mahoney 1988; Ross et al. 2005).

The geochemical stratigraphy of the central-western part of the Deccan Traps was established in the eighties (Najafi et al. 1981; Cox and Hawkesworth 1985; Beane et al. 1986; Subbarao et al. 1988). Eleven formations (from Jawhar to Panhala) and several members within each formation are distinguished in the Western Ghats; the thickest stratigraphic sequence is estimated to be roughly 3,500 m (Mahoney 1988). Other magmatic types with no equivalents in the Western Ghats are the rocks of the Pavagadh section in Gujarat (Sheth and Melluso 2008), the potassic rocks of the Rajpipla area (Krishnamurthy and Cox 1980; Mahoney et al. 1985; Melluso et al. 1995) and the abundant low-Ti tholeiites of Gujarat, apparently chemical equivalents of the Bushe basalts of the Western Ghats (Melluso et al. 1995, 2006a). The petrogenesis of the Deccan basalts is not simple, as the rocks change in their geochemical and isotopic characteristics from basalt types nearly free of crustal contamination, resembling transitional mid-ocean ridge basalts (the Ambenali types), to rocks which have suffered very strong crustal contamination with upper or lower crustal lithologies (e.g. Neral, Bushe, Mahableshwar basalts), to incompatible element enriched basalts which are not so different isotopically and geochemically from those of the present-day Reunion basalts (those in the Pavagadh section and Narmada–Tapti; cf. Melluso et al. 2006). More details about the petrogenesis of the Deccan rocks can be found in Peng et al. (1994), Melluso et al. (1995, 2004, 2006a), Sen (2001), Chatterjee and Bhattacharji (2008), Mahoney et al. (2000) and Sano et al. (2001). Magma types equivalent to some found in the Indian mainland are well known in the Seychelles archipelago (Devey and Stephens 1991, 1992).

The Deccan Traps are currently the subjects of various investigations, including:

- The age and the volume of the erupted products in their context of as the primary cause of extinction events that occurred at the end of Mesozoic, probably as a

result of a massive release of aerosols, ash, and potentially toxic elements such as sulfur over very short duration (e.g. Chenet et al. 2008 and references therein; Jay and Widdowson 2008; Self et al. 2008).
- The eruption sites, which were likely located in areas of major un-oriented or strongly oriented (E–W and N–S) dykes, are still not well constrained, and a major effort is presently devoted to linking the chemical and isotopic composition of the dykes to those of the lavas (e.g. Vanderkluysen et al. 2006; Bondre et al. 2006).
- The composition of the lavas has long been known to be influenced by various mantle and crustal components (e.g. Lightfoot et al. 1990; Peng et al. 1994), but their relative interplay, and the ultimate influence of mantle plumes in the petrogenesis of the basalts, are still poorly understood (e.g. Melluso et al. 2006a; Sheth 2005; Chatterjee and Bhattacharji 2008).

The chemical variation in the mineral phases of flood basalts is important for the following reasons:

- Minerals are useful for recognising genetic relationships between rocks with different degrees of magmatic evolution (e.g., basalts, basaltic andesites) or discriminating magmatic series affinity (e.g. alkali or tholeiitic basalts; Dal Negro et al. 1982).
- Mineral compositions may assist long-range correlations between rock types, and can be also useful for tephrostratigraphy (e.g., Harangi et al. 2005).
- Minerals provide useful information about intensive variables governing the crystallization (temperature, pressure, oxygen fugacity).
- Paragenesis and mineral composition are useful in the search for primitive magmas, which retain information on the mantle sources, as well as on magmatic and volcanic processes.

This paper summarizes the available data on minerals found in the Deccan Trap rocks, and places them in context. Despite the apparently monotonous whole-rock major element chemical composition of the Deccan (>90% tholeiitic basalts), we show that the phases record significant compositional ranges as a consequence of significant changes in crystallization conditions. Pre- or post-Deccan alkaline rocks (e.g., Bose 1973; Gwalani et al. 1993; Melluso et al. 2002; Karmalkar et al. 2005) are not considered in this review.

7.2 Data Set and Analytical Techniques

A total of about 1,700 microprobe analyses (of which about 400 are still unpublished or not used in diagrams of earlier papers) from different chemical types of Deccan basalts are used in the present review. Even though a significant number of analyses relate to the Gujarat volcanic rocks (Melluso 1992; Melluso et al. 1995) and the Tapti dykes (Melluso et al. 1999), an effort was made to include the rocks of the Western Ghats using literature and unpublished data (e.g. Sen 1986; Sethna and Sethna 1988; Krishnamurthy et al. 2000). Prominence was given to poorly known outcrops and to minerals, such as chromiferous spinels, which are still poorly known in the Deccan Trap literature. Details about the location of the samples are given in the tables.

Mineral analyses were obtained with three different microprobes: an ARL-SEMQ WDS microprobe operating at Cagliari University, a Cameca SX-50 WDS-EDS microprobe operating at CNR, IGAG, Rome, and a JEOL JSM EDS microprobe operating at CISAG, University of Napoli. The different analytical conditions have already been described elsewhere (e.g. Melluso et al. 1995, 1999, 2008). Bias between the different data sets has been examined and is negligible for the data in this paper.

7.3 Summary of General Chemical and Petrographic Characteristics

Picritic basalts are found in several areas of Deccan, but they are particularly frequent in Gujarat and in the northern part of the Western Ghats, around Igatpuri and Mahableshwar (Cox and Hawkesworth 1985; Beane and Hooper 1988; Melluso et al. 1995; Sano et al. 2001). They are strongly olivine-phyric basalts, and it is well known that many have excess olivine phenocrysts (Beane and Hooper 1988; Melluso et al. 1995; Krishnamurthy et al. 2000). Calcic plagioclase and/or clinopyroxene may be companion phenocrysts. Olivine is often the host for tiny spinel inclusions. Ca-rich clinopyroxene is augite in tholeiitic types, and the typical and distinctive titaniferous diopside in the more alkaline types. Groundmass minerals may also include quartz, as well as sanidine, Na-plagioclase, Fe–Ti oxides and Fe-rich pyroxenes. Intrusive variants are picritic gabbros, found among other outcrops at Khapoli (Lonavala, Western Ghats) and in the Saurashtra peninsula (Gujarat). These rocks contain primocrysts of cumulus olivine with spinel inclusions, and plagioclase+ pyroxene(s). Interstitial oxides complete the paragenesis.

Basalts are by far the most abundant magma type of the Deccan Traps. A particular textural variant includes the peculiar giant plagioclase basalts that have already been described in several papers (e.g. Pattanayak and Shrivastava 1999; Higgins and Chandrasekharam 2007). The phases include olivine (rare with respect to the picritic basalts, and, when Fe-rich, later than plagioclase in the crystallization sequence), plagioclase, pyroxenes (usually augite and pigeonite, with the former usually as phenocryst phase) and Fe–Ti oxides, with minor accessory phases and glass.

Basaltic andesites are rare; they are the olivine-free variants of basalts, with mostly plagioclase and clinopyroxene (± pigeonite) phenocrysts and interstitial oxides.

More evolved rocks, such as *andesites, dacites and rhyolites* are uncommon, though the latter are frequent in the Saurashtra. The dacite D66 (Table 7.1) shows weakly porphyritic texture with phenocrysts of zoned plagioclase, Fe-rich augite with rims of more Mg-rich augite, pigeonite and magnetite, set in a granophyric groundmass. The Pavagadh rhyolitic pitchstone D89 has phenocrysts of ferroaugite, pigeonite, Na-plagioclase, Fe–Ti oxides and olivine in an abundant glassy matrix. In many rhyolites, phenocryst quartz may be present with Na-plagioclase or alkali feldspar and altered mafic minerals.

Representative major oxide chemical analyses are provided in Table 7.1, and more information can be found in Melluso et al. (1995, 1999).

7.4 The Compositions of the Primary Phases

7.4.1 Olivine

Olivine is most commonly found in the picritic basalts and basalts and, more rarely, in rhyolites. This mineral has a very large range of composition, from Fo_{92} to Fo_{11} (Fig. 7.2a). This range is observed in the basalts, whereas rare olivine in the rhyolites has a far more restricted range (Fo_{27}–Fo_{28}) (Table 7.2). The frequency histogram shows a marked peak of abundance at Fo content of 90–70, with a secondary peak at Fo content of 60–50, suggesting a minor bimodality. Minor elements such as Mn have good negative correlation with decreasing MgO, whereas CaO does not show any significant correlation with MgO, as could be expected in olivine crystallization at low-pressure (average CaO = 0.34 ± 0.09 wt% over 323 analyses). Ni content in olivine is available only for a few samples (see Krishnamurthy et al. 2000) and have maximum values of about 0.4 wt% NiO. Our measurements of Ni in olivine reach values as high as 0.7 wt% NiO in a Gujarat low-Ti picritic basalt. The large compositional range of olivine found in many lava samples is noteworthy (cf. Fig. 7.2b).

7.4.2 Chrome-Bearing Spinel

It is one of the most abundant accessory phases of the olivine-bearing basalts. It occurs as inclusions in olivine or, very rarely, in plagioclase and clinopyroxene. Chemical zoning is observed in the largest spinel crystals, and we have frequently noted chemically different grains within the same olivine crystals, that are also zoned. Olivine compositions that include chrome spinels range in composition from Fo_{91} to Fo_{58}. Olivine grains with lower forsterite contents very rarely contain (low-Cr, low-Al) spinel inclusions.

Table 7.1 Whole rock analyses of Deccan Trap rocks of which mineral data have been provided. T Egg. = 18MgO + 1050 °C (Whole-rock geothermometer of Eggins 1993)

Type	Area	Sample	SiO_2	TiO_2	Al_2O_3	Fe_2O_3	MnO	MgO	CaO	Na_2O	K_2O	P_2O_5	Mg#	T Eggins
gb	Khapoli	BU6	42.56	0.80	5.87	16.61	0.21	26.92	5.98	0.90	0.01	0.14	78	
LTiA	Saurashtra	D108	45.27	0.89	14.54	12.80	0.16	15.07	9.43	1.36	0.30	0.14	73	1,321
HTi	Pavagadh	D81	45.81	1.90	11.85	12.71	0.14	14.30	11.12	1.28	0.62	0.27	72	1,307
LTiA	Saurashtra	D57	46.05	0.79	14.22	12.04	0.18	14.14	10.31	1.83	0.35	0.10	73	1,305
LTiA	Saurashtra	D101	45.57	1.22	13.19	13.10	0.19	14.03	10.13	2.03	0.42	0.13	71	1,303
HTi	Pavagadh	D97	45.69	2.46	13.30	12.63	0.15	13.02	9.83	1.75	0.79	0.35	70	1,284
LTi	Saurashtra	D130	47.50	1.16	13.83	11.88	0.18	12.40	10.79	1.75	0.40	0.11	70	1,273
HTi	Narmada	D216	44.14	2.47	13.38	13.47	0.19	12.41	11.71	1.12	0.52	0.55	67	1,273
HTi	Saurashtra	D44	46.08	2.30	13.45	12.27	0.16	10.70	11.98	1.93	0.65	0.46	66	1,243
Bushe	Western Ghats	BU4	48.15	1.09	14.08	12.49	0.19	10.40	10.78	1.98	0.65	0.18	65	1,237
Bushe	Western Ghats	BU2	51.48	1.18	12.95	10.72	0.15	10.25	10.56	1.97	0.54	0.19	68	1,234
ThL	Western Ghats	D333	49.51	1.76	12.37	12.52	0.18	10.15	10.90	1.85	0.59	0.17	65	1,233
gb	Saurashtra	D30	46.45	0.97	16.60	12.48	0.18	10.14	10.50	2.21	0.35	0.12	65	
LTiA	Saurashtra	D56	46.47	1.06	15.49	13.13	0.19	10.05	10.77	2.32	0.42	0.12	63	1,231
Bushe	Western Ghats	BU1	50.56	1.47	13.60	11.44	0.20	9.58	10.78	1.91	0.24	0.22	65	1,223
LTi	Saurashtra	D45	48.61	1.13	14.33	12.22	0.19	8.26	12.58	2.01	0.46	0.21	60	1,199
LTi	Rajpipla	D208	49.93	0.86	15.49	11.12	0.18	8.23	11.70	1.96	0.37	0.15	62	1,198
Mah	Western Ghats	MEL27	51.51	2.00	13.90	12.59	0.18	7.04	10.10	2.32	0.76	0.22	56	1,177

LTi	Saurashtra	D47	50.03	0.85	14.25	13.29	0.21	6.98	12.09	2.06	0.11	0.12	54	1,176
dyke	Tapti	Bom122	51.41	1.80	16.35	10.65	0.20	6.52	9.54	2.57	0.39	0.58	58	1,167
HTi	Pavagadh	D79	48.24	2.87	16.32	12.49	0.17	6.31	9.20	2.89	1.24	0.28	53	1,164
dyke	Tapti	Bom151	50.83	1.98	16.50	11.57	0.20	5.71	9.20	2.96	0.28	0.78	53	1,153
dyke	Tapti	RN24	51.74	0.83	17.43	10.52	0.15	5.30	10.37	2.44	1.03	0.19	53	1,145
dyke	Tapti	Bom207	50.91	1.08	18.02	10.15	0.15	5.13	11.38	2.65	0.24	0.29	53	1,142
dyke	Tapti	TAP3	49.52	3.32	14.88	13.94	0.18	4.73	9.66	2.81	0.54	0.42	43	1,135
dyke	Tapti	TAP15	50.56	3.61	15.18	14.19	0.17	3.30	8.88	2.87	0.64	0.60	34	1,109
dacite	Saurashtra	D66	66.32	0.96	13.37	6.08	0.10	2.80	4.84	2.46	2.83	0.19	51	1,100
rhyolite	Saurashtra	D4	72.89	0.65	12.56	4.76	0.07	0.66	2.18	2.70	3.79	0.12	24	1,062
rhyolite	Pavagadh	D89	70.05	0.67	13.24	5.53	0.13	0.63	2.16	4.35	3.17	0.08	20	1,061

gb, gabbro; LTi, low Titanium basalts (A if alkaline); HTi, high-Titanium basalts; ThL, Thakurvadi Fm.; Mah, Mahableshwar Fm.
Mg# = atomic 100Mg/(Mg + Fe^{2+}) with Fe_2O_3/FeO = 0.15

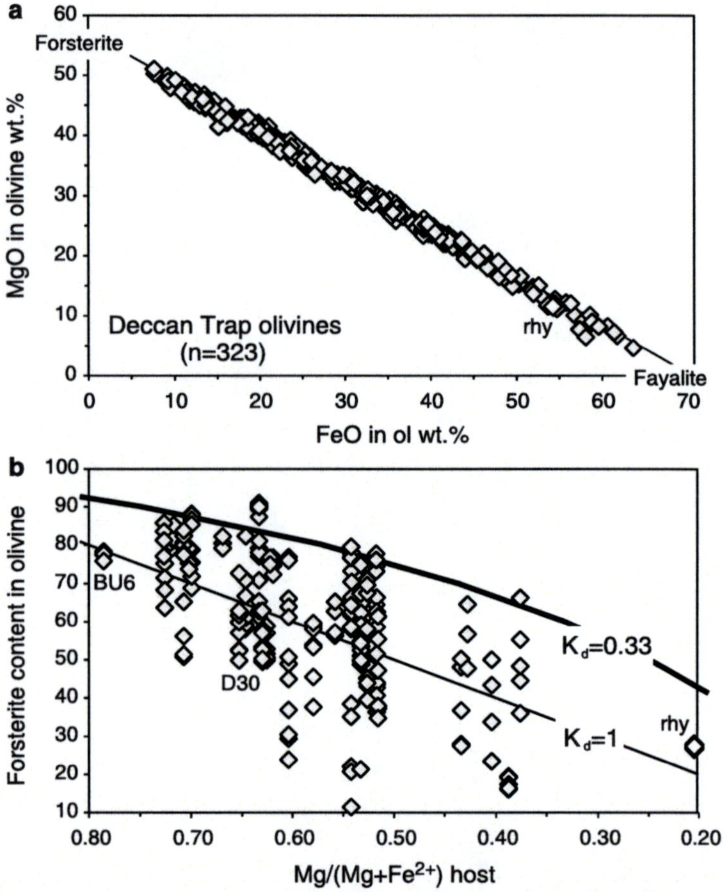

Fig. 7.2 (**a**) MgO versus FeO (in wt%) for the Deccan olivines. Symbols: olivine in basalts and gabbros: grey rhombs; olivine in rhyolite: white rhombs; (**b**) whole rock Mg# versus Fo content of olivines. The *bold line* represent the Fe–Mg partition coefficient between olivine and liquid (K_d) = 0.33 (Roeder and Emslie 1970) the *thinner line* is that with K_d = 1 (perfect equilibrium crystallization)

Chrome-bearing spinel shows a very large range of composition (Fig. 7.3 and b), varying almost continuously from Al-spinel [Al_2O_3 = 53.2 wt%; Cr# = 12; Cr# = molar 100Cr/(Cr + Al)], to Cr-spinel (Cr_2O_3 = 50.7 wt%; Cr# = 84) and to chromiferous Ti-magnetite [Cr# = 5–19; Mg# = 5–13; TiO_2 = 17–21%; Mg# = molar 100Mg/(Mg + Fe^{2+})]. Sample D56 has the largest range of compositions (Cr_2O_3 from 5 to 37 wt%; Al_2O_3 from 6.5 to 53 wt%), and it represents one of the few cases where spinel is deep brown. Some of the most Al-rich spinels of this sample overlap in chemical composition with those in spinel lherzolite mantle xenoliths found in the Kutch region (Krishnamurthy et al. 1988). The most Cr-rich spinels have been found in a Bushe picritic basalt (sample BU1) and in the Pavagadh basalts (sample D97;

Table 7.2 Representative compositions of olivine and chromiferous spinel in the Deccan Trap rocks

Type	Area	Sample	SiO_2	FeO	MnO	MgO	CaO	sum		Fo	
Olivine											
gb	Saurashtra	D30	35.78	34.73	0.61	27.97	0.30	99.4		59	
LTiA	Saurashtra	D56	40.87	8.84	0.07	49.18	0.33	99.3		91	
LTiA	Saurashtra	D101	39.69	16.29	0.28	42.00	0.38	98.6		82	
LTiA	Saurashtra	D101	36.07	36.05	0.97	25.70	0.54	99.3		56	
LTi	Saurashtra	D47	38.96	19.15	0.29	41.46	0.33	100.2		79	
LTi	Saurashtra	D47	36.51	31.01	0.47	31.07	0.36	99.4		64	
LTi	Saurashtra	D47	30.70	63.67	1.02	4.51	0.53	100.4		11	
LTi	Saurashtra	D47	31.77	58.72	0.93	8.49	0.47	100.4		20	
rhyolite	Pavagadh	D89	32.00	53.81	1.72	11.57	0.28	99.4		28	
Mah	Western Ghats	MEL27	37.16	32.15	0.44	30.03	0.36	100.1		62	
Bushe	Western Ghats	BU4	38.05	25.30	0.17	37.24	0.28	101.0		72	
Bushe	Western Ghats	BU4	34.67	32.35	0.43	30.54	0.39	98.4		63	
gb	Khapoli	BU6	39.02	20.49	0.49	39.42	0.43	99.9		77	
			TiO_2	Al_2O_3	FeO	MnO	MgO	Cr_2O_3	sum	Mg#	Cr#
Cr-spinel											
LTiA	Saurashtra	D101	8.62	10.98	50.46	0.76	5.22	24.78	100.8	21	60
LTiA	Saurashtra	D108	0.58	38.77	23.47	0.30	14.53	22.31	100.0	61	25
gb	Saurashtra	D30	4.18	6.63	62.68	0.31	1.66	17.91	93.4	8	30
LTiA	Saurashtra	D56	0.22	25.18	28.38	0.08	8.82	32.40	95.1	43	42
LTiA	Saurashtra	D56	0.18	53.16	14.12	0.12	20.93	11.91	100.4	81	12
LTiA	Saurashtra	D57	7.04	13.39	50.40	0.46	2.75	23.26	97.3	12	39
HTi	Pavagadh	D97	1.54	12.19	25.05	0.36	10.69	47.19	97.0	52	72
HTi	Pavagadh	D97	12.65	7.15	56.59	0.59	5.36	13.85	96.2	21	57
gb	Khapoli	BU6	1.31	12.25	33.52	0.39	5.98	47.18	100.6	29	72
Bushe	Western Ghats	BU1	2.15	6.54	32.65	0.93	3.16	50.65	96.1	16	84
Bushe	Western Ghats	BU2	1.63	15.69	32.85	0.66	6.54	38.49	95.9	33	62

Fo = 100Mg/(Mg + Fe); Mg# = Mg/Mg + Fe^{2+}); Cr# = 100Cr/(Cr + Al)

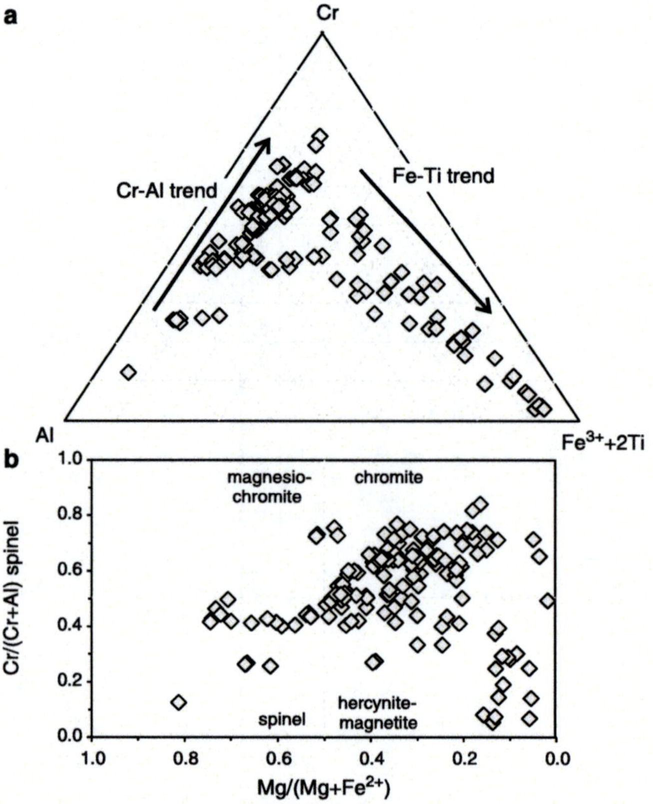

Fig. 7.3 (**a**) Composition of the Deccan chromiferous spinels included in olivine when plotted in the Cr–Al–Fe^{3+} + 2Ti diagram. Symbol as in Fig. 7.2; (**b**) Mg/(Mg + Fe^{2+}) vs Cr/(Cr + Al) in the same spinels. The decrease of Mg# and the general increase in Cr# indicates the preferred coupled substitution Mg + Al->Fe^{2+} + Cr

see also Krishnamurthy and Cox 1977, for data on spinels in other high-Ti basalts of Gujarat) (Table 7.2). The new data added in this paper considerably enlarge the known compositional field of spinels towards the Cr-rich side, but still well within the range of the terrestrial spinels considered by Barnes and Roeder (2001). Indeed, they follow the two typical trends: one mostly along the Al-spinel-chromite side of Fig. 7.3a (the Al–Cr trend of Barnes and Roeder 2001) and the other towards the magnetite + ulvöspinel apex (the Fe–Ti trend of Barnes and Roeder 2001).

7.4.3 *Pyroxene*

Four types of pyroxene have been found in the Deccan Trap volcanic rocks: Ca-rich clinopyroxene of the diopside-hedenbergite series, Ca-rich pyroxene of the augite

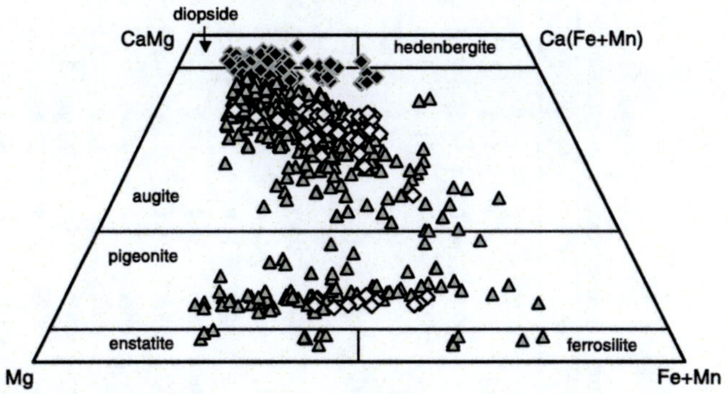

Fig. 7.4 Composition of Deccan pyroxenes in the standard Ca–Mg–Fe_t diagram. Symbols: *black rhombs*: clinopyroxene of mildly alkaline basalts; *grey triangles*: pyroxene of tholeiitic samples; *white rhombs*: pyroxene of evolved tholeiitic rocks (basaltic andesites, andesites, dacites and rhyolites)

series, Ca-poor pyroxene of pigeonite composition and orthopyroxene (Fig. 7.4; Table 7.3). Ca-Na-rich clinopyroxenes of the strongly alkaline rocks of Amba Dongar, Murud-Janjira and in intrusions of Saurashtra are not considered here.

Ca-rich clinopyroxene with broadly diopside/salite composition or close to the diopside-augite limit, occur in the transitional rocks of the Pavagadh section (e.g. Sheth and Melluso 2008) and in some Gujarat basalts with slight alkaline affinity and low-Ti geochemistry (e.g. Melluso et al. 1995, 2006a). This type of pyroxene is not associated with low Ca-pyroxene and has a tendency to Ti- Na- and Al-enrichment in Fe-rich, evolved, compositions (Fig. 7.5a–c). This enrichment is higher in the Pavagadh high-Ti basalts, but TiO_2 reaches values as high as 2.78 wt% also in the Gujarat low-Ti basalts, and is accompanied by similarly high Al_2O_3 contents (Al_2O_3 up to 3.4 wt%). Na contents are generally low, and increase with decreasing Mg# (Fig. 7.4b).

Ca-rich clinopyroxene of the augite series is the most widespread mafic mineral in the Deccan Traps. It occurs as a phenocryst or groundmass phase in rocks of most formations and chemical types in the Western Ghats and elsewhere. This is a further evidence of the common tholeiitic nature of most Deccan Traps magmas. It has a compositional range from $Ca_{39}Mg_{49}Fe_{10}$ (Bushe basalt) to $Ca_{25}Mg_{16}Fe_{59}$ (Tapti dyke). As typical of tholeiitic augites worldwide, minor elements such as Ti, Al and Na first increase and then decrease with Mg# suggesting strong control of co-crystallizing phases (plagioclase and Fe–Ti oxides) for these elements (Fig 7.5a–c). On the other hand, Mn has a negative correlation with Mg#. It is interesting to note that the largest compositional range is found in the basalts; more evolved rocks (basaltic andesites to rhyolites) have low TiO_2, more scattered Na_2O (Fig. 7.5a–c) and a more marked tendency to increase in Mn with decreasing Mg#. A few dacites of Gujarat (sample D66) have rare reverse zoning in the crystals, and a few compositions more typical of pyroxene of tholeiitic basalts, that can be taken as evidence of magma mixing. The frequency histogram of Mg# in the Ca-rich pyroxenes has a marked peak at Mg#=80–60 (Fig. 7.5d).

Table 7.3 Representative compositions of pyroxene in the Deccan Trap rocks. Analyses of sample SH51 are taken from Chandrasekharam et al. (2000)

Type	Area	Sample	SiO_2	TiO_2	Al_2O_3	FeO	MnO	MgO	CaO	Na_2O	Cr_2O_3	tot	Ca	Fe*	Mg
Ca-rich px															
gb	Saurashtra	D30	50.70	0.99	2.37	8.26	0.22	14.82	21.75	0.30	0.14	99.6	44.4	13.5	42.1
LTiA	Saurashtra	D56	49.41	1.46	5.07	8.58	0.14	14.32	20.13	0.40		99.5	43.0	14.5	42.5
LTiA	Saurashtra	D101	49.33	1.65	3.14	12.34	0.59	11.26	21.24	0.62		100.2	45.2	21.5	33.3
HTi	Saurashtra	D44	50.07	0.87	3.62	5.29	0.05	15.83	23.06	0.29	0.14	99.2	46.8	8.5	44.7
HTi	Saurashtra	D44	47.73	1.93	5.39	7.84	0.17	13.40	22.37	0.36	0.02	99.2	47.3	13.2	39.4
HTi	Saurashtra	D44	44.28	3.58	6.84	9.76	0.15	11.65	22.29	0.38	0.10	99.0	48.2	16.7	35.1
LTi	Saurashtra	D47	50.79	0.42	1.89	20.98	0.33	14.03	10.34	0.21		99.0	22.2	35.6	42.0
HTi	Pavagadh	D97	52.46	0.65	2.31	4.27	0.11	16.45	22.42	0.23	1.01	99.9	46.0	7.0	47.0
HTi	Pavagadh	D97	48.13	3.03	4.02	10.57	0.31	13.01	20.04	0.37		99.5	43.0	18.2	38.8
gb	Khapoli	BU6	52.04	0.83	2.03	8.09	0.31	16.33	18.24	0.32	0.36	98.6	38.4	13.8	47.8
gb	Khapoli	BU6	52.94	0.45	1.58	6.90	0.22	17.36	19.05	0.18	0.99	99.7	39.1	11.4	49.5
Bushe	Western Ghats	BU2	50.99	0.66	1.25	13.22	0.40	15.64	16.04	0.17		98.4	33.1	22.0	44.9
Mah	Western Ghats	MEL27	51.26	0.91	2.79	8.15	0.40	15.52	19.61	0.29	0.72	99.7	41.0	13.9	45.1
Mah	Western Ghats	MEL27	50.83	1.03	1.55	14.30	0.37	13.63	17.49	0.15	0.07	99.4	36.5	23.9	39.6
dyke	Tapti	B122	51.12	1.07	1.74	16.62	0.45	15.31	13.33	0.26		99.9	27.8	27.7	44.4
dyke	Tapti	B122	48.01	0.55	0.51	34.01	0.70	5.12	11.40	0.12		100.4	25.0	58.9	15.6
dyke	Tapti	TAP15	49.25	0.76	1.06	26.06	0.55	8.17	14.17	0.15		100.2	30.6	44.6	24.5
dacite	Saurashtra	D66	50.02	0.63	1.94	14.98	0.53	13.93	17.26	0.17		99.5	35.4	24.8	39.8
rhyolite	Pavagadh	D89	50.08	0.48	0.92	18.46	0.82	10.69	17.85	0.18		99.5	37.4	31.4	31.1
rhyolite	Pavagadh	D89	51.39	0.35	0.93	18.93	0.93	9.50	17.18	0.59		99.8	37.4	33.6	28.8

Ca-poor px															
dyke	Tapti	B122	48.37	0.40	0.97	34.78	0.81	9.37	4.94	0.14		99.8	10.8	60.2	28.5
dyke	Tapti	B122	51.40	0.46	1.14	21.91	0.41	19.49	4.99	0.02		99.8	10.1	35.0	54.7
gb	Khapoli	BU6	56.24	0.46	0.50	15.87	0.24	26.95	1.77	0.13	0.24	102.4	3.4	24.3	72.3
gb	Khapoli	BU6	54.03	0.97	1.29	15.35	0.28	25.83	2.20	0.24	0.23	100.4	4.4	24.2	71.4
gb	Khapoli	BU6	55.12	0.28	0.73	12.93	0.27	25.14	4.30	0.26	0.29	99.3	8.7	20.8	70.5
gb	Khapoli	BU6	54.27	0.34	1.04	14.15	0.26	25.10	4.07	0.10	0.44	99.8	8.1	22.4	69.5
Bushe	Western Ghats	BU2	51.56	0.79	0.71	22.79	0.45	17.95	4.78	0.17	0.06	99.3	10.0	37.9	52.1
dacite	Saurashtra	D66	51.52	0.30	0.63	23.75	1.07	18.37	4.62	0.02		100.3	9.3	38.9	51.6
dacite	Saurashtra	D66	49.02	0.42	0.66	28.66	0.76	14.46	4.37	0.03		98.4	9.2	48.1	42.4
dyke	Tapti	SH51	51.44	0.19	0.73	25.91	0.66	19.13	1.20	0.03		99.3	2.5	42.7	54.8
dyke	Tapti	SH51	51.75	0.30	0.70	25.56	0.65	18.77	1.96			99.7	4.0	42.2	53.8
rhyolite	Saurashtra	D4	49.38	0.23	0.49	36.47	0.73	11.34	1.41			100.1	3.1	62.8	34.1

Ca, Mg, Fe* (Fe + Mn) in mol%

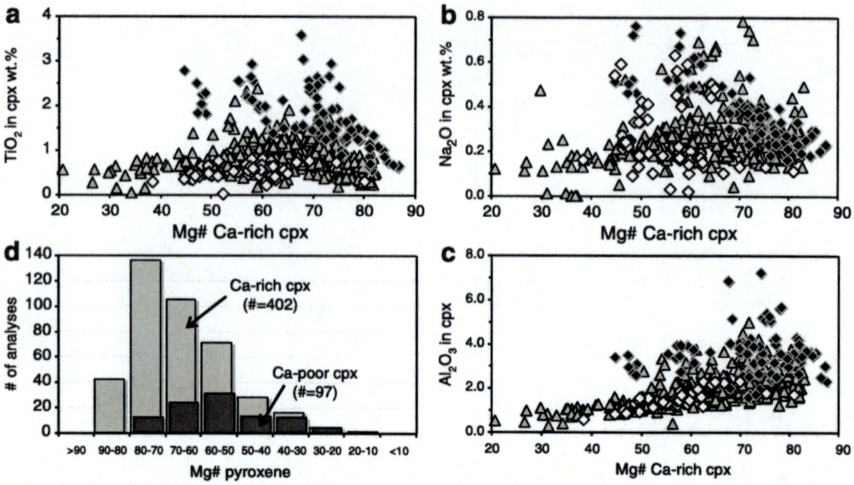

Fig. 7.5 (**a–c**), TiO_2–Mg#, Na_2O–Mg#, Al_2O_3–Mg# diagrams for the Ca-rich pyroxene compositions. (**d**) Frequency histogram of Ca-rich and Ca-poor pyroxenes. The shifted maximum peaks and the lack of the highest Mg# in pigeonite and orthopyroxene indicates the slightly delayed crystallization of pigeonite and orthopyroxene with respect to augite and diopside. Symbols as in Fig. 7.4

Pigeonite is by far the most abundant low-Ca pyroxene, as typical of tholeiitic magmas worldwide. It ranges from $Ca_8Mg_{71}Fe_{21}$ to $Ca_9Mg_{18}Fe_{73}$ with the most Mg-rich compositions found in the Khapoli gabbro and the most Fe-rich compositions (and, actually, the largest compositional range) found in the Tapti dykes. The frequency histogram of Ca-poor pyroxenes (Fig. 7.5d) has the peak displaced towards lower Mg# (Mg# = 60–50) than that of coexisting Ca-rich augite. It is worth noting that coexisting Ca-rich and Ca-poor pyroxene in the evolved rocks (basaltic andesites to rhyolites) have a smaller range in Mg# with respect to the range observed in the basalts (Fig. 7.5a–c), and this range is clearly displaced towards lower Mg# in both augite (Fe–augite) and pigeonite (or orthopyroxene).

Orthopyroxene is rarely found: it has been analyzed in the Khapoli picritic gabbro (at the base of the Lonavala escarpment), where it reaches the most Mg-rich compositions observed to date in the Deccan volcanic rocks ($Ca_4Mg_{72}Fe_{24}$). It is also found in the Tapti dykes (Chandrasekharam et al. 2000; Melluso et al. 1999; R. Ray, unpublished data 2009) and, with more Fe-enrichment, in a few rhyolites in Saurashtra ($Ca_3Mg_{34}Fe_{63}$; Melluso et al. 1995) (Fig. 7.4).

7.4.4 Feldspar

Plagioclase (Table 7.4) is the most abundant phase of the Deccan volcanic rocks, and ranges in composition from bytownite An_{84}, found in a Tapti dyke (Melluso et al. 1999) to almost pure albite (mostly in dacites and rhyolites of

Table 7.4 Representative compositions of feldspar, amphibole, biotite, Ti-magnetite and ilmenite in the Deccan Trap rocks. An, Ab, Or in mol%; Ulvöspinel and ilmenite in mole fractions. Mg# = 100Mg/(Mg + Fe^{2+})

Type	Area	Sample	SiO_2	Al_2O_3	FeO	CaO	Na_2O	K_2O	tot		An	Or	Ab
feldspar													
HTi	Pavagadh	D81	51.96	30.03	0.69	12.65	3.98	0.33	99.6		62.5	1.9	35.6
HTi	Pavagadh	D81	65.09	19.88	0.32	1.16	5.60	7.91	100.0		5.6	45.5	48.9
gb	Khapoli	BU6	51.04	29.32	0.33	13.90	3.61	0.20	98.4		67.2	1.2	31.6
gb	Khapoli	BU6	51.85	29.11	0.46	13.47	3.74	0.15	98.8		66.0	0.9	33.1
LTiA	Saurashtra	D56	48.06	32.61	0.79	15.69	2.59	0.11	99.9		76.5	0.6	22.9
HTi	Pavagadh	D79	64.94	19.81	0.29	1.23	4.79	8.95	100.0		6.0	51.8	42.2
HTi	Pavagadh	D79	63.00	21.63	0.48	3.06	6.09	5.57	99.8		14.8	32.0	53.2
LTi	Rajpipla	D208	64.98	18.93	0.70	0.52	4.45	10.00	99.6		2.5	58.1	39.3
dyke	Tapti	B207	46.68	33.69	0.58	16.89	1.78	0.05	99.7		83.7	0.3	16.0
dyke	Tapti	TAP13	63.27	21.93	0.69	3.05	8.05	2.74	99.7		14.6	15.6	69.8
dyke	Tapti	B151	64.75	18.40	0.15	0.07	0.52	16.04	99.9		0.3	95.0	4.7
HTi	Pavagadh	D97	51.26	30.64	0.64	13.38	3.63	0.33	99.9		65.8	1.9	32.3
HTi	Pavagadh	D97	58.38	25.00	0.72	7.00	6.35	2.09	99.5		33.4	11.9	54.8
HTi	Pavagadh	D97	54.90	27.53	1.43	9.93	5.39	0.50	99.7		49.0	2.9	48.1
			TiO_2	Al_2O_3	Fe_2O_3	FeO	MnO	MgO		Cr_2O_3sum		Ulr	Ilm
Fe–Ti oxides													
Bushe	Western Ghats	BU4	15.53	1.92	35.24	43.09	0.22	1.03		97.0		44.8	
Bushe	Western Ghats	BU4	17.03	2.78	31.23	44.61	0.27	0.99		96.9		48.9	
Bushe	Western Ghats	BU4	51.50	0.13	2.75	39.55	0.64	3.44		98.0			97.2
Bushe	Western Ghats	BU4	49.62	0.05	7.42	39.29	0.46	2.74		99.6			93.0
dyke	Tapti	B122	20.07	3.42	23.81	48.66	0.46	0.06		96.5		57.9	
dyke	Tapti	B122	50.45		3.99	44.23	0.54	0.34		99.5			96.2

(continued)

Table 7.4 (continued)

			TiO_2	Al_2O_3	Fe_2O_3	FeO	MnO	MgO		Cr_2O_3sum	Ulr	Ilm
gb	Saurashtra	D30	13.75	2.45	39.41	39.73	0.48	2.23		98.0	38.9	
gb	Saurashtra	D30	49.48	0.12	7.23	37.25	0.46	3.81		98.4		93.0
HTi	Pavagadh	D97	16.20	2.27	34.11	44.72	1.14	0.12	0.33	98.6	46.1	
HTi	Pavagadh	D97	48.15		10.56	37.61	0.62	2.85		99.8		90.1
LTiA	Saurashtra	D101	16.85	1.47	34.68	45.54	1.40		0.01	100.1	47.7	
LTiA	Saurashtra	D101	49.07	0.11	8.57	40.64	0.82	1.50	0.05	100.8		91.8
LTi	Saurashtra	D130	52.25		5.47	34.95	0.57	6.44	0.55	100.2		94.5
HTi	Rajpipla	D216	48.79	0.17	5.94	38.66	4.88	0.16		99.4		93.3
ThL	Western Ghats	D333	15.19	1.71	35.14	42.87	0.87	0.37	0.14	96.3	37.6	
ThL	Western Ghats	D333	48.89	0.03	6.81	41.29	0.44	1.26	0.08	98.8		93.3
dyke	Tapti	B207	20.03	1.55	24.60	46.59	0.99	0.34		94.1	59.7	
dyke	Tapti	B207	50.66	0.09	4.78	43.26	0.58	0.97		100.3		95.3
pseudob	Saurashtra	D57	54.18		31.52	9.51	0.62	4.00		99.8		
pseudob	Saurashtra	D30	51.72		30.47	8.59	0.54	3.85		95.2		

Ulvöspinel and ilmenite in mol%

Type	Area	Sample	SiO_2	TiO_2	Al_2O_3	FeO	MnO	MgO	CaO	Na_2O_2	K_2O	tot	Mg#
amphibole													
LTi	Saurashtra	D45	45.51	1.03	5.16	24.76	0.58	8.05	9.81	1.94	0.64	97.5	36.1
dyke	Tapti	B122	44.56	1.49	6.22	25.15	0.27	7.28	9.87	1.93	0.95	97.7	33.8
rhyolite	Saurashtra	D4	44.69	0.14	6.49	26.11	0.50	7.80	9.95	1.84	1.08	98.6	34.3
biotite													
dyke	Tapti	TAP3	36.96	2.88	11.30	28.37	0.10	8.06	0.15	0.28	7.46	95.6	33.5

Mg# = 100Mg/(Mg + Fe)

Gujarat), with a marked frequency peak at labradorite-bytownite compositions (Fig. 7.6a). Zoning is dominantly normal, and very large compositional ranges are often observed in the same sample, passing from cores of phenocrysts to rims and then to groundmass compositions. Elements such as Fe are very scattered, though this element decreases slightly with Ca towards Na-rich compositions (Fig. 7.6b). This is probably the result of Fe–Ti oxide fractionation extracting iron from the melts.

Alkali feldspar is common in the groundmass of many rocks, including different basalt types, and ranges in composition from sanidine to anorthoclase, the latter more frequent in the high-Ti basalts of Pavagadh. The basaltic andesites to rhyolite rocks have a slightly more restricted range than that observed in the basalts.

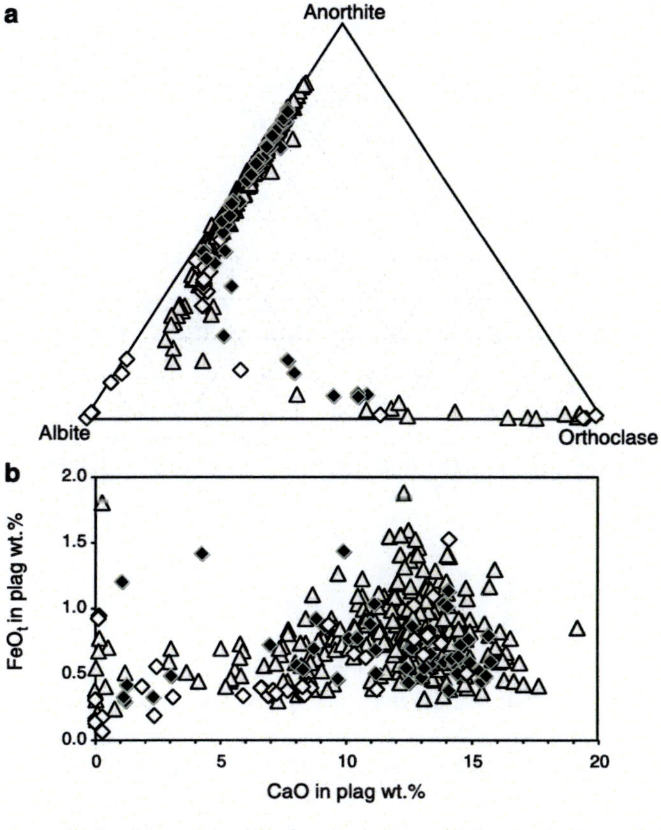

Fig. 7.6 (**a**) Ca–Na–K diagram (mol%) for the Deccan feldspars. Symbols as in Fig. 7.4. (**b**) FeO_t–CaO diagram (both in wt%) for the Deccan feldspars. Symbols as in Fig. 7.4

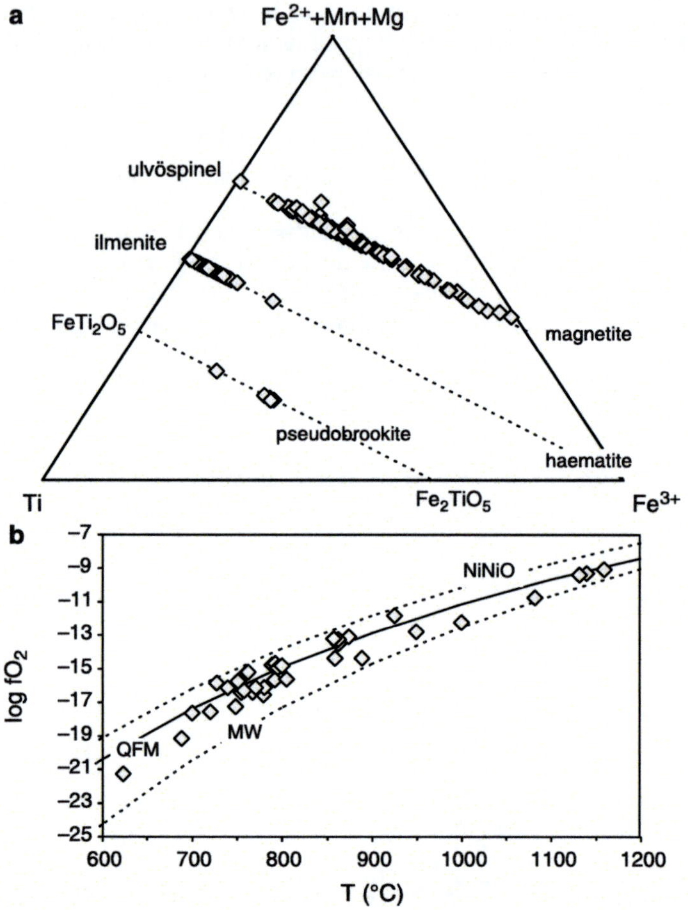

Fig. 7.7 (**a**) Diagram Fe–Ti–Mg–Mn (atoms) for the Deccan groundmass oxides (spinels, rhombohedral oxides and pseudobrookites); (**b**) T–fO_2 estimates for coexisting Fe–Ti oxides in the Deccan Traps. QFM = quartz–fayalite–magnetite buffer; WM = wustite–magnetite buffer, NiNiO = nickel–nickel oxide buffer

7.4.5 Fe–Ti Oxides

Titaniferous magnetite and ilmenite (Table 7.4; Fig. 7.7a) are common phases in the Deccan rocks (De 1964; Murari et al. 1993). They are found in the groundmass or are interstitial in basalts and basaltic andesites, and are often included in other minerals (pyroxene, feldspar) in more evolved rock types (andesites, dacites and rhyolites). The two minerals have distinctly different ranges of composition: Ti-magnetite has ulvöspinel contents ranging from 7.3 to 92.8 mol%, whereas ilmenite has a more limited range ($ilm_{99.3}$ to $ilm_{89.7}$). There is a broadly continuous range of compositions between chrome-bearing and Fe–Ti spinels. Ilmenite reaches MgO values as high as 6.4 wt% (in picritic basalt D130), and MnO as high as 4.9 wt% (in picritic basalt D216). Pseudobrookite is a rare accessory of many basalt types (Fig. 7.7a), though it is already known in other flood basalt provinces (Bowles 1988). The chemical composition is slightly variable, with TiO_2 ranging from 51 to 58 wt% and total iron as FeO from 34.7 to 39.3 wt%.

Coexisting spinel–ilmenite pairs gave a range of equilibration temperatures from 1,159°C to 624°C, with the highest values found in the picritic basalt D101 and oxygen fugacities from –9 to –21 in log fO_2 units. Most of the data plot between 700°C and 800°C, indicating that the composition of the coexisting minerals is close to subsolidus re-equilibration. The data tightly cluster around the quartz-fayalite-magnetite (QFM) synthetic oxygen buffer at low pressure (Fig. 7.7b), as is typical of most continental flood basalt provinces (e.g., Melluso et al. 2006b). Earlier papers on the Deccan Traps also indicate T-fO_2 conditions close or slightly lower than those pertinent to the QFM buffer (cf. Melluso et al. 1995, 1999; Sano et al. 2001; Sen 1986; Sethna and Sethna 1988).

7.4.6 *Hydrous Phases and Other Minerals*

Calcic amphibole and biotite are rare interstitial phases of many basalt types. Amphiboles are mostly Fe–hornblende, with low Mg# (31–45), and TiO_2 (0–2 wt%).

Brown mica has been found at Pavagadh and in the interstices of Tapti basaltic dykes. It varies in composition from phlogopite at Pavagadh (Mg# = 84–80) to biotite in the Tapti dykes (Mg# = 52–33) (Table 7.4).

7.5 Discussion

The compositional range of the Deccan Trap minerals is not different from that already observed in rocks of other flood basalt provinces such as the Ferrar (Antarctica) or Madagascar (Brotzu et al. 1988; Melluso et al. 2001, 2005, 2006a; Mahoney et al. 2008). The Ferrar rocks are mostly made up of low-Ti tholeiitic types, thus very rich in augite and pigeonite (or orthopyroxene), whereas the Madagascar LIP is made up of tholeiitic and slightly alkaline magma types, thus having closer resemblance to the Deccan Traps (cf. Melluso et al. 2001, 2005; Mahoney et al. 2008).

The general paucity of very Mg-rich olivines (excluding a few examples; Krishnamurthy et al. 2000), and the relative abundance of more Fe-rich variants can be linked to at least two causes: the paucity of very primitive liquids in the Deccan Traps (see above), and the crystallization conditions, with relatively low oxygen fugacity in many samples, thus stabilizing olivine instead of Fe–Ti oxides. When the forsterite content of olivines is plotted against the Mg# of the host rock (Fig. 7.2b) several interesting features are noted, including the remarkable range of composition of many samples, and the difficulty in finding olivines typically in equilibrium with the host rock according to the Roeder and Emslie (1970) Fe–Mg partitioning between olivine and melt. In Fig. 7.2b, most olivines plot below the equilibrium curve at $Kd_{Fe/Mg} = 0.33$. This curve seems to represent a reasonable upper limit for the most Mg-rich olivine compositions. Considering the olivine compositional trends within each sample, if the Fe-rich olivine compositions of basalts are in continuous equilibrium with a changing melt composition (thus obeying to the same partitioning of Fe and Mg between melt and olivine of Fig. 7.2b), olivine must have been in equilibrium with liquids with very high Fe/Mg ratios, similar to, or even higher than, those found in rhyolites; therefore these crystals should have been in equilibrium with the entire basalt-to-rhyolite liquid sequence, found in the same lava flow or dyke. We consider this hypothesis very difficult to verify. The compositions of olivine in rhyolite D89 also plot away from the curve at $K_d = 0.33$ (Fig. 7.2b). Another possibility is that the most Fe-rich olivines (usually rims or groundmass crystals) crystallized in a non-equilibrium environment. Olivine of the Khapoli gabbro (BU6) and Gujarat gabbro (D30) plot close to the curve $Kd_{Fe/Mg} = 1$, testifying once more that the compositions of these intrusive samples cannot be considered as representative of liquid compositions (Fig. 7.2b). This also renders problematic the use of olivine-liquid geothermometers in the Deccan rocks (Roeder and Emslie 1970; Ford et al. 1983). Whole-rock geothermometers (e.g. Eggins, 1993) indicate maximum liquidus (=olivine-in) temperatures of about 1,370°C, but most samples with MgO close to 10% have calculated liquidus temperatures in the range 1,240–1,220°C. The use of olivine-bulk rock thermometers is seriously biased when the bulk rocks cannot be considered as liquid compositions.

The spinels included in olivine also have a very wide compositional range, which is broadly similar to that found in other spinel suites in flood basalt sequences (Karoo, Madagascar, Siberian Traps, Emeishan). The causes of such wide chemical variation in the Deccan spinels could be the consequence of varying pressure of crystallization as suggested by the Al–Cr trend shown in Fig. 7.3a (cf. Dick and Bullen 1984), the degree of melt-depletion of the source as reflected in the chemical composition of the host magmas, or local crystallization conditions, whereas the Fe–Ti trend towards a high ulvöspinel content of the Cr–Al-poor spinel inclusions could be the result of local non-equilibrium conditions (Barnes and Roeder 2001). These non-equilibrium conditions can be triggered by possible cation substitutions in the spinel lattice (such as Cr-> Fe^{3+}; Fe^{2+}+2Cr-> $2Fe^{2+}$+Ti) favoured by the interstitial Fe–Ti-rich tholeiitic melt compositions, and by the partitioning of Al in other co-crystallizing phases such as plagioclase.

The highest Mg# of diopside and diopside-augite in the mildly alkaline rocks are higher than the highest Mg# of the augites in the tholeiites. This suggests a higher temperature of crystallization of clinopyroxene in the more alkaline rocks. This is in agreement with the higher abundance of Ca-rich clinopyroxene as a phenocryst phase in the Gujarat high-Ti rocks.

The marked chemical differences between the two Ca-rich clinopyroxene types are not only seen in the classification diagram, but also in the minor element compositional trends. The generally high and increasing Ti content of clinopyroxene in mildly alkaline rocks (high $CaTiAl_2O_6$) is the reverse of that found in Ca-rich clinopyroxene of the tholeiites, where $CaTiAl_2O_6$ smoothly decreases with Mg#. The decreasing Na, Ti and Al in the augites (Fe-augites), while being a typical feature of tholeiitic augites worldwide, can be interpreted as derived from decreasing extent of partitioning of these elements in a structure increasingly rich in enstatite-ferrosilite compared with diopside-hedenbergite (the Ca-bearing components). This is also enhanced by the strong competition of increasingly Na-rich plagioclase, the total lack of a tendency towards peralkaline conditions (increasing aegirine components in clinopyroxene) and Fe–Ti oxides in determining the minor element budget of Ca-rich clinopyroxene.

The low pressure of crystallization of Ca-rich clinopyroxene in the Deccan Traps is evident from their low-to-negligible Al^{VI} contents. Pyroxene barometry (Nimis 1999) indicates equilibration pressures not more than 3 kbar for the most Mg-rich compositions (Mg# > 80). Pigeonite thermometry (Ishii 1975) indicates temperatures ranging from 1,160°C to 882°C, similar to the values obtained by two-pyroxene thermometry (maximum values <1,200°C; e.g. Lindsley 1983).

The large overlap in the Ca-rich pyroxene and plagioclase compositions among different basalt types including Mahableshwar and low-Ti Gujarat basalts precludes the use of Ca-rich pyroxene and plagioclase compositions as markers of different tholeiitic basalts. This has the consequence that the Deccan tholeiites crystallized under similar physico-chemical conditions throughout the province, thus implying similar feeding systems throughout the history of the Deccan eruption. On the other hand, the presence of more Ca-rich clinopyroxene without Ca-poor pyroxene is used to identify more alkaline magmas in parts of the Deccan. The presence of more alkaline magmas has no direct relationship with their degree of incompatible element enrichment, as indicated by the presence of more alkaline types in the low-Ti, incompatible element depleted basalts in Gujarat (the low-TiA types of Melluso et al. 1995, 2006a).

7.6 Conclusions and Future Developments

The main minerals found in the Deccan Trap volcanic rocks record a very wide range of compositions as a result of changing temperatures and chemical compositions of the melts with which the minerals were in equilibrium. Most minerals in the basalts have compositions typical of those observed in tholeiitic or mildly alkaline rocks with contrasting degrees of magmatic evolution. More detailed work on the Deccan phases is required to correlate volcanic units (for instance, the Deccan-related dolerites of Seychelles and dyke-sill-lava flow units) by matching chemical and isotopic characteristics of the minerals, to identify the geochemical affinity of magmas from trace element and isotopic composition in key minerals, to understand magma chamber processes and the timing of eruption events from the zonation of the parageneses and chemical composition of minerals, including trace element distributions, and to establish compositional ranges in minerals within given formations. This paper hopefully provides a basis for future detailed mineral chemical (including trace elements) and isotopic work within these contexts.

Acknowledgments This work is our humble tribute to Prof. Mihir Bose. His work on the Deccan rocks was greatly influential during the development of the Ph.D. thesis of the first author. We gratefully thank Carlo Garbarino, Marcello Serracino and Roberto de' Gennaro for their essential technical and scientific support for the microprobe work over several years, first at Cagliari and then at Rome and Napoli, and for their good humour and kindness that made the tedious work enjoyable. We also thank all the people involved or interested in our Deccan Trap work over the years, in particular L. Beccaluva, P. Brotzu, A. Khateeb, P. Javeri, J.J. Mahoney, B. Sethna, H. Sheth and the late Keith Cox. We apologize if major papers dealing with a substantial data set on mineral chemistry of the Deccan Traps were not taken into account. The comments of three anonymous reviewers were very helpful in improving contents and style of an earlier version of this manuscript, and so was the further reading of a very kind Ed Stephens. Funds for this project were provided by "Fondi per la Ricerca Dipartimentale 2008" to Leone Melluso.

References

Barnes S-J, Roeder PL (2001) The range of spinel compositions in terrestrial mafic and ultramafic rocks. J Petrol 42:2279–2302

Beane JE, Hooper PR (1988) A note on the picrite basalts of the Deccan Traps. In Subbarao KV (ed) Deccan Flood basalts. Geol Soc India Mem 10:117–134

Beane JE, Turner CA, Hooper PR, Subbarao KV, Walsh JN (1986) Stratigraphy, composition and form of the Deccan Basalts, Western Ghats, India. Bull Volcanol 48:61–83

Bondre NR, Hart WK, Sheth HC (2006) Geology and geochemistry of the Sangamner mafic dyke swarm, western Deccan volcanic province, India: implications for regional stratigraphy. J Geol 114:155–170

Bose MK (1972) Deccan basalts. Lithos 5:131–145

Bose MK (1973) Petrology and geochemistry of the igneous complex of Mount Girnar, Gujarat, India. Contrib Miner Petrol 39:247–266

Bowles JFW (1988) Definition and range of naturally occurring minerals with the pseudobrookite structure. Am Miner 73:1377–1383

Brotzu P, Capaldi G, Civetta L, Melluso L, Orsi G (1988) Jurassic Ferrar dolerites and Kirkpatrick basalts in northern Victoria Land (Antarctica): stratigraphy, geochronology and petrology. Soc Geol It Mem 43:97–116

Chandrasekharam D, Vaselli O, Sheth HC, Keshav S (2000) Petrogenetic significance of ferro-enstatite orthopyroxene in basaltic dikes from the Tapi rift, Deccan flood basalt province, India. Earth Planet Sci Lett 179:469–476

Chatterjee N, Bhattacharji S (2008) Trace element variations in Deccan basalts: roles of mantle melting, fractional crystallization and crustal assimilation. J Geol Soc India 71:171–188

Chenet AL, Fluteau F, Courtillot V, Gerard M, Subbarao KV (2008) Determination of rapid Deccan eruptions across the Cretaceous-Tertiary boundary using paleomagnetic secular variation: results from a 1200-m-thick section in the Mahabaleshwar escarpment. J Geophys Res 113:B04101. doi:10.1029/2006JB004365

Cox KG, Hawkesworth CJ (1985) Geochemical stratigraphy of the Deccan Traps at Mahabaleshwar, Western Ghats, India, with implication for open system processes. J Petrol 26:355–387

Dal Negro A, Molin GM, Cundari A, Piccirillo EM (1982) Intracrystalline cation distribution in natural clinopyroxenes of tholeiitic, transitional and alkaline basaltic rocks. In: Saxena SK (ed) Advances in physical geochemistry, vol 2. Springer, New York, pp 117–150

De A (1964) Iron-titanium oxides and silicate minerals of the alkali olivine basalts, tholeiites and acidic rocks of the Deccan Trap series and their significance. International geological congress New Delhi, Rep 22nd session, III, pp 126–138

Devey CW, Stephens WE (1991) Tholeiitic dykes in the Seychelles and the original spatial extent of the Deccan. J Geol Soc Lond 148:979–983

Devey CW, Stephens WE (1992) Deccan-related magmatism west of the Seychelles–India rift. In: Storey BC, Alabaster T, Pankhurst RJ (eds) Magmatism and the causes of continental break-up. Geological Society of London Special Publications 68, London, pp 271–291

Dick HJB, Bullen T (1984) Chromian spinel as a petrogenetic indicator in abyssal and alpine-type peridotites and spatially associated lavas. Contrib Miner Petrol 86:54–76

Duncan RA, Pyle DG (1988) Rapid eruption of the Deccan flood basalts at the Cretaceous/Tertiary boundary. Nature 333:841–843

Eggins SM (1993) Origin and differentiation of picritic arc magmas, Ambae (Aoba), Vanuatu. Contrib Miner Petrol 114:79–100

Ford CE, Russell DG, Craven JA, Fisk MR (1983) Olivine-liquid equilibria: temperature, pressure and composition dependence of the crystal/liquid cation partition coefficients for Mg, Fe^{2+}, Ca and Mn. J Petrol 24:256–265

Gwalani LG, Rock NMS, Chang W-J, Fernandez S, Allegre CJ, Prinzhofer A (1993) Alkaline rocks and carbonatites of Amba Dongar and adjacent areas, Deccan Igneous Province Gujarat, India: 1. Geology, petrography and petrochemistry. Miner Petrol 47:219–253

Harangi S, Mason PRD, Lukacs R (2005) Correlation and petrogenesis of silicic pyroclastic rocks in the Northern Pannonian Basin, Eastern-Central Europe: in situ trace element data of glass shards and mineral chemical constraints. J Volcanol Geotherm Res 143: 237–257

Higgins MD, Chandrasekharam D (2007) Nature of sub-volcanic magma chambers, Deccan Province, India: evidence from quantitative textural analysis of plagioclase megacrysts in the Giant Plagioclase basalts. J Petrol 48:885–900

Ishii T (1975) The relations between temperature and composition of pigeonite in some lavas and their application to geothermometry. Miner J 8:48–57

Jay AE, Widdowson M (2008) Stratigraphy, structure and volcanology of the SE Deccan continental flood basalt province: implications for eruptive extent and volumes. J Geol Soc Lond 165:177–188

Karmalkar NR, Rege S, Griffin WL, O'Reilly SY (2005) Alkaline magmatism from Kutch, NW India: implications from plume-lithosphere interaction. Lithos 81:101–119

Krishna KS, Gopala Rao D, Sar D (2006) Nature of the crust in the Laxmi Basin (14°-20° N), western continental margin of India. Tectonics 25: TC1006. doi:10.1029/2004TC001747

Krishnamurthy P, Cox KG (1977) Picrite basalts and related lavas from the Deccan Traps of Western India. Contrib Miner Petrol 62:53–75

Krishnamurthy P, Cox KG (1980) A potassium-rich alkalic suite from the Deccan Traps, Rajpipla, India. Contrib Miner Petrol 73:179–189

Krishnamurthy P, Gopalan K, MacDougall JD (2000) Olivine compositions in picrite basalt and the Deccan Volcanic cycle. J Petrol 41:1057–1069

Krishnamurthy P, Pande K, Gopalan K, Macdougall JD (1988) Upper mantle xenoliths in alkali basalts related to Deccan Trap volcanism. In Subbarao KV (ed) Deccan Flood basalts. Geol Soc India Mem 10:53-67

Lindsley DA (1983) Pyroxene thermometry. Am Miner 68:477–493

Lightfoot PC, Hawkesworth CJ, Devey CW, Rogers NW, Van Calsteren PWC (1990) Source and differentiation of Deccan Trap lavas: implications of geochemical and mineral chemical variations. J Petrol 31:1165–1200

Mahoney JJ (1988) Deccan Traps. In: Macdougall JD (ed) Flood basalts. Kluwer, Dordrecht, pp 151–194

Mahoney JJ, Macdougall JD, Lugmair GW, Gopalan K, Krishnamurthy P (1985) Origin of contemporaneous tholeiitic and K-rich alkalic lavas: a case study from the northern Deccan Plateau, India. Earth Planet Sci Lett 72:39–53

Mahoney JJ, Saunders AD, Storey M, Randriamanantenasoa A (2008) Geochemistry of the Volcan de l'Androy basalt-rhyolite complex, Madagascar Cretaceous igneous province. J Petrol 49:1069–1096. doi:10.1093/petrology/egn018

Mahoney JJ, Sheth HC, Chandrasekharam D, Peng ZX (2000) Geochemistry of flood basalts of the Toranmal section, northern Deccan Traps, India: implications for regional Deccan stratigraphy. J Petrol 41:1099–1120

Melluso L (1992) Caratteristiche petrologiche e geochimiche di vulcaniti continentali del Deccan, India. Ph.D. dissertation, Ferrara University, 140 pp

Melluso L, Barbieri M, Beccaluva L (2004) Petrogenesis and regional chemical correlations of the flood basalt sequence in the Central Deccan Traps, India. Proc Indian Acad Sci 114:587–603

Melluso L, Beccaluva L, Brotzu P, Gregnanin A, Gupta AK, Morbidelli L, Traversa G (1995) Constraints on the mantle sources of the Deccan Traps from the petrology and geochemistry of the basalts of Gujarat State (Western India). J Petrol 36:1393–1432

Melluso L, Morra V, Brotzu P, Mahoney JJ (2001) The Cretaceous igneous province of central-western Madagascar: evidence for heterogeneous mantle sources, crystal fractionation and crustal contamination. J Petrol 42:1249–1278

Melluso L, Morra V, Brotzu P, Tommasini S, Renna MR, Duncan RA, Franciosi L, d'Amelio F (2005) Geochronology and petrogenesis of the Cretaceous Antampombato-Ambatovy complex and associated dyke swarm, Madagascar. J Petrol 46:1963–1996

Melluso L, Mahoney JJ, Dallai L (2006a) Mantle sources and crustal input as recorded in High-Mg Deccan Trap basalts of Gujarat (India). Lithos 89:259–274

Melluso L, Morra V, Fedele L (2006b) An overview of phase chemistry and magmatic evolution in the Cretaceous flood basalt province of northern Madagascar. Period Miner 75:175–188

Melluso L, Sethna SF, Morra V, Khateeb A, Javeri P (1999) Petrology of the mafic dyke swarm of the Tapti river in the Nandurbar area (Deccan Volcanic Province, India). Geol Soc India Mem 43:735–755

Melluso L, Sethna SF, D'Antonio M, Javeri P, Bennio L (2002) Geochemistry and petrogenesis of sodic and potassic mafic alkaline rocks in the Deccan Volcanic province, Mumbai area (India). Miner Petrol 74:236–254

Melluso L, Cucciniello C, Petrone CM, Lustrino M, Morra V, Tiepolo M, Vasconcelos L (2008) Petrology of Karoo volcanic province magmatism in the southern part of the Libombos Monocline, Mozambique. J Afr Earth Sci 52:139–151. doi:10.1016/j.jafrearsci.2008.06.002

Murari R, Krishnamurthy P, Tikhonenko PI, Gopalan K (1993) Magnesian ilmenites in picrite basalts from Siberian and Deccan Traps – additional mineralogical evidence for primary melt compositions (?). Miner Mag 57:733–735

Najafi SJ, Cox KG, Sukheswala RN (1981) Geology and geochemistry of the basalt flows (Deccan Traps) of the Mahad-Mahableshwar section, India. Geol Soc India Mem 3:300–315

Nimis P (1999) Clinopyroxene geobarometry of magmatic rocks. Part 2. Structural geobarometers for basic to acid, tholeiitic and mildly alkaline magmatic systems. Contrib Miner Petrol 135:62–74

Pande K (2002) Age and duration of the Deccan Traps, India: a review of radiometric and paleomagnetic constraints. Proc Indian Acad Sci 111:115–123

Pattanayak SK, Shrivastava JP (1999) Petrography and major oxide geochemistry of basalts from the eastern Deccan Volcanic province India. Geol Soc India Mem 43:233–270

Peng ZX, Mahoney JJ, Hooper PR, Harris C, Beane JE (1994) A role for lower continental crust in flood basalt genesis? Isotopic and incompatible element study of the lower six formations of the Western Deccan Traps. Geochim Cosmochim Acta 58:267–288

Ray R, Shukla AD, Sheth HC, Ray JS, Duraiswami RA, Vanderkluysen L, Rautela CS, Mallik J (2008) Highly heterogeneous Precambrian basement under the central Deccan Traps, India: direct evidence from xenoliths in dykes. Gondwana Res 13:375–385

Roeder PL, Emslie RF (1970) Olivine-liquid equilibrium. Contrib Miner Petrol 29:275–289

Ross P-S, Ukstins Peate I, McClintock MK, Xu YG, Skilling IP, White JDL, Houghton BF (2005) Mafic volcaniclastic deposits in flood basalt provinces: a review. J Volcanol Geotherm Res 145:281–314

Sano T, Fuji T, Deshmukh SS, Fukuoka T, Aramaki S (2001) Differentiation processes of Deccan Trap basalts: contribution from geochemistry and experimental petrology. J Petrol 42:2175–2195

Self S, Blake S, Sharma K, Widdowson M, Sephton S (2008) Sulfur and chlorine in Late Cretaceous Deccan magmas and eruptive gas release. Science 319:1654–1657

Sen G (1986) Mineralogy and petrogenesis of the Deccan Trap lava flows around Mahabaleshwar, India. J Petrol 27:627–663

Sen G (2001) Generation of Deccan Trap magmas. Proc Indian Acad Sci (Earth Planet Sci) 110:409–431

Sethna SF, Sethna BS (1988) Mineralogy and petrogenesis of Deccan Trap basalts from Mahabaleshwar, Igatpuri, Sagar and Nagpur areas, India. In Subbarao KV (ed) Deccan Flood basalts. Geol Soc India Mem 10:69–90

Sheth HC (2005) From Deccan to Reunion: no trace of a mantle plume. In: Foulger GR, Natland JH, Presnall DC, Anderson DL (eds) Plates plumes and paradigms. Geological society of American special paper 388, Boulder, pp 477–501

Sheth HC, Melluso L (2008) The Mount Pavagadh volcanic suite, Deccan Traps: geochemical stratigraphy and magmatic evolution. J Asian Earth Sci 32:5–21

Subbarao KV, Bodas MS, Hooper PR, Walsh JN (1988) Petrogenesis of Jawhar and Igatpuri Formations, Western Deccan basalt Province. In Subbarao KV (ed) Deccan Flood basalts. Geol Soc India Mem 10:253–280

Vanderkluysen L, Mahoney JJ, Hooper PR, Sheth HC, Ray R (2006) Location and geometry of the Deccan Traps feeder system inferred from dyke geochemistry. EOS Trans Am Geophys Union 87:V13B–0681

Chapter 8
Recycling of Flow-Top Breccia Crusts into Molten Interiors of Flood Basalt Lava Flows: Field and Geochemical Evidence from the Deccan Traps

Hetu C. Sheth, Jyotiranjan S. Ray, P. Senthil Kumar, Raymond A. Duraiswami, Rudra Narayan Chatterjee, and Trupti Gurav

Abstract Thick flood basalt lava flows cool conductively inward from their tops and bases, usually developing columnar jointing. Although relatively rapid cooling in such flows due to meteoric water circulation has been previously demonstrated, mixing of the surface crust with the interior – as observed in active lava lakes – has not been shown. Here we report large radial columnar jointing structures (rosettes) with cores of highly brecciated, weathered and amygdaloidal material within Deccan flood basalt lava flows. The morphology of such breccia-cored rosettes, petrographic observations, and geochemical data, particularly Nd–Sr isotopic ratios, all suggest that the features formed due to the sinking of the flow-top breccia crusts into these flows' molten interiors and the resultant warping of isotherms around these "cold anchors". Thus, cooling in some thick flood basalt lava flows may be accelerated by sinking of cooler upper crusts into hotter, molten interiors.

H.C. Sheth (✉), R.N. Chatterjee, and T. Gurav
Department of Earth Sciences, Indian Institute of Technology Bombay (IITB), Powai, Mumbai 400076, India
e-mail: hcsheth@iitb.ac.in

J.S. Ray
Physical Research Laboratory (PRL), Navrangpura, Ahmedabad 380009, India
e-mail: jsray@prl.res.in

P.S. Kumar
National Geophysical Research Institute (NGRI), Council of Scientific and Industrial Research (CSIR), Uppal Road, Hyderabad 500007, India
e-mail: senthilngri@yahoo.com

R.A. Duraiswami
Department of Geology, University of Pune, Pune 411007, India
e-mail: raymond_d@rediffmail.com

J. Ray et al. (eds.), *Topics in Igneous Petrology*, DOI 10.1007/978-90-481-9600-5_8,

8.1 Introduction

Active lava lakes, because of continuous supply of hot magma from below and rapid cooling at the surface, are characterized by vigorous convection and multiple events of crust formation and overturn (Duffield 1972; Harris et al. 1999; Stovall et al. 2009). Cooling of thick flood basalt lava flows, on the other hand, has generally been considered to be slow and conductive, with little role for convection, or for recycling of solidified upper crusts of the flows into their molten interiors. Some studies suggest enhanced cooling due to circulation of meteoric water (e.g., Long and Wood 1986) but it is usually assumed that the crust and interior of a cooling flood basalt lava flow remain independent and do not undergo wholesale remixing.

Such a perception, however, is primarily because of paucity of evidence. Indeed, historic eruptions have rarely, if at all, produced flows as thick and laterally extensive as some flood basalt flows. Evidence for recycling processes is therefore not easy to obtain, and the processes that occur in cooling flood basalt flows must be inferred from the morphology and internal structures of these flows. Here we present field, petrographic and geochemical evidence, from several exposures in the Deccan Traps of India, that suggests that flow-top breccia crusts of thick flood basalt flows sometimes get recycled into their molten interiors, and strongly affect the temperature distribution within, leading to the formation of radial columnar jointing structures (rosettes) around them. We call such structures "breccia-cored rosettes" to distinguish them from simple rosettes (radial columnar jointing, without obvious cores) described from several areas (e.g., Spry 1962; Scheidegger 1978).

8.2 The Breccia-Cored Rosettes: Field Observations

Bondre et al. (2004a, b) and Sheth (2006) provide general overviews of the morphology and emplacement of Deccan flood basalt lava flows that currently occupy ~500,000 km^2 in western and central India (Fig. 8.1a). The breccia-cored rosettes we describe are restricted to the "simple" flows from this province, which tend to be single cooling units of greater spatial dimensions than typical pahoehoe lobes constituting the compound pahoehoe flows (Bondre et al. 2004a, b). Several simple flows are characterized as rubbly pahoehoe (Duraiswami et al. 2008), i.e., flows with extensively brecciated flow-top crusts, although flows with preserved tops are also found. Two of the three field exposures we discuss in this study are in the Koyna–Satara region of the Deccan, east of the Western Ghats escarpment, where some of the younger stratigraphic formations (Poladpur, Ambenali and Mahabaleshwar) are exposed at increasing elevations above sea level. The elevations of our exposures mean that the flows belong to the ~550-m-thick Ambenali Formation, one of the thickest Deccan formations (e.g., Lightfoot et al. 1990).

The first exposure is in the abandoned Koyna quarry, just south of the Koyna Dam at Koynanagar (Fig. 8.1b). GPS coordinates of the quarry are N 17° 24′ 26.6″ and E 73° 44′ 39″, and the altitude 652 m, which means that the flow belongs to uppermost one-fifth part of the Ambenali Formation (the Kumbharli Ghat stratigraphic section of

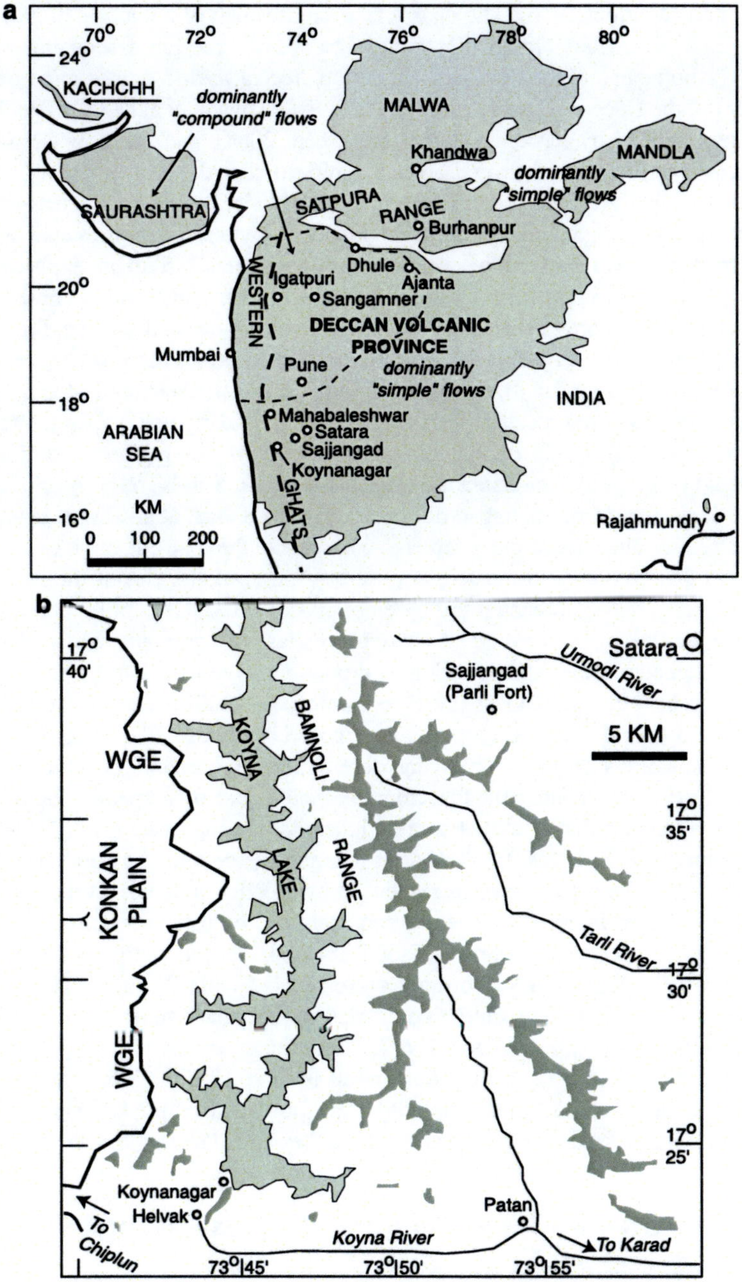

Fig. 8.1 (**a**) Sketch-map of the Deccan Traps (*shaded*) showing the areas dominated by compound pahoehoe and simple flows, and some key localities. Based on Bondre et al. (2004a, b). (**b**) Map of the Satara-Sajjangad-Koynanagar area, showing the Koyna reservoir and the Bamnoli Range with its basalt-capping ferricrete cap (medium gray outcrops forming a dendritic pattern). WGE is the Western Ghats escarpment (Simplified from Ollier and Sheth 2008 and references therein)

Devey and Lightfoot 1986). The quarry face exposes part of a thick (~20 m exposed) rubbly pahoehoe flow. No smaller flow lobes are observed in association with this flow. The upper part of the flow shows flow-top breccia, as do many simple flows from this part of the Deccan (Bondre et al. 2004a, b; Duraiswami et al. 2008). The flow-top breccia is capped by ferricrete (Ollier and Sheth 2008), and the flow's base is not exposed. This flow shows several breccia-cored rosettes, the best developed of which is shown in Fig. 8.2. The breccia core is vertically elongated and consists of highly weathered and amygdaloidal material (welded with secondary zeolites and calcite). It is surrounded by a garland of well-developed columns, ~5 m in exposed length (Fig. 8.2a–c). The basalt forming the columns is non-vesicular and is chilled against and welded to the core as seen in the quarry face (hazardous to approach, Fig. 8.2a) as well as fallen blocks (Fig. 8.3a, b). Rosettes at other locations in the quarry seem to be similarly cored (Fig. 8.3c, d), though none of these is as spectacular as the first. Apart from the breccia-cored rosettes, the flow is rather irregularly and poorly jointed.

A second well-developed breccia-cored rosette in this region is exposed near Sajjangad (Fig. 8.1b). The exposure (Fig. 8.4a) is in a 5-m-high, N–S-oriented road cut, ~3 km down from Sajjangad hill on the road to Satara, at an approximate elevation of 775 m. The core of the rosette is >6 m wide in the road cut and has an exposed height of 3 m. Its top is truncated, but it crops out on both sides of the road, and is evidently large. It is characterized by close-spaced cuboidal jointing (Fig. 8.4b, c). The angular fragments show small white plagioclase feldspar laths a few millimetres long, contained in a much finer dark gray groundmass with a bluish sheen, suggesting a considerable amount of glass (cf. Duraiswami et al. 2008). The rosette consists of well-developed columns with an average exposed height of ~1.5 m above road level. The columns are subvertical below the core but nearly horizontal on the side of the core, following the general rule that columns should develop perpendicular to cooling surfaces or contacts and parallel to the highest thermal gradient (e.g., Budkewitsch and Robin 1994; Lyle 2000). The basalt forming the columns is non-vesicular, and chilled against the breccia core as evidenced by its finer grain size at the contact.

The third breccia-cored rosette outcrop examined by us is in a simple basalt flow exposed in a road cut ~18 km NNE of Burhanpur, on the National Highway 6 to Khandwa (Fig. 8.1a). GPS coordinates are N 21° 27′ 01.7″ and E 76° 16′ 44.3″. The breccia core is highly altered basalt rich in secondary zeolites and calcite, and roughly ellipsoidal with a major axis of 2 m and minor axis of 1 m (Fig. 8.4d, e). The contact between the core and the host basalt is sharp, and the basalt lava is chilled against the inclusion. The host simple flow has subvertical columns everywhere in the exposure, except around the inclusion, where they rotate toward the inclusion.

8.3 The Breccia-Cored Rosettes: Previous Interpretations and Suggested Formation Mechanisms

The rosettes in the Koyna quarry flow have been noted for a long time and interpreted in different ways. Because of the brecciated material at the core of the main Koyna quarry rosette (Fig. 8.2), Agashe and Gupte (1971) called it a "volcanic vent".

Fig. 8.2 (**a–c**) Outcrop photographs of the Koyna quarry main breccia-cored rosette. FTBC denotes flow-top breccia crust. Note the gradual widening of columns with distance away from the inclusion. Note persons for scale in (**a**, **b**), and GPS, 11 cm long, in (**c**). Locations of samples taken are shown

Chandrasekharam and Parthasarathy (1978) mentioned that "Around Koyna huge blocks of zeolitized tuff surrounded by radiating columns of massive basalts are commonly encountered". Later Misra (2002) has considered the same structure to be a filled lava tube. We discuss below these previous interpretations and suggested formation mechanisms and why we consider them untenable.

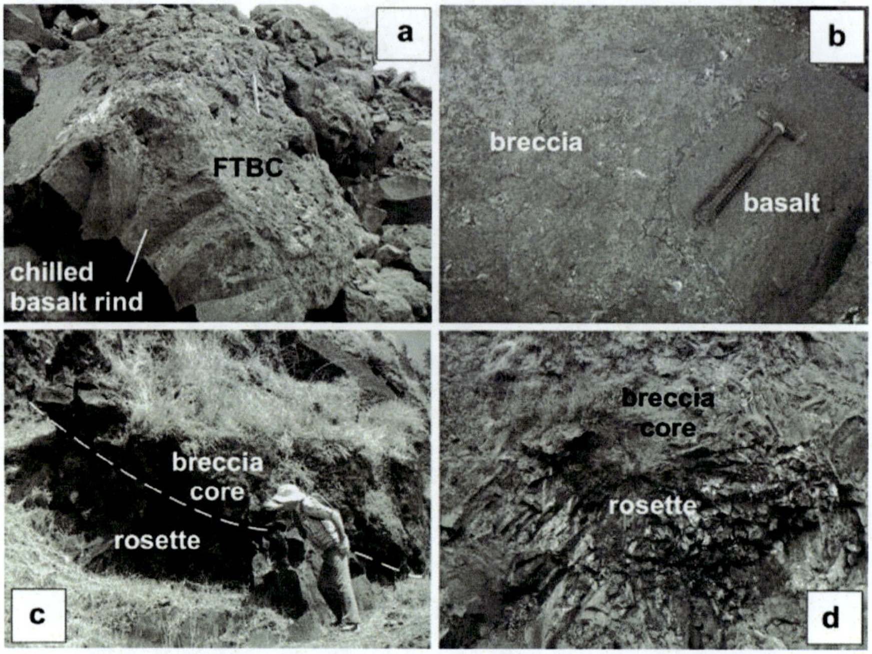

Fig. 8.3 (**a**) The contact of flow-top breccia crust (FTBC) and chilled host basalt, as observed in a toppled block on the quarry floor. Pen for scale is 14 cm long. (**b**) A second toppled block of breccia with chilled basalt. (**c**) Another example of breccia-cored rosette on the quarry floor, between the quarry face with the main rosette and the Koyna dam. After this photograph was taken, this particular example is no longer accessible due to the construction of a high fence between the quarry face and the dam. (**d**) Partially developed rosette visible in the quarry face at about middle height. It is not accessible, and the height of view is 7–8 m. Its breccia core has no visible connection to the flow-top breccia crust of the flow

8.3.1 Water Circulation into Lava Flow Producing Unequally Thick Flow-Top Breccia Crust and Rosettes

Ingress of meteoric water along cooling joints has been considered as one explanation for the formation of rosette jointing (e.g., Long and Wood 1986). Lyle (2000) discusses how water can percolate along the irregular contact between the flow-top breccia and flow interior, leading to the formation of radial jointing, whereas DeGraff and Aydin (1987) have described lobes of radial jointing that are often associated with fractures penetrating through the flow crust. Philpotts and Dickson (2002) illustrate how thick flood basalt flows can have a highly undulating and cuspate boundary between the upper crust and interior as a result of seepage of surface water into the flow along fractures, so that the crust can *grow* to a considerable depth into the flow interior. This can be particularly true for rubbly pahoehoe flows (Gilbaud et al. 2005; Duraiswami et al. 2008). In such cases, the protuberances would serve to locally warp the solidification fronts and cause chevrons, fans and rosettes of columns to develop (e.g., see Fig. 11 in Lyle 2000).

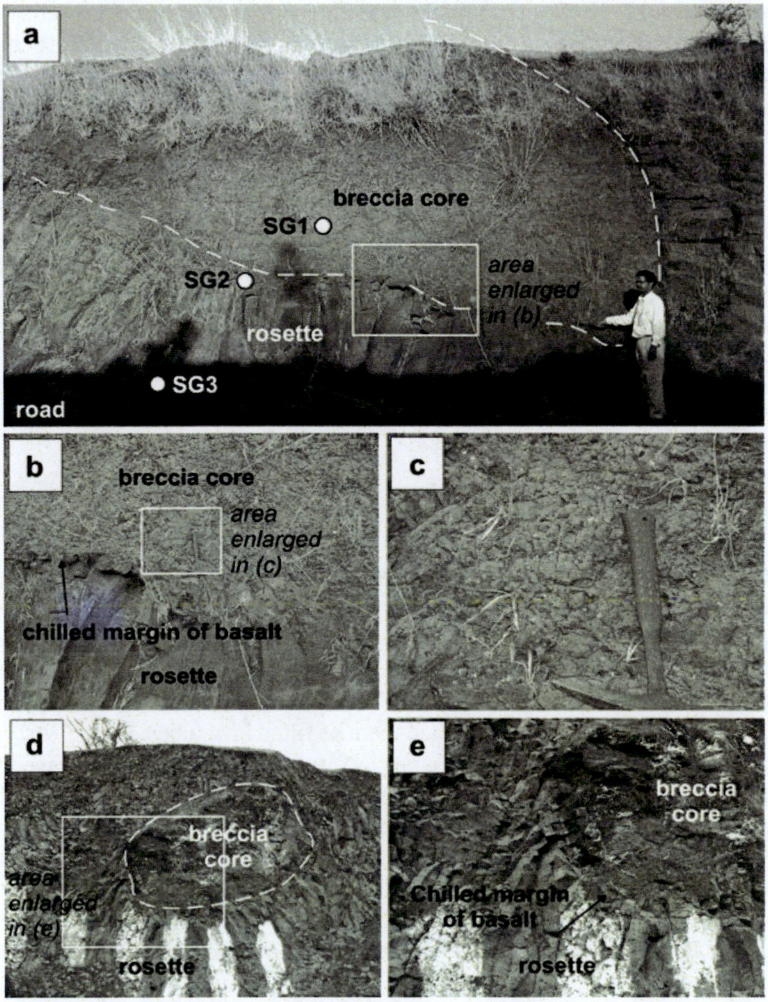

Fig. 8.4 Outcrop photographs of the Sajjangad and Burhanpur breccia-cored rosettes. In (**a**), note the columns being vertical below the breccia core and becoming horizontal on its side. Note also the *inward-turning* boundary of the breccia core above the person (*white dashed line*), showing that, though the top of the core is truncated and the exposure incomplete, the core is definitely suspended into the flow, and cannot, for example, represent a locally thicker part of the flow-top breccia crust in situ. (**b, c**) Close-up of the chilled contact between the breccia core and the basalt columns. In (**c**), note bluish sheen of the breccia material. (**d, e**) Outcrop photographs of the Burhanpur breccia-cored rosette, where the core is suspended into the flow and severed from any original flow-top breccia crust. Boundary of the breccia core is marked by *white dashed line*. *Thicker white patches* are lime used during road construction work

However, the Koyna quarry breccia-cored rosette in Fig. 8.2 could not simply have formed around an unusually thick part of the flow-top breccia crust, because (1) several such cored rosettes occur in this flow at various heights, and unlike the case with the main rosette, their cores do not seem to be physically connected with the upper crust. (2) No fractures are seen to extend from these cored rosettes to the upper crust. (3) The core of the Sajjangad rosette, its truncated upper part notwithstanding, is suspended into the flow, as its inward-curving upper boundary shows (Fig. 8.4a). (4) The core of the Burhanpur rosette also has the appearance of being loose and suspended into the flow, as the core is surrounded on all sides, including the upper side, by the host lava flow (Fig. 8.4d, e). (5) Simple chevrons or fans developed due to water circulation into a cooling flow will have the columns meeting at a point (or a narrow zone). Here, the rosettes have sizeable cores of brecciated, highly vesicular (now amygdaloidal) material in their middle. Therefore, taken together, these observations suggest that, though meteoric water ingress from the flow-top breccia into the molten interior may indeed have occurred, the breccia-cored rosettes cannot have formed simply by such water ingress, or cannot be related to cuspate lower boundaries of upper crusts.

8.3.2 Rosettes as Lava Tubes

As regards the simple rosettes (apparently lacking inclusions or cores) described in the literature, common interpretation has been that they represent filled lava tubes (e.g., Waters 1960; Spry 1962; Bear and Cas 2007). Waters (1960) observed several large (tens of metres across) radial jointing structures in the Columbia River flood basalts, termed them "war bonnets" (referring to the radial feathers on a native American's headdress), and interpreted them as filled lava tubes. Spry (1962) interpreted the radial arrangement of columns as resulting from isotherms shaped like a series of concentric cylinders, resulting from imperfect focused flow of lava through tubes or tunnels.

Misra (2002) has considered the Koyna quarry's main rosette in Fig. 8.2 to be a filled lava tube, one of many he claims in the Deccan. Besides problems with his other examples (see Dole et al. 2002; Duraiswami et al. 2005; Sheth 2006), this structure cannot be a lava tube for several reasons: (1) For a lava tube to have a vertically elongated cross-section as seen in the Koyna quarry rosette, lava flowing within the tube would have had to thermally erode its floor (Greeley et al. 1998). There is no evidence for this. (2) In none of the breccia cores of these Deccan rosettes are concentric bands of vesicles apparent, a feature commonly observed in filled tubes from younger provinces (e.g., Keszthelyi and Self 1998). (3) Most important, in a lava tube that progressively shrinks towards the centre before shutting off, the central portion is the last to cool. Examination of all Koyna quarry breccia-cored rosettes, as well as those of the other two localities described here, leaves no doubt that their central portions were cold compared to the host lava, which got chilled at the contact with the core, and roughly circular solidification fronts that propagated outwards into the surrounding lava created radial jointing. Observe how (Fig. 8.2a, c), as in a bunch of radially arranged pencils, there are several columns with small cross-sectional areas close to the core, indicating rapid nucleation on faster cooling (see Grossenbacher and McDuffie 1995). The columns become fewer in number and widen outward from the core by laterally merging with each other, owing to slower cooling. Eventually the columns merge with the rest of the lava flow without any break.

Waters (1960) considered the war bonnets, which he interpreted as tubes, to have progressively frozen shut from the boundaries inwards, but if this were the case the columns would have had narrower cross-sections at the boundaries of the structures and wider cross-sections at the centre (with the most gradual cooling). His sketch shows that the columns are *wider* in radially outward directions. Greeley et al. (1998) noted the absence of concentric arrangement of vesicle zones in the war bonnet structures, the textural uniformity of the basalt forming the columns with the rest of the host flow, as well as the grading of the columns into the host rock without any break. They concluded that these structures cannot be filled lava tubes. We agree with that view and, based on our own field observations of the Deccan rosettes, conclude that the commoner rosettes, and certainly the Deccan breccia-cored rosettes, do not represent lava tubes.

8.4 Inferences from Field Observations, Present Methodology, and Results

At all three breccia-cored rosette outcrops we have examined, field observations suggest the introduction of sizeable masses of relatively cold, brecciated, highly vesicular (now amygdaloidal) material into the molten interior of a cooling lava flow. The material is definitely basaltic, though highly weathered, and cemented by zeolites and calcite. Where permitted by exposure, as in the Koyna quarry, we identify this breccia with the flow-top breccia of the flow. The teardrop shape in cross-section, of the breccia core of the Koyna quarry main rosette (Fig. 8.2a, b), strongly suggests that a portion of the solidified, brecciated upper crust foundered into the interior at some stage during the cooling of the flow. This sinking crustal mass could have been sheet-like in three dimensions. This suggests that the cores of the rosettes formed well within the flow interior have been derived by sinking of the flow-top breccia crust forming the roof parts.

8.4.1 Samples

To test our outcrop-based interpretations and to get further clues to the origin of these intriguing structures from mineralogical and textural information, as well as geochemical "fingerprinting", we sampled breccia cores and enclosing lava at the Koyna quarry and Sajjangad. At Koyna quarry, we sampled the breccia core of the main rosette at its lower boundary (sample KY1) and a large column under it. Sample KY2 comes from the top of the column, just below the contact with the breccia core, and sample KY3 comes from the same column, 2 m below KY2 (Fig. 8.2b). We also took a breccia sample (KY4) from one of the many large, toppled blocks on the quarry floor; it appears that these blocks come not from the breccia core preserved in the rosette itself, but the flow-top breccia crust preserved in the upper part of the flow. Thus, we have two samples (KY1 and KY4) that represent the flow-top breccia crust.

At the Sajjangad outcrop, we sampled the breccia core ~0.5 m above its lower boundary (sample SG1). We took two samples of an enclosing lava column, one (SG2) just at the contact with the breccia core, and the other (SG3) 1.5 m below (Fig. 8.4a). We carried out petrographic studies of the breccia and host basalt column samples, analyzed these samples for the major elements, several key trace elements, as well as Nd–Sr isotopic ratios, and determined their LOI (loss on ignition) values.

8.4.2 Petrography

Figure 8.5 shows the essential petrographic features of the samples. Glass is common and widespread. Samples KY1 (Fig. 8.5a) and KY4 show clusters of plagioclase phenocrysts, groundmass glass, as well as a considerable amount of secondary minerals. Samples KY2 and KY3 (Fig. 8.5b–d) show phenocrysts (~1 mm long) and microphenocrysts of plagioclase, showing well-developed twinning, as well as smaller amounts of clinopyroxene and olivine. All these are generally euhedral or subhedral in shape. The groundmass is fine-grained, with minute grains of opaque oxides in a background of glass, dark between crossed nicols.

Sample SG1 shows clusters of closely packed, euhedral plagioclase and olivine microphenocrysts surrounded by a considerable amount of glass (Fig. 8.5e). Samples SG2 and SG3 show euhedral, well-twinned plagioclase phenocrysts up to several millimetres across (Fig. 8.5f–h). The interesting feature of SG2 is that the plagioclase phenocrysts poikilitically enclose well-developed olivine crystals (Fig. 8.5f, g), indicating that olivine crystallized earlier than plagioclase. The large olivine grain in Fig. 8.5f is euhedral, with both its characteristic randomly oriented fractures and one set of parallel cleavage planes, and is partly enclosed within the large plagioclase grain. An interesting aspect of this sample (SG2) is that the plagioclase grain appears to have suffered partial resorption along its right edge (Fig. 8.5f). Such corrosion may be produced by the delivery of fresh, hotter melt around the crystal.

8.4.3 Geochemistry: Major and Trace Elements

The seven rock samples were analyzed for the major and several trace elements, by inductively coupled plasma atomic emission spectrometry at IIT Bombay and by X-ray fluorescence spectrometry at PRL Ahmedabad. Loss on ignition (LOI)

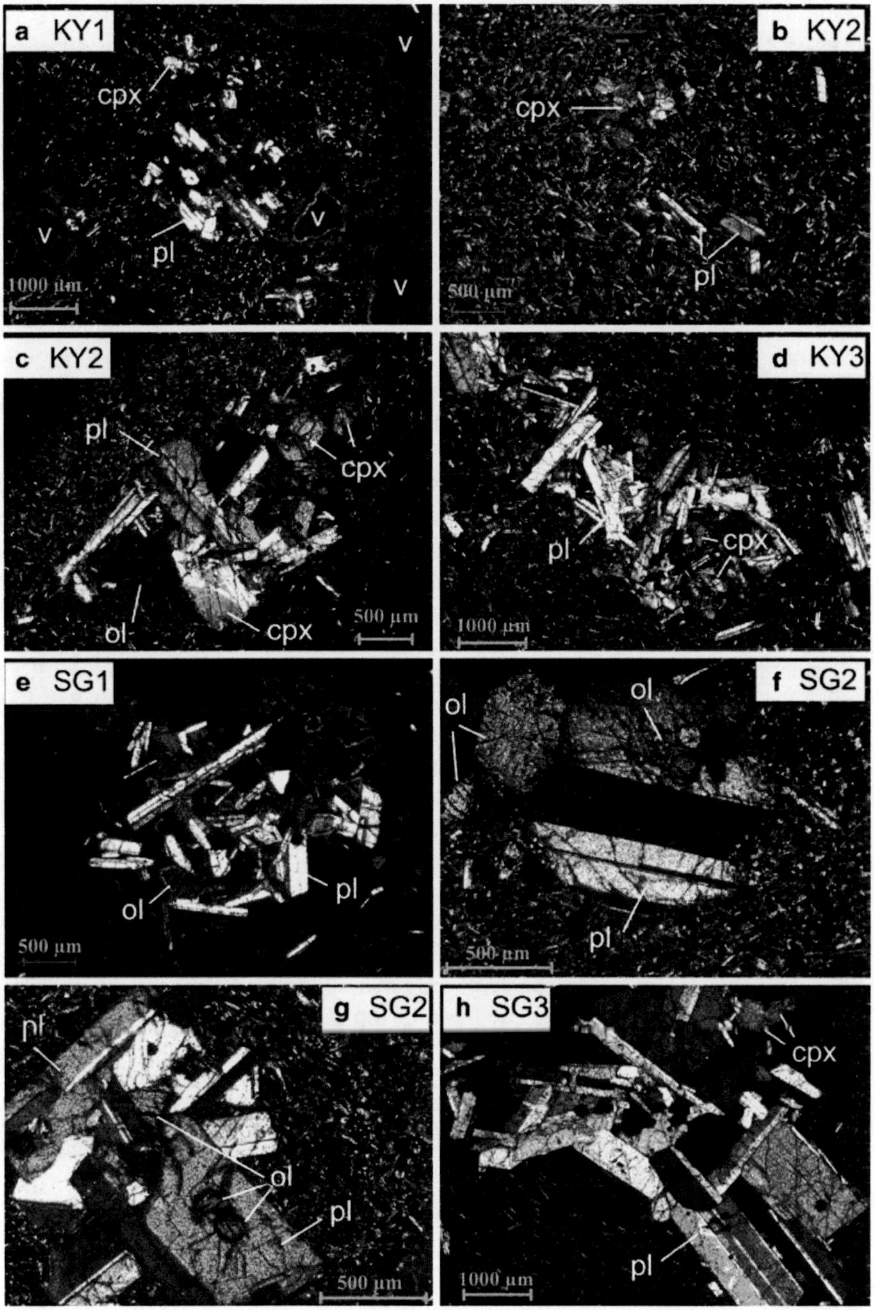

Fig. 8.5 The Koyna quarry and Sajjangad flow-top breccia, breccia core and host basalt column samples observed in thin section. All views are between crossed nicols. Abbreviations used are: v, vesicle or secondary mineral filling a vesicle; pl, plagioclase; ol, olivine; cpx, clinopyroxene. The opaque oxide grains are too minute to label

values of the samples were also determined by drying the sample powders at 100°C overnight and igniting them at 950°C. The data are presented in Table 8.1. Also given are the rock names and Mg number values obtained after adjusting the data on an LOI-free basis with the program SINCLAS (Verma et al. 2002). All samples are quartz-normative (5.70–13.5 wt% normative quartz) subalkalic basalt or basaltic andesite, as computed by SINCLAS.

Figure 8.6 shows binary plots of some key trace elements and element ratios that help to identify the formational affinities of the studied samples, and which have been used in many previous studies of the Deccan geochemical stratigraphy (e.g., Cox and Hawkesworth 1985; Devey and Lightfoot 1986; Mitchell and Widdowson 1991; Sheth et al. 2004; Jay and Widdowson 2008). Except for Ba, which becomes mobilized at moderate levels of alteration, the other elements used on the plots are all alteration-resistant. The Zr/Nb ratio remains unaffected by crystal fractionation or even significant alteration (Mitchell and Widdowson 1991), which makes it a powerful "tracer" in assigning even weathered samples to their correct stratigraphic formations.

The three plots of Fig. 8.6 do not reveal any systematic difference between the Koyna quarry and the Sajjangad samples, or, more important, between the breccia core and host basalt column samples from either locality. All samples can be confidently ascribed to the Ambenali Formation (as inferred already from the position of these exposures in the established stratigraphy). A value of 100 ppm Ba constitutes the upper limit for the Ambenali Formation and lower limit for the Mahabaleshwar Formation (Mitchell and Widdowson 1991). Data for samples KY1 and KY4 plot outside the Ambenali Formation field and the samples appear to resemble the Mahabaleshwar Formation because of their elevated Ba and Ba/Y. We believe that such high Ba values are due to their severe alteration and secondary mineralization (either calcite or zeolites would explain the high barium levels). Note that both samples nevertheless retain the Ambenali-like Zr/Nb ratios, and have nearly Ambenali-like Sr contents. Data for the Sajjangad breccia core sample (SG1) plot within the Ambenali field in all three plots.

The striking similarity in trace element concentrations and element ratios between the breccia core samples and the basalt lava columns at both field sites is significant. Such similarity indicates that the cores are parts of the same flow and are not accidental inclusions, whether picked up from the top of an underlying flow, or of rubble falling into a new, cooling, valley-confined lava flow from the valley walls. As regards such possible valley-confined flows, arguments for palaeotopography control on Deccan lava flows have been presented by Kale et al. (1992), and have been reiterated more recently by Jay and Widdowson (2009) who on different grounds suggest up to 80 m of topographic relief during eruptions in this very region.

8.4.4 Geochemistry: Nd and Sr Isotopic Ratios

Bulk rock Nd and Sr isotopic ratio measurements were performed at PRL Ahmedabad. Powders made from rock chips picked free of secondary minerals

Table 8.1 Major and trace element and Nd–Sr isotopic data for the Koyna (KY) and Sajjangad (SG) breccia core and surrounding basalt column samples

Rock name	Basaltic andesite	Basaltic andesite	Subalkalic basalt	Subalkalic basalt	Subalkalic basalt	Subalkalic basalt	Basaltic andesite	Recom.	Meas.
	Breccia core	Lava column	Lava column	FTBC boulder	Breccia core	Lava column	Lava column		
Sample No.	**KY1**	**KY2**	**KY3**	**KY4**	**SG1**	**SG2**	**SG3**	**BHVO-2**	**BHVO-2**
SiO_2	50.95	48.87	47.48	51.15	47.80	45.15	49.95	49.95 ± 1.00	49.23
TiO_2	3.95	3.01	3.07	3.91	3.55	3.41	3.20	2.73 ± 0.16	2.70
Al_2O_3	9.41	12.50	13.21	13.26	14.34	13.01	12.93	13.38 ± 0.26	13.81
Fe_2O_{3T}	18.59	14.78	16.04	13.45	17.16	16.73	15.31	12.23 ± 0.48	12.07
MnO	0.18	0.23	0.21	0.12	0.19	0.21	0.19	0.17 ± 0.02	0.16
MgO	7.13	6.18	5.62	7.98	5.74	5.20	5.69	7.23 ± 0.28	7.58
CaO	9.18	10.63	12.01	8.38	9.69	11.39	9.97	11.45 ± 0.46	11.56
Na_2O	1.24	2.33	2.23	1.81	2.03	2.19	2.34	2.14 ± 0.18	2.36
K_2O	0.47	0.18	0.39	0.36	0.12	0.50	0.47	0.51 ± 0.04	0.52
P_2O_5	0.19	0.26	0.29	0.27	0.29	0.31	0.27	0.22 ± 0.02	0.21
Sum	101.29	98.97	100.55	100.69	100.91	98.10	100.31		
LOI	5.60	0.45	1.31	9.83	2.90	1.80	0.94		
Mg No.	49.1	51.3	45.0	58.1	43.9	42.1	48.3		
Cr_{ICP}	187	296	289	95	145	208	242	280 ± 38	294
Sr_{ICP}	286	235	230	541	143	223	240	389 ± 46	417
Y_{ICP}	21.2	37.2	35.4	32.8	67.9	42.9	41.2	26 ± 4	28
Zr_{ICP}	143	163	157	151	234	178	188	177 ± 22	174
Ba_{ICP}	201	56.8	85	438	72.5	54.5	124	130 ± 26	134
Rb_{XRF}	5.5	2.1	14.2	7.8	1.8	4.8	10.8	9.8 ± 2	8.6
Sr_{XRF}	183	217	216	235	207	227	219	389 ± 46	392
Y_{XRF}	28	26	27	26	43	35	32	26 ± 4	25
Zr_{XRF}	200	164	159	188	202	206	189	177 ± 22	176
Nb_{XRF}	18.5	14.8	14.9	17.7	15.7	16.4	14.8	18 ± 4	19
$(^{143}Nd/^{144}Nd)_p$	0.512854	0.512834	0.512838	0.512835	0.512833	0.512825	0.512839		

(continued)

Table 8.1 (continued)

Rock name	Basaltic andesite	Basaltic andesite	Subalkalic basalt	Subalkalic basalt	Subalkalic basalt	Subalkalic basalt	Basaltic andesite
	Breccia core	Lava column	Lava column	FTBC boulder	Breccia core	Lava column	Lava column
Sample No.	**KY1**	**KY2**	**KY3**	**KY4**	**SG1**	**SG2**	**SG3**
ε_{Ndp}	4.2	3.8	3.9	3.8	3.8	3.6	3.9
$(^{87}Sr/^{86}Sr)_p$	0.70433	0.70422	0.70439	0.70447	0.70440	0.70444	0.70436
$(^{87}Rb/^{86}Sr)_p$	0.087	0.028	0.190	0.096	0.025	0.061	0.143
$(^{87}Sr/^{86}Sr)_{66}$	0.70425	0.70419	0.70421	0.70438	0.70438	0.70438	0.70423

Notes: Major oxide determinations (reported in wt%) on two samples (KY3 and SG2) are by ICP-AES at IIT Bombay, and on the remaining five samples by XRF at PRL Ahmedabad. Trace elements (in ppm) were determined by ICP, at IIT Bombay, and by XRF, at PRL. Concentrations of the same element measured in both laboratories (e.g., Sr) are not necessarily expected to agree because the measurements are on chips picked free of secondary minerals independently in both laboratories and, noting the extremely pervasive alteration and secondary mineralization in especially the breccia samples, the picked chips might still contain minor amounts of these. Nevertheless, some measurements do agree closely, e.g., the Zr concentrations at both laboratories for samples KY2 and KY3. The high accuracy of each method is indicated by the recommended and measured values on the USGS reference material BHVO-2 (analytical errors shown are 2σ, and the reference values are of Wilson 1997). BHVO-2 reference values for trace elements are those of Raczek et al. (2001). Mg number = [atomic $Mg/(Mg + Fe^{2+})$] × 100.

The $(^{143}Nd/^{144}Nd)_p$ and $(^{87}Sr/^{86}Sr)_p$ values are present-day ratios. Nd and Sr isotope ratios were corrected for instrumental fractionation using $^{146}Nd/^{144}Nd = 0.7219$ and $^{86}Sr/^{88}Sr = 0.1194$, respectively. For calculating ε_{Ndp} the chondritic average $^{143}Nd/^{144}Nd$ value used is 0.512638. Within-run 2σ errors for the Nd and Sr data are 0.000008 and 0.00001, respectively, consistently on all samples, corresponding to a 2σ uncertainty of 0.2 units on ε_{Nd}. BHVO-2 analyzed for Nd and Sr isotopic ratios gave values of 0.512975 ± 0.000009 and 0.70346 ± 0.00002, respectively, comparable to the values of Weis et al. (2005): $^{143}Nd/^{144}Nd = 0.512983 \pm 0.000010$ (relative to La Jolla Nd value of 0.511853 ± 0.000016), and $^{87}Sr/^{86}Sr = 0.703481 \pm 0.000020$ (relative to NBS987 Sr value of 0.71025 ± 0.00001; errors are $2\sigma_m$). The average isotopic ratios for JNdi Nd and NBS987 Sr analyzed over a period of 3 years are $^{143}Nd/^{144}Nd = 0.512105 \pm 0.000008$ (−10.4 ± 0.2 ε_{Nd} units) and $^{87}Sr/^{86}Sr = 0.71023 \pm 0.00001$ (the errors are $2\sigma_m$). The value of $^{143}Nd/^{144}Nd = 0.512105$ for JNdi corresponds to a value of 0.511848 for the widely used La Jolla Nd standard (Tanaka et al. 2000). The Sr isotopes have been age-corrected for 66 Ma using $^{87}Rb/^{86}Sr$ ratios calculated from the whole-rock XRF Rb and Sr abundances. Procedural blanks for Sr and Nd are <500 pg each.

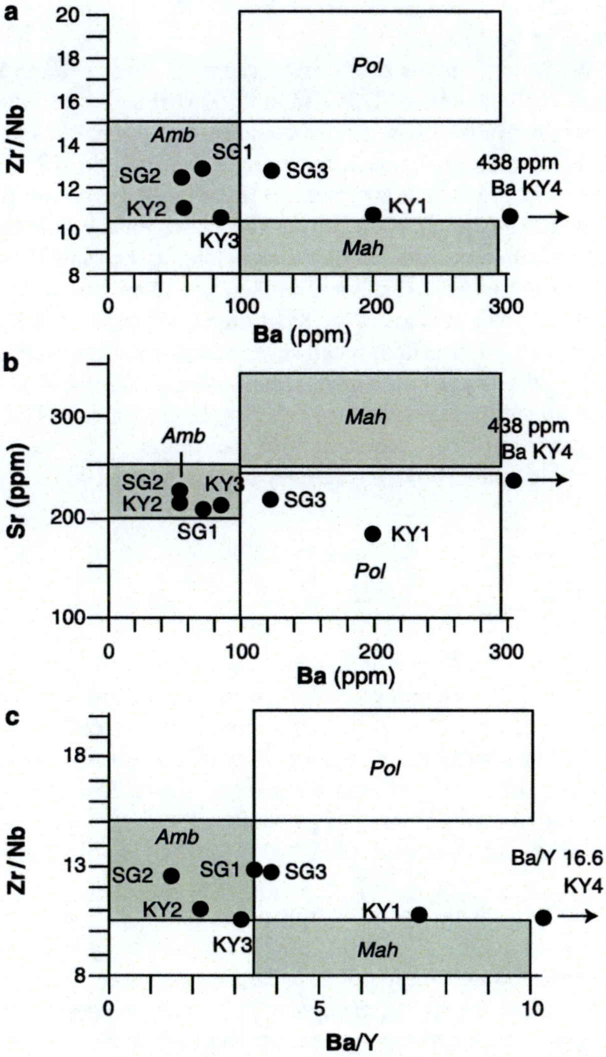

Fig. 8.6 Plots of key elements and element ratios for the Deccan flood basalts, useful in identifying three major, younger stratigraphic formations that outcrop in the study region. Boundaries between the formations are those given by Devey and Lightfoot (1986) and the plots are as used by Jay and Widdowson (2008). Formation names are abbreviated as follows: Pol, Poladpur; Amb, Ambenali; Mah, Mahabaleshwar

were dissolved using a standard HF–HNO_3–HCl dissolution procedure for silicate rocks. After Sr was separated by conventional cation-exchange column chemistry, Nd was separated from the other rare earth elements using Ln-specific resin from Eichrom with dilute HCl as the elutant. Isotopic ratio measurements were carried out in static multi-collection mode on an ISOPROBE-T mass spectrometer. The data are reported in Table 8.1.

Measured Nd isotopic ratios of all seven samples, covering both localities, are identical within analytical error. This similarity is striking, especially considering the level of alteration in the breccia core samples. Age-corrections to the ratios for a ~66 Ma age being minor and equal, there are no Nd isotopic differences between the breccia cores and host basalt at either locality, as well as *between* the two localities. The identical Nd isotopic ratios of the breccia cores and host basalt columns provide the significant confirmation that they belong to the same flow and, at each locality, the flow-top breccia crust was broken up, mobilized and carried into a largely molten flow interior. This is the most important result of this study.

The Sr isotopic ratios measured on all seven samples are also very close, and the computed initial Sr isotopic ratios have a narrow range from 0.70419 to 0.70438, the latter value shared by three samples covering both localities. The Nd–Sr isotopic ratios are Ambenali-like, consistent with indications of the trace element data and previous stratigraphic work (e.g., Devey and Lightfoot 1986).

8.5 Discussion

Lava flows similar to the Koyna quarry flow show evidence of having been thick, extensive cooling units. Vigorous convection similar to that in active lava lakes (e.g., Duffield 1972; Stovall et al. 2009) almost certainly did not occur in these flows; however, our observations suggest at least a single episode of sinking of colder, denser crust into the interior in the manner of a "cold anchor". A somewhat analogous process involving sinking of gravitationally unstable crystal mush is suggested by Philpotts and Dickson (2000, 2002) for the Holyoke flow in the northeastern USA.

In thick sheet flows with preserved upper crusts, it is difficult to imagine spontaneous initiation of sinking of the crust. Although solid basalt is ~10% denser than liquid basalt (Philpotts and Ague 2009, p. 22), the upper crust may be buoyed up by its vesicular nature. However, there may be a small temperature–density–time window during which part of the upper crust can detach and sink into the molten interior. Sinking of crustal fragments might be easier in rubbly pahoehoe flows such as these with fragmented and brecciated crusts (Duraiswami et al. 2008). The process might be aided by injection of new pulses of basalt lava into the molten flow interior (as must occur during the inflation of such flows), such pulses being less dense on account of being both hotter and less evolved (lower Fe/Mg) than the resident lava. Possible evidence for this is seen in the sample SG2, where relatively large, well-formed plagioclase crystals show corrosion and partial redissolution (Fig. 8.5f).

Whereas small pieces of flow-top breccia crust might get resorbed within the hot molten interior, larger blocks would strongly warp isotherms by inducing radial cooling. The breccia cores were stationary enough within a largely or completely molten interior (reflected in the chilled basalt rinds around them) for the well-developed columnar rosettes to form. Possibly, the increased viscosity of the host lava getting chilled against them "captured" them in place. Physical modelling of the mechanism(s) by which the flow-top breccia crusts might have sunk into the molten interiors is an interesting topic for future work.

A further interesting implication of the combined trace element and Nd–Sr isotopic data, the elevations of the Koyna and Sajjangad exposures (~650 and ~775 m, respectively), and the rosettes themselves, is that *both exposures, some 32 km apart in a NE-SW direction, may represent the same lava flow*. In the region southwest of Satara, the area of this study, the lava pile has a regional dip of 0.5° exactly due south (Subbarao and Hooper 1988). If so, a simple apparent dip calculation shows that Sajjangad, 32 km NE of the Koyna quarry, should be 140 m higher. Noting that we do not see either the top or the base of the particular flow, at either locality, and that the flow is at least 20 m thick as seen in the Koyna quarry, the match of the expected and observed heights is striking. We consider it as confirming the outcropping of the same flow in both localities. There is evidently something special about this Koyna quarry-Sajjangad flow, a fairly extensive flow that has recycled its upper crust everywhere along its lateral extent.

According to V. Bhosale (Groundwater Survey and Development Agency, Govt. of Maharashtra, 2007, personal communication) cored rosettes are common in many sections through the simple flows of the southern Deccan Traps. In fact, the other example of a breccia-cored rosette illustrated here (Fig. 8.4d, e) comes from the Satpura region of the Deccan that abounds in simple flows (De 1996; Sengupta and Ray 2006).

Finally, rosette jointing is not uncommon in basaltic provinces (e.g., Spry 1962; Scheidegger 1978; De 1996) but the breccia-cored rosettes described here from the Deccan have not been reported from any other basaltic province, to our knowledge. Our results mean that it would be worth to closely study the war bonnet structures of the Columbia River flood basalt province, as well as the rosettes described from other basaltic provinces, to determine whether they too contain cores of flow-top breccia crusts.

8.6 Conclusions

Large rosettes (radial columnar jointing structures) are found in some of the rubbly pahoehoe flows of Deccan Traps which, unlike the rosettes described from other basaltic provinces, contain large (several metres across) breccia cores. Field observations, petrographic data, and geochemical data including Nd–Sr isotope data, all indicate that the breccia cores of these rosettes are blocks of the flow-top breccia crusts that became unstable and sank into these flows' molten interiors. This strongly affected the temperature distribution within the molten flow interiors, warped the isotherms around the relatively cold masses, and led to the formation of the breccia-cored rosettes. This study shows that, whereas lava lakes actively recycle their upper crusts, multiple times, at least one such cycle of crustal overturn and "subduction" can occur, under suitable conditions, in large and thick flood basalt lava flows.

Acknowledgments We thank Vilas Bhosale, Vinit Phadnis and Cliff Ollier for company and assistance on various field trips, and Deepak Joshi, Alok Kumar, Anil D. Shukla and Shilpa Netrawali for help with the isotopic, XRF and ICP measurements. The XRF used at PRL is a part of the PLANEX facilities of the Indian Space Research Organization (ISRO). Very intensive discussions with Ninad Bondre are greatly appreciated, and discussions with Stephen Self have also been helpful. We thank Jyotisankar Ray for his kind invitation to contribute this paper to this volume. The paper was significantly improved and tightened by the reviews of Gautam Sen and Surendra Pal Verma. Senthil Kumar thanks the Director, NGRI for permission to publish this paper.

References

Agashe LV, Gupte RB (1971) Mode of eruption of the Deccan Trap basalts. Bull Volcanol 35:591–601

Bear AN, Cas RAF (2007) The complex facies architecture and emplacement sequence of a Miocene submarine mega-pillow lava flow system, Muriwai, North Island, New Zealand. J Volcanol Geotherm Res 160:1–22

Bondre NR, Duraiswami RA, Dole G (2004a) Morphology and emplacement of flows from the Deccan volcanic province, India. Bull Volcanol 66:29–45

Bondre NR, Duraiswami RA, Dole G (2004b) A brief comparison of lava flows from the Deccan volcanic province and the Columbia-Oregon Plateau flood basalts: implications for models of flood basalt emplacement. In: Sheth HC, Pande K (eds) Magmatism in India through time. Proc Ind Acad Sci (Earth Planet Sci) 113:809–817

Budkewitsch P, Robin P-Y (1994) Modelling the evolution of columnar joints. J Volcanol Geotherm Res 59:219–239

Chandrasekharam D, Parthasarathy A (1978) Geochemical and tectonic studies on the coastal and inland Deccan Trap volcanics and a model for the evolution of Deccan Trap volcanism. Neues Jahrb Miner Abh 132:214–229

Cox KG, Hawkesworth CJ (1985) Geochemical stratigraphy of the Deccan Traps at Mahabaleshwar, Western Ghats, India, with implications for open system magmatic processes. J Petrol 26:355–377

De A (1996) Entablature structure in Deccan Trap flows: its nature and probable mode of origin. Gondwana Geol Mag 2:439–447

DeGraff JM, Aydin A (1987) Surface morphology of columnar joints and its significance to mechanics and direction of joint growth. Geol Soc Am Bull 99:605–617

DePaolo DJ (1987) Neodymium isotope geochemistry. Springer, New York

Devey CW, Lightfoot PC (1986) Volcanological and tectonic control of stratigraphy and structure in the western Deccan Traps. Bull Volcanol 48:195–207

Dole G, Bondre N, Duraiswami RA, Kale VS (2002) Discussion on 'Arterial system of lava tubes and channels within Deccan volcanics of western India' by K. S. Misra. J Geol Soc Lond 60:597–599

Duffield WA (1972) A naturally occurring model of global plate tectonics. J Geophys Res 77:2543–2555

Duraiswami RA, Bondre NR, Dole G (2005) The Songir structure: inflated lava flow or tube? J Geol Soc Ind 65:357–365

Duraiswami RA, Bondre NR, Managave S (2008) Morphology of rubbly pahoehoe (simple) flows from the Deccan volcanic province: implications for style of emplacement. J Volcanol Geotherm Res 177:822–836

Gilbaud M, Self S, Thordarson Th, Blake S (2005) Morphology, surface structures, and emplacement of lavas produced by Laki, A. D. 1783–1784. In: Manga M, Ventura G (eds) Kinematics and dynamics of lava flows. Geol Soc Am Spec Pap 396:81–102

Greeley R, Fagents SA, Harris RS, Kadel SD, Williams DA (1998) Erosion by flowing lava: field evidence. J Geophys Res 103:27325–27345

Grossenbacher KA, McDuffie SM (1995) Conductive cooling of lava: columnar joint diameter and stria width as functions of cooling rate and thermal gradient. J Volcanol Geotherm Res 69:95–103

Harris AJL, Flynn LP, Rothery DA, Oppenheimer C, Sherman SB (1999) Mass flux measurements at active lava lakes: implications for magma recycling. J Geophys Res 104:7117–7136

Jay AE, Widdowson M (2008) Stratigraphy, structure and volcanology of the SE Deccan continental flood basalt province: implications for eruptive extent and volumes. J Geol Soc Lond 165:177–188

Jay AE, Widdowson M (2009) New palaeomagnetic data from the Mahabaleshwar Plateau, Deccan flood basalt province, India: implications for the volcanostratigraphic architecture of continental flood basalt provinces. J Geol Soc Lond 166:13–24

Kale VS, Kulkarni HC, Peshwa VV (1992) Discussion on a geological map of the southern Deccan Traps, India and its structural implications. J Geol Soc Lond 149:473–475

Keszthelyi LP, Self S (1998) Some physical requirements for the emplacement of long basaltic lava flows. J Geophys Res 103:27447–27464

Lightfoot PC, Hawkesworth CJ, Devey CW, Rogers NW, van Calsteren PWC (1990) Source and differentiation of Deccan Trap lavas: implications of geochemical and mineral chemical variations. J Petrol 31:1165–1200

Long PE, Wood BJ (1986) Structures, textures, and cooling histories of Columbia River basalt flows. Geol Soc Am Bull 97:1144–1155

Lyle P (2000) The eruption environment of multi-tiered columnar basalt lava flows. J Geol Soc Lond 157:715–722

Misra KS (2002) Arterial system of lava tubes and channels in Deccan volcanics of western India. J Geol Soc Ind 59:115–124

Mitchell C, Widdowson M (1991) A geological map of the southern Deccan Traps, India and its structural implications. J Geol Soc Lond 148:495–505

Ollier CD, Sheth HC (2008) The High Deccan duricrusts of India and their significance for the 'laterite' issue. J Earth Syst Sci 117:537–551

Philpotts AR, Dickson LD (2000) The formation of plagioclase chains during convective transfer in basaltic magma. Nature 406:59–61

Philpotts AR, Dickson LD (2002) Millimeter-scale modal layering and the nature of the upper solidification zone in thick flood-basalt flows and other sheets of magma. J Struct Geol 24:1171–1177

Philpotts AR, Ague JJ (2009) Principles of igneous and metamorphic petrology, 2nd edn. Cambridge University Press, New York, 667 pp

Raczek I, Stoll B, Hofmann AW, Jochum KP (2001) High-precision trace element data for the USGS reference materials BCR-1, BCR-2, BHVO-1, BHVO-2, AGV-1, AGV-2, DTS-1, DTS-2, GSP-1 and GSP-2 by ID-TIMS and MIC-SSMS. Geostand Newslett: J Geostand Geoanal 25:77–86

Scheidegger AE (1978) The tectonic significance of joints in the Canary Islands. Rock Mech 11:69–85

Sengupta P, Ray A (2006) Primary volcanic structures from a type section of Deccan Trap flows around Narsingpur-Harrai-Amarwara, central India: implications for cooling history. J Earth Syst Sci 115:631–642

Sheth HC (2006) The emplacement of pahoehoe lavas on Kilauea and in the Deccan Traps. J Earth Syst Sci 115:615–629

Sheth HC, Mahoney JJ, Chandrasekharam D (2004) Geochemical stratigraphy of flood basalts of the Bijasan Ghat section, Satpura Range, India. J Asian Earth Sci 23:127–139
Stovall WK, Houghton BF, Harris AJL, Swanson DA (2009) A frozen record of density-driven crustal overturn in lava lakes: the example of Kilauea Iki 1959. Bull Volcanol 71:313–318
Spry A (1962) The origin of columnar jointing, particularly in basalt flows. J Austr Geol Soc 8:191–216
Subbarao KV, Hooper PR (Compilers) (1988) Reconnaissance map of the Deccan Basalt Group in the Western Ghats, India. In: Subbarao KV (ed) Deccan flood basalts. Geol Soc Ind Mem 10:enclosure
Tanaka T, Togashi S, Kamioka H, Amakawa H, Kagami H, Hamamoto T, Yuhara M, Orihashi Y, Yoneda S, Shimizu H, Kunimaru T, Takahashi K, Yanagi T, Nakano T, Fujimaki H, Shinjo R, Asahara Y, Tanimizu M, Dragusanu C (2000) JNdi-1: a neodymium isotopic reference in consistency with La Jolla neodymium. Chem Geol 168:279–281
Verma SP, Torres-Alvarado IS, Sotelo-Rodriguez ZT (2002) SINCLAS: standard igneous norm and volcanic rock classification system. Comp Geosci 28:711–715
Waters AC (1960) Determining directions of flow in basalts. Am J Sci 258A:350–366
Wilson SA (1997) Data compilation for USGS reference material BHVO-2, Hawaiian basalt. US Geological Survey Open File Report
Weis D, Kieffer B, Maershalk C, Pretorius W, Barling J (2005) High-precision Pb-Sr-Nd-Hf isotopic characterization of USGS BHVO-1 and BHVO-2 reference materials. Geochem Geophys Geosyst 6. doi: 1029/2004GC000852

Chapter 9
Giant Plagioclase Basalt from Northern Part of Jhabua District, Madhya Pradesh, Central India

Biswajit Ghosh

Abstract A brief petrological account of six basalt flows from northern part of Jhabua district, Madhya Pradesh, central India is presented. The lowest flow with a thick red bole horizon at the bottom resting over the Precambrian Aravalli Supergroup of rocks is a giant plagioclase basalt (GPB) flow. The second flow is distinctly porphyritic containing augite phenocrysts. The other four flows are also porphyritic with phenocrysts of mainly plagioclase and microphenocryst of augite. Major element chemistry indicates that lowest two flows (Flows-1 and 2) belong to one chemical type (CT-I) while the rest four flows (Flows-3–6) belong to other (CT-II). Flows-1 and 2 appear to be two different flow units of a single compound flow. It is interpreted here that filtration mechanism after a period of magma chamber dormancy might have played which caused the difference in phenocryst mineralogy.

9.1 Introduction

The lava flows in the Deccan volcanic province are termed "giant plagioclase basalt" or "giant phenocryst basalt" (GPB), which are characterized by plagioclase megacrysts (1.5–10 cm long) (Karmarkar et al. 1972; Bodas et al. 1985; Beane et al. 1986; Hooper et al. 1988; Pankov et al. 1994; Sen 2001; Higgins and Chandrasekharam 2007). GPB flows occur at various stratigraphic levels and these appear to mark cyclicity in the eruption of volcanic sequence (Misra et al. 2001). They serve as good marker horizons in the lower part of Deccan stratigraphy (Kalsubai Subgroup) (Table 9.1). Desmukh et al. (1977) gave a detailed account of four GPB horizons in the Western Ghats lava sequence. Later Beane (1988) from his study added at least two more GPB flows.

B. Ghosh (✉)
Department of Geology, University of Calcutta, 35, Ballygunge Circular Road, Kolkata 700019, India
e-mail: bghosh_geol@hotmail.com

J. Ray et al. (eds.), *Topics in Igneous Petrology*, DOI 10.1007/978-90-481-9600-5_9,

Table 9.1 Simplified stratigraphy of the Deccan Basalt Group in the Western Ghats showing the position of GPBs at the top of each formation (After Hooper et al. 1988)

Group	Subgroup	Formation	Member/flow
Deccan Basalt	Wai	Desur	
		Panhala	
		Mahabaleshwar	
		Ambenali	
		Poladpur	
	Lonavala	Bushe	
		Khandala	
	Kalsubai	Bhimashankar	Monkey Hill GPB
			Giravili GPB
		Thakurvadi	Manchar GPB
		Neral	Tunnel-5 GPB
		Igatpuri	Kashele GPB
		Jawahar	Thalghat GPB

GPB flows are generally highly fractionated evolved lavas (low MgO) and their chemical compositions contrast sharply with the surrounding more primitive flows. These flows mark the formational boundaries with a shift in the chemical trend at or near the top of each of them (Bodas et al. 1988; Misra et al. 2001). Hooper et al. (1988) demonstrated that despite having similarity in appearance to each other between the six GPB flows in the Kalsubai Subgroup, each GPB possesses a unique combination of chemical characters. In the present contribution, I describe the petrography and geochemistry of six basalt flows including one GPB from northern part of Jhabua district, Madhya Pradesh. The Deccan traps in this area of Jhabua district form a part of the Malwa Plateau and overlie the rocks of the Precambrian Aravalli Supergroup towards north (Fig. 9.1). The conclusions arrived in this paper are based mainly on the field evidences supported by few mineral composition and whole rock geochemical data followed by comparisons with other well studied GPB flows in other parts of India.

9.2 Petrography and Mineralogy

Deccan traps in the presently investigated area consist of six horizontally-disposed flows (Flows-1–6) horizontally disposed that can be traced over large distances. The lower most flow with a thick red bole horizon at the bottom resting over Aravalli Supergroup of rocks is a GPB flow having average length of plagioclase phenocryst of 1.8 cm. Representative field photographs and photomicrographs of this GPB have been given in Fig. 9.2. The overall texture of the rock is porphyritic with the groundmass minerals commonly showing intergranular texture. Plagioclase megacrysts do not show any appreciable zoning and are more calcic when compared with those occurring in groundmass

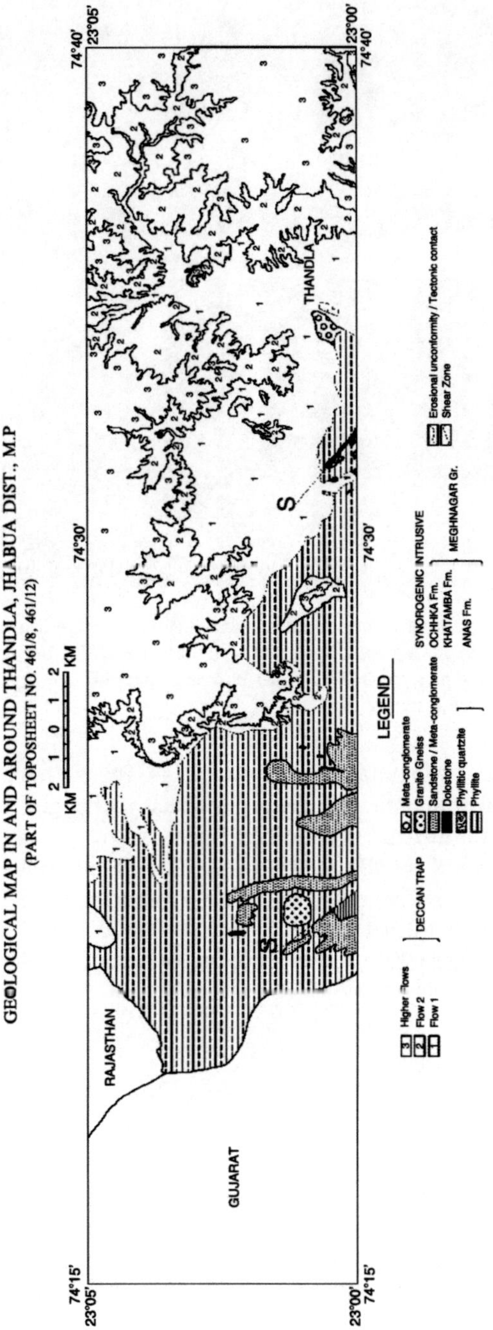

Fig. 9.1 Geological map of northern part of Jhabua district, Madhya Pradesh (mapped by the author)

Fig. 9.2 (**a**) Field photograph of giant plagioclase basalt (*GPB*) resting over red bole horizon in a dug well section; (**b**) field photograph of plagioclase laths of the GPB flow; (**c**) photomicrograph of the GPB flow showing plagioclase laths; (**d**) photomicrograph showing rounded rersorbed phenocrysts of augite (Cpx) in flow-2

(Table 9.2, Fig. 9.3a). One notable feature of this GPB is the presence of alkali feldspar occupying the interstices in the groundmass. Other groundmass minerals are clinopyroxene (augite) and titano-magnetite.

The contact between this GPB and the upper flow (Flow-2) is ill defined in the field. This second flow is distinctly porphyritic with phenocrysts of augite. Under microscope, it shows well developed flow texture (Fig. 9.2d). The next four flows (Flows-3–6) are also porphyritic with phenocryst of mainly plagioclase and micro-phenocryst of augite.

9.3 Geochemistry

Major elements for representative rock samples (Table 9.3) were analysed by XRF at G.S.I., Central Geochemical Laboratory, Kolkata, India. In the FeO^t–MgO–(Na_2O + K_2O) diagram, plots for all the flows fall well in the tholeiitic field (Fig. 9.3b). Inspite of having a common tholeiitic chemistry, two compositional types (CT) are well recognized when the major element chemistry of all the flows are plotted on bivariate ($Na_2O + K_2O$) versus SiO_2 diagram after Irvine and Baragar (1971) and Cox et al. (1979)

Table 9.2 Representative electron microprobe analyses of constituent minerals of the GPB flow

Ref. Nos.	1	2	3	4	5	6	7	8	9	10
SiO_2	53.36	53.67	56.50	55.78	64.87	65.21	50.54	50.54	0.10	0.05
TiO_2	0.08	0.12	0.21	0.09	0.10	0.21	1.29	0.98	22.58	22.61
Al_2O_3	27.76	27.72	25.70	26.23	20.08	19.01	1.70	0.94	1.51	1.44
Cr_2O_3	0.00	0.00	0.00	0.01	0.00	0.03	0.09	0.00	0.56	0.69
FeO^t	0.50	0.69	0.72	0.58	0.36	0.41	10.74	14.42	69.93	69.84
MnO	0.00	0.01	0.02	0.11	0.00	0.00	0.31	0.42	0.50	0.62
MgO	0.12	0.11	0.04	0.03	0.03	0.01	13.82	11.75	1.11	1.02
CaO	12.11	11.50	8.72	8.62	1.74	0.99	20.27	20.11	0.02	0.09
Na_2O	4.68	4.86	6.05	6.42	6.52	5.86	0.33	0.33	0.00	0.00
K_2O	0.57	0.55	0.64	0.72	6.15	7.35	0.02	0.02	0.03	0.00
Total	99.18	99.23	98.6	98.59	99.85	99.08	99.11	99.51	95.68	95.66
Number of atoms										
O	8	8	8	8	8	8	6	6	32	32
Si	2.448	2.458	2.582	2.556	2.918	2.961	1.919	1.942	0.030	0.015
Ti	0.003	0.004	0.007	0.003	0.003	0.007	0.037	0.028	5.103	5.112
Al	1.501	1.497	1.384	1.415	1.065	1.017	0.076	0.042	0.535	0.510
Cr	0.000	0.000	0.000	0.000	0.000	0.001	0.003	0.000	0.133	0.164
Fe^t	0.019	0.027	0.028	0.022	0.014	0.015	0.341	0.463	–	–
Fe^{2+}	–	–	–	–	–	–	–	–	12.502[a]	12.483[a]
Fe^{3+}	–	–	–	–	–	–	–	–	5.067[a]	5.072[a]
Mn	0.000	0.001	0.001	0.004	0.000	0.000	0.010	0.014	0.127	0.158
Mg	0.008	0.008	0.003	0.002	0.002	0.001	0.782	0.673	0.497	0.457
Ca	0.595	0.564	0.427	0.423	0.084	0.048	0.824	0.828	0.006	0.029
Na	0.416	0.432	0.536	0.570	0.569	0.516	0.025	0.025	0.000	0.000
K	0.033	0.032	0.038	0.042	0.353	0.426	0.001	0.001	0.000	0.000

Ref. Nos. 1–2: Plagioclase megacryst; 3–4: Groundmass plagioclase; 5–6: Groundmass alkali feldspar; 7–8: Groundmass augite and 9–10: Groundmass titano-magnetite

[a] Fe^{2+} and Fe^{3+} are after charge-balance

(Fig. 9.3c and d). The recognition of these two CTs is also reflected on the behaviour of some other important oxides like TiO_2, P_2O_5, CaO and MgO of the lavas. The geochemical characteristics of these two CTs are summarized below.

CT-I: High TiO_2 (~3.40 wt%), high P_2O_5 (~0.50 wt%), low CaO (<10.00 wt%) and low MgO (<5.50 wt%).

CT-II: Low TiO_2 (<2.70 wt%), low P_2O_5 (<0.30 wt%), high CaO (>10.45 wt%) and high MgO (>5.60 wt%).

Thus, Flows-1 and 2 belong to the CT-I and the rest of the flows (Flows-3–6) belong to CT-II. Although first two flows belong to the same chemical type having comparable TiO_2 and P_2O_5 content, close observation reveals that there is a chemical difference between these two flows as evident from MgO and CaO values. This difference can be ascribed to crystal–liquid fractionation which has been discussed later.

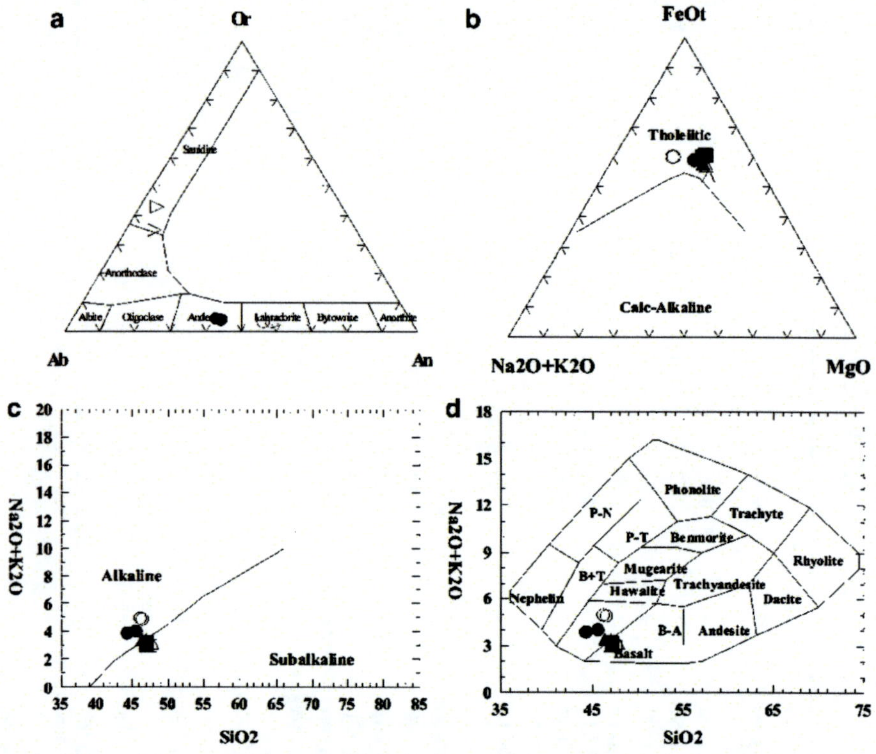

Fig. 9.3 (**a**) Ternary Ab–An–Or feldspar plots of the GPB flow. *Open circle*: megacryst plagioclase, *solid circle*: groundmass plagioclase, *open triangle*: groundmass alkali feldspar. (**b**) AFM plots of the six Deccan flows. Dividing line is after Irvine and Baragar (1971). *Open circle*: flow-1, *solid circle*: flow-2, *open triangle*: flow-3, *solid triangle*: flow-4, *open square*: flow-5 and *solid square*: flow-6; (**c**) total alkali versus silica (TAS) diagram after Irvine and Baragar (1971). Symbols are same as in (**b**); (**d**) TAS diagram of Cox et al. (1979) for volcanic rocks. Symbols are same as in **b**

Table 9.3 Major element compositions of basalt flows

Ref. Nos.	1	2	3	4	5	6	7	8	9	10	11	12
SiO_2	46.50	46.20	44.30	45.60	47.00	47.80	47.00	46.40	47.00	47.50	47.20	47.20
TiO_2	3.41	3.41	3.42	3.50	2.56	2.57	2.67	2.68	2.29	2.24	2.15	2.21
Al_2O_3	15.74	15.81	13.81	14.07	14.33	14.88	13.85	14.00	13.65	13.68	13.67	13.72
Fe_2O_3	7.52	8.10	7.00	7.08	6.47	6.13	7.07	7.17	7.14	7.40	6.93	7.27
FeO	6.12	5.76	7.02	7.11	5.94	4.95	6.30	6.30	7.29	6.84	7.38	7.20
MnO	0.18	0.19	0.21	0.21	0.19	0.19	0.20	0.20	0.23	0.22	0.22	0.22
MgO	3.59	3.58	5.20	5.30	5.71	5.82	5.75	5.83	5.89	5.78	5.64	5.76
CaO	8.71	8.73	9.95	10.12	11.16	11.32	10.47	10.59	10.98	10.82	10.73	10.98
Na_2O	3.25	3.29	2.82	2.90	2.40	2.47	2.61	2.65	2.51	2.52	2.16	2.48
K_2O	1.63	1.66	1.09	1.09	0.66	0.67	0.74	0.75	0.44	0.44	1.10	0.52
P_2O_5	0.46	0.46	0.55	0.56	0.29	0.28	0.28	0.29	0.22	0.22	0.23	0.23
LOI	2.06	2.00	3.65	1.46	2.39	2.16	2.18	2.21	1.40	1.37	1.31	1.31
Total	99.17	99.19	99.02	99.00	99.10	99.24	99.12	99.07	99.04	99.03	98.72	99.10
D.I.	44.25	44.13	32.34	32.78	29.16	30.20	31.07	30.37	27.77	29.09	28.98	28.28
CIPW normative minerals												
Quartz	6.02	5.39	0.57	1.02	4.13	4.59	3.78	2.65	3.35	4.60	3.55	3.69
Orthoclase	9.92	10.09	6.75	6.60	4.03	4.08	4.51	4.58	2.66	2.66	6.67	3.14
Albite	28.32	28.64	25.02	25.16	21.00	21.53	22.78	23.15	21.75	21.83	18.76	21.46
Anorthite	24.23	24.13	22.85	22.70	27.26	28.35	24.63	24.86	25.26	25.29	24.99	25.31
Diopside	0.00	0.00	19.81	19.70	22.03	21.46	21.27	21.54	23.15	22.36	22.48	22.98
Hypersthene	9.17	9.14	6.14	6.07	6.04	5.05	6.47	6.47	8.16	7.31	8.41	7.72
Magnetite	10.73	9.56	10.64	10.52	9.70	9.15	10.57	10.73	10.60	10.99	10.31	10.78
Hematite	0.34	1.74	0.00	0.00	0.00	0.00	0.00	0.00	0.00	0.00	0.00	0.00
Ilmenite	6.67	6.66	6.81	6.82	5.03	5.03	5.23	5.25	4.45	4.36	4.19	4.29
Apatite	1.03	1.03	1.26	1.25	0.65	0.63	0.63	0.65	0.49	0.49	0.52	0.51

Ref. Nos. 1–2: Flow-1; 3–4: Flow-2; 5–6: Flow-3; 7–8: Flow-4; 9–10: Flow-5 and 11–12: Flow-6

9.4 Discussion

GPBs represent flows in which the plagioclase phenocrysts accumulate by flotation in highly differentiated evolved magmas as a result of their buoyancy and concentrates near the roofs of magma chambers. During a period of magma chamber dormancy such plagioclase crystals grow to large sizes. Finally a GPB eruption perhaps occurs when a new batch of magma enters the chamber from below and sweeps through the plagioclase mush, mingles and mixes, then squeezes the mush out of the chamber to form the GPB flows (Higgins and Chandrasekharam 2007). Thus each GPB eruption represents the culmination of an eruptive cycle of a magma chamber (Sen 2001).

Red bole horizons are commonly associated with GPBs, with the GPBs often erupting on a red bole horizon. From this observation Sen (2001) inferred that while red bole (in situ or transported clay rich soil) was forming at the surface, the subsurface magma chamber was undergoing a dormant (non-eruptive) phase during which the magma was differentiating and plagioclase crystals already present in it were growing to large sizes.

Hooper et al. (1988) have demonstrated that each of the six GPB of the Kalsubai Subgroup recorded from Western Ghat sequence possesses a unique combination of chemical characters. The major element data of the presently studied GPB of Jhabua district, Madhya Pradesh are compared with the data given by Hooper et al. (1988). The comparison shows that the presently investigated GPB matches well with the Thalghat GPB which is characterized by >0.35% P_2O_5, >3.27% TiO_2 and >1.00% K_2O.

In the present study, there is a distinct chemical dissimilarity between two compositional types. Here again, although flows-1 and 2 belonging to CT-I have comparable TiO_2 and P_2O_5 content, there is an appreciable difference in their MgO and CaO content which is the reflection of the difference in their phenocryst mineralogy, flow-1 having plagioclase as phenocryst (GPB) while flow-2 having clinopyroxene as phenocryst.

From the petrographic and chemical characters it is inferred that during the dormancy of CT-I magma, while the plagioclase crystals were growing and accumulating near the roof of the chamber, pyroxene crystals were sinking and accumulating near the base of the chamber. Finally eruption occurred when next batch of magma (CT-II) entered the chamber from below and squeezed out the plagioclase rich magma first followed by pyroxene rich magma. It seems that flows-1 and 2 which were recognized in the field as two different flows with ill defined contact simply represent two different flow units of a single compound flow (CT-I).

Desmukh (1988) also envisaged a similar filtration mechanism to explain the phenocryst-variation in different flow units of a compound flow near Ellora caves where the lava sequence is marked by alternating non-porphyritic and strongly porphyritic flows. Flow units in some compound flows contain large plagioclase phenocryst up to 5 cm in length constituting about 40 vol.% of the rock (Desmukh 1988).

Summarising, major element chemistry indicates that lower most two flows (Flows-1 and 2) belong to one chemical type (CT) while the rest four flows

(Flows-3–6) belong to other. Flows-1 and 2 appear to be two different flow units of a single compound flow. Filtration mechanism after a period of magma chamber dormancy might have played an important role which caused the difference in phenocryst mineralogy.

Acknowledgments The work was carried out as a part of field-session program (2004–2005) of Geological Survey of India (GSI). Mr. B. Chattopadhyay and Mr. S. Sengupta of GSI, EPMA Lab., Central Petrological Laboratory, Kolkata are sincerely thanked for assistance with the microprobe analysis. Special thanks are due to Dr. Abhinaba Roy, GSI, Eastern Region, Kolkata for his patient review and insightful comments on an early version of the manuscript.

References

Beane JE (1988) Flow stratigraphy, chemical variation and petrogenesis of Deccan flood basalts, Western Ghats, India. Ph.D. dissertation, Washington State University, Pullman, 576 pp

Beane JE, Turner CA, Hooper PR, Subbarao KV, Walsh JN (1986) Stratigraphy, composition and form of the Deccan Basalts, Western Ghats, India. Bull Volcano 48:61–83

Bodas MS, Khadri SFR, Subbarao KV, Hooper PR, Walsh JN (1985) Flow stratigraphy of a part of the western Deccan basalt province – a preliminary study. In: 5th international geological congress, Washington, DC, 1985, pp 339–347

Bodas MS, Khadri SFR, Subbarao KV (1988) Stratigraphy of the Jawhar and Igatpuri Formations, Western Ghat lava pile, India. In: Subbarao KV (eds) Deccan flood basalts. Memoir of the Geological Society of India 10:235–252

Cox KG, Bell JD, Pankhrust RJ (1979) The interpretation of igneous rocks. George Allen & Unwin, London, p 450

Desmukh SS (1988) Petrographic variations in compound flows of Deccan Traps and their significance. In: Subbarao KV (eds) Deccan flood basalts. Memoir of the Geological Society of India 10:305–319

Desmukh SS, Aramaki S, Shimizu N, Kurasawa N, Konda T (1977) Petrography of the basalt flows exposed along Mahabaleshwar and Amboli sections in Western Ghats, India. Record Geol Surv Ind 108:81–103

Higgins MD, Chandrasekharam D (2007) Nature of Sub-volcanic magma chambers, Deccan Province, India: evidence from quantitative textural analysis of plagioclase megacrysts in the giant plagioclase Basalts. J Petrol 48:885–900

Hooper PR, Subbarao KV, Beane JE (1988) The Giant Plagioclase Basalts (GPBs) of the Western Ghats, Deccan Traps. In: Subbarao KV (eds) Deccan flood basalts. Memoir of the Geological Society of India 10:135–144

Irvine TN, Baragar WRA (1971) A guide to the chemical classification of the common volcanic rocks. Can J Earth Sci 8:523–548

Karmarkar BM, Kulkarni SR, Marathe SS, Sowani PV, Peshwa VV (1972) Giant phenocryst basalt in the Deccan Trap. Bull Volcano 35:965–974

Misra KS, Sabale AB, Bhutani R (2001) Lava channel/tube network in western Maharashtra: study of their morphology, formation and transportation of Deccan lava. Geological Survey. India. Special Publications 64, pp 503–510

Pankov V, Oleinikov BV, Krishnamurthy P, Murari R, Gopalan K (1994) Mineral and melt inclusions in the Giant Plagioclase phenocrysts of Deccan Basalts, Western Ghats, India: some comparisons with plagioclases of intrusive Siberian Traps and implications on the physicochemical conditions during magmatic evolution. In: Subbarao KV (ed) Volcanism. Geological Society of India, Bangalore, pp 187–199

Sen G (2001) Generation of Deccan Trap magmas. Proc Ind Acad Sci 110:409–431

Chapter 10
Petrogenesis of Flood Basalts of the Narsingpur–Harrai–Amarwara–Lakhnadon Section of Eastern Deccan Province, India

Piyali Sengupta and Jyotisankar Ray

Abstract The present area of investigation falls in the Eastern Deccan segment where available field and geochemical data are rather limited. In the present study, five distinctive structural zones namely Lower Vesicular zone, Lower Colonnade Zone, Entablature Zone, Upper Colonnade Zone and Upper Vesicular Zone have been identified in different lava flows which are helpful for the purpose of stratigraphic correlation. All the lavas under investigation can be classified into three distinct geochemical groups (viz. Group I, Group 2 and Group 3) on the basis of their Mg′ number and TiO_2 wt%. The Group 1 lavas are marked by low Mg′ No (0.39–0.45) and high TiO_2 (2.9–3.6 wt%) while Group 2 lavas are characterized by moderate Mg′ No (0.45–0.50) and moderate TiO_2 (2.4–2.9 wt%). On the other hand, high Mg′ No (0.49–0.54) and low TiO_2 (1.91–2.31 wt%) characterize the Group 3 basalts. Interestingly, the basaltic flows corresponding to three distinct groups are intermixed with each other; in other words, lava flows belonging to several groups are juxtaposed with one another with increasing height.

Evolutionary trends of the entire basaltic lava flows of the study area are also attempted based on the major and trace element (including REE) chemistry. All the investigated basaltic lava flows in general show a restricted compositional range and they fall in the basalt field in the total alkali versus silica diagram. They have, in general, a tholeiitic imprint (with a tendency of becoming high iron tholeiite). Closer examination of several variation diagrams involving different major elements versus Mg′ No plot indicates that the three basaltic groups have distinct and secluded geochemical trends suggestive of their derivation from discrete magma pulses. However, each investigated group of basalts shows evidence of intra-group differentiation. Variation diagrams involving plot of several compatible elements/incompatible elements versus Mg′ number also indicate discrete or

P. Sengupta and J. Ray (✉)
Department of Geology, University of Calcutta, 35, Ballygunge Circular Road, Kolkata 700019, India
e-mail: jsray65@hotmail.com

J. Ray et al. (eds.), *Topics in Igneous Petrology*, DOI 10.1007/978-90-481-9600-5_10,

independent derivation of the three investigated basalt groups. The trace element variation diagrams also reveal subsequent low to moderate degree of internal fractionation within each investigated group. This observation (regarding derivation of three groups corresponding to different magmatic pulses and subsequent internal differentiation) has also been corroborated from plots of several major and trace elements versus Zr plot. Chondrite normalized rare earth element plots for the three investigated group of basalts indicate that their fundamental patterns for the REE variation are essentially the same showing most primitive pattern for the Group 3 basalts and the most evolved pattern for the Group 1 basalts. Further it has been found that $\Sigma(REE)_N$ for Group 3 basalts is the least while the same value corresponding to Group 1 basalts is the highest. The $\Sigma(REE)_N$ for Group 2 basalts is intermediate between those of Group 1 and Group 3. Several trace element-ratios namely Zr/Nb, La/Nb, Ba/Nb, Ba/Th, Rb/Nb, K/Nb, Th/Nb, Th/La and Ba/La were successfully employed to constrain the source characteristics of the investigated basalt- groups. On the basis of these trace element ratios, it has been found that in general, Group 1 basalts have affinity either to EMI/EMII or HIMU sources while Group 2 and Group 3 basalts show a close proximity towards HIMU parentage although affinity towards EMI and EMII was also hinted. Thus a distinct enriched mantle character has been documented from due consideration of several critical trace element ratios. Consideration of other trace element ratios like Lu/Hf, La/Sm and Sm/Yb brings out an enriched garnet peridotite plume source material which subsequently underwent different degrees of partial melting corresponding to different ambient conditions. It has been shown by theoretical calculations that a relatively low degree of melting of the initial plume source generated the enriched type of basalts (Group 1 basalts) while a significantly higher degree of melting have produced the primitive group of basalts (Group 3). The intermediate group of basalts has geochemical characteristics intermediate between Group I and Group 3 and therefore might owe their origin to moderate degree of mantle melting. In each of the cases, the generated melt underwent different degrees of crystal cumulation, mixing of the supernatant liquids and the settled crystals to give rise to heterogeneous porphyritic basalts. It is also possible that the generated melts (corresponding to different degrees of mantle melting) might have mixed up with one another as evidenced by the petrographic analyses of the different group of basalts. The possibility of effusion of the parent magma along the Narmada–Tapti rift and related lineament-reactivation/subsequent rifting due to interaction of hot mantle plume with the lithospheric weak planes has also been critically viewed.

10.1 Introduction

The term continental flood basalt provinces or CFBs are large areas of the continents covered by considerable thickness of laterally extensive lava flows which are considered to have been erupted in comparatively short period of time from fissures at various periods in the geological past. Continental flood basalts are variants of

Large Igneous Province, commonly referred to as LIPs, which are voluminous occurrences of dominantly mafic igneous rocks in oceans and continents not directly related to plate-tectonic processes (Carlson 1991; Coffin and Eldholm 1994). Most of the CFB provinces include relatively evolved tholeiitic basalts which have isotopic and incompatible element ratios within the range of present-day oceanic basalts, particularly oceanic island basalts (MacDougall 1988; Carlson 1991). The rapid melt production rate documented by the eruption of huge volumes of magma in a short interval of time (in a million year or less) is persistent with the hypothesis that flood basalts are triggered by the arrival of new mantle plume at the base of the continental lithosphere (Morgan 1972; Richards et al. 1989; Campbell and Griffiths 1990).

Some of the major CFB provinces are the Siberian Traps of Asia, the Parana province of Brazil with its smaller equivalent in the Etendeka area of Namibia, the Deccan Traps of India, the Karoo dolerites of South Africa, the Ferrar dolerites of Antarctica, dolerites of Tasmania, the North Atlantic province of Europe and the Columbia River Plateau and Snake River Plain of USA.

Deccan volcanism with a tremendous burst of volcanic activity marks a unique episode in Indian geological history covering nearly two-thirds of peninsular India. The magmatism of Deccan Trap volcanic province of India has attracted the attention of geologists because of their eruption around 65 million years ago, coinciding with the Cretaceous–Tertiary boundary (Duncan and Pyle 1988; Courtillot et al. 1988; Venkatesan et al. 1993; Bakshi 1994; Sheth et al. 1997). This was apparently a magmatic event generally ascribed to the fragmentation of the Gondwanaland with the opening of Indian Ocean. Some authors emphasize that this huge volcanism occurred during the passage of India over an extremely vigorous mantle plume or hot spot which is presently located below Reunion Island, as suggested from plate-tectonic reconstruction (Morgan 1972). The isotopic studies of the picritic flows in eastern Saurashtra, first described by West (1958) and later by Krishnamurthy and Cox (1977) show a close isotopic similarity between some basalts and picritic flows with recent eruptions from the Reunion hotspot (Peng and Mahoney 1995).

Field and geochemical data of the Eastern Deccan is relatively limited. Some sporadic sections in the north and south of the Narmada (Fermor and Fox 1916; Crookshank 1936; Alexander and Paul 1977; De 1996; Deshmukh et al. 1996a, b; Yedekar et al. 1996; Nair et al. 1996; Peng et al. 1998; Pattanayak and Shrivastava 1999; Mahoney et al. 2000) have been studied, but till date no standard stratigraphic column for the Eastern Deccan could be erected. Therefore definitive correlation of different flows (belonging to the Eastern Deccan) with the well-established formations of Western Deccan is rather difficult. Some of the important works of Eastern Deccan already carried out by previous workers are summarized below for ready reference for the readers.

Crookshank (1936) geologically mapped an area lying to the northern slopes of the Satpura Hills between Morand and Sher rivers which includes portions of the present area of investigation. However Crookshank (1936) could not establish the flows separately. In addition, he also showed presence of a number of dykes and sills adjacent to the Deccan Trap lavas. Alexander and Paul (1977) reported the occurrence of tholeiites of low-K content among the basalt flow of Sagar area in Madhya Pradesh. Alexander (1999) observed that the low-K tholeiites are peripheral in occurrence to the Deccan Trap region around Sagar. De (1996) studied Deccan Trap flows in Satpura region of Madhya Pradesh and showed that the Entablature zone have developed in Deccan Trap flows in such a manner that their structures are very helpful in identifying individual lava flows occurring in a pile of lavas. The highly fragmented nature of the Entablature zone requires that the configuration of the isotherms should have had a near vertical disposition when the interior part of the lava reached the joint causing isotherm.

Deshmukh et al. (1996a, b) identified 36 and 15 basalt flows within 625 and 385 m thick lava piles in Chikaldara and Behramghat sections from the eastern part of the Deccan Trap province and studied their geology and chemical stratigraphy.

Yedekar et al. (1996) studied geochemical characteristics of the basalts of Chhindwara–Jabalpur–Seoni–Mandla (CJSM) sector of the eastern Deccan volcanic province that lies in the eastern side of the present study area. From trace element studies, they established that these basalts were derived by less than 10% partial melting of mantle, followed by olivine and orthopyroxene fractionation.

Sano and Fuji (1998) showed on the basis of trace element ratios (Ba/Nb and Ba/Sr) that some particular basalt from central India is similar to Ambenali basalts in the Western Ghats area.

Peng et al. (1998) carried out extensive geochemical work on north-eastern Deccan. On the basis of their studies, they have indicated close geochemical similarities between northeastern and southwestern Deccan lavas. Isotopic similarities indicate that the Ambenali formation and perhaps a few Khandala or Poladpur and Bushe Formation flows (all belonging to well established southwestern Deccan) may be present in the northeastern Deccan as well. It was observed that nearly all of the Khandala and Poladpur-like northeastern Deccan lava analyses have higher isotopic ratios than any southwestern Khandala or Poladpur formation basalts, which suggests that these lavas must have traversed different routes through the crust than the southwestern flows and thus issued from different feeder dykes.

Pattanayak and Shrivastava (1999) studied the 900 m thick lava pile in the Eastern Deccan outlier around Mandla, which they termed as Mandla lobe and identified 37 distinct basaltic lava flows within it. On the basis of petrography, field disposition and major oxide chemistry they subdivided these flow units into eight chemical groups. Recently, Mahoney et al. (2000) studied tholeiitic basalts forming a 850 m thick flood basalt sequence at Toranmal in the northern Deccan Traps. These lavas have large range in isotopic ratios [$\epsilon Nd(t)$ =+2.1 to –15.7, $(^{87}Sr/^{86}Sr)_t$ = 0.70457–0.71416, $^{206}Pb/^{204}Pb$ = 16.999–20.245], similar to those of Mahabaleshwar area indicating continental lithospheric influence in the petrogenesis of these basalt lavas.

The present work involves a detailed field, petrological and geochemical investigations in the Deccan Traps of eastern parts of Satpura areas of Madhya Pradesh in the south of the Narmada river which forms a part of Toposheet No. 55N (including 55 N/1, N/2, N/3, N/5, N/6, N/10). The mapped area lies between latitude 22°20′ and 22°50′, longitude 79°10′ and 79°36′ (Fig. 10.1).The major town within the area is Narsingpur (lat 22°57′N, long 79°11′ E), located in the extreme north of the area,

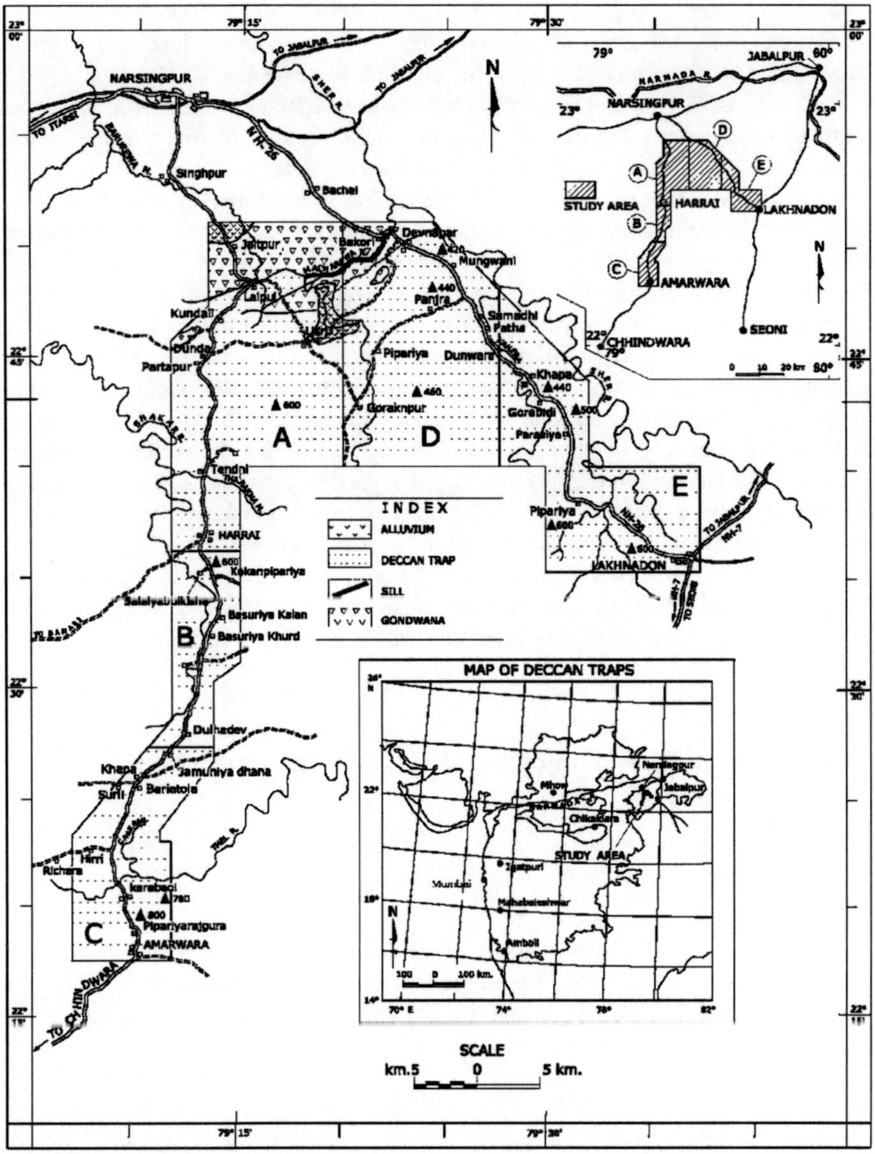

Fig. 10.1 Geological map showing generalized distribution of Investigated Deccan Trap Lava Flows and other related rocks (mapped by the first author). Important localities in the study area have also been shown. The lava flows were studied along two sectors viz. Narsingpur–Harrai–Amarwara (N-H-A) sector and Narsingpur–Lakhnadon (N-L) sectors. The *top-right inset* map shows different sub-sectors within the study area (**a**–**e**). The *bottom-inset* shows the location of the investigated area in the map of India. The sill encountered along the Macharewa river has been excluded in the present study; contour heights are in metres

which is situated at a distance of 100 km southwest of Jabalpur. The other two important towns in the area of investigation are Amarwara and Lakhnadon, which are situated at the south and east of the present area respectively (Fig. 10.1).

The area of the present research work has been specially selected as it reaches high altitudes (up to 885 m) and exposes a large number of flows in the eastern part of the Satpura area. There are however no detailed investigations in the Deccan Traps of Eastern Satpuras of Madhya Pradesh. Moreover, investigation on flow-to-flow correlation using volcanic structures is very few in the Eastern Deccan Trap area.

Available data acquired to date on the chemistry of the lava flows in the eastern Deccan, in general, are insufficient. To meet the gaps in existing knowledge, the present work involves erection of stratigraphic succession of the area by correlation of the different lava flows which was hitherto unattended. The different basdaltic flows have been identified and correlated in the field with the help of primary volcanic structures. Attempts have been made to identify the different chemical types in the stratigraphic succession of the present area and study the detailed petrology of those different groups. The nature of the original magma (or magmas), the mechanism of crystallization, cooling history and petrogenesis of the flows have been established from extensive geochemical and petrologic studies. However, petrological details of intrusive sill (occurring along Macharewa river; Fig. 10.1) have not been attempted in this paper and will be addressed separately.

10.2 Sampling, Petrographic Methods and Analytical Procedures

10.2.1 Sampling

It is to be mentioned that primary volcanic structures are useful in characterizing individual lava flows and have been used successfully in the study on the Columbia River basalts, U.S.A., the Hawaiian basalts and in other areas. In case of Deccan basalts, it has been demonstrated by De (1974) that a three-tier classification of primary volcanic structures is useful in characterization and recognition of individual lava flows in the field. The three divisions are Lower Colonnade Zone, which is overlain by Entablature Zone and the Upper Colonnade Zone. Therefore, in the present investigation, individual flows have been distinguished in the field by the characteristics of these structural zones. Flow stratigraphy constructed at suitable locations of the area has been correlated in order to get complete succession of Deccan Trap lava flows in the region. Samples were collected systematically from all the representative structural zones of individual lava flows. Field sections (measured) of local flow stratigraphy have also been prepared at suitable areas showing well-exposed primary volcanic structures of different flows at different heights. This aspect has been discussed in detail later in this paper.

10.2.2 Petrographic Methods

Thin sections of rocks have been prepared from different structural zones of different basaltic flows. Detailed study of the thin sections from different zones of a single flow has helped to know the petrographic, mineralogical and textural variations during different stages of cooling of a particular flow. The important characters (shape, size, zoning, twinning, pattern of inclusions etc.) of phenocryst minerals (olivine, augite and plagioclase) of every flow have been studied in details in order to differentiate the mineralogical characters of different flows. Modal analyses of representative rock samples were carried out by point counting method of Chayes (1949).

10.2.3 Analytical Procedures (Whole Rock Analyses)

For whole rock chemical analyses of investigated basalt samples, each sample was carefully sawed to trim off all weathered portions. Then, each sample was lapped to grind off the sawn surfaces; to ensure that, the sample was ground far enough to get rid of all striation-marks formed during rock-cutting. The sample was then cleaned with tap water and dried. Next the sample was cleaned with ultrasonic baths for three times (each time for 25 min) to remove all unwanted materials caught up during grinding and lapping of the sample. The sample was next placed inside of a heavy-duty freezer bag on a sturdy flat surface. The sample was next crushed into peanut-sized chips by using a hammer. But in order to eliminate contact with metal, the metal hammer was covered with cloth-duct tape. The peanut-sized chips were next placed in a pyrex beaker and covered with 18 MΩ (ultrapure) water and next cleaned ultrasonically for 30 min. The sample was then air-dried and only fresh grains were finally selected (chips with alteration were discarded). The sample was then put into agate-mill so that they fill the gap between ball and the chamber. On turning the mill on, the quality of the powdered sample was checked frequently (after every 5 min) (with new pair of gloves on for each different sample) to have a feel of the powdered sample. When the sample was powdered very finely (talcum powder like feel) the completion of grinding was marked. The milled powder was tipped onto a clean piece of paper and then placed into a clean, new zip lock plastic bag. The mill was next cleaned thoroughly so that no contamination could take place while using the next sample.

Major element analysis of ten oxides (SiO_2, Al_2O_3, Fe_2O_3 total, MnO, MgO, CaO, Na_2O, K_2O, TiO_2, P_2O_5) of selected specimens from every flow have been analyzed at the Geological Survey of India, Calcutta, by XRF method (model: Philips).

The major elements were analyzed using fused disks. For each sample, two fused disks were prepared (with the sample powder and the flux mixture). One fused disk was kept preserved for double-checking the accuracy of the output data. The fused disk was transferred into Platinum-crucible into pre-heated fusion furnace using W–alloy tongs. After preparation of the fusion disks, those were kept on the hot plate to anneal for at least 45 min. After preparation of the fused glass disks, the platinum crucibles were cleaned using HCl and double-distilled water. LOI (loss on ignition) was determined by noting difference of weights of the ignited samples (before and after ignition) before preparation of the discs. The machine set up and other analytical details were followed using the procedure adopted by Norrish and Chappell (1977). Precision of analytical data is same as given in Mahoney et al. (1998).

Trace element and REE analyses were done from Rochester University, USA. The powdered samples were dissolved with HF/HNO_3 treatment (in each case 50 mg of rock was dissolved in 100 ml of 5% HNO_3). Internal standards were prepared (for different elements) using 5% HNO_3 stock solutions of matrix standards (A and B). Each matrix standard was prepared by adding 3 g of a stock solution to a test tube followed by 3 g. of internal standard stock. The matrix standard concentration for each element in the final solution was close to those prescribed values in the laboratory. The stock solution of the spike was next prepared. Two sets of tubes were prepared for each concentration-standard, each standard reference material (BHVO) and each unknown sample. During the analyses, the matrix standard, concentration standard, standard reference (BHVO), acid blanks and unknown samples were sequenced in a definite order. Between each test tube, a three-stage wash sequence was run: (a) 10% HNO_3 + one drop of HF per 150 ml of nitric acid (b) 10% HNO_3 and (c) 5% HNO_3. Each run was continued for 90 seconds. A standard reference (BHVO) (unspiked and spiked) at the beginning, middle and end of the run was employed to maintain consistency of the data-output. At the close of sequence, matrix standard A, matrix standard B then the concentration standards of 1, 10 and 100 ppb (unspiked and spiked) were run. The data reduction was done taking due consideration of blank correction, interference correction, matrix factor and machine drift correction. Procedures followed during the analyses were according to the conventional broad-steps outlined by McGinnis et al. (1997). Precision of analytical data is same as given in Neal et al. (1997).

10.3 Primary Volcanic Structures of the Deccan Trap Flows of the Present Area

Twenty Deccan Trap flows have been identified in the Narsingpur–Harrai–Amarwara sector (N-H-A sector) while nine Deccan Trap Flows have been identified in the Narsingpur–Lakhnadon sector (N-L sector) (Fig. 10.1) and correlated in the field on the basis of primary volcanic structures. The height-wise occurrences of each lava flow and corresponding thickness for both N-H-A and N-L sectors have been shown in the table for ready reference.

Each and every Deccan Trap flow of the present area is characterized by some distinctive structural features. Some of the flows are characterized by huge Lower and Upper Colonnade Zones while some show variable nature of Entablature Zone like fanning columns, intersecting, inclined columns etc. Lower or Upper Vesicular Zones also characterize some flows showing variable nature of shape, size, distribution of vesicles etc. In order to make a coherent description of the different lava flows encountered in N-H-A and N-L sectors, it is necessary to devise criteria to systematize them in a meaningful manner. In order to make such systematization possible, emphasis was laid on chemical basis of the investigated lava flows rather than on physical aspects; because chemical characters are more incisive to decipher groupings of lava flows because they yield perceptible spectrum. On the other hand, the physical zonations of lava flows such as presence of Lower and Upper Colonnade Zones, Entablature Zone etc. are meaningful to recognize different lava flows in the field with sufficient ease. But in a volcanic terrain (as in the present study) yielding a large number of successive lava flows, it is more important to group the same on the basis of inherent chemical characters (Cox and Hawkesworth 1985; Lightfoot and Hawkesworth 1988; Pattanayak and Shrivastava 1999).

On the basis of foregoing discussions, an attempt has been made to classify the basaltic lava flows of the presently studied sectors on the basis of Mg′ No. and TiO_2 content. The Mg′ No. of the investigated flows can be categorized into low Mg′ No. (0.39–0.45), moderate Mg′ No. (0.45–0.50) and high Mg′ No. (0.49–0.54). TiO_2 content of the flows also show similar variations so that they can be classified into high TiO_2 (2.9–3.6 wt%), moderate TiO_2 (2.4–2.9 wt%) and low TiO_2 (1.91–2.31 wt%) lavas. Accordingly, in order to classify the investigated lava flows, their compositions were plotted in TiO_2 versus Mg′ No. diagram (Fig. 10.2a). Figure 10.2a clearly shows that all the investigated lava flows can be classified into following three distinct groups. The details are under:

Groups	Character	Constituent flows	Fig. no.
1	Low Mg′ number, High TiO_2 flows	Flow IV, V, VI, XX of N-H-A sector and Flow I,II, III,VIII of N-L sector	10.2a
2	Moderate Mg′ number, moderate TiO_2 flows	Flow VII, VIII, X, XI, XIII, XIV, XV of N-H-A sector and Flow IV, V, VI, VII of N-L sector	10.2a
3	Very high Mg′ number, low TiO_2 flows	Flow IX, XII, XVI, XVII, XVIII, XIX of N-H-A sector and Flow IX of N-L sector	10.2a

Concomitant to the above distinct variations of TiO_2 and Mg′ No., it is noteworthy to mention that certain other major element oxides namely P_2O_5, CaO and Al_2O_3 show marked ranges corresponding to each group. As it will be discussed later in this paper, trace element and rare earth contents of these three groups of basaltic lavas show distinct range of values and confirm the validity of these chemical groupings. Flows I, II and III in the N-H-A sector and the sill in the N-L sector show a completely different chemical character and are not therefore included within these groups.

10.4 Correlative Stratigraphy of the Lava Flows

As mentioned earlier, 20 Deccan Trap flows of N-H-A sector and nine Deccan Trap lava flows of the N-L sector have been established on the basis of distinct physical characters. Further all these lava flows (belonging to both the sectors) have been

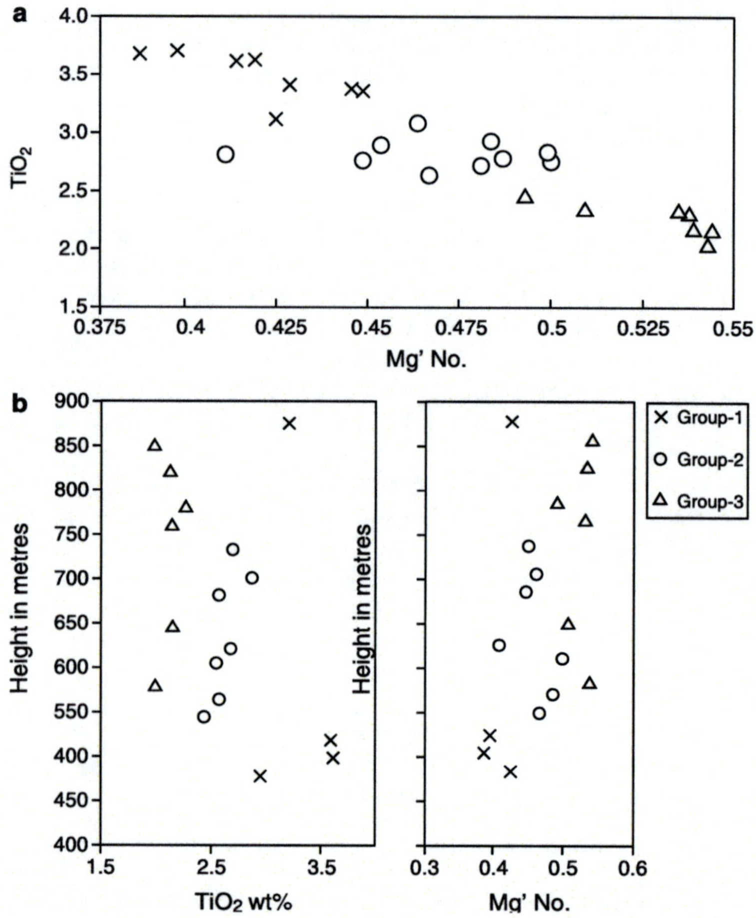

Fig. 10.2 (**a**) *Top panel*: TiO_2 versus Mg′ No. Plot in an attempt to classify the investigated basaltic rocks (*cross* – Group-1, *circle* – Group-2 and *triangle* – Group-3 basalts). (**b**) *Bottom panel*: Variation of TiO_2 and Mg′ No. with increasing heights in the present area of investigation. Explanations of symbols are same as in (**a**)

classified into three distinct groups as detailed earlier. It is interesting to note that the basaltic lavas corresponding to three different groups are intermixed with each other and this character can be well documented in terms of variation diagrams involving TiO_2 (wt%) and Mg′ No. plotted with respect to increasing height (Fig. 10.2b).

Field criteria, physical characteristics of the lava flows and corresponding chemical groups have been used to erect a standard stratigraphic succession of the investigated area and subsequent presentation of correlative stratigraphy between the Deccan Trap lava flows (including intertrappean beds) occurring in N-H-A sector

Table 10.1 Stratigraphic succession of N-H-A sector

Flow no.	Thickness of flow (m)	Height above mean sea level (m)
Basalt Flow XX	25	860
Basalt Flow XIX	20	840
Basalt Flow XVIII	30	810
Basalt Flow XVII	30	780
Basalt Flow XVI	20	760
Basalt Flow XV	30	730
Basalt Flow XIV	33	697
Basalt Flow XIII	22	675
Basalt Flow XII	33	642
Basalt Flow XI	22	620
Basalt Flow X	19	601
Intertrappean Bed	1	600
Basalt Flow IX	20	580
Basalt Flow VIII	18	562
Basalt Flow VII	20	542
Basalt Flow VI	22	520
Basalt Flow V	18	502
Basalt Flow IV	22	480
Basalt Flow III	18	462
Basalt Flow II	25	437
Basalt Flow I	32	405
Gondwanas	Base not seen	

and N-L sector. The details are shown in Tables 10.1 and 10.2. Correlative stratigraphy of the Deccan Trap lava flows of the study area has been presented in Table 10.3.

10.5 Resume of Petrographic Characters of the Lava Flows

The basaltic flows represent inequigranular phenocrystal basalts characterized by development of phenocrystal phases like olivine, pyroxene and plagioclase while the groundmass is composed of tiny grains of olivine, plagioclase, pyroxene, opaque and interstitial glass. Olivine phenocrysts are mostly pseudomorphed by iddingsite and chlorite leaving relicts of smaller olivine grains in some cases. Pyroxene phenocrysts mostly occur as prismatic and elongate grains of augite with few pigeonites. Plagioclase phenocrysts are mostly lath- like grains with some tabular shaped varieties. The largest plagioclase phenocryst is 1 cm in longest diameter which is comparable with Giant Plagioclase Basalts (GPBs) of the western Deccan segment (Hooper et al. 1988). A petrographic note on the basaltic lava flows of the present area of investigation was earlier published by Sengupta and Ray (2007).

Table 10.2 Stratigraphic succession of N-L sector

Flow no.	Thickness of flow (m)	Height above mean sea level (m)
Basalt Flow IX	20	580
Basalt Flow VIII	30	550
Intertrappean Bed	5	545
Basalt Flow VII	25	520
Basalt Flow VI	20	500
Basalt Flow V	18	482
Basalt Flow IV	20	462
Basalt Flow III	22	440
Basalt Flow II	18	422
Basalt Flow I	22	400
Gondwanas	10	390
The Sill	5 m exposed	385
Gondwanas	Base not seen	

Table 10.3 Correlative stratigraphy of the Deccan Trap lava flows of the study area

N-H-A sector	N-L sector	Corresponding chemical group
Flow XX	No	1
Flow XIX	Corresponding	3
Flow XVIII	representative	3
Flow XVII	lava flows	3
Flow XVI		2
Flow XV		2
Flow XIV		2
Flow XIII		3
Flow XII		2
Flow XI		2
Flow X		
Intertrappean Bed		
Flow IX	Flow IX	3
	Flow VIII	2
No	Intertrappean Bed	
Equivalent Flows	Flow VII	2
	Flow VI	2
Flow VIII	Flow V	2
Flow VII	Flow IV	2
Flow VI	Flow III	1
Flow V	Flow II	1
Flow IV	Flow I	1
Flow III Flow II Flow I	SILL (equivalent to flow I, II, III of N-H-A sector)	
Gondwanas	Gondwanas	

Table 10.4 Variation of major element oxides and different related parameters of the three basaltic groups

Group 1									Group 2									
Serial No.	1	2	3	4	5	6	7	8	9	10	11	12	13	14	15	16	17	18
Sp. No.	PSG231	PSG236	PSG239	PSG274	PSG76	PSG221	PSG178	PSG213	PSG31	PSG124	PSG600	PSG1	PSG255	PSG258	PSG302	PSG203	PSG183	PSG207
Flow No	IV	V	VI	XX	I	II	III	VIII	VII	VIII	X	XI	XIII	XIV	XV	IV	V	VII
Sector	N-H-A	N-H-A	N-H-A	N-H-A	N-L	N-L	N-L	N-L	N-H-A	N-H-A	N-H-A	N-H-A	N-H-A	N-H-A	N-H-A	N-L	N-L	N-L
SiO_2	47.53	51.28	50.07	48.27	50.64	48.29	50.36	47.57	47.47	47.00	48.52	49.86	48.69	49.73	48.78	47.62	48.02	48.28
TiO_2	2.97	3.62	3.60	3.24	3.47	3.19	3.47	3.17	2.47	2.61	2.59	2.71	2.61	2.91	2.73	2.56	2.75	2.66
Al_2O_3	11.41	12.87	12.87	12.02	11.32	12.05	11.32	11.81	13.63	12.88	12.63	13.39	13.81	13.32	13.02	12.74	13.04	13.28
Fe_2O_3	2.22	2.35	2.14	2.26	2.18	2.28	2.18	2.3	2.05	2.12	1.97	2.06	2.14	1.90	2.11	2.12	2.10	1.93
FeO	12.3	12.99	11.86	12.47	12.06	12.6	12.06	12.7	11.34	11.72	10.87	11.40	11.82	10.48	11.65	11.69	11.62	10.69
MnO	0.21	0.22	0.19	0.21	0.19	0.21	0.19	0.21	0.21	0.21	0.21	0.17	0.21	0.18	0.19	0.19	0.19	0.19
MgO	5.36	4.82	4.60	5.52	5.14	5.98	5.02	6.11	5.85	6.57	6.41	4.69	5.69	5.35	5.71	6.39	6.42	6.27
CaO	9.42	9.23	8.62	10.17	9.31	10.16	9.41	10.22	11.03	10.58	10.65	10.23	10.64	10.08	10.56	10.62	10.83	11.05
Na_2O	1.58	2.65	2.50	2.17	2.17	2.27	2.15	2.05	2.32	1.99	1.94	2.47	2.32	2.36	2.27	2.01	2.19	2.14
K_2O	0.09	0.53	1.05	0.21	0.41	0.28	0.39	0.22	0.19	0.13	0.14	0.25	0.37	0.34	0.27	0.15	0.19	0.17
P_2O_5	0.31	0.30	0.34	0.28	0.31	0.27	0.31	0.24	0.17	0.22	0.22	0.24	0.24	0.24	0.24	0.21	0.25	0.21
LOI	5.56	nd	1.24	2.01	1.73	1.9	2.08	2.06	nd	2.38	2.94	1.96	nd	1.97	1.53	2.61	1.67	2.15
Total	98.96	100.86	99.08	98.83	98.93	99.48	98.94	98.66	96.73	98.41	99.09	99.43	98.54	98.86	99.06	98.91	99.27	99.02
Mg'No.	0.43	0.39	0.40	0.43	0.42	0.45	0.42	0.45	0.47	0.49	0.50	0.41	0.45	0.47	0.46	0.48	0.49	0.50
FeO(t)	14.30	15.11	13.79	14.50	14.02	14.65	14.02	14.77	13.19	13.63	12.64	13.25	13.75	12.19	13.55	13.60	13.51	12.43
Na_2O+ K_2O	1.67	3.18	3.55	2.38	2.58	2.55	2.54	2.27	2.51	2.12	2.08	2.72	2.69	2.70	2.54	2.16	2.38	2.31
D.I	21.72	30.61	32.20	23.54	28.91	23.28	28.58	21.34	21.00	18.92	21.23	27.10	23.11	26.88	23.51	19.97	20.88	21.32
S.I.	24.87	20.65	20.77	24.39	23.41	25.54	23.03	26.13	26.90	29.16	30.05	22.47	25.47	26.19	25.94	28.58	28.51	29.58
CIPW norms																		
Qz	7.82	5.06	4.84	3.94	8.13	2.42	8.08	2.7	0.25	1.31	3.99	4.72	1.29	4.90	2.70	2.07	1.23	2.21
Or	0.53	3.13	6.21	1.24	2.42	1.65	2.31	1.3	1.12	0.77	0.83	1.48	2.19	2.01	1.60	0.89	1.12	1.00
Ab	13.37	22.42	21.15	18.36	18.36	19.21	18.19	17.34	19.63	16.84	16.41	20.90	19.63	19.97	19.21	17.01	18.53	18.11

(continued)

Table 10.4 (continued)

Group 1									Group 2									
Serial No.	1	2	3	4	5	6	7	8	9	10	11	12	13	14	15	16	17	18
Sp. No.	PSG231	PSG236	PSG239	PSG274	PSG76	PSG221	PSG178	PSG213	PSG31	PSG124	PSG600	PSG1	PSG255	PSG258	PSG302	PSG203	PSG183	PSG207
Flow No	IV	V	VI	XX	I	II	III	VIII	VII	VIII	X	XI	XIII	XIV	XV	IV	V	VII
Sector	N-H-A	N-H-A	N-H-A	N-H-A	N-L	N-L	N-L	N-L	N-H-A	N-H-A	N-H-A	N-H-A	N-H-A	N-H-A	N-H-A	N-L	N-L	N-L
An	23.78	21.66	20.80	22.44	19.94	21.87	20.09	22.38	26.22	25.83	25.34	24.71	26.18	24.75	24.54	25.30	25.20	26.14
Di	17.43	18.54	16.47	21.78	20.15	22.23	20.45	22.2	22.71	20.85	21.49	20.42	20.80	19.60	21.81	21.52	22.32	22.61
Hy	20.88	19.07	17.64	18.98	17.73	20.22	17.27	20.76	18.74	21.90	19.80	16.55	19.83	16.82	18.88	21.09	20.38	18.53
Ol	–	–	–	–	–	–	–	–	–	–	–	–	–	–	–	–	–	–
Mt	3.23	3.41	3.11	3.27	3.16	3.3	3.16	3.33	2.97	3.07	2.86	2.99	3.10	2.75	3.06	3.07	3.04	2.80
Ilm	5.64	6.87	6.84	6.15	6.59	6.06	6.59	6.02	4.69	4.96	4.92	5.15	4.96	5.53	5.18	4.86	5.22	5.05
Ap	0.73	0.71	0.81	0.66	0.73	0.64	0.73	0.57	0.40	0.52	0.52	0.57	0.57	0.57	0.57	0.50	0.59	0.50

Group 3												
Serial No.	19	20	21	22	23	24	25					
Sp. No.	PSG 66	PSG250	PSG263	PSG303	PSG293	PSG289	PSG280					
Flow No	IX	XII	XVI	XVII	XVIII	XIX	IX					
Sector	N-H-A	N-H-A	N-H-A	N-H-A	N-H-A	N-H-A	N-L	Max.	Min.	X(n = 25)	S..D.	Avg..Deccan[a]
SiO_2	49.13	49.86	47.47	48.16	48.26	48.15	48.89	51.28	47	48.72	1.13	50.56
TiO_2	2.03	2.19	2.18	2.31	2.16	2.01	1.91	3.62	1.91	2.72	0.51	2.57
Al_2O_3	14.29	14.33	14.01	14.17	13.78	13.95	13.19	14.33	11.32	13.01	0.92	13.83
Fe_2O_3	1.75	1.8	1.85	2.06	1.85	1.79	1.88	2.35	1.75	2.06	0.17	1.93
FeO	9.66	9.96	10.21	11.36	10.21	9.87	10.39	12.99	9.66	11.36	0.96	10.68
MnO	0.18	0.16	0.19	0.2	0.19	0.17	0.19	0.22	0.16	0.19	0.02	0.17
MgO	6.64	6.1	6.99	6.51	6.99	6.92	7.25	7.25	4.6	5.97	0.76	5.12
CaO	11.24	10.75	11.04	11.89	11.35	11.66	11.24	11.89	8.62	10.48	0.81	9.62

Na_2O	2.05	2.36	1.32	2.18	2.02	1.89	2.01	2.65	1.58	2.16	0.23	2.65
K_2O	0.14	0.19	0.14	0.11	0.15	0.12	0.14	1.05	0.09	0.25	0.20	0.93
P_2O_5	0.19	0.18	0.18	0.18	0.17	0.16	0.16	0.34	0.16	0.23	0.05	0.22
LOI	1.65	1.58	3.22		2.07	2.63	1.87	5.56	1.24	2.23	0.90	–
Total	98.95	99.46	99.30	99.13	99.20	99.32	99.12	100.86	96.73	99.02	0.66	98.28
Mg′No.	0.539	0.51	0.535	0.494	0.538	0.544	0.543	0.544	0.387	0.47	0.05	
FeO(t)	11.24	11.58	11.88	13.21	11.88	11.48	12.08	15.11	11.24	13.21	1.11	12.42
$Na_2O + K_2O$	2.19	2.55	1.96	2.29	2.17	2.01	2.15	3.70	1.67	2.41	0.43	3.58
D.I	20.97	23.78	17.86	19.33	19.01	18.34	19.40	32.2	17.86	22.91	3.98	30.47
S.I.	32.81	29.89	33.27	29.30	32.94	33.61	33.46	33.609	20.651	27.48	3.95	24.03
CIPW norms												
Qz	2.51	2.98	1.81	0	1.21	1.52	1.56	8.13	0	3.17	2.30	2.55
Or	1.12	0.83	0.65	0.89	0.71	0.83	0.83	6.21	0.53	1.51	1.18	5.5
Ab	17.34	19.97	15.4	18.44	17.09	15.99	17.01	22.42	13.37	18.24	1.97	22.42
An	29.23	28.11	29.74	28.44	28.19	29.17	26.57	29.74	19.94	25.06	2.89	23.1
Di	20.53	19.9	19.62	24.28	22.27	2.69	23.22	24.28	2.69	20.24	4.06	19.25
Hy	19.72	18.9	21.5	17.7	20.49	19.73	21.39	21.9	16.55	19.38	1.51	17.27
Ol	–	–	–	1.63	–	–	–	–	–	–	–	–
Mt	2.54	2.61	2.68	2.99	2.68	2.6	2.73	3.41	2.54	2.98	0.25	2.8
Ilm	3.86	4.16	4.14	4.39	4.1	3.82	3.63	6.87	3.63	5.18	0.97	4.88
Ap	0.45	0.43	0.43	0.43	0.4	0.38	0.38	0.81	0.38	0.55	0.13	0.52

X = average S.D. = standard deviation
[a]Thompson et al. 1983

10.6 Whole Rock Geochemistry

10.6.1 Major Element Variations

In this section, an attempt is being made to present the major element variations of all the different basaltic groups as a whole and to trace out any systematic pattern existing amongst them. The major element analyses of different rock samples belonging to all the three groups and their average composition have been furnished in Table 10.4 along with respective CIPW norms.

The correlation co-efficient (r) between different elements or element-oxides as determined through the regression methods (Dixon and Massey 1957; Chayes 1962) are attempted in case of selected major element oxides on the premise that correlation coefficients can help in establishing whether a prospective suite of rocks has resulted due to single stage close-system process or more than one magmatic upsurge. The ideal single stage close-system variation would result in production of a series of magma by separation of phases, e.g. fractionation of one or more solid phases. Involvement of an open system may impart introduction of a new compositional component (e.g. from external contaminant source), which may affect array of plots in variation diagram affecting the initial or main trend. Sudden cumulation or residue left after partial melting can also affect the geometry of the variation curve eventually affecting the value of correlation coefficients. If an ideal single stage magmatic process has controlled the petrogenesis of an assemblage, the members will show inter element correlation co-efficient varying from +1 to −1. In reality, however a number of factors control the correlation coefficients. Occasionally the variation trend shows curved course of liquid line of descent in biaxial variation diagram (which is actually two dimensional projection of the actual data space) and correlation coefficient (r) may vary over the region particularly when precipitation of the solid phase in the melt has variation in composition and or proportion. Because of the close-table effect, significant negative correlation may be enhanced and the resulting value may be far more negative. Therefore great care has been taken in the present study to furnish correlation co-efficient data which are free from close-table effect. In many cases, insignificant correlation may have been depicted within a particular group of basaltic flows in respect to some chemical attributes but among the different basaltic groups, when the same pair of chemical attributes are being applied, gives significant correlation values.

Silica: SiO_2 content of the investigated lava flows varies from 47.00 to 51.28 but often there is overlapping SiO_2 values in case of different investigated groups. The average SiO_2 content of the entire suite is 48.72, which is found to be slightly lower than that of average Deccan composition (Thompson et al. 1983). The generalized silica oversaturated character of the investigated basalts has been documented by the presence of normative quartz in its average composition (normative quartz 3.17%), which closely corresponds to that of average Deccan basalt of Thompson et al. (1983). However in one sample belonging to primitive (with high Mg'No.) group, a slight silica undersaturation (normative olivine 1.63 wt%) is reflected.

SiO_2 contents of the investigated basalts play an important role in putting them into systematic nomenclature. On this basis, total alkali contents ($Na_2O + K_2O$ wt%) of different samples were plotted against SiO_2 (wt%) values in total alkali versus silica diagram (Le Bas et al. 1986). All the investigated data plots belonging to different groups characteristically fall into the basaltic field (Fig. 10.3a) indicating almost similar degree of silica content in all the analyzed samples.

Alumina: Alumina shows moderate range of variation in the investigated samples ranging from 11.32 to 14.33 wt%. The average alumina content of all the samples is 13.01 wt% which is almost similar to that of the average Deccan basalt composition (13.83 wt%) (Thompson et al. 1983). Due to their very limited range of variation, Al_2O_3 content of the investigated samples do not show any perceptible patterns in

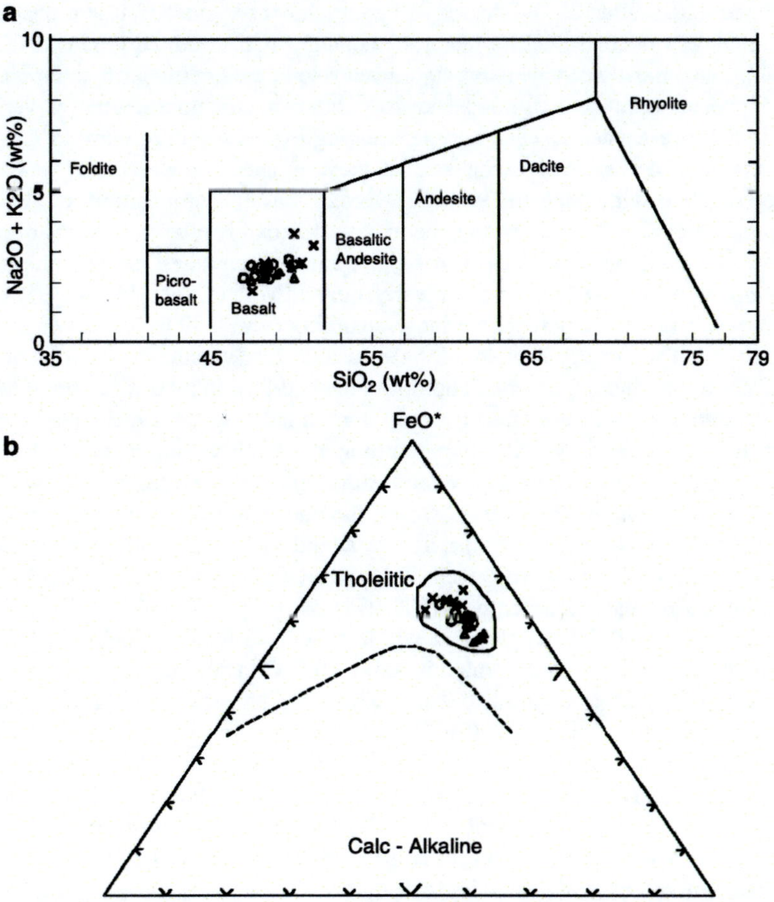

Fig. 10.3 **(a)** *Top panel*: Total alkali versus SiO_2 diagram for the investigated basalts (Le Bas et al. 1986). **(b)** *Bottom panel*: Plot of investigated samples in FeO^*–($Na_2O + K_2O$)– MgO diagram. Dividing line is after Irvine and Baragar (1971). Symbols are same as in Fig. 10.2

any of the variation diagrams constructed. The limited range of alumina variation may be ascribed to restricted variation of corresponding crystallizing phases namely plagioclase and/or pyroxene.

Total Iron Oxides (FeO total): The FeO total contents of any investigated suite of igneous rocks depict the degree of evolution of the magma and concomitant increasing trend of iron-enrichment. The FeO (t) content of the investigated basalts have a span from 15.11 to 11.24 wt% with an average of 13.21 FeO (t) %. The average value of the presently studied rocks indicates a comparable FeO (t) content with that of the average Deccan composition (12.42 wt%)(Thompson et al. 1983). This suggests a similar or comparable trend of Fe-enrichment in the studied basaltic lava flows.

FeO (t) content of the investigated lavas have been depicted in FeO*–MgO–(Na_2O + K_2O) diagram (Fig. 10.3b) where all the investigated data-plots show a limited range of distribution and fall well within the tholeiite field. However, even within the restricted data scatter, a mild degree of Fe-enrichment is evident. It is probable that Fe-enrichment is controlled by magnesia content in all the three groups of investigated basalts; mostly olivine plays significant role in calibrating such variation.

Magnesia: Like FeO(t), magnesia also shows a moderate degree of variation within the investigated Deccan samples ranging from 4.60 (minimum) to 7.25 wt% (maximum). The average MgO content of all the investigated samples is 5.97 wt% which matches well with that of average Deccan basalt composition (5.12 wt%) (Thompson et al. 1983). Thus, in general, a very broad similarity in terms of primitiveness of the investigated Deccan basalts and the average Deccan composition (Thompson et al. 1983) has been clearly hinted. Mg′No. (Mg/Mg + Fe^{2+}) of the studied basalts show a perceptible range spanning from 0.39 to 0.54 having a average of 0.47. The average Mg′No. of these basalts clearly matches with that of average Deccan basalt composition (Thompson et al. 1983). In fact, variation of Mg′No in a given mafic lava suite often serves as a good indicator for bracketing degree of internal differentiation within the parent magma. Such variations of Mg′No often have sympathetic or antipathetic relationship with other oxides or oxide-ratios in the same given magmatic suite. However, the nature of variation (between several oxides/oxide-ratios and Mg′No.) may not attend the same significance level. On this premise, variation of TiO_2 wt%, P_2O_5 wt%, CaO wt% and FeO(t) wt% have been plotted against the respective Mg′No (Fig. 10.4a–d). The calculated correlation coefficients (for all the samples as a whole as well as for those belonging to individual groups) have been presented in Table 10.5. Considering the entire spectrum of samples (n = 25), variation of TiO_2 against Mg′No. shows a highly significant negative correlation (r = –0.91) (Fig. 10.4a). However, in case of Group 1 basalts, r = –0.79 assumes a significant value. In case of Group 2 basalts the value of correlation coefficient (r = –0.23) becomes non significant while in case of Group 3 basalts the correlation coefficient (r = –0.81) again assumes significance. Hence for TiO_2 versus Mg′No. correlation (in respect of individual groups) the level of significance of correlation coefficients differ widely attesting the attainment of different degrees of internal fractionation. Figure 10.4b shows P_2O_5 versus Mg′No. variation wherein a highly significant negative correlation for the entire spectrum

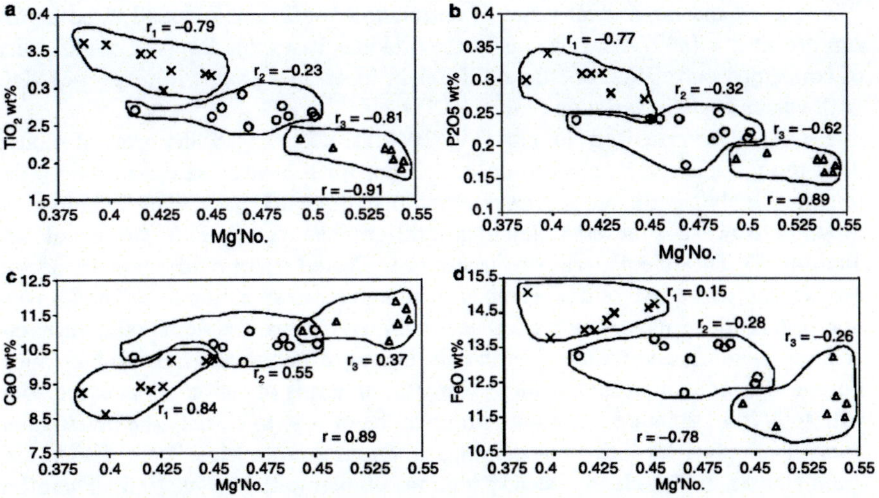

Fig. 10.4 **(a–d)** Plot of TiO_2 (wt%), P_2O_5 (wt%), CaO (wt%), and FeO (wt%) against Mg'No for the investigated basalts. Symbols are same as in Fig. 10.2

Table 10.5 Selected correlation coefficient data among different major element oxides and Mg'No.[a]

Parameter	Fig no.	Correlation coefficient for the entire spectrum (r)	r_1 (for Group 1)	r_2 (for Group 2)	r_3 (for Group 3)
TiO_2 vs Mg'No.	4a	–0.91 (H.S.)	–0.79 (S)	–0.23 (N. S.)	–0.81 (S)
P_2O_5vs Mg'No.	4b	–0.89 (H.S.)	–0.77 (S)	–0.32 (N. S.)	–0.62 (S)
CaO vs Mg'No.	4c	0.89 (H.S.)	0.84 (H. S.)	0.55 (N. S.)	0.37 (N. S.)
FeO (t) vs Mg'No.	4d	–0.78 (H.S.)	0.15 (N.S.)	–0.28 (N. S.)	0.26 (N. S.)

H.S. = Highly significant
S = Significant
N.S. = Non-significant
[a]Correlation coefficients have been tested at the 5% and 1% levels of significance (taking degrees of freedom = 23) [data for testing significance – levels have been taken from Snedecor and Cochran 1967, p. 557]

of rocks (r = –0.89) is evident. However, in case of different groups (namely Group 1, Group 2 and Group 3) the levels of significance of concerned correlation coefficient (r) differ widely, again suggesting variable degree of internal fractionation. Variation of CaO when plotted against Mg'No. shows a clear positive correlation

(Fig. 10.4c) and the overall value of correlation coefficient is found to be highly significant (r = 0.89). However, as in the previous cases, the levels of significance of concerned correlation coefficient differ in individual groups in respect of CaO wt% versus Mg′No. variation (Table 10.5).

FeO(t) wt% versus Mg′No. plot (Fig. 10.4d) reflects a bit wider array of plotted data, though the overall correlation coefficient for the entire samples (r = –0.78) remains highly significant. However, attainment of significance levels in individual group in this case is far less (attending non significant correlations) (for details see Table 10.5). Thus the patterns of variation of selected major oxides with respect to Mg′No. are suggestive of discrete magma pulses (corresponding to different investigated basaltic groups) which ultimately suffered internal fractionation in different degrees yielding correlation coefficients having different levels of significance. Further, Mg content in mafic suite (expressed in terms of cation %) in association with total iron, titanium and aluminium may be of help to classify the investigated rocks. Data plots for individual samples when referred to Mg–(Fe* + Ti)–Al diagram (Jensen 1976), clearly occupy high iron tholeiite field (Fig. 10.5). Therefore slight enrichment of iron for the investigated basalts (mainly at the expense of iron) is evident from the diagram.

Lime: CaO content in the investigated suite of rocks has a moderate scatter ranging from 8.62 to 11.89 wt% with an average of 10.48 wt% (Table 10.4). This average value is slightly higher than that of average Deccan basalt composition (Thompson et al. 1983). The lime content has been manifested in terms of presence of normative anorthite and diopside in all samples. Further, the lime contents go hand in hand along with Mg′No. in the investigated lava groups which speaks in favor of progressive depletion of Ca with increasing degree of differentiation. Nevertheless,

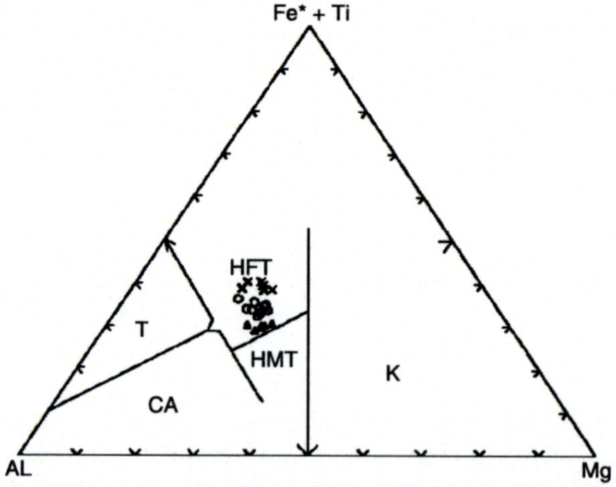

Fig. 10.5 Plot of investigated basalts in Fe* + Ti–Al–Mg diagram (After Jensen 1976). HMT: High-Mg tholeiite, HFT: High-Fe tholeiite. Symbols are same as in Fig. 10.2

the patterns of CaO depletion (with advent of magmatic crystallization) for three different groups are quite distinct (Fig. 10.4c).

Total Alkali: In the investigated basaltic lava groups, total alkali ranges from 3.7 to 1.67 wt% (Table 10.4); the average being 2.41 wt%. This average (Na_2O + K_2O) is slightly less than that of average Deccan Basalt composition (Thompson et al. 1983). However, it must be mentioned that the degree of variation of total alkalies in different lava groups as a whole is not too wide so that the data plots of the investigated basalts when referred to (Na_2O + K_2O) versus SiO_2 diagram (Fig. 10.3a) are clearly restricted well within the basalt field as has been stated earlier. Moreover the total alkali contents of all the samples lie well within the limit of normal tholeiitic basalts as documented from other parts of the world (Cox et al. 1979).

Titania: TiO_2 contents of the investigated basaltic samples belonging to three different groups range within wide limit namely from 1.91 to 3.62 wt% with an average of 2.72 wt%. This average is bit higher than that of average Deccan basalt composition of Thompson et al. 1983 (Table 10.4). However, titania contents of different investigated groups are distributed in distinct patterns so that a clear demarcation of the lava flows into three broad groups can be recognized. These are high TiO_2 (2.97–3.62 wt%), moderate TiO_2 (2.47–2.91 wt%) and low TiO_2 (1.91–2.31 wt%) basalts. In fact, such a systematic variation of TiO_2 contents and corresponding variations of MgNo′. (Fig. 10.2a) help to systematize the spatially distributed lava flows in the present study.

10.7 Variation of Trace Elements and Element Ratios

Trace element abundances of different investigated samples belonging to three different groups have been presented in Table 10.6. Trace element composition of average Deccan Basalt (Thompson et al. 1983) has also been cited for the purpose of comparison.

Strontium: Strontium content for the entire spectrum of the investigated basalt ranges from 202 to 345.44 ppm with an average value of 239.52 ppm. This average (239.52 ppm) is slightly higher than the Sr content of average Deccan Basalt composition (Thompson et al. 1983). So far as the individual basaltic groups are concerned, Sr shows overlapping values. In case of Group 3 basalts (primitive group) Sr shows relatively lesser values, possibly due to lesser concentration of plagioclase in the group. Ca/Sr ratio in the investigated samples show a range from 190.89 to 389.5 with an average of 308.63 which is very close to that of average Deccan Basalt composition. Like absolute Sr contents, Ca/Sr ratio also shows almost overlapping values for different basalt groups.

Rubidium: Rubidium content in the investigated basaltic samples shows a wide scatter (ranging from as low as 0.79 ppm to as high as 143 ppm). The average concentration of Rb is found to be 19.18 ppm, which is slightly higher than the Rb content of the average Deccan basalt composition (15 ppm) (Thompson et al. 1983). On a closer scan of Rb contents in respect to individual groups, it is revealed

Table 10.6 Trace element (in ppm) and element ratios of the investigated flows and average Deccan basalts

Group 1						Group 2			Group 3							
Serial no.	1	2	3	4	5	6	7	8	9	10	11	12				
Sp. No.	PSG 231	PSG 236	PSG 274	PSG 76	PSG 178	PSG 183	PSG 207	PSG 302	PSG 250	PSG 263	PSG 293	PSG 280	Min.	Max.	Avg. n = 12	Avg. Deccan
Ni	42.81	32	31	31.66	31.68	58.04	57.2	44.23	64.3	64.81	65.19	54.74	31	65.19	48.14	44
Cr	95.98	39.2	38.9	40.15	43.55	99.62	99.8	80.2	210.7	188.9	163.82	113.67	38.9	210.7	101.21	44
Sc	28.51	25.04	25	25.1	26.64	25	25. 4	26. 49	27.7	27.5	27.6	28.89	25	28.89	26.57	–
V	336.63	346.25	349	353.7	362.84	310.33	320	313.64	294	298	294	285	285	362.84	321.95	–
Ba	188.07	229.6	115	274.9	220	81	91.81	135.29	58	39	58	84	39	274.93	131.23	239
Rb	0.79	20.8		7.2	16.59	1.33	143	7.2	3.53	2.91	2.12	5.55	0.79	143	19.18	15
Sr	345. 44	232.68	234	259	256.27	240	243	226.3	209	215.6	211	202	202	345. 44	293.52	219
Zr	225.77	267.03	198	261.3	272.74	150.03	143	162. 41	123	121	127.74	115.17	115.2	272.74	180.60	203
Y	43	50	38	45.91	51	33.52	32	34. 41	30	29	30	29	29	51	37.15	50
Nb	17.08	19.53	16.7	19.39	19.86	14.38	13	12.93	10.96	10.99	10.51	9.78	9.78	19.86	14.59	15.9
Ta	1.01	1.17		1.16	1.19	0.98		0.77			0.63	0.59	0.59	1.19	0.94	1.39
Hf	6.12	7.06		7.27	7.28	4		4.52			3.5	3.21	3.21	7.28	5.37	
Th	3. 49	5. 44		5.27	5.35	1.39		1.67			1.18	1.35	1.18	5. 44	3.14	2.12
Ga	13.37	15.62		16.39	16.7	13.57		13.67			12.93	12.66	12.66	16.7	14.36	24
Cu	168.75	175.89		182.3	193.05	141.18		140.3			120.13	107.86	107.9	193.05	153.68	202
Zn	98.87	106.55		106.7	112.56	84.2		87.25			74. 48	73.68	73.68	112.56	93.03	149
Ca (ppm)	65,940	64,610	71,1090	65,170	65,870	70,210	77,350	73,920	75,250	77,280	79,450	78,680	64,610	79,450	72,076.67	67,340
Ca/Sr	190.89	277.68	304.231	251.7	257.0336	292.54	318.313	326.65	360.048	358.442	376.54	389.505	190.9	389.5	308.63	307.4886
Zr/Nb	13.218	13.673	11.8563	13.48	13.73313	10.433	11	12.561	11.2226	11.01	12.15	11.7761	10.43	13.733	12.18	12.7673
Zr/Y	5.2505	5.3406	5.21053	5.692	5.347843	4.4758	4.46875	4.72	4.1	4.17241	4.258	3.97138	3.971	5.6922	4.75	4.06
Nb/Y	0.3972	0.3906	0.43947	0.422	0.389412	0.429	0.40625	0.3758	0.36533	0.37897	0.35033	0.33724	0.337	0.4395	0.39	0.318
Sc/Ni	0.666	0.7825	0.80645	0.793	0.840909	0.4307	0.44406	0.5989	0.43079	0.42432	0.4233	0.52777	0.423	0.8409	0.60	–

that Rb content is highly fluctuating even within the same group; e.g. in Group 2 basalts, Rb content shows strong fluctuations starting with very low values (1.33 ppm) which shoots up to as high as 143 ppm. The normal way of explaining Rb concentration in a magmatic suite is to correlate its value with modal feldspar content. But even with that assumption, on the basis of petrographic studies, it is very difficult to explain such abrupt variation of Rb content manifested in a single basaltic group of the present study. It is therefore probable, that Rb suffered mobilization from the parent magma immediately before crystallization of the rocks or at a subsequently later geological time.

Barium: Barium is an important incompatible LIL element. Like Rb, Ba also shows strong fluctuations in respect of its composition having a minimum value of 39 ppm which increases up to a maximum of 274.93 ppm. The average Ba concentration is found to be 131.23 ppm which is lesser than that of average Deccan basalt composition (239 ppm) (Thompson et al. 1983). Within the investigated samples, the Group 1 basalts show a weak tendency to contain higher concentration of Ba which may be attributed to its enriched character relative to its primitive counterpart.

Nickel: Nickel has intermediate ionic size between Mg^{2+} and Fe^{2+}. Among the Mg, Fe and coherent heavy trace elements, the sequence of entry into crystal lattice is considered to be Mg, Ni, Fe^{2+} (Taylor 1965) which has been established from the inherent electronegativity relations (Ringwood 1955). In the present study, the Ni content shows a minimum value of 31 ppm which rises to a maximum value of 65.19 ppm. The average Ni content of the investigated basalt is found to be 48.14 ppm which is closely comparable to Ni content (44 ppm) of the average Deccan basalts (Thompson et al. 1983). Figure 10.6a shows a variation of Ni with Mg′ No. in the investigated basalts. From this figure it is evident that Group 1 basalts have a very close range of Ni content. However, in general, for the investigated flows, an increasing concentration of Ni is evident with rising Mg′ No. The maximum Ni content, as it is expected, is observed to be in the primitive (Group 3) type.

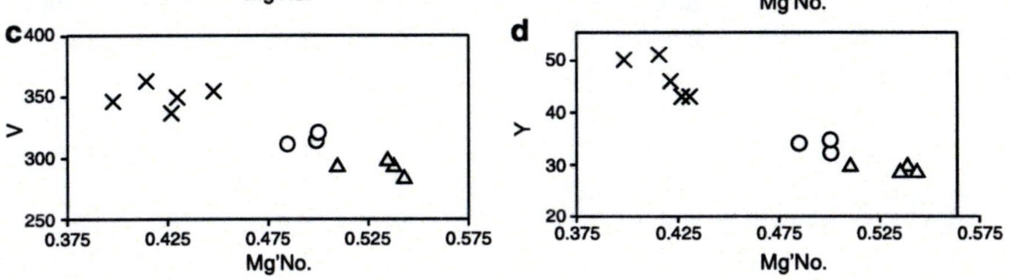

Fig. 10.6 **(a–d)** Plot of Ni (ppm), Cr (ppm),V (ppm) and Y (ppm) of the investigated basalts against respective Mg′No. Symbols are same as in Fig. 10.2

Chromium: Chromium is a ubiquitous heavy trace metal present in all the basaltic flows in varying concentration. The chromium content ranges from 38.9 to 210.7 ppm with an average value of 101.21 ppm. This value is slightly higher than the chromium content of the average Deccan basalt composition of Thompson et al. 1983 (44 ppm). In the variation diagram (Fig. 10.6b), the plots of chromium against Mg′No. shows a clear array of points with positive slope. However, the general sympathetic correlation between chromium and Mg′ No. is not identically manifested in all the three groups and the pattern of variation of chromium (with rising Mg′No.) is found to be slightly dissimilar in case of three different groups.

Vanadium: Vanadium is mostly removed from a magma by magnetite in which it proxies for Fe^{3+}. The relatively lesser values of electronegativity and higher crystal field stabilization energy generally compel vanadium to be concentrated in early formed magnetite. However, significant concentration of vanadium is also found in pyroxenes having a higher value of Fe^{3+} content. The vanadium content in the investigated basaltic samples range from 285 to 362.84 ppm; the average being 321.95 ppm (Table 10.7). In order to judge whether vanadium concentration has some preferential incorporation in any of the three basaltic groups, a variation diagram involving vanadium (ppm) versus Mg′No. has been constructed (Fig. 10.6c). This figure shows that the enriched group (Group 1 basalt) shows a relatively higher concentration of vanadium than its primitive counterparts (Group 3 basalt). It would imply that apart from magnetite, a mafic-silicate phase (pyroxene) has played a significant role in accumulating vanadium in the enriched type (Group 1 basalt).

Yttrium: The yttrium content of the investigated basaltic samples varies within a very short range (with a minimum value of 29 ppm while the maximum being 51 ppm). The average yttrium concentration is found to be 37.15 as against Y = 50 ppm for the average Deccan basalt composition (Thompson et al. 1983). Figure 10.6d shows variation of yttrium content with varying Mg′No. It is evident from this figure that yttrium contents assume highest value in Group 1 or enriched type of basalt while yttrium content is relatively low for the primitive type (Group 3 basalt). This feature reflects relatively higher concentration of yttrium in the more evolved or enriched type.

Zirconium: The combination of high charge and comparatively higher radius (0.79 Å) makes zirconium apart from any of the major elements in common magmatic systems. The zirconium represents a high field strength incompatible element and generally its concentration rises with increasing degree of evolution of the magma. In the present study, zirconium content of the investigated samples ranges from 115.17 to 272.74 ppm (Table 10.7). The average zirconium concentration is found to be 180.60 ppm which is relatively less than the zirconium content (203 ppm) of the average Deccan basalt composition (Thompson et al. 1983). Considering zirconium to be an effective indicator of the enrichment of the concerned magma types, several variation diagrams were constructed in the present study with respect to relevant zirconium concentrations. Figure 10.7a–e represents variation of several major element oxides namely TiO_2, MgO, Al_2O_3, CaO and P_2O_5 with respect to zirconium. TiO_2 and P_2O_5 show increasing trends while MgO, Al_2O_3 and

Table 10.7 Rare earth element (in ppm) and chondrite normalised values of the investigated flows and average Deccan basalts (minimum, maximum and average values for the entire spectrum of rocks have also been shown)

	La	Ce	Pr	Nd	Sm	Eu	Gd	Tb	Dy	Ho	Er	Tm	Yb	Lu
Group 1					**Group 2**		**Group 3**							
Serial no.	1	2	3	4	5		6	7	8					
Sp. No.	PSG 231	PSG 236	PSG 76	PSG 178	PSG 183		PSG302	PSG 293	PSG 280	Min.	Max.	Avg.	Avg. Deccan	C1 Chondrite
La	21.87	27.15	25.9	25.56	12.68		14.94	10.65	11.51	10.65	27.15	18.78	19.30	0.237
Ce	50.44	61.02	57.84	56.62	30.87		35.07	25.43	26.64	25.43	61.02	42.99	43.00	0.612
Pr	6.92	8.13	7.75	7.58	4.43		4.9	3.66	3.63	3.63	8.13	5.88	5.20	0.095
Nd	31.94	36.39	35.12	34.11	21.2		23.16	17.54	16.84	16.84	36.39	27.04	27.60	0.467
Sm	8.56	9.47	8.85	8.75	5.94		6.42	5.14	4.68	4.68	9.47	7.23	7.60	0.153
Eu	2.68	2.8	2.66	2.58	2.13		2.19	1.83	1.64	1.64	2.80	2.31	2.47	0.058
Gd	9.04	9.77	9.01	8.95	6.39		6.97	5.77	5.05	5.05	9.77	7.62	8.40	0.2055
Tb	1.5	1.57	1.48	1.43	1.05		1.13	0.95	0.84	0.84	1.57	1.24	–	0.0374
Dy	8.44	8.93	8.93	9	5.8		6.57	5.41	4.86	4.86	9.00	7.24	8.18	0.254
Ho	1.64	1.69	1.73	1.76	1.12		1.27	1.06	0.96	0.96	1.76	1.40	1.55	0.0566
Er	4.34	4.51	4.48	4.5	2.91		3.32	2.75	2.61	2.61	4.51	3.68	4.20	0.1655
Tm	0.59	0.61	0.63	0.63	0.39		0.47	0.38	0.37	0.37	0.63	0.51		0.0255
Yb	3.81	3.98	4.1	3.97	2.53		3.09	2.41	2.38	2.38	4.10	3.28	3.63	0.17
Lu	0.58	0.59	0.6	0.59	0.37		0.46	0.36	0.36	0.36	0.60	0.49	0.53	0.025
La_N	92.28	114.56	109.28	107.85	53.50		63.04	44.94	48.57	44.94	114.56	79.25	81.43	
Ce_N	82.42	99.71	94.51	92.52	50.44		57.30	41.55	43.53	41.55	99.71	70.25	70.26	
Pr_N	72.84	85.58	81.58	79.79	46.63		51.58	38.53	38.21	38.21	85.58	61.84	54.74	
Nd_N	68.39	77.92	75.20	73.04	45.40		49.59	37.56	36.06	36.06	77.92	57.90	59.10	
Sm_N	55.95	61.90	57.84	57.19	38.82		41.96	33.59	30.59	30.59	61.90	47.23	49.67	
Eu_N	46.21	48.28	45.86	44.48	36.72		37.76	31.55	28.28	28.28	48.28	39.89	42.59	
Gd_N	43.99	47.54	43.84	43.55	31.09		33.92	28.08	24.57	24.57	47.54	37.07	40.88	
Tb_N	40.11	41.98	39.57	38.24	28.07		30.21	25.40	22.46	22.46	41.98	33.26	–	

(continued)

Table 10.7 (continued)

	La	Ce	Pr	Nd	Sm	Eu	Gd	Tb	Dy	Ho	Er	Tm	Yb	Lu
Dy_N	33.23	35.16	35.16	35.43	22.83		25.87	21.30	19.13	19.13	35.43	28.51	32.20	
Ho_N	28.98	29.86	30.57	31.10	19.79		22.44	18.73	16.96	16.96	31.10	24.80	27.39	
Er_N	26.22	27.25	27.07	27.19	17.58		20.06	16.62	15.77	15.77	27.25	22.22	25.38	
Tm_N	23.14	23.92	24.71	24.71	15.29		18.43	14.90	14.51	14.51	24.71	19.95	0.00	
Yb_N	22.41	23.41	24.12	23.35	14.88		18.18	14.18	14.00	14.00	24.12	19.32	21.35	
Lu_N	23.20	23.60	24.00	23.60	14.80		18.40	14.40	14.40	14.40	24.00	19.55	21.20	
$\Sigma(REE)_N$	659.36	740.66	713.31	702.03	435.87		488.74	381.32	367.04	367.04	740.66	561.04	526.19	
Ce_N/Yb_N	3.68	4.26	3.92	3.96	3.39		3.15	2.93	3.11	2.93	4.26	3.55	3.29	
La_N/Sm_N	1.65	1.85	1.89	1.89	1.38		1.50	1.34	1.59	1.34	1.89	1.64	1.64	
Sm_N/Yb_N	2.50	2.64	2.40	2.45	2.61		2.31	2.37	2.18	2.18	2.64	2.43	2.33	
La/Sm	2.554907	2.86695	2.926554	2.9211	2.13468		2.327103	2.071984	2.459402	2.07	2.93	2.53	2.01	
Lu/Hf	0.094771	0.08357	0.082531	0.081	0.0925		0.10177	0.102857	0.11215	0.08	0.11	0.09	–	
Average Chondrite Normalised REE data for Group 1, Group 2, Group 3 basalts (normalized with respect to C1 Chondrite, Sun and McDonough 1989)														
Average Group 1	105.99	92.28	79.94	73.64	58.21	46.20	44.73	39.97	34.74	30.12	29.65	26.93	24.11	23.32
Average Group 2	85.29	75.82	66.74	62.41	50.82	42.77	39.78	35.63	29.92	25.86	23.27	20.66	20.12	20.06
Average Group 3	46.75	42.54	38.37	36.81	32.09	29.91	26.33	23.93	20.22	17.84	16.19	14.71	14.09	14.17

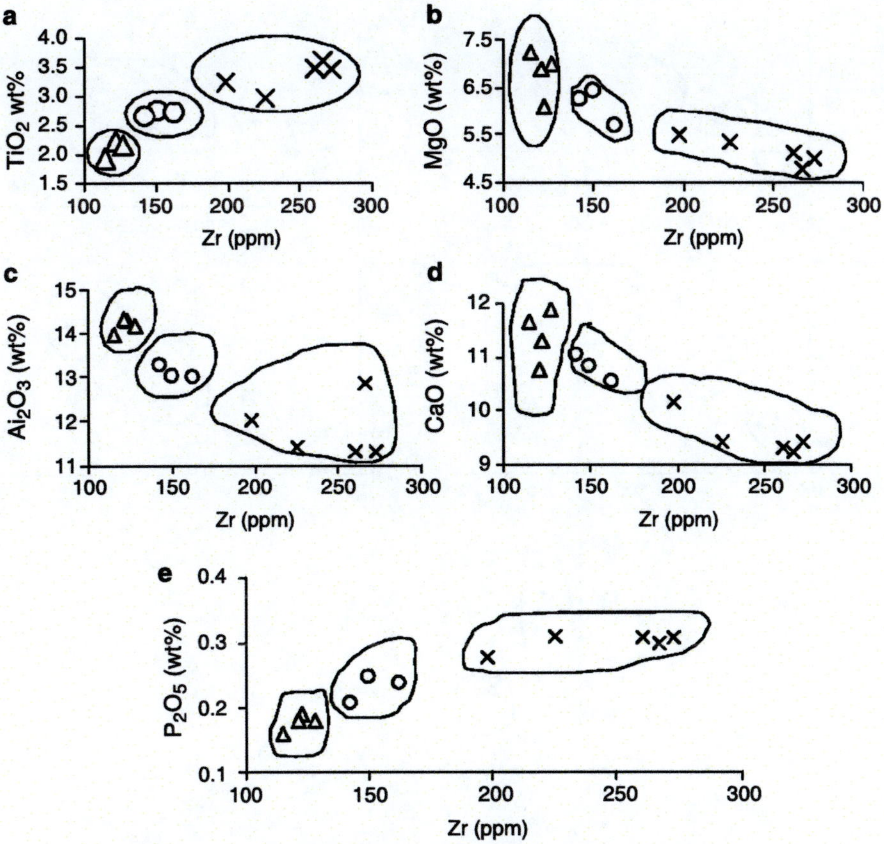

Fig. 10.7 **(a–e)** Plot of TiO_2 (wt%), MgO (wt%), Al_2O_3 (wt%), CaO (wt%), and P_2O_5 (wt%) against Zr (ppm) of the investigated basalts. Symbols are same as in Fig. 10.2

CaO show declining trends with rising zirconium concentration in the investigated samples. These indicate relatively higher concentration of TiO_2 and P_2O_5 in the enriched type of basalts. On the other hand, MgO, Al_2O_3 and CaO assume relatively lower values in the enriched type being controlled by the distinct magmatic impulses and/or subsequent mineral fractionation. Even with the maintenance of a generalized pattern, Fig. 10.7 shows clear seclusion corresponding to different basaltic groups, which might be linked to the evolution of the parental magma types. Figure 10.8a–e depicts variation of several trace elements with varying zirconium concentrations. From Figs. 10.8a–d it is clear that vanadium, yttrium, niobium and barium increase steadily with rising zirconium contents. This speaks strongly in favor of increased amount of vanadium, yttrium, niobium and barium contents being controlled by corresponding degree of magmatic evolution. Here again (in Fig. 10.8a–e), clear seclusion of data plots (corresponding to different magmatic ambience) is evident. Variation of chromium, when plotted against

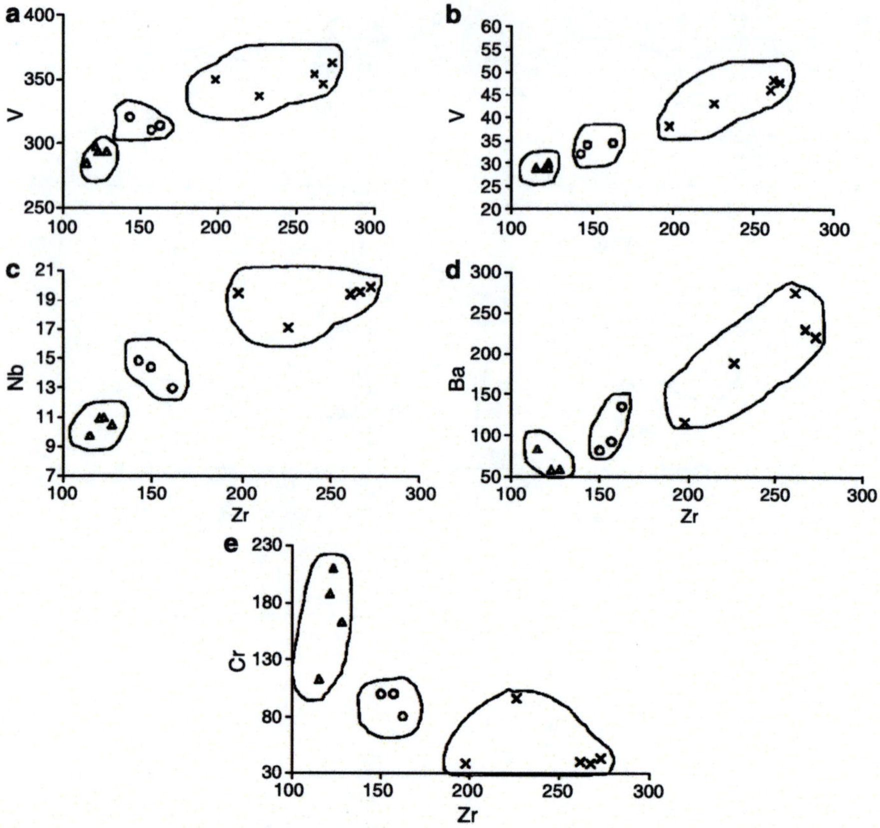

Fig. 10.8 (**a–e**) Plot of V (ppm), Y (ppm), Nb (ppm), Ba (ppm), and Cr (ppm), against Zr (ppm) of the investigated basalts. Symbols are same as in Fig. 10.2

respective zirconium concentration (Fig. 10.8e) shows a rough tendency of falling trend with increasing zirconium, which also corresponds to discrete evolutionary history of the magma pulses.

10.7.1 Rare Earth Elements

The rare earth element (REE) composition of the investigated lava flows have been given in Table 10.7. The REE data of the flows belonging to three different groups have been normalized against chondrite (Sun and McDonough 1989). The normalized patterns for the different flows have been depicted in Fig. 10.9a. From Table 10.7 it appears that in general $\sum(\mathrm{REE})_N$ for Group 3 flows is the least while

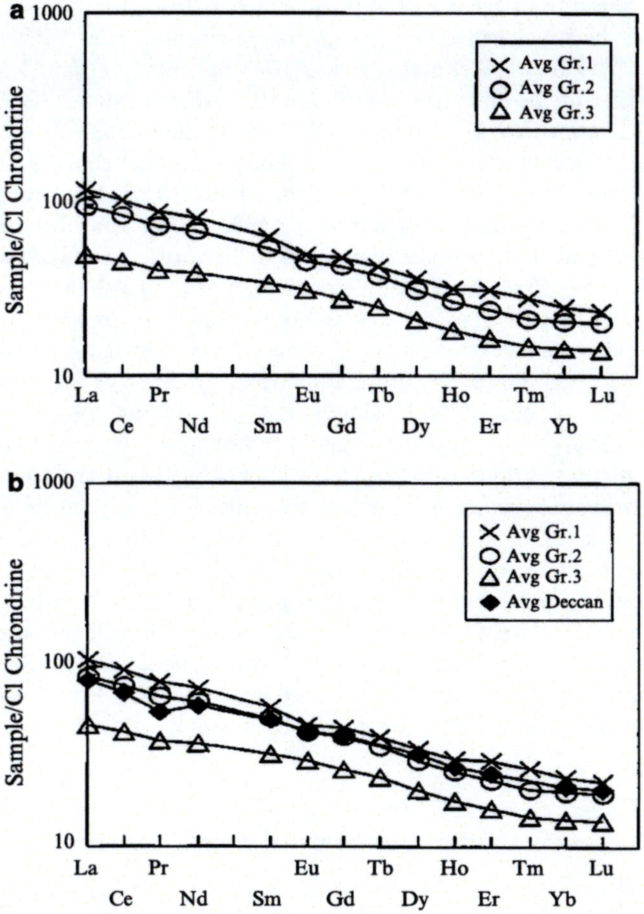

Fig. 10.9 (**a**) Chondrite-normalized REE pattern for the investigated basaltic rocks (*cross*: Average of Gr 1 basalts, *circle*: Average of Gr.2 basalts and *triangle*: average of Gr.3 basalts). (**b**) Chondrite-normalized REE pattern for the investigated basaltic rocks (*cross*: Average of Gr 1 basalts, *circle*: Average of Gr.2 basalts and *triangle*: average of Gr.3 basalts). *Solid diamond*: Average Deccan Trap Basalts (Thompson et al. 1983)

the same value corresponding to Group 1 basalts is the highest. The $\sum (\mathrm{REE})_N$ data for Group 2 member are intermediate between those of Group 1 and Group 3 and therefore reflect REE variation almost intermediate between those of Group 1 and Group 3. Figure 10.9a represents rock/chondrite plot for the average Group 1, Group 2 and Group 3 flows. This figure clearly shows that for all the three investigated groups, the fundamental patterns for the REE variation are essentially the same showing most primitive pattern for the Group 3 basalts and most evolved pattern for the Group 1 basalts. The similar slopes for the three distinct groups can also be visualized from Table 10.7 as documented by $(Ce)_N/(Yb)_N$ ratios. In case of Group 3, data have a close $(Ce/Yb)_N$ ratios (ranging from 2.93 to 3.11) whereas the Group 1 basalts have distinct higher ratios {$(Ce /Yb)_N$ ranging from 3.68 to 4.26}. Group 2 basalts, on the other hand (see Table 10.7), show moderate $(Ce/Yb)_N$ ratios (3.15–3.39) intermediate between those of Group 1 and Group 3. As mentioned earlier, these almost similar ratios have resulted in identical chondrite normalized slope in all the three members in Fig. 10.9a. Figure 10.9b shows rock/chondrite pattern of Deccan basalts (Thompson et al. 1983) along with those of Group1, Group2 and Group3 basalts under investigation. The overall slope of the chondrite normalized pattern of all the three investigated groups and average Deccan basalts are found to maintain a more or less similar slope in the relevant diagram (Fig. 10.9b). It is often customary to depict the pattern of variation of trace element concentrations (including REE) of basaltic groups in terms of mantle normalized diagram devised by Sun and McDonough (1989). Useful petrogenetic information using appropriate geochemical tools have been obtained from volcanic rocks of Pioneer Formation of Southwest British Columbia by Dostal and Church (1994). On this premise, average mantle normalized values for several trace elements for all the investigated groups have been depicted in Fig. 10.10. As it is expected, the mantle normalized diagram for all the three groups show a more or less uniform pattern with different degrees of enrichment. Figure 10.9a shows distinct geochemical polarity for the investigated Group 3 and Group 1 basalts whereas Group 2 basalts reveal values intermediate between those of Group 1 and Group 3. This figure further corroborates relatively enriched patterns of the Group 1 basalts and a relatively pristine character for the Group 3 basalts.

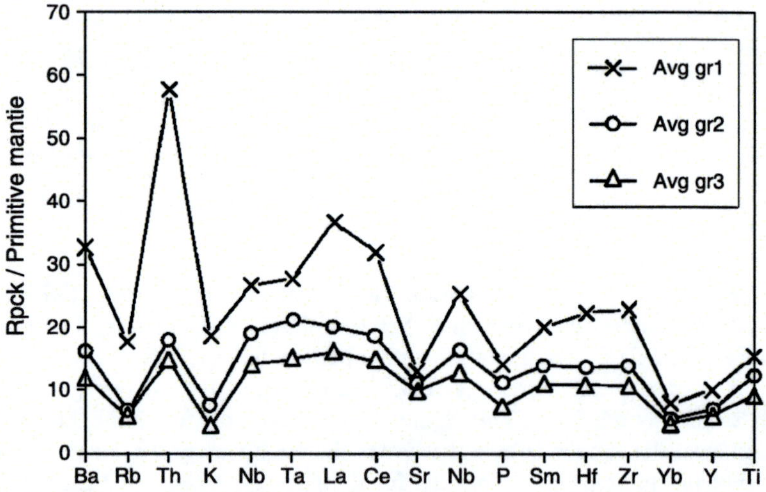

Fig. 10.10 Primitive mantle normalized trace element pattern for three investigated basalt-groups (average composition) [For details, see text]

10.8 Petrogenesis

10.8.1 Constraining Source Characteristics

In order to arrive at a meaningful conclusion to constrain the source characters, attention was paid to several critical trace element ratios for bracketing the source characteristics. The useful ratios in this context are Zr/Nb, La/Nb, Ba/Nb, Ba/Th, Rb/Nb, K/Nb, Th/Nb, Th/La and Ba/La. All these trace element ratios maintain distinctive values corresponding to different source characteristics (Weaver 1991). On this basis, these elemental ratios for all the three investigated basaltic groups (with range of variation and average values) have been presented in Table 10.8. From this table, it is evident that Group 1 basalts, on consideration of majority of the trace element ratios correspond either to EMI or EMII or HIMU sources. Group 2 basalts, on the other hand, (Table 10.8) show in general a more proximity towards HIMU sources; although affinity towards EMI and EMII sources cannot also be overruled. Group 3 basalts (Table 10.8) in majority of the cases show affinity towards HIMU sources, although in some cases, EMI sources are also hinted. Therefore, whatever may be the actual source characteristics (EMI or EMII or HIMU or any possible combination amongst them), a distinctive parental source corresponding to enriched mantle can be documented. The EMI source was originally from a mantle that had been enriched by small volume-melts and metasomatic fluids and it is having characteristically low Sr^{87}/Sr^{86} ratios. The other type of enriched mantle source (EMII) refers to recycled subducted sediments protolith and thereby imparting significant enrichment in the mantle. Characteristically EMII sources have high Sr^{87}/Sr^{86} ratios. The other enriched mantle (HIMU) refers to source with high U^{238}/Pb^{204} ratio.

10.8.2 Mantle Melting Model

In order to have a generalized idea to generate the parental melt (giving rise to different basaltic flows of the investigated area), plume source may be a good candidate (Watson and Mc Kenzie 1991). It is generally believed that the plume responsible to produce the melt has a central potential temperature of 1,558°C corresponding to a mechanical boundary layer of 72 km thickness. The melt-producing region has a vertical extent of 55 km and a radial extent of approximately 130 km to produce the melt in the scale of 0.16 km^3 per year. As the mantle source characteristics for the present flows have been deduced to be an enriched one (Table 10.8), it will be worthwhile to assess the melt characterization assuming an enriched mantle source. Figure 10.11a shows plots of Lu/Hf versus La/Sm, when plots for investigated basaltic flows indicate melting of garnet peridotite source component; the enriched group (Group 1) corresponds to relatively lower degree of melt production. Figure10.11b involving plot of $(Sm)_n/(Yb)_n$ versus $(La)_n/(Sm)_n$

Table 10.8 Several critical trace element ratios of the basaltic flows of the study area to constrain source characteristics

Group 1 basalts										Group 2 basalts						
						Min	Max	Avg	Source[a]				Min	Max	Avg	Source
Zr/Nb	13.218	13.673	11.716	13.478	13.733	11.716	13.733	13.164	EMI	10.433	11.000	12.561	10.433	12.561	11.331	EMI
La/Nb	1.280	1.390	–	1.336	1.287	1.280	1.390	1.323	EMI	0.882	–	1.155	0.882	1.155	1.019	EMI
Ba/Nb	11.011	11.756	6.805	14.179	11.078	6.805	14.179	10.966	EMII	5.633	7.062	10.463	5.633	10.463	7.719	EMII
Ba/Th	53.888	42.206		52.169	41.121	41.121	53.888	47.346	HIMU	58.273	–	81.012	58.273	81.012	69.643	HIMU
Rb/Nb	0.046	1.065	–	0.371	0.835	0.046	1.065	0.579	EMII	0.092	–	0.557	0.092	0.557	0.325	HIMU
K/Nb	43.724	225.185	103.110	175.458	162.949	43.724	225.185	142.085	HIMU	109.638	108.511	173..273	108.511	173.273	130.474	HIMU
Th/Nb	0.204	0.279	–	0.272	0.269	0.204	0.279	0.256	Akin to EMII	0.097	–	0.129	0.097	0.129	0.113	EMII
Th/La	0.160	0.200	–	0.203	0.209	0.160	0.209	0.193	EMII	0.110	–	0.112	0.110	0.112	0.111	HIMU
Ba/La	8.599	8.457	–	10.615	8.607	8.457	10.615	9.070	EMII OIB	6.388	–	9.056	6.388	9.056	7.722	HIMU

Group 3 basalts								
					Min	Max	Avg	Source[a]
Zr/Nb	11.223	11.010	12.154	11.776	11.010	12.154	11.541	EMI
La/Nb	–	–	1.013	1.177	1.013	1.177	1.095	EMI
Ba/Nb	5.292	3.549	5.519	8.589	3.549	8.589	5.737	HIMU
Ba/Th	–	–	49.153	62.222	49.153	62.222	55.687	HIMU
Rb/Nb	0.322	0.265	0.202	0.567	0.202	0.567	0.339	HIMU
K/Nb	143.850	105.705	118.428	118.783	105.705	143.850	121.692	HIMU
Th/Nb	0.000	0.000	0.112	0.138	0.000	0.138	0.063	HIMU
Th/La	–	–	0.111	0.117	0.111	0.117	0.114	HIMU
Ba/La	–	–	5.446	7.298	5.446	7.298	6.372	HIMU

EMI: Enriched mantle components (~ slightly modified bulk earth component)
EMII: Enriched mantle components (recycled subducted sedimentary protolith)
HIMU: Enriched mantle with high U^{238}/Pb^{204} ratio
[a]The source characteristics have been constrained using the ratios given by Weaver (1991)

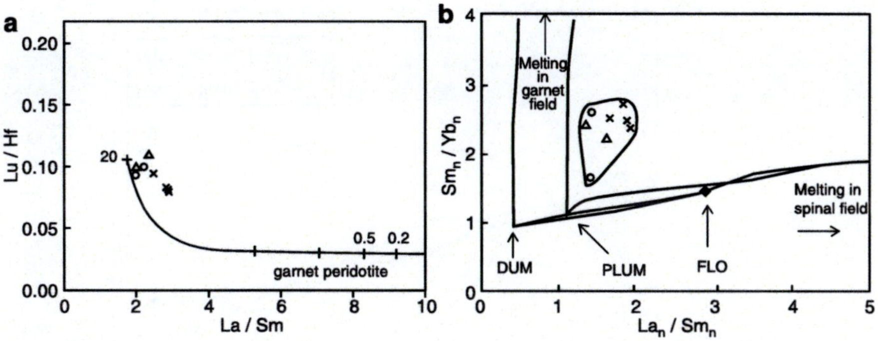

Fig. 10.11 (**a**) Lu/Hf versus La/Sm plot for the investigated basalt-groups of the study area. Garnet peridotite field is after Regelous et al. (2003). *Tick marks* indicate progressive degree of partial melting. Explanations of symbols are same as in Fig. 10.2. (**b**) Plot of Sm_n/Yb_n versus La_n/Sm_n for investigated basalt groups. Symbols are same as in Fig. 10.2 Different melting fields have been incorporated from Harpp and White (2001)

depicts fields for depleted upper mantle and plume with respective melting domains in garnet peridotite and spinel peridotite field. The plots for the investigated flows (Fig. 10.11b) are clearly secluded in the melting domain of garnet peridotite source. The apparent spread of the plotted data indicates varying degree of melting of the garnet peridotite source (for details see Harpp and White 2001). With this basic assumption, an attempt has been made to put forward a quantitative model for the generation of the three investigated groups of basalts. The relevant REE compositions of the initial plume source has been incorporated from Storey et al. (1997) and the corresponding data have been furnished in Table 10.9. This table also incorporates respective values of distribution coefficients (for different elements in respect of constituent phases) used in the melting exercises.

As three investigated groups correspond to different degrees of primitiveness (or enrichment) (Fig. 10.9a), it is obvious that they correspond to discrete degrees of melting of the plume which was followed by magma-upsurge in a rising plume environment. With this backdrop of knowledge, it is expected that a relatively low degree of melting of the initial plume source might have generated the enriched type (Group1 basalts) while a significantly higher degree of melting might have produced primitive group of basalts (Group 3). The intermediate groups of basalts (Group 2) have a geochemical characteristics intermediate between Group 1 and Group 3 and therefore might owe their origin to moderate degree of melting. The observed and calculated patterns of the Group 1, Group 2 and Group 3 basalts have been furnished in Figs. 10.12, 10.13 and 10.14 respectively. Summary observations worked out from critical theoretical calculations in respect of generation of Group 1, Group 2 and Group 3 basalts have been given in Fig. 10.15. The data for REE- modeling for different groups of the investigated basalts have been furnished in Table 10.10. It has to be noted that all the flows (corresponding to three different groups) basically represent phenocrystal basalts with different degrees of dominance of constituent phases like olivine, clinopyroxene and plagioclase

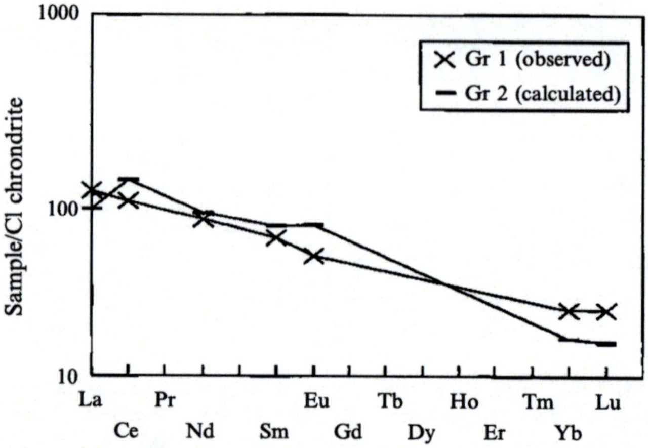

Fig. 10.12 Rock/Chondrite patterns for Group 1 basalts. *Cross*: average composition (observed). *Solid bars*: calculated composition

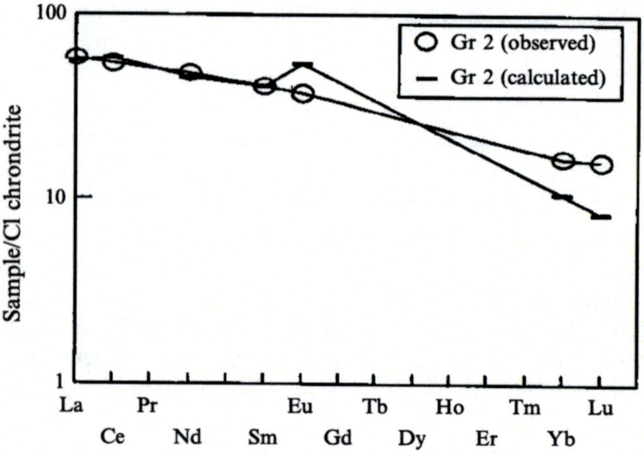

Fig. 10.13 Rock/Chondrite patterns for Group 2 basalts. *Circle*: average composition (observed). *Solid bars*: calculated composition

depending upon individual group-characteristics, controlling structural parameters and inherent petrography. Therefore all the investigated basaltic flows represent varying degrees of mechanical mixtures between ambient fluids and solid fractionates. It is interesting to note that theoretical calculations reveal that all the three groups of basalts suffer a little degree (~1–2%) of late stage apatite fractionation. Such apatite fractionation is often suggestive of mixing between enriched and slightly depleted melt compositions (Farley et al. 1992). In case of plume

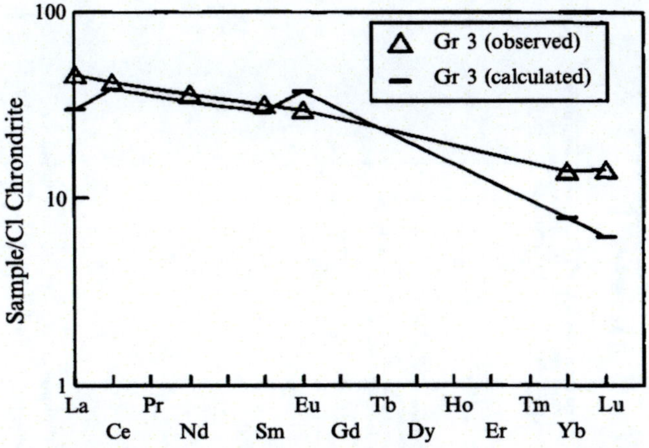

Fig. 10.14 Rock/Chondrite patterns for Group 3 basalts. *Triangle*: average composition (observed). *Solid bars*: calculated composition

Table 10.9 Data of Rare Earth Element composition of the initial plume source and respective distribution co-efficient for different element in respect of constituent phases

Data in (ppm)							
	La	Ce	Nd	Sm	Eu	Yb	Lu
Plume source[a]	32	65	35	7.8	2.6	2.26	0.32
Distribution co-efficient (K_d) of different elements for several constituent phases used for partial melting and fractional crystallisation							
	La	Ce	Nd	Sm	Eu	Yb	Lu
Olivine	–	0.009	0.009	0.009	0.008	0.034	0.11
Plagioclase	0.14	0.14	0.08	0.08	0.32	0.07	0.08
Orthopyroxene	–	0.02	0.05	0.05	0.05	0.34	0.11
Clinopyroxene	0.08	0.34	0.6	0.9	0.9	1	0.8
Garnet	0.05	0.05	0	0.6	0.9	30	35
Apatite	–	31	50	54	27	21	17

Data for Kd values have been taken from Henderson 1984, p. 26
[a]Data from Storey et al. 1997

derived HIMU type of enriched basalts, fractionation of apatite in the magma chamber has been documented by Zhang et al. (1999). Therefore, the late stage apatite fractionation testifies binary mixing of the derived melts from the plume components. Evidences of generation of multiple magmatic pulses and subsequent mixing between them have also been documented by House et al. (1997) and Burnard et al. (1998). In the present case, apart from apatite fractionation, evidences of open system melt-mixing (subsequent to upsurge of magma-pulses) have been imprinted on phase mineralogy (details have been given by Sengupta and Ray 2007).

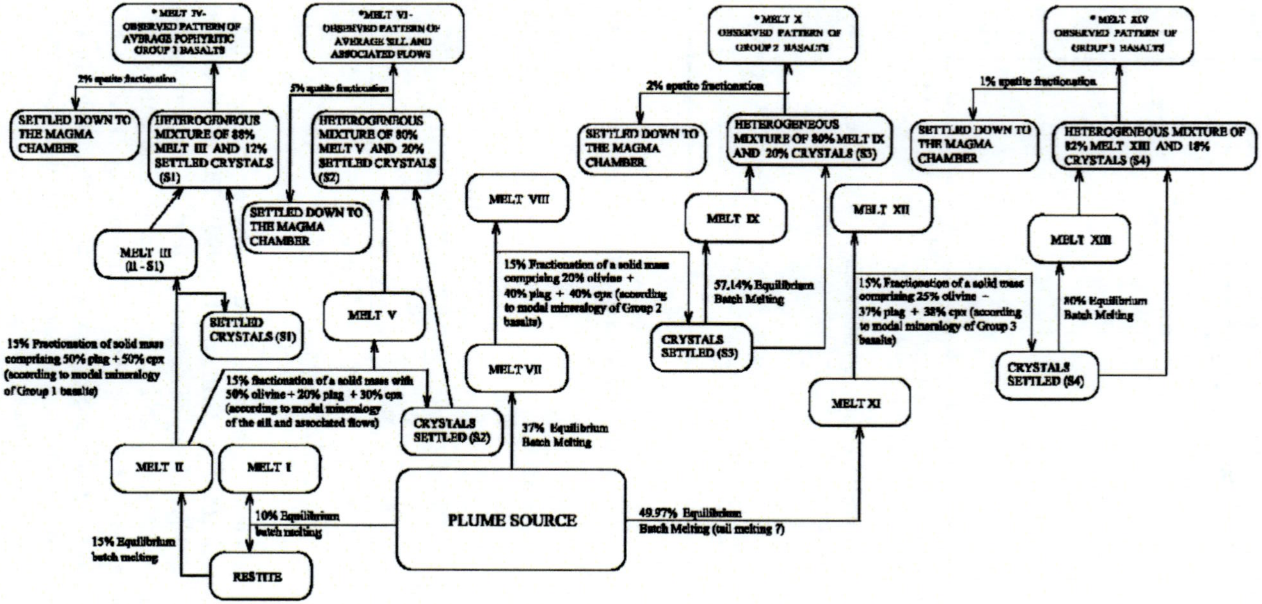

Fig. 10.15 Flow chart showing petrogenetic evolution of the investigated basaltic groups starting with an enriched plume source. It is to be noted that different melt sources might have been mingled together upon upwelling. The details of petrogenesis of the intrusive sill (encountered at Macharewa river, Fig. 10.1) has not been dealt with in this present paper but included in the summary of the flow chart for having a cogent view

Table 10.10 Data for REE -modelling for three basaltic groups of the study area (REE data in ppm)[a]

Group 1 basalt	Modal data of source composition (plume)	Melting mode	Restite composition (left after 10% equilibrium batch melting of the plume source)	Composition of Melt II	Composition of Melt III	Composition of Melt IV (calculated composition)	REE data for melt (observed)
	Ol-58.00	Ol-16.00	La-2.98	La-18.78	La-21.93	La-19.91	La-25.12
	Opx-26.00	Opx-42.00	Ce-23.02	Ce-115.72	Ce-130.93	Ce-74.36	Ce-56.48
	Cpx-15.00	Cpx-41.50	Nd-17.41	Nd-72.64	Nd-81.02	Nd-37.67	Nd-34.39
	Gt-1.00	Gt-1.00	Sm-4.71	Sm-21.20	Sm-22.74	Sm-10.36	Sm-8.91
			Eu-1.58	Eu-5.75	Eu-6.37	Eu-3.99	Eu-2.68
			Yb-2.08	Yb-3.77	Yb-4.05	Yb-2.73	Yb-3.96
			Lu-0.290	Lu-0.511	Lu-0.557	Lu-0.393	Lu-0.590
				[melting mode of restite]			
				Ol-15.80			
				Opx-42.00			
				Cpx-41.50			
				Gt-0.700			
Group 2 basalt	**Modal data of source composition (plume)**	**Melting mode**	**Composition of Melt VII**	**Composition of Melt VIII**	**Composition of Melt IX**	**Composition of Melt X (calculated composition)**	**REE data for melt (observed)**
	O1-58.00	Ol-22.00	La-85.89	La-99.50	La-14.24	La-13.39	La-13.81
	Opx-26.00	Opx-44.00	Ce-169.37	Ce-192.59	Ce-61.04	Ce-35.18	Ce-32.97
	Cpx-15.00	Cpx-33.10	Nd-89.42	Nd-100.33	Nd-45.97	Nd-21.30	Nd-22.18
	Gt-1.00	Gt-0.900	Sm-19.11	Sm-21.02	Sm-13.88	Sm-6.19	Sm-6.18
			Eu-6.34	Eu-6.87	Eu-5.07	Eu-3.11	Eu-2.16
			Yb-3.49	Yb-3.81	Yb-2.76	Yb-1.81	Yb-2.81
			Lu-0.460	Lu-0.510	Lu-0.312	Lu-0.210	Lu-0.410

(continued)

Table 10.10 (continued)

Group 3 basalt	Modal data of source composition (plume)	Melting mode	Composition of Melt XI	Composition of Melt XIII	Composition of Melt XIV (calculated composition)	REE data for melt (observed)
	O1-58.00	Ol-35.00	La-64.00	La-7.44	La-7.26	La-11.08
	Opx-26.00	Opx-34.00	Ce-129.41	Ce-32.00	Ce-23.93	Ce-26.03
	Cpx-15.00	Cpx-30.00	Nd-69.70	Nd-23.60	Nd-15.44	Nd-17.19
	Gt-1.00	Gt-0.980	Sm-15.27	Sm-7.04	Sm-4.50	Sm-4.91
			Eu-5.08	Eu-2.78	Eu-2.18	Eu-1.73
			Yb-3.24	Yb-1.64	Yb-1.33	Yb-2.39
			Lu-0.43	Lu-0.190	Lu-0.160	Lu-0.360
				[melting mode of crystals already settled down (S4)]		
				Ol-29.00		
				Pl-35.00		
				Cpx-36.00		

[a]Please also see relevant Figs. 10.12 to 10.15

10.9 Discussion

Critical evaluation of the available published petro-mineralogical and geochemical data of the Deccan basalts (for both western and eastern segments) indicate that in majority of the cases, enriched mantle or sub-continental lithospheric mantle or plume source was involved. In fact synchronity of magmatism at the western and eastern Deccan indicates that the diameter of the Reunion-Deccan plume head might have been at least ~800 km (Sen and Cohen 1994). Moreover by noting the geometry of the plateau-uplifts and associated magmatism, Mc Getchin et al. (1980) suggested following idealized steps which explains the interrelation between the two: (i) thermal expansion of the lithosphere due to deep mantle plume or hot spot and (ii) final ending at detachment and collapse of lithosphere into large pieces beneath the continents and flow of mantle material in the asthenosphere to replace it. According to McKenzie (1984) the most important consequence of the intrusion of great thicknesses of magma into the lower crust is the associated surface uplift. These arguments therefore suggest that large igneous intrusions are a major cause of uplift. McKenzie (1984) also opined that another cause of uplift is hot upwelling mantle. The corresponding amplitude for uplift in continent is about 700 m. Whatever be the nature of plume source, it has been agreed upon that in general, early alkaline magmatism in Deccan Traps corresponds to the period of plume incubation (Sheth and Chandraseharam 1997). Further, if one has to accept the plume model for Deccan basalt magma generation, Reunion plume is the most acceptable candidate (Peng and Mahoney 1995; O'Neil et al. 2003; Bhattacharji et al. 2004). Incubation and upwelling of the Reunion plume has caused enormous outpouring of Deccan lavas (mean age ~65–66 million years). However, Mahoney et al. (2002) opined that the location, ages and geochemical characteristics of marine volcanic rocks preserved in the south Tethyan Suture zone of Pakistan suggest that the Reunion hot spot was active off northwestern greater India well before the emplacement, far to the south, of the Deccan flood basalts the great bulk of which was erupted at 65–66 million years and are widely believed to be associated to the hot spot's plume-head phase (Fig. 10.16).

Now, this is the opportune moment to discuss implications of obtained petrological and geochemical data in the light of knowledge already existing for Deccan basalts or continental flood basalts in general. First, the investigated lava piles do not show a clear gradation of chemical characters from bottom upward. In other words, different chemical groupings of the investigated lavas are not systematically arranged so far as their spatial relations are concerned. On the other hand, in a vertical section of the lava pile, their chemical characteristics (or progressive evolution of magma) are not systematically related while going from the base to the top of the sequence. Thus, the basal portion of the basaltic lava pile shows enriched chemistry which may be followed upward by basaltic flows having primitive or intermediate chemistry. This fact was already cited while systematizing the basaltic lava-characters on the basis of major element chemistry. Close juxtaposition of basaltic flows (in terms of mix-up of several chemical groups in a vertical section) is also evident

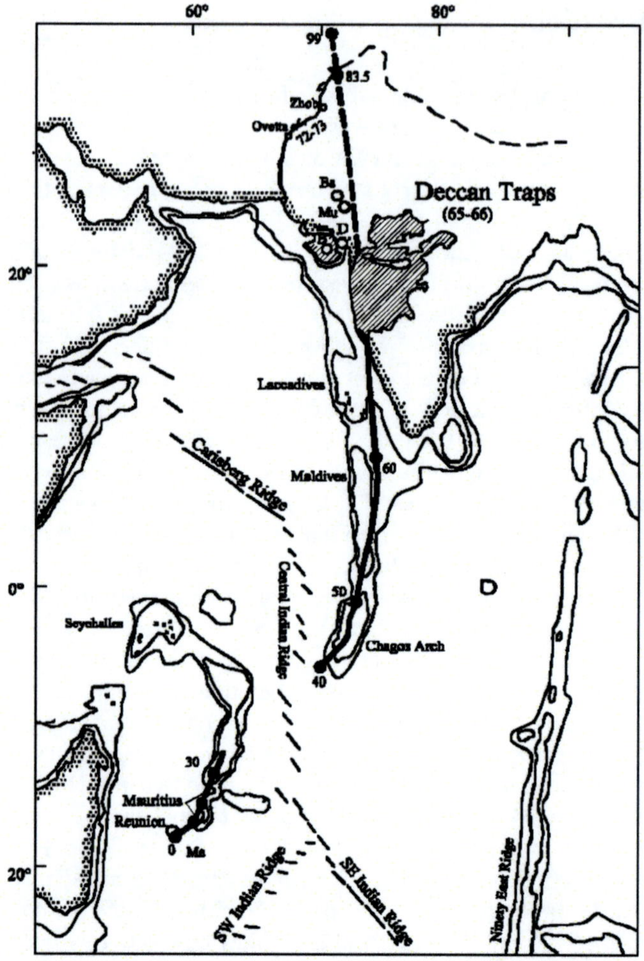

Fig. 10.16 The heavy line indicates the model 0–99 Ma track of the Reunion hotspot (Mahoney et al. 2002) assuming a stationary hotspot and that plate reconstruction are accurate. The dashed pre-Deccan portion is anchored by the 83.5 and 99 Ma plate reconstructions of Kent et al. (2002). The south Tethyan suture zone of Pakistan passes north and east into the Indus–Yarlung suture zone indicated schematically by the *light dashed line*. D: Dhandhuka, Ba: Botad, Ba: Barmer, Mu: Mundwara

in terms of their trace element chemistry, expressed in terms of several trace element ratios (Fig. 10.17). Thus, this situation in the present case, reasonably corresponds to the model of Cox and Hawkesworth (1985) highlighting the idea of periodical replenishment of parental magmas operative in a large scale. Second, critical consideration of trace element chemistry of the investigated basalts point to distinctly enriched mantle source and thus enriched mantle source was taken up while modelling the evolution of three groups of investigated basalts in the present

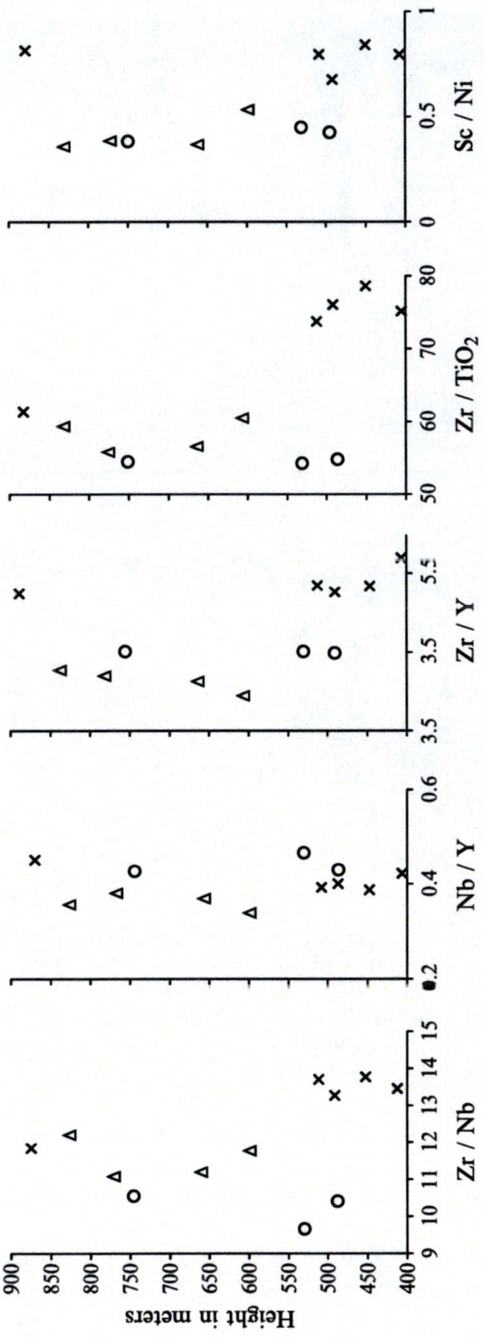

Fig. 10.17 Variation of different trace element ratios of the three basaltic groups with increasing height. Explanations of symbols are same as in Fig. 10.2

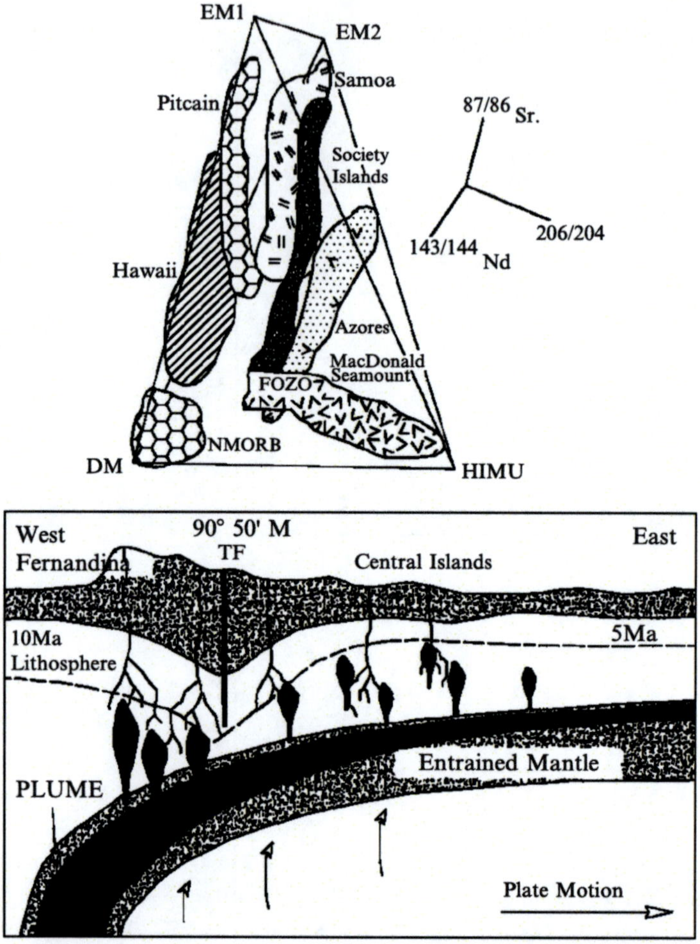

Fig. 10.18 *Top panel*: Tetrahedral diagram showing positions of DM, EMI, EMII, HIMU and FOZO (for details see text). *Bottom panel*: Model of entrainment of plume-head (from a plume source originating at West Fernandina, Galapagos) far into eastern part of Galapagos (Harpp and White, 2001). This model can be applied for the present study involving the Reunion plume having a large plume-head (over 800 km in diameter). This model explains the possibility of the plume-head located far off to the present area (lying at the western Deccan segment)

case. It is of particular attention that several types of enriched sources namely EMI, EMII and HIMU sources were indicated in the present study. When viewed in Nd–Sr–Pb isotopic space, arrays defined by two and three component mixing lines involving DM (depleted mantle), HIMU, EMI, EMII (see Condie 2001, p. 165) occupy at the corners of a tetrahedron (see Fig. 10.18a) However, there is a notable lack of mixing arrays joining EMI, EMII and HIMU which seems to rule out random mixing in the mantle (Hart et al. 1992; Hauri et al. 1994; Condie 2001). The mixing arrays are systematically oriented, originating from points along the EMI-HIMU and EMII-HIMU joins and converging in a region within the lower part of the tetrahedron, but not at depleted mantle (DM). This convergence suggests existence of another mantle component that appears in the mantle plumes. This component, which is called FOZO (or C) after focal zone from the converging arrays, is a depleted mantle component (Hart et al. 1992). The absence of any mixing arrays emanating from the EMI-EMII join or from the EMI-EMII-HIMU face of the tetrahedron indicates that there was some process that caused juxtaposition of EM components with HIMU that predates the mixing with FOZO. The probability that most or all hot spots (or plumes) do not mix with the shallow depleted mantle (DM) is consistent with most plume entrainment occurring in the lower mantle. To explain the existence of 'depleted' and 'enriched' mantle, scientists have put forward the model of compositional stratification in the deep-mantle (see Kellog et al. 1999; Helffrich and Wood 2001). Model of Helffrich and Wood (2001) clearly shows that mantle convection is layered, with the deeper mantle largely isolated from the upper mantle and further the mechanism of whole-mantle convection provided that the observed heterogeneities are remnants of oceanic and continental crust that make up about 16% and 0.3%, respectively of mantle volume. [However the nature of differentiation of the mantle (see Hoffman 1997) has been viewed differently by Meiborn and Anderson (2003) where they have supported the view that statistical distribution of lithologic components and sampling theory should replace the concept of distinct, isolated geochemical reservoirs, and extensive convective stirring prior to sampling]. Absence of any depleted mantle-signature in the present study can therefore be explained in terms of EMI, EMII, DM and HIMU-tetrahedron.

Third, the investigated basalts show presence of both high and low TiO_2 basalts. This feature has been observed in many other flood basalt provinces of the world (Condie 2001). Geochemical studies indicate that the two basalt-groups cannot be related by fractional crystallization, there is considerable debate as to whether they represent different degrees of melting of the same source or contamination of plume derived magmas by sub-continental lithosphere (Sharma 1997; Hawkesworth et al. 2000). Although the origin of high and low Ti flood basalts is still not resolved, most trace element and isotopic data demand two sources for the two groups (Sharma 1997). Condie (2001) has summarized different views regarding the origins of high and low Ti basalts. According to Condie (2001), high Ti magmas appear to have been generated by low degrees of melting of garnet lherzolite, leaving garnet in the restite, whereas low Ti magmas reflect a high degree of melting of the mantle source. Quantitative trace element modelling, attempted in the present study clearly reveals that different degrees of melting (of an enriched garnet peridotite mantle source) generated investigated basaltic flows corresponding to different chemical groups. Present study indicates that Group 1 basalts (with high TiO_2) generates only on low degree (~10%) garnet peridotite mantle-melting while the other polar end or Group 3 basalts (with low TiO_2) owe their origin to a high degree (~50%) of garnet peridotite mantle-melting. This may be taken as gross-analog of plume-melting model (head and tail melting) proposed by Campbell et al. (1989).

Therefore, outcome of the present study conforms to the models proposed by Fodor (1987) and Sharma (1997).

Fourth, how did involvement of Reunion plume take place in the generation of the present Deccan basalt magmatism is an important question (related to the plume-model for LIPS); do uplift and extensional tectonics precede magmatism as predicted by numerical models? This factor assumes more importance particularly when some counter-arguments (opposed to plume model) do exist for Deccan Traps, as for example, that championed by model of Sheth (1999). According to Sheth's model, continental mantle delamination, crack-propagation through oceanic lithosphere and crack-controlled melting and magma focusing can explain well observations generally ascribed to plumes. In fact, Sheth's (1999) model is following generalized alternative model proposed by Anderson et al. (1992) considering presence of enriched perisphere to explain presence of enriched basalts. However, in the absence of definite geodynamic evidences of crack-propagation and lack of numerical support, for Deccan basalts, model proposed by Sheth (1999) is still in its infancy and needs vigorous field and petrogeochemical support before coming up as a viable and realistic model. Obviously, the amount of uplift is controlled by plume buoyancy and by lateral variations in the thickness and rheology of the lithosphere. In the simplest case, plume heads with high viscosity, a maximum uplift of 500–1,000 m should be attained when the plume is at depths of 100–200 km (Griffiths and Campbell 1990). The uplift should begin 25 million years before the time of maximum uplift and precede major basaltic volcanism by 3–30 million years. Hence, unless the lithosphere is already thinned from forces in the plates, melting in the top of the plume should not occur until uplift has ceased. Uplift is followed by subsidence, which results from some combination of following three factors: (i) as the plume head thins and spreads laterally, it looses buoyancy and begins to sink over the plume axis; (ii) the removal of magma from the top of the plume results in deflation, causing subsidence of overlying lithosphere and (iii) gradual heat losses from the surface over hundreds of millions of years also results in some subsidence. In the present study, actual delineation of the magma generated-centre is difficult to assess but presence of sill, spatially associated with the enriched type of basaltic Group (or Group 1), may suggest remnant centre of probable magma-generation. This dyke/sill activity was possibly related to extensional tectonics (Bose 1996; Sheth 2000) and eruption of the investigated Deccan basaltic flows. But if the plume head (or generation of the magma-centre) was located far off to the present area (lying at the western Deccan segment) then a large plume head (having over 800 km diameter) corresponding to Reunion plume (see Sen and Cohen 1994; Bhattacharji et al. 2004) may be envisaged. Model of entrainment of plume head (from a plume source originating at West Fernandina, Galapagos) entraining far into eastern part of Galapagos has been invoked by Harpp and White (2001) (Fig. 10.18b). Similar situation may be envisaged for Eastern Deccan segment as well. However, a possibility may also lie that effusion of mafic bodies took place along the Narmada–Tapti rift are related to lineament reactivation and subsequent rifting due to interaction of hot mantle plume with the lithospheric weakness (lineaments).

Acknowledgments The present paper forms a part of the Ph.D. thesis of the first author (PSG). PSG is indebted to late Prof. A. De for providing erudite guidance during the early part of this research-work. The authors are thankful to Prof. A.R. Basu of Rochester University, USA for kindly arranging trace element analyses by ICP-MS at his laboratory. The authors are extremely grateful to Prof. John Mahoney, Department of Geology and Geophysics, University of Hawaii, USA for some stimulating discussion on problems of Eastern Deccan volcanism. A preliminary draft of this paper was kindly examined by Prof. Jarda Dostal of St. Mary's University, Canada. The authors also thank the Head, Department of Geology, University of Calcutta for providing necessary laboratory facilities. Sincere thanks are due to Sohini Ganguly, Mousumi Banerjee and Abhishek Saha (research- scholars of the Department of Geology, University of Calcutta) for their kind help rendered to the authors during preparation of the final version of this paper.

References

Alexander PO (1999) Rare earth abundances in low-K basalts at Sagar, India. In: Srivastava RK, Chandra R (eds) Magmatism in relation to diverse tectonic settings. A.A. Balkema, Rotterdam, pp 415–427

Alexander PO, Paul DK (1977) Geochemistry and Sr isotopic composition of basalts from eastern Deccan basaltic province, India. Min Mag 41:165–172

Anderson DL, Zhang Y, Tanimoto T (1992) Plume- heads, continental lithosphere, flood basalts and tomography. Geol Soc Lond Spec Publ 68:99–124

Bakshi AK (1994) Geochronological studies of whole-rock basalts, Deccan traps, India: evaluation of the timing of volcanism relative to the K-T boundary. Earth Planet Sci Lett 121:43–56

Bhattacharji S, Sharma R, Chatterjee N (2004) Two and three dimensional gravity modeling along western continental margin and intraplate Narmada–Tapti rifts: its relevance to Deccan flood basalt volcanism. Proc Indian Acad Sci (Earth Planet Sci Lett) 113(4):771–784

Bose MK (1996) Deccan basalts, mantle activity and extensional tectonics. Gondwana Geological Magazine 2:267–281

Burnard PG, Farley KA, Mclnnes B (1998) Multiple fluid phases in a Samoan harzburgite. Chem Geol 147:99–111

Campbell IH, Griffiths RW (1990) Implications of mantle plume structure for the evolution of flood basalts. Earth Planet Sci Lett 99:79–93

Campbell IH, Griffiths RW, Hill RI (1989) Melting in an Archean mantle plume: head it's basalts, tail it's komatiites. Nature 339:697–699

Carlson RW (1991) Physical and chemical evidences on the cause and source characteristics of flood basalt volcanism, Australian. J Earth Sci 38:525–544

Chayes F (1949) A simple point counter for thin section. Am Miner 36:1–11

Chayes F (1962) Numerical correlation and petrographic variation. J Geol 70:440–452

Coffin MF, Eldholm O (1994) Large igneous provinces: crustal structure, dimensions and external consequences. Rev Geophys 32:1–36

Condie KC (2001) Mantle plumes and their record in earth history. Cambridge University Press, U.K., p 306

Courtillot V, Feraud G, Maluski H, Vandamme D, Moreau MG, Besse J (1988) Deccan flood basalts and the Cretaceous/ Tertiary boundary. Nature 333:843–846

Cox KG, Hawkesworth CJ (1985) Geochemical stratigrapghy of the Deccan Traps at Mahabaleswar, western Ghats, India with implications for open system magmatic processes. J Petrol 26:355–377

Cox KG, Bell JD, Pankhurst RJ (1979) The interpretation of igneous rocks. George Allen & Unwin, London, p 450

Crookshank H (1936) Geology of the northern slopes of the Satpuras between Morand and Sher River. Geol Surv India, Mem 66(2):276–349

De A (1974) Short and long distance correlation of Deccan Trap lava flows. Abs Bull Geol Min Met Soc India 47:50

De A (1996) Entablature structure in Deccan Trap flows: its nature and probable mode of origin. Gond Geol Mag 2:439–447

Deshmukh SS, Sano T, Fujii T, Nair KKK, Yedekar DB, Umino S, Iwamori H, Aramaki S (1996a) Chemical stratigraphy and geochemistry of the basalt flows from the central and eastern parts of the Deccan Volcanic Province of India. Gondwana Geological Magazine 2:145–170

Deshmukh SS, Sano T, Nair KKK (1996b) Geology and chemical stratigraphy of the Deccan basalts of Chikaldara and Behramghat sections of the eastern part of the Deccan Trap province, India. Gondwana Geological Magazine 2:1–22

Dixon WJ, Massey FJ (1957) Introduction to statistical analyses. Mc Graw Hill, New York, p 488

Dostal J, Church BN (1994) Geology and geochemistry of the volcanic rocks of the Pioneer Formation, Bridge river area, SW British Columbia. Geol Mag 31:245–253

Duncan RA, Pyle DG (1988) Rapid eruption of the Deccan flood basalts at the Cretaceous/Tertiary boundary. Nature 333:841–843

Farley KA, Natland JH, Craig H (1992) Binary mixing of enriched and degassed (primitive?) mantle components (He, Sr, Nd, Pb) in Samoan lavas. Earth Planet Sc, Lett 111:183–199

Fermor LL, Fox CS (1916) Deccan Trap flows of Linga. Chhindwara District, Central Province. Geol Surv Ind Rec 47:81–136

Fodor RV (1987) Low- and high-TiO_2 flood basalts of southern Brazil: origin from picritic parentage and a common mantle source. Earth and Planetary Sci Lett 84:423–430

Griffiths RW, Campbell IH (1990) Stirring and structure in mantle plumes. Earth Planet Sci Lett 99:66–78

Harpp KS, White WM (2001) Tracing a mantle plume: isotopic and trace element variations of Galapagos seamounts. Geochem Geophys Geosyst 2:1525–2027

Hart SR, Hauri EH, Oschmann LA, Whitehaed JA (1992) Mantle plumes and entrainment: isotopic evidences. Science 256:517–519

Hauri EH, Whitehead JA, Hart SR (1994) Fluid dynamic and geochemical aspects of entrainment in mantle plumes. J Geophys Res 99(24):275–300

Hawkesworth CJ, Gallagher K, Kisten L, Mantovani MSM, Peate DW, Tuner SP (2000) Tectonic controls on magmatism associated with continental break- up: an example from the Parana-Etendeka Province. Earth Planet Sci Lett 179:335–349

Helffrich GR, Wood BJ (2001) The earth's mantle. Nature 412:501–507

Henderson P (1984) General geochemical properties and abundance of the rare earth elements. In: Henderson P (ed) Rare earth element geochemistry. Elsevier, Amsterdam, pp 1–32

Hoffman AW (1997) Mantle geochemistry: the message from oceanic volcanism. Nature 385:219–229

Hooper PR, Subbarao KV, Beane JE (1988) The Giant plagioclase Basalts (GPBs) of the western Ghats, Deccan Traps. Mem Geol Soc India 10:135–144

House MA, Wernicke BP, Farley KA, Dumitru TA (1997) Cenozoic thermal evolution of the central Sierra Nevada, Cal Th/ He thermochonometry. Earth Planet Sci Lett 151:167–179

Irvine TN, Baragar WRA (1971) A guide to the chemical classification of common volcanic rocks. Can J Earth Sci 8:523–548

Jensen LS (1976) New cationic plots for classifying subalkalic volcanic rocks. Ontario Divisions of Mines. MP66.22

Kellog LH, Hager BH, Vander Hilst RD (1999) Compositional stratification in the deep mantle. Science 283:1881–1884

Kent RW, Pringle MS, Muller RD, Saunders AD, Ghose NC (2002) $^{40}Ar/^{39}Ar$ geochronology of the Rajmahal basalts, India, and their relationships to the Kerguelen plateau. J Petrol 43:1141–1153

Krishnamurthy P, Cox KG (1977) Picrite basalts and related lavas from the Deccan Traps of western India. Contrib Miner Petrol 62:53–75

Le Bas MJ, Maitre RW, Streckeisen A, Zanattin B (1986) A chemical classification of volcanic rocks based on the total alkali and silica diagram. J Petrol 27:745–750

Lightfoot PC, Hawkesworth CJ (1988) Origin of Deccan Trap lavas: evidence from combined trace element and Sr, Nd and Pb isotopic studies. Earth Planet Sci Lett 91:89–104

MacDougall JD (1988) Continental flood basalts and MORB: a brief discussion of similarities and differences in their petrogenesis. In: Macdougall JD (ed) Continental flood basalts. Kluwer, Dordrecht, The Netherlands, pp 331–341

Mahoney JJ, Frei R, Tejada MLG, Mo XX, Leat PT, Näggler TF (1998) Tracing the Indian Ocean mantle domain through time: isotopic results from old Western Indian, East Tethyan and South Pacific sea floor. J Petrol 39:1285–1306

Mahoney JJ, Sheth HC, Chandrashekharam D, Peng ZX (2000) Geochemistry of flood basalts of the Toranmal section, Northern Deccan Traps, India: implications for regional Deccan stratigraphy. J Petrol 41(7):1099–1120

Mahoney JJ, Duncan RA, Khan W, Gnos E, Cormick M (2002) Cretaceous volcanic rocks of the South Tethyan Suture Zone, Pakistan: implications for the Reunion hotspot and Deccan Traps. Earth Planet Sci Lett 203:295–310

Mc Getchin TR, Burke KC, Thompson GA, Young RA (1980) Mode and mechanisms of plateau uplifts. In: Bally AW, Bender PL, Mc Getchin TR, Walcot RI (eds) Dynamics of plate interior. American Geophysical Union and Geological Society of America, Boulder, CO, pp 99–110

McGinnis CE, Jain JC, Neal CR (1997) Characterization of Memory effects and development of an effective wash-protocol for the measurement of petrogenetically critical trace elements in Geological samples by ICS-MS. Geostandards Newslett 21(2):289–305

McKenzie DP (1984) The generation and compaction of partially molten rock. J Petrol 25:713–765

Meiborn A, Anderson DL (2003) The statistical upper mantle assemblage. Earth Planet Sci Lett 271:123–139

Morgan WJ (1972) Plate motions and deep mantle convection. Geol Soc Am Mem 132:7–22

Nair KKK, Chatterjee AK, Sano T (1996) Stratigraphy and geochemistry of the Deccan basalts along Toranmal section, western Satpura region. In: Deshmukh SS, Nair KKK (eds) Deccan Basalts, Gondwana Geological Magazine, Special volume 2:23–48

Neal CR, Mahoney JJ, Kroenke LW, Duncan RA, Petterson MG (1997) The Ontong java plateau. In: Mahoney JJ, Coffin M (eds) Large igneous provinces: continental, oceanic and planetary flood volcanism. Monograph, American Geophysical Union, Washington DC, pp 183–216

Norrish K, Chappell BW (1977) X-ray fluorescence spectrometry. In: Zussman J (ed) Physical methods in determinative mineralogy, 2nd edn. Academic, New York, pp 201–272

O'Neil C, Dietmar M, Steinberger B (2003) Geodynamic implications of moving Indian Ocean Hotspots. Earth Planet Sci Lett 215:151–168

Pattanayak SK, Shrivastava JP (1999) Petrography and major-oxide geochemistry of basalts from the Eastern Deccan Volcanic Province, India. In: Subbarao KV (ed) Deccan volcanic province, West Volume. Geological Society of India, Memoir 43(1):233–270

Peng ZX, Mahoney JJ (1995) Drillhole lavas from the north-western Deccan Traps, and the evolution of Reunion hotspot mantle. Earth Planet Sci Lett 134:169–185

Peng ZX, Mahoney JJ, Hooper PR, Macdougall JD, Krishnamurthy P (1998) Basalts of the northeastern Deccan traps, India: isotopic and elemental geochemistry and relation to southwestern Deccan stratigraphy. J Geophys Res 103 (B12):29,843–29,866

Regelous M, Hoffman AW, Aboucham WF, Galer SJG (2003) Geochemistry of lavas from the Emperor Seamounts and the Geochemical evolution of Hawaiian magmatism from 85 to 42 Ma. J Petrol 44(1):113–140

Richards MA, Duncan RA, Courtillot VE (1989) Flood basalts and hotspot tracks: plume heads and tails. Science 246:103–107

Ringwood AE (1955) The principles governing trace element distribution during magmatic crystallization. Pt. 1. Influence of electronegativity. Geochim Cosmochim Acta 7(3–4):189–202

Sano T, Fuji T (1998) Chemical variation of the uncontaminated (Ambenali-like) basalts of Deccan Traps. Gondwana Geological Magazine 2:301–310

Sen G, Cohen TH (1994) Deccan intrusion, crustal extension, doming and the size of the Deccan-Reunion plume head. In: Subbarao KV (ed) Volcanism (Radhakrishna vol). Wiley Eastern Ltd., New Delhi, pp 201–216

Sengupta P, Ray J (2007) Mineral chemistry of Basaltic Lava Flows from Narsingpur-Harrai-Amarwara-Lakhnadon areas of eastern Deccan, Central India. In: Ray J, Bhattacharyya C (eds) Igneous petrology: 21st century perspective. Allied publishers, New Delhi, p 37

Sharma M (1997) Siberian Traps. Am Geophys Union Monogr 100:273–295

Sheth HC (1999) Flood basalts and large igneous provinces from deep mantle plumes: fact, fiction and fallacy. Tectonophysics 311:1–29

Sheth HC (2000) The timing of crustal extension, diking and eruption of the Deccan Flood Basalt. Int Geol Rev 42(11):1007–1016

Sheth HC, Chandraseharam D (1997) Early alkaline magmatism in the Deccan Traps: implications for plume incubation and lithospheric rifting. Phys Earth Planet Inter 104(4):371–376

Sheth HC, Duncan RA, Chandrasekharam D, Mahoney JJ (1997) Early alkaline magmatism in the Deccan Traps: implications for plume incubation and lithospheric rifting. Phys Earth Planet Inter 104:371–376

Snedecor GW, Cochran WG (1967) Statistical methods. Oxford & IBH Publishing Co, Calcutta, Bombay, New Delhi, p 593

Storey M, Mahoney JJ, Saunders AD (1997) Cretaceous basalts in Madagascar and the transition between plume and continental lithosphere mantle sources. Geophys Monogr 100:95–122

Sun SS, McDonough WF (1989) Chemical and isotopic systematic of oceanic basalts: Implications for mantle composition and processes. In: Saunders A, Norrey MJ (eds) Magmatism in the ocean basins. Geological Society Special Publications 42, London, pp 313–345

Taylor SR (1965) The application of trace element data to problems of petrology. In: Arhens LH (ed) Physics and chemistry of the earth, vol 13. Pergamon Press, London, pp 133–213

Thompson RN, Morrison MA, Dickin IL, Hendry GL (1983) Continental flood basalts ....archinids rule OK? In: Hawkesworth CJ, Norry MJ (eds) Continental basalts and mantle xenoliths. Natwich, Shiva, pp 158–185

Venkatesan TR, Pande K, Gopalan K (1993) Did Deccan volcanism pre-date the Cretaceous/Tertiary transition? Earth Planet Sci Lett 119:181–189

Watson S, Mc Kenzie D (1991) Melt generation by plumes: a study of Hawaiian volcanism. J Petrol 32(3):501–537

Weaver BL (1991) The origin of ocean island basalt end-member compositions: trace elements and isotopic constraints. Earth Planet Sci Lett 104:381–397

West WD (1958) The petrography and petrogenesis of forty eight flows of Deccan Traps penetrated by borings in western India. Trans Ind Natl Inst Sci 4:1–56

Yedekar DB, Aramaki S, Fujii T, Sano T (1996) Geochemical signature and stratigraphy of the Chhindwara–Jabalpur–Seoni–Mandla sector of the Eastern Deccan Volcanic Province and problems of its correlation. Gondwana Geological Magazine Special 2:49–68

Zhang M, Oireilly SY, Chen D (1999) Pacific- and Indian-MORB mantle as source reservoirs for the Cenozoic basalts in eastern Australia: Pb-Sr-Nd isotope evidence. Geology 27:39–43

Arc Volcanism

Chapter 11
The Intra-oceanic Barren Island and Narcondam Arc Volcanoes, Andaman Sea: Implications for Subduction Inputs and Crustal Overprint of a Depleted Mantle Source

Martin J. Streck, Frank Ramos, Aspen Gillam, Dhanapati Haldar, and Robert A. Duncan

Abstract The active Barren Island volcano and the Pleistocene Narcondam volcano, located ~140 km to the North, are the only two subaerially exposed Andaman arc volcanoes, which rise from the 1,000 to 2,300 m deep seafloor of the Andaman Sea, and are associated with the subduction of the Indian plate beneath the Burma plate. Lavas at Barren Island range in composition from basalt to andesite while lavas from Narcondam volcano range from andesite to silicic andesite/dacite. Similarities in the geochemistry of both lava suites include strong and comparable depletions in Nb and Ta (K_2O/Nb ~0.7; Ba/Nb 130–250), low MORB-like Nb/Zr (0.01–0.03) and nearly constant U/Th (0.15–0.22). These characteristics suggest a genetic link between both magma suites, but there are geochemical differences such as elevated trace element abundances and ratios in Narcondam lavas compared to Barren Island lavas. These include elevated Ba, Rb, and U concentrations and higher Ba/Zr and Nb/Zr. Additionally, isotopic ratios of Barren Island lavas are $^{87}Sr/^{86}Sr$ ~0.7039–0.7041, $^{143}Nd/^{144}Nd$ ~0.51285–0.51296, and $^{206}Pb/^{204}Pb$ ~18.063–18.309. For Narcondam, isotope ratios are $^{87}Sr/^{86}Sr > 0.705$, $^{143}Nd/^{144}Nd$ ~0.51270, and $^{206}Pb/^{204}Pb$ ~18.565–18.617. Additional geochemical parameters (e.g., Sr/Y, Zr/Y, Th/La, U/La, Ba/La) of Narcondam lavas positively correlate with increasing SiO_2 but are anchored at the mafic end within compositions observed at Barren Island volcano.

M.J. Streck (✉) and A. Gillam
Department of Geology, Portland State University, Portland, OR 97207, USA
e-mail: streckm@pdx.edu

F. Ramos
Department of Geological Sciences, New Mexico State University, Las Cruces, NM 88003, USA
e-mail: framos@nmsu.edu

D. Haldar†
Geology Department, Presidency College, Kolkata, 700 073, India

R.A. Duncan
College of Oceanic and Atmospheric Sciences, Oregon State University, Corvallis, OR 97331, USA
e-mail: rduncan@coas.oregonstate.edu

†Deceased

J. Ray et al. (eds.), *Topics in Igneous Petrology*, DOI 10.1007/978-90-481-9600-5_11,

Similarities in the geochemistry of both lava suites, of which some (e.g. high Th/Nb) stand out in comparison to other intra-oceanic arc systems, almost certainly originate early during the generation of these magmas, most-likely in the source of magmas that do not change subsequently. On the other hand, generally more incompatible element enriched silicic Narcondam magmas are best explained by amphibole-dominated fractionation of a Barren Island-type basalt. Such a scenario would be consistent with an increase of Sr/Y correlated with increased silica and the presence of greater amphibole in samples with higher Sr/Y. The shift in isotopic values from typical Barren Island values to those of Narcondam samples is likely caused by assimilation of extended continental crust and/or sediments from the fan associated with the Irrawaddy Delta at the Myanmar continental margin.

11.1 Introduction

Little has been published regarding the geochemistry and petrography of Andaman arc volcanoes related to the subduction of the Indian plate beneath the Burma plate. However, Barren Island and Narcondam volcanoes offer highly contrasting signatures that offer insights into sources, residence, and processes affecting arc-related magmas. Limited prior work includes a cursory focus on Barren Island and Narcondam volcanoes (e.g. Luhr and Haldar 2006; Haldar et al. 2007; Pal et al. 2007) and general overview data (Shanker et al. 2001), although more but rather limited contributions exist in Indian journals (see Haldar et al. 2007 for list of references). In addition, recently acquired, high quality major and trace element data and isotopic ratios exist only for Barren Island volcano lavas (Luhr and Haldar 2006) but data for Narcondam volcano is sparse and incomplete (cf. Haldar et al. 2007). This work intends to close this gap and directly compares the geochemistry of Barren Island lavas with Narcondam volcano lavas, highlighting similarities and distinct differences of volcanic products that erupted only 140 km apart.

11.2 Geotectonic Framework

Narcondam and Barren Island volcanoes are part of the Neogene Inner Island Arc (Fig. 11.1) of the Andaman and Nicobar Islands in the Andaman Sea (Shanker et al. 2001). Narcondam Island is located ~140 km NNE from Barren Island. The Pliocene-Holocene Narcondam and active Barren Island volcanoes are the only subaerial, intra-oceanic arc volcanoes associated with the subduction of the Indian Plate beneath the Burma plate. Narcondam and Barren Island volcanoes are narrow volcanic constructions built on the Andaman seafloor (see below). Subduction of the Indian Plate occurs obliquely at a rate of 40–50 mm/yr (Thorne et al. 2005).

Narcondam and Barren Island constitute the northern part of the Sunda arc and lie between the Holocene Sumatran volcanic chain to the south and the Holocene volcanoes of the central Myanmar Lowlands to the north (Luhr and Haldar 2006).

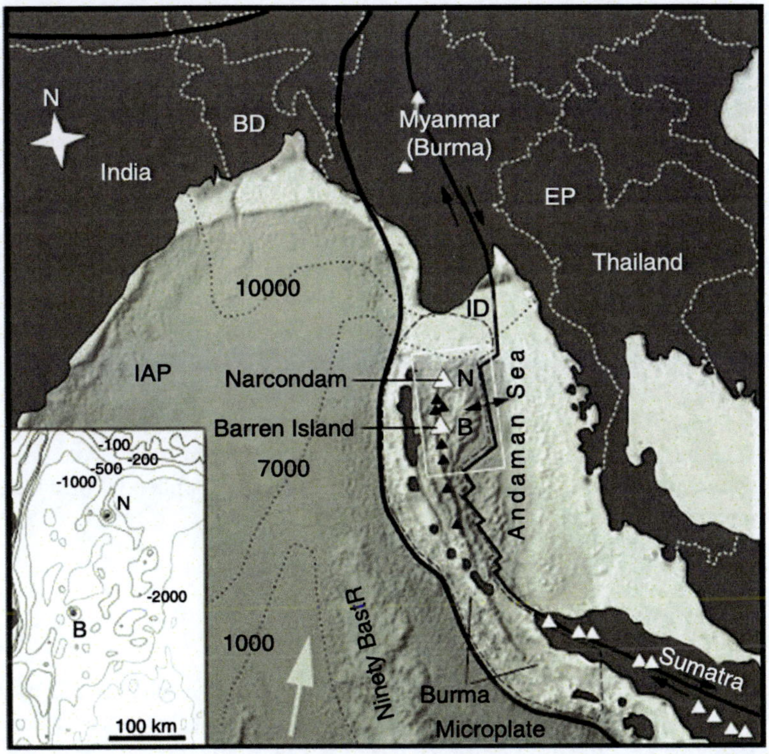

Fig. 11.1 Barren Island (B) and Narcondam (N) volcanoes with respect to geotectonic and regional geographic framework. Thickest solid line is convergent plate boundary between Eurasien plate (EP) and Indian-Australian plate (IAP). Thinner dashed line outlines the Burma microplate (Outline after http://www.lib.utexas.edu/maps/middle_east_and_asia/sumatra_earthquake_2004.pdf, accessed March 2009) and Solid line to the east of the Burma plate is a transform fault north and south of the Andaman Sea and a spreading center within the Andaman Sea. Convergence direction of the IAP towards the convergent plate boundary is indicated by single *gray arrow* after Luhr and Haldar (2006). *White triangles* are subaerial volcanoes and black triangles are some of the volcanic seamounts (After Rodolfo 1969). *Dotted lines* are sediment thickness in meters (Simplified after http://www.ngdc.noaa.gov/mgg/sedthick/sedthick.html, accessed Feb 2009). Other abbreviations include: ID = Irrawaddy Delta; BD = Bangladesh. *Inset*, highlighted by *white rectangle*, shows the bathmetry of this part of the Andaman Sea; *lighter gray lines* indicate greater water depth in meters (After Office of Naval Research 2002 accessed, Feb. 2009)

The top of the eastward dipping Indian plate is thought to lie ~90 km below the magmatic arc (Dasgupta and Mukhopadhyay 1993). This intra-oceanic arc is situated in a ~300–400 km wide, 1,000–3,000 m deep back-arc rift basin characterized by active seafloor spreading in its eastern part about ~100 km east of the axis between Narcondam and Barren Island volcanoes (Curray 2005). The Andaman Sea results from intraplate extension and rifting in the overriding plate that began at ~11 Ma but true seafloor spreading with generation of new oceanic crust started only at ~4 Ma (Raju et al. 2004; Khan and Chakraborty 2005). Prior to the onset of spreading, the Andaman sea experienced extension and rifting along the plane joining regionally important transform faults, the Sagaing fault in the north and the Semengko fault in the south (Raju et al. 2004; Khan and Chakraborty 2005). Given the young age of active backarc spreading, Narcondam and Barren Island volcanoes are unlikely to sit on newly formed oceanic crust but rather these volcanoes more likely reside on stretched and thinned crust of a hybrid nature as is suggested by the rough bathymetry immediately surrounding the volcanoes and by petrological data discussed below (Fig. 11.1).

Two large sediments fans occur regionally from materials supplied from the Himalayas, and surroundings, to local oceanic basins. One, the Bengal Fan and its main delivery system the Ganges river, has covered the entire NE part of the Indian Ocean seafloor with a thick sediment accumulation (Fig. 11.1) (Gasparotto et al. 2000). Estimates suggest a sediment thickness in excess of 3.5 km that overlies the subducting Indian Plate at the Andaman trench. This enormous thickness of sediment at a convergent plate boundary is nearly unrivaled worldwide (cf. Plank and Langmuir 1998). The other sediment fan is located in the northern part of the Andaman Sea where the Irrawaddy Delta (southern coast of Myanmar) supplies Himalayan sediments directly to the Andaman Sea (Luhr and Haldar 2006).

11.3 Barren Island Volcano

Barren Island volcano rises from a water depth of ~2,300 m to a height of 355 m above sea level (Luhr and Haldar 2006). The submarine volume is estimated at ~400 km^3 and the subaerial part is only 1.3 km^3 of the island that is 3 km across (Luhr and Haldar 2006). Volcanic rocks of Barren Island can be subdivided into five eruptive periods. The oldest samples predate the first historic eruption in 1787. Attempts to date the oldest rocks by $^{40}Ar/^{39}Ar$ geochronology have had limited success and ages on two samples have large errors due to young ages and low potassium contents (see below).

The eruptive period that began in 1787 continued intermittently until 1832. Subsequent eruptions occurred in 1991, 1994–1995, and Barren Island also began erupting again in 2006–2009 (Luhr and Haldar 2006; https://lists.asu.edu/cgi-bin/wa?LIST=VOLCANO). Over these periods, Barren Island volcano has erupted porphyritic basaltic to andesitic lavas dominated by plagioclase phenocrysts. However, historic activity is mostly basaltic to basaltic andesite in composition (Table 11.1) (Luhr and Haldar 2006).

11.4 Narcondam Island Volcano

Narcondam volcano forms a small island, ~3.5 km across and 710 m high, that rises isolated from its immediate ~17 km wide base at 1,000 m beneath the sea, although the nearby ocean floor is as deep as ~2,000 m (Fig. 11.1). Estimated submarine and subaerial volumes are 95 and 4 km^3, respectively, based on depth contours of the Office of Naval Research (2002). Last activity is likely Holocene but subaerial lavas have been dated as old as 550 and 700 ka (see below).

Table 11.1 Major and trace element data of Barren Island and Narcondam volcanoes

Panel A:								
ID #	9/21/5/92	F-1/1	X/F-1	X/L/1/B	P-1	12/F-2	L-1/B	L-2/B/F-2
Period	1	1	1	1	1	2	2	2
XRF		Normalized major elements (wt%):						
SiO2	51.47	51.42	50.44	52.92	59.28	53.01	53.55	54.74
TiO2	0.671	0.672	0.689	0.964	1.171	0.967	0.997	1.347
Al2O3	18.28	18.13	19.00	21.42	16.99	21.33	20.72	17.17
FeO*	7.53	7.66	7.46	7.06	7.56	7.07	7.30	9.54
MnO	0.156	0.157	0.147	0.146	0.189	0.146	0.151	0.199
MgO	8.10	8.23	8.01	2.91	2.73	2.94	3.02	3.84
CaO	10.66	10.61	11.53	10.52	6.33	10.45	10.04	8.43
Na2O	2.72	2.71	2.38	3.41	4.80	3.42	3.54	3.99
K2O	0.32	0.32	0.28	0.53	0.78	0.53	0.56	0.60
P2O5	0.084	0.084	0.070	0.128	0.177	0.129	0.133	0.150
ppm, ICP-MS (see footnote)								
Cs	0.06	0.04	0.05	0.40	0.69	0.40	0.43	0.48
Rb	5.5	5.4	4.7	14.2	17.3	14.2	15.2	14.2
Ba	64	63	52	103	137	104	109	115
Sr	181	176	175	230	173	228	227	190
Th	0.63	0.61	0.63	2.03	1.55	2.04	2.16	1.61
U	0.13	0.13	0.11	0.30	0.30	0.31	0.33	0.28
Nb	0.48	0.46	0.41	0.74	1.06	0.77	0.80	1.08
Ta	0.05	0.05	0.04	0.06	0.09	0.06	0.06	0.10
Zr	53	52	44	72	114	72	78	91
Hf	1.56	1.55	1.37	2.07	3.36	2.12	2.24	2.72
Y	19.86	19.75	19.36	26.73	43.11	26.85	28.26	36.04
Pb	2.61	2.16	1.29	2.52	4.16	2.45	2.45	3.20
Sc	36.1	35.3	37.3	29.0	30.1	29.5	30.7	41.8

(continued)

Table 11.1 (continued)

Panel A:

ID #	9/21/5/92	F-1/1	X/F-1	X/L/1/B	P-1	12/F-2	L-1/B	L-2/B/F-2
Ni, X	129	134	122	19	10	15	15	14
Cr, X	321	310	290	14	18	16	16	14
Co, I	31.34	n.a.	34.14	16.16	13.03	14.81	10.82	22.28
V, X	226	224	236	298	212	303	305	435
Zn, X	72	71	59	71	94	70	77	96
Ga, X	17	15	15	20	19	20	19	19
Cu, X	49	51	38	93	159	92	55	128
La	3.25	3.18	2.80	6.73	7.12	6.70	6.93	6.81
Ce	8.12	7.99	7.15	15.65	17.91	15.55	16.04	16.38
Pr	1.21	1.18	1.06	2.16	2.60	2.17	2.22	2.36
Nd	6.23	6.14	5.64	10.59	13.45	10.66	10.90	12.16
Sm	2.17	2.13	2.02	3.31	4.64	3.37	3.48	4.12
Eu	0.84	0.83	0.79	1.19	1.61	1.21	1.22	1.47
Gd	2.75	2.68	2.64	4.04	5.99	3.95	4.19	5.09
Tb	0.52	0.51	0.50	0.71	1.12	0.72	0.75	0.94
Dy	3.49	3.44	3.41	4.75	7.42	4.72	5.00	6.39
Ho	0.75	0.75	0.74	1.01	1.61	1.03	1.06	1.35
Er	2.09	2.08	2.10	2.83	4.59	2.83	3.03	3.80
Tm	0.32	0.32	0.31	0.42	0.69	0.43	0.44	0.57
Yb	1.99	2.01	1.95	2.68	4.40	2.65	2.86	3.63
Lu	0.33	0.32	0.32	0.42	0.71	0.43	0.46	0.58

Panel B:

ID #	FG/16/5/91	FG/91/L	3/21/5/92	8/9/292	20/24/1/95	29/8/3/95	13D/8/3/95	13B/8/3/95	3C/8/3/95	1/11/5/95	17/6/10/95	17/6/10/95
Period	3	3	3	3	4	4	4	4	4	4	4	4
XRF			Normalized major elements (wt%):									
SiO2	56.48	55.39	55.21	54.19	51.66	51.60	51.53	51.70	51.65	51.71	51.50	51.50
TiO2	1.188	1.222	1.191	1.094	0.839	0.830	0.832	0.839	0.833	0.833	0.820	0.820
Al2O3	17.52	17.68	18.00	18.52	21.52	21.50	21.51	21.42	21.57	21.33	21.49	21.49
FeO*	8.62	8.82	8.64	8.33	7.06	7.10	7.14	7.12	7.05	7.10	7.14	7.14
MnO	0.192	0.194	0.191	0.178	0.146	0.146	0.146	0.147	0.146	0.147	0.145	0.145
MgO	3.30	3.61	3.64	4.11	3.95	4.01	4.00	4.00	3.93	4.17	4.05	4.05
CaO	7.54	8.10	8.31	9.08	11.35	11.33	11.38	11.28	11.36	11.32	11.41	11.41
Na2O	4.23	4.13	4.00	3.76	2.98	2.98	2.97	2.98	2.97	2.91	2.95	2.95
K2O	0.74	0.68	0.65	0.59	0.40	0.40	0.40	0.40	0.40	0.39	0.39	0.39
P2O5	0.178	0.164	0.159	0.144	0.100	0.099	0.099	0.100	0.100	0.099	0.097	0.097
ppm, ICP-MS (see footnote)												
Cs	0.60	0.51	0.50	0.43	0.26	0.25	0.24	0.25	0.24	0.24	0.35	0.35
Rb	20.4	18.3	17.9	15.6	10.4	10.1	10.2	10.4	10.2	10.2	9.9	9.9
Ba	141	130	130	111	78	75	76	78	75	75	75	75
Sr	194	197	202	198	218	216	218	220	218	217	215	215
Th	2.85	2.55	2.52	2.16	1.33	1.30	1.30	1.37	1.36	1.32	1.27	1.27
U	0.43	0.39	0.39	0.32	0.21	0.21	0.20	0.20	0.21	0.20	0.19	0.19
Nb	1.10	0.95	0.98	0.83	0.54	0.52	0.54	0.55	0.54	0.53	0.50	0.50
Ta	0.09	0.07	0.07	0.07	0.04	0.04	0.04	0.05	0.04	0.05	0.04	0.04
Zr	104	95	94	83	59	58	58	59	58	58	56	56
Hf	3.05	2.77	2.73	2.42	1.74	1.74	1.71	1.75	1.72	1.71	1.66	1.66
Y	37.82	35.86	35.45	31.61	23.43	23.13	23.27	23.90	23.18	23.46	22.62	22.62
Pb	3.59	3.15	3.11	2.82	2.23	1.91	1.89	1.80	1.87	1.96	2.10	2.10

(continued)

Table 11.1 (continued)

Panel B:												
Sc	34.5	36.3	36.2	35.0	30.7	30.4	30.9	30.7	30.4	31.9	30.2	30.2
Ni, X	12	13	16	27	26	29	31	31	32	34	29	29
Cr, X	6	8	13	45	43	46	42	41	41	78	46	46
Co, I	18.23	20.14	19.55	21.44	19.92	20.93	20.53	20.37	20.06	21.29	20.97	20.97
V, X	318	357	354	329	271	269	270	272	272	276	264	264
Zn, X	95	94	92	85	70	69	68	68	67	70	66	66
Ga, X	19	18	20	21	18	18	19	18	18	18	18	18
ID #	FG/16/5/91	FG/91/L	3/21/5/92	8/9/292	20/24/1/95	29/8/3/95	13D/8/3/95	13B/8/3/95	3C/8/3/95	1/11/5/95	17/6/10/95	17/6/10/95
Cu, X	71	114	49	111	87	84	83	67	77	58	77	77
La	9.13	8.32	8.28	7.10	4.73	4.56	4.60	4.76	4.70	4.56	4.53	4.53
Ce	21.31	19.57	19.33	16.81	11.25	11.09	11.06	11.38	11.17	10.96	10.95	10.95
Pr	2.98	2.72	2.69	2.35	1.60	1.59	1.58	1.63	1.60	1.56	1.57	1.57
Nd	14.47	13.44	13.47	11.64	7.98	8.01	7.98	8.23	8.14	7.87	7.81	7.81
Sm	4.57	4.28	4.31	3.74	2.70	2.63	2.70	2.73	2.64	2.68	2.58	2.58
Eu	1.54	1.48	1.46	1.31	1.01	0.96	0.99	1.02	0.98	0.99	0.96	0.96
Gd	5.43	5.20	5.12	4.53	3.28	3.27	3.26	3.35	3.28	3.23	3.24	3.24
Tb	0.99	0.95	0.94	0.83	0.61	0.60	0.61	0.62	0.61	0.61	0.59	0.59
Dy	6.58	6.27	6.19	5.51	4.05	3.94	4.03	4.15	4.02	4.00	3.96	3.96
Ho	1.41	1.36	1.32	1.17	0.88	0.87	0.88	0.89	0.87	0.87	0.86	0.86
Er	3.96	3.78	3.74	3.37	2.46	2.41	2.45	2.50	2.43	2.45	2.41	2.41
Tm	0.59	0.56	0.56	0.50	0.37	0.36	0.36	0.37	0.36	0.36	0.36	0.36
Yb	3.74	3.59	3.56	3.17	2.30	2.27	2.33	2.34	2.30	2.30	2.27	2.27
Lu	0.60	0.57	0.57	0.51	0.37	0.37	0.37	0.38	0.37	0.37	0.36	0.36

Panel C:

ID #	N/87	13/20/2/87 inclusion	20/23/2/87	25/22/2/87	4/20/2/87	5/22/2/87(A)	8/22/2/87	19/22/2/87
XRF		Normalized major elements (wt %):						
SiO2	62.47	53.72	60.40	62.49	61.69	57.56	57.66	59.28
TiO2	0.468	0.856	0.523	0.468	0.534	0.608	0.611	0.597
Al2O3	18.00	19.94	18.33	17.93	18.22	19.11	18.90	18.17
FeO*	4.50	6.44	5.03	4.34	4.87	5.76	5.94	5.52
MnO	0.106	0.122	0.108	0.103	0.118	0.126	0.128	0.111
MgO	2.78	5.16	3.24	2.84	2.46	4.19	4.22	3.72
CaO	6.27	9.65	7.28	6.56	6.71	8.23	8.16	7.81
Na2O	3.69	2.99	3.48	3.56	3.75	3.24	3.20	3.37
K2O	1.61	1.02	1.50	1.58	1.54	1.08	1.08	1.34
P2O5	0.114	0.089	0.111	0.124	0.114	0.103	0.105	0.085
ppm, ICP-MS (see footnote)								
Cs	0.97	1.94	2.79	4.42	2.45	2.57	2.60	1.27
Rb	50.9	29.6	46.2	60.1	53.0	42.0	43.0	44.1
Ba	345	225	312	341	308	293	274	400
Sr	314	272	377	403	331	308	305	286
Th	6.32	3.36	5.66	6.42	5.77	4.53	4.50	4.69
U	1.22	0.70	1.06	1.24	1.16	0.95	0.94	1.02
Nb	2.23	1.24	2.29	2.29	2.07	1.48	1.52	1.48
Ta	0.19	0.10	0.18	0.19	0.17	0.12	0.13	0.13
Zr	78	62	76	92	76	71	71	67
Hf	2.19	1.96	2.20	2.52	2.17	2.08	2.09	1.98
Y	13.78	22.32	14.22	13.34	17.76	21.63	26.48	23.52
Pb	6.23	6.19	5.55	6.47	5.81	8.27	7.65	5.15
Sc	13.8	37.9	19.0	14.8	18.1	22.8	22.5	25.2

(continued)

Table 11.1 (continued)

Panel C:								
Ni, X	17	33	12	16	15	30	32	22
Cr, X	29	40	20	39	21	49	45	69
Co, I	13.63	25.05	16.30	12.84	15.09	21.09	20.68	19.48
V, X	116	253	145	108	131	166	168	158
Zn, X	44	55	43	41	45	54	56	51
Ga, X	16	16	18	17	17	17	18	17
Cu, X	23	49	21	24	46	36	40	25
La	12.46	9.39	11.63	14.99	12.87	13.49	16.21	20.38
Ce	22.32	15.08	21.36	25.24	21.35	19.39	19.72	19.93
Pr	2.41	2.26	2.37	2.88	2.55	2.76	3.19	4.26
Nd	9.59	10.21	9.65	11.10	10.60	11.66	13.66	16.79
Sm	2.25	3.09	2.40	2.43	2.61	2.95	3.41	3.89
Eu	0.75	1.01	0.79	0.80	0.84	0.98	1.05	1.10
ID #	N/87	13/20/2/87	20/23/2/87	25/22/2/87	4/20/2/87	5/22/2/87(A)	8/22/2/87	19/22/2/87
Gd	2.27	3.56	2.44	2.30	2.76	3.32	3.67	3.92
Tb	0.38	0.63	0.40	0.37	0.47	0.54	0.60	0.64
Dy	2.40	4.00	2.58	2.29	2.95	3.41	3.93	4.08
Ho	0.50	0.86	0.53	0.49	0.63	0.74	0.84	0.83
Er	1.41	2.30	1.48	1.36	1.76	2.05	2.32	2.33
Tm	0.22	0.34	0.22	0.21	0.26	0.29	0.33	0.34
Yb	1.39	2.11	1.42	1.36	1.66	1.84	2.07	2.20
Lu	0.24	0.34	0.23	0.22	0.27	0.31	0.34	0.35

A, B: Barren Island lavas; C: Narcondam lavas; X denotes values obtained by XRF analysis and I from Instrumental Neutron Activation; period in Panels A, B indicates eruptive period, with the following designations 1 = pre-historic, 2 = 1787–1832, 3 = 1991, 4 = 1995 activity
Prenormalization totals range from 99.6 to 98% for Barren Island samples and from 99.2 to 97.9% for Narcondam samples

Subaerial lavas are coarsely porphyritic medium-K andesites and dacites with typical volcanic arc chemical signatures (Haldar et al. 2007; this study). Andesites and dacites often contain fine-grained cognate inclusions which record slightly more mafic compositions as primitive as basaltic andesite (Haldar et al. 2007; Table 11.1). Mineral assemblages of Narcondam andesites and dacites are dominated by phenocrysts of plagioclase followed by hornblende, orthopyroxene and oxide minerals.

11.5 Samples and Analytical Methods

The Barren Island suite includes 19 samples encompassing all but the most recent 2006–2009 eruptions while the Narcondam island suite is composed of eight samples. All samples were analyzed for major and trace elements by X-ray fluorescence (XRF) and trace elements by inductively-coupled mass spectrometry (ICP-MS) at the GeoAnalytical Laboratory at Washington State University. Precision estimates are found in Johnson et al. (1999) and Knaack et al. (1994). In addition, all samples were analyzed by instrumental neutron activation analysis (INAA) at Portland State University. Analytical methods and accuracies are similar to those reported by Streck and Grunder (1997). Agreement between methods for the same suite of trace elements is excellent. Trace elements reported in Table 11.1 are mostly ICP-MS values with some exceptions as noted.

Amphibole compositions were determined using the CAMECA SX50 electron microprobe at Oregon State University. Analytical conditions included an accelerating voltage of 15 kV, a beam current of 50 nA, and a slightly defocused (~5 μm diameter) beam. Collection times on peak and background positions were 10 and 5 s for Na and K, Mg, Mn, Si, and Ca; collection times were 20 s (peak) and 10 s (background) for Al, Ti, Fe, Cr, F, and Cl.

About 150 mg of rock powder was dissolved for Sr, Nd, and Pb isotope analyses. Sr was separated using cation exchange resin and analyzed using a thermal ionization mass spectrometer (TIMS) with five Faraday collectors in dynamic mode with $^{86}Sr/^{88}Sr$ ratios normalized to 0.1194. Nd isotopes were also measured using TIMS with seven Faraday collectors in dynamic mode with $^{146}Nd/^{144}Nd$ normalized to 0.7219 (see Ramos 1992 for analytical details). Pb isotopes were analyzed using multi-collector inductively coupled plasma mass spectrometry (MC-ICPMS) using five collectors in static mode with Tl-doping (Wolff and Ramos 2003) and normalized to $^{205}Tl/^{203}Tl = 0.41891$ (May and Watters 2004).

11.5.1 Age Determinations

Crystallization ages of Barren Island and Narcondom volcano lavas were determined using the $^{40}Ar/^{39}Ar$ incremental heating method at the Noble Gas Mass Spectrometry lab at Oregon State University. Small cores (5 mm diameter) were cut from the freshest interiors of whole rocks, then sliced into disks of about 150 mg mass. We loaded the labeled disks with small quantities of monitor mineral FCT-3 biotite (28.04 Ma; Renne et al. 1998) in quartz vials and irradiated them for 6 h in the 1MW TRIGA reactor at Oregon State University. Samples were heated in a double-vacuum, thermocouple-controlled, resistance furnace. After each heating step, followed by gas cleanup, the isotopic composition of Ar was analyzed using a MAP 215–50 mass spectrometer. We reduced the isotopic data and made age calculations using the ArArCALC software provided by Koppers (2002). Further details of the analytical procedures are described in Duncan and Keller (2004) and on the laboratory website (http://www.coas.oregonstate.edu/research/mgg/chronology.html).

Ages for two samples of the stratigraphically oldest sequence of Narcondom lavas and two samples of Barren Island volcano were determined (Table 11.2). Barren Island samples yielded weighted plateau ages (±2 s.d. errors) of 33 ± 70 and 32 ± 88 ka. For these rocks, proportions of radiogenic Ar were very low due to their youth and low K-contents (primitive mafic character). Resulting analytical uncertainties were larger than the measured ages. Although pre-historic, these samples are most likely no older than Holocene. In contrast, both Narcondam samples yielded reasonably precise weighted plateau ages of 548 ± 22 and 707 ± 170 ka, and therefore constrain the activity period of subaerial cone building to ≤700,000 years.

11.6 Petrography

Phenocryst abundances are based on visual estimates using model graphs (e.g. "Models for Visual Estimation of Percentage Composition", Williams et al. 1982) and are thought to be within 5% at percentages between 10 and 30% and better than 5% below 10% phenocryst content. We consider phenocrysts to be larger than 300 μm and microphenocrysts between ~50 and 300 μm but both are set in a matrix of finer crystals (microlites). Furthermore, the term phenocryst is used as a reference to size only (cf., Streck 2008).

Lavas from Barren Island volcano have variable abundances of phenocryst-sized minerals. Seventeen out of 19 samples, regardless of age, contain 6–30% phenocrysts (Fig. 11.2) while two samples contain as little as 1% phenocrysts, most of which are also rather small (<300 μm). Independent of phenocryst proportions, mineral assemblages (Fig. 11.2) are dominated by plagioclase with lesser amounts of olivine, clinopyroxene and oxides (cf., Luhr and Haldar 2006; Haldar et al. 2007). Among basalts and basaltic andesites of our sample suite, characteristic differences in the abundance and types of phenocrysts among eruptive periods exist. Pre-historic basalt samples contain the highest abundance (>20%) of phenocrysts and the highest abundance of large (>500 μm across) and complexly zoned clinopyroxenes, in addition to large plagioclase and olivine phenocrysts (Fig. 11.2). Basalt samples of 1787–1832 and 1995 eruptions are similar; both contain fewer phenocrysts (~13–18%) dominated by plagioclase

Table 11.2 40Ar–39Ar radiometric ages for Barren Island and Narcondam Volcano, Andaman Islands

Sample	Material	Total fusion age (ka)	2σ error	Plateau age (ka)	2σ error	N	MSWD	Isochron age (ka)	2σ error	40Ar/36Ar initial	2σ error	J
Barren Island												
F/1/1	whole rock	94	98	33	70	7/7	0.73	6	79	299	5	0.00161
9/21/5/92	whole rock	42	107	32	88	5/5	0.10	17	108	297	5	0.00165
Narcondam Volcano												
N/87	plagioclase	4325	174	707	170	4/6	0.45	789	336	260	106	0.00175
13/20/2/87	whole rock	611	38	548	22	6/7	0.67	542	31	298	9	0.00174

Ages calculated using biotite monitor FCT-3 (28.04 Ma) and the total decay constant l = 5.530E-10/year. N is the number of heating steps (defining plateau/total); MSWD is an F-statistic that compares the variance within step ages with the variance about the plateau age. J combines the neutron fluence with the monitor age

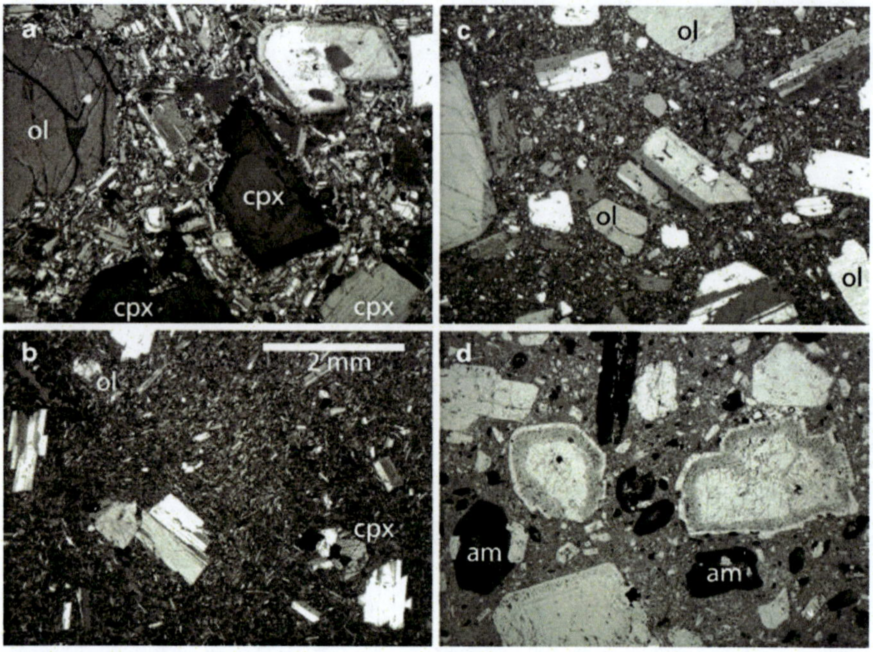

Fig. 11.2 Petrographic characteristics of basaltic Barren Island and intermediate Narcondam lavas. All images are at the same scale; ol = olivine, cpx = clinopyroxene, am = pseudomorph after amphibole; all unlabeled phenocrysts are plagioclase except in **d** where all dark patches are pseudomorphs after amphibole; (**a**) Barren Island basalt samples under cross polarized light of prehistoric eruptive period with large plagioclase, zoned clinopyroxene and olivine; (**b**) Barren Island basalt of the 1991 eruptive period with plagioclase phenocrysts and relative small pyroxene and olivine; (**c**) Barren Island basalt of the 1995 eruptive period with large plagioclase and olivine phenocryst; (**d**) Narcondam lava under plain light showing abundant plagioclase and amphibole phenocrysts; amphibole phenocrysts are now mostly pseudomorphs that partially to fully degraded yet preserve characteristic amphibole shapes

and olivine, with limited or no clinopyroxene. Basaltic andesite samples of the 1991 eruptions have the lowest phenocryst contents (6–10%). Exceptions are one pre-historic andesite (P-1) and one 1787–1832 period basaltic andesite (L-2/B/F-2) that both contain only 1% phenocrysts (Fig. 11.2). Phenocrysts in 1991 samples are composed of mostly smaller plagioclase (~400 μm). Furthermore, small olivines are similar in abundance to pyroxene that occurs either as single grains or in glomerocrysts. All samples at 55 wt% SiO_2 or higher contain the fewest phenocrysts and are texturally closest to samples of the 1991 eruptive phase. Petrographic differences observed here were in part also observed by Luhr and Haldar (2006) and Haldar et al. (2007) including the lower abundance of clinopyroxene as phenocyrst in all historic eruptions compared to pre-historic ones.

Samples from Narcondam volcano are also rich (~20–30%) in phenocryst-sized minerals of which plagioclase is the dominant mineral type (Figs. 11.2d, 11.3 and 11.4).

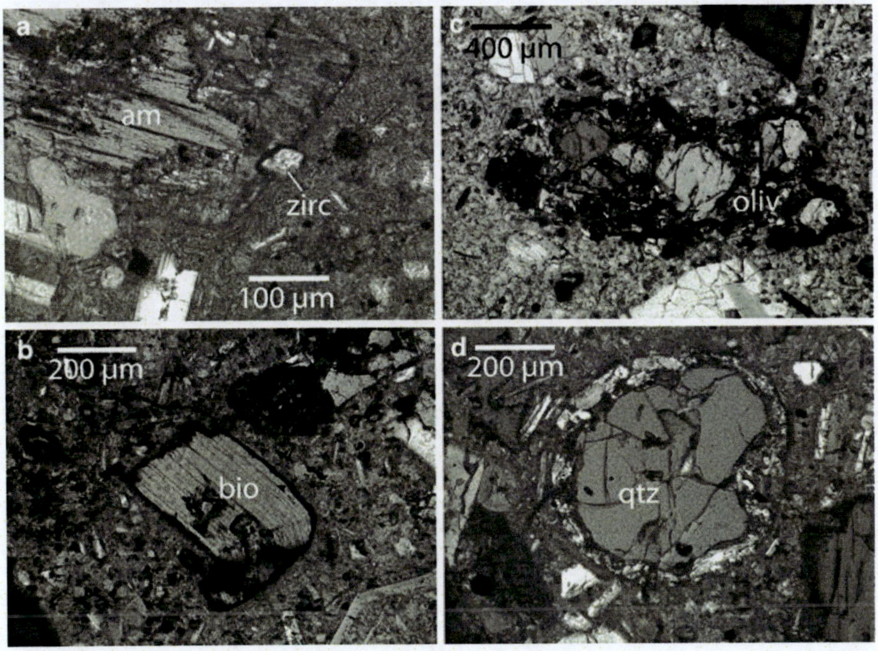

Fig. 11.3 Narcondam sample N87 (62.5 wt% SiO_2) with striking disequilibrium phenocrystic assemblage of olivine (oliv), partially decomposed hornblende (am) with zircon inclusion (zirc), biotite (bio), and quartz (qtz) with reaction rim

Samples with phenocryst contents of less than 10% have not been observed. The second most abundant phenocryst-sized phase is amphibole (Table 11.3). Amphibole varies from perfectly euhedral crystals without signs of decomposition to hornblende crystals that have been completely transformed to a fine-grained mineral aggregate or to pyroxene pseudomorphs. Amphiboles in individual samples can reflect the entire textural range but in some samples all hornblendes exist only as pseudomorphs after hornblende (Figs. 11.2d, 11.3 and 11.4). The opposite, a sample where all hornblende phenocrysts are pristine, has not been observed. Identification of pseudomorphs as former amphibole crystals in samples where only pseudomorphs are present is still possible because at least a few pseudomorphs show characteristic crystal shapes of amphibole (Figs. 11.2d and 11.4d). Orthopyroxene phenocrysts are third in abundance, followed by clinopyroxene and biotite. Biotite, similarly to hornblende, varies in texture from unaltered to largely decomposed. Among mafic silicates, least abundant minerals include occasional olivine crystals that also appear as glomerocrysts and that can be rimmed with orthopyroxene. All eight Narcondam samples contain phenocryst-sized quartz crystals that can be nearly euhedral, rounded and embayed, or rounded with addition of a reaction rim. Abundances of quartz in some samples can be as high as 2%, between orthopyroxene to clinopyroxene contents. Accessory phases include titanomagnetite, ilmenite, apatite and zircon.

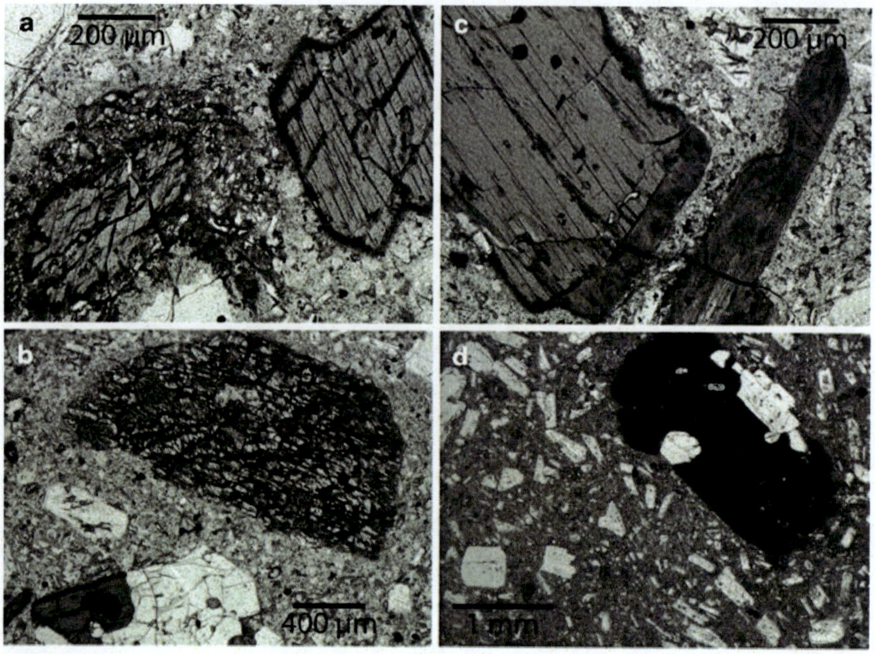

Fig. 11.4 Amphibole texture in dacitic to andesitic Narcondam samples; Pictures **a–c** in 62.5 wt% SiO_2 sample and picture **d** of 57.7 wt% sample. (**a**) Amphibole phenocrysts with and without thick coarse grained reaction rim; (**b**) orthopyroxene pseudomorph after amphibole; (**c**) amphibole phenocrysts degraded into very fine reaction products; (**d**) pseudomorphs after amphibole consisting of opaque mass that likely consists of a fine-grained mineral aggregate; note small pseudomorph along left side with characteristic amphibole shape

Although many petrographic features are similar to those described by Pal et al. (2007), there are also significant differences. Pal et al. (2007) describe samples that do not contain amphibole while we find amphibole in all samples (either as fresh crystals or as decomposed pseudomorphs with clear amphibole outlines). We found phenocryst-sized quartz also in samples of andesitic composition. We have not found Cummingtonite. And finally, zircon was not an identified accessory phase by Pal et al. (2007).

11.7 Geochemistry

11.7.1 Major and Trace Elements

Subaerial Barren Island lavas are basalts and andesites while lavas at Narcondam volcano are basaltic andesites and dacites (cf., Luhr and Haldar 2006; Haldar et al. 2007; Pal et al. 2007). Barren Island lavas are low-K, tholeitiic to medium-K calc-alkaline in character (Fig. 11.5) while Narcondam lavas are medium-K, calc-alkaline (after Gill 1981).

Table 11.3 Representative amphibole analyses

Label	13/28/2/87	13/28/2/87	N/87	N/87	N/87	N/87	N/87
Crystal ID	amph3	amph4	amph2	amph2	amph2	amph3	amph6
Location	Middle	Center	Center	Middle	Rim	Center	Center
SiO_2	40.12	46.76	46.86	46.64	46.96	46.51	47.07
TiO_2	1.85	1.13	1.01	1.08	1.04	0.91	0.84
Al_2O_3	14.32	8.67	7.82	8.09	7.46	7.74	7.82
FeO*	12.33	11.93	12.90	12.98	12.64	13.01	12.75
MnO	0.16	0.33	0.42	0.45	0.45	0.40	0.47
MgO	14.04	16.27	15.85	15.84	16.10	15.85	15.93
CaO	11.34	10.43	10.28	10.33	10.37	10.35	10.24
Na_2O	2.21	1.64	1.20	1.28	1.15	1.21	1.11
K_2O	0.40	0.27	0.28	0.31	0.29	0.28	0.22
F	0.06	0.10	0.12	0.07	0.08	0.15	0.14
Cl	0.02	0.07	0.13	0.12	0.12	0.09	0.07
Total	96.84	97.60	96.87	97.18	96.66	96.50	96.66
–O=F, Cl	0.02	0.05	0.08	0.07	0.07	0.07	0.06
Total	96.82	97.55	96.79	97.11	96.60	96.43	96.60

More specifically, the earliest Barren Island lavas (i.e. of pre-historic age) range in composition from basalts with 8 wt% MgO and Cr and Ni concentrations of ~300 and 130 ppm, respectively, to andesites with up to 59 wt% SiO_2 (Fig. 11.5, Table 11.1; Luhr and Haldar 2006). All historic to recent basalts are noticeably more evolved and have MgO contents ranging from ~3 to 4.5 wt% and Cr and Ni concentrations of mostly 15–50 and 15–30 ppm, respectively (Fig. 11.5, Table 11.1; Luhr and Haldar 2006). Particularly low Cr and Ni contents are indications for extensive fractionation. Element–element diagrams of Barren Island compositions, irrespective of age, define mostly linear trends (Figs. 11.5 and 11.6) of increasing or decreasing concentrations typical for a tholeiitic liquid line of descent (i.e., increasing concentrations of FeO*, TiO_2, and P_2O_5; Naslund 1989). Elemental concentrations from the primitive to evolved endmembers of the suite change by a factor of 2–4 for the most incompatible elements. The following are notable points: (1) the inflection point in the enrichment trends of Fe, Ti, and P towards more differentiated compositions (i.e., trends that are distinctive for tholeiitic evolution) appears to be at 56 wt% SiO_2; (2) enrichment of Na is about equal to K, and to other incompatibly behaving elements; (3) Rare Earth elements (REE's) are enriched to about the same degree from La to Lu, and thus REE normalized patterns are simply parallel but offset to higher values (Fig. 11.7) with the exception of Eu which is enriched at slightly lower rates than neighboring REE's leading to decreasing Eu/Eu* from 1.05 to 0.93; (4) Eu/Eu* ratios exceeding 1 cannot be simply ascribed to accumulated plagioclase (cf. Luhr and Haldar 2006) because Barren Island lavas with relatively low Al_2O_3 of ~18 wt% encompass the entire range of Eu/Eu* while those with the highest Al_2O_3 of 21 wt% have Eu/Eu* of 1 (Table 11.1); and (5) also notable is the conservative behavior of Sr that remains near ~200 ppm throughout the Barren Island suite (Fig. 11.6) which is also

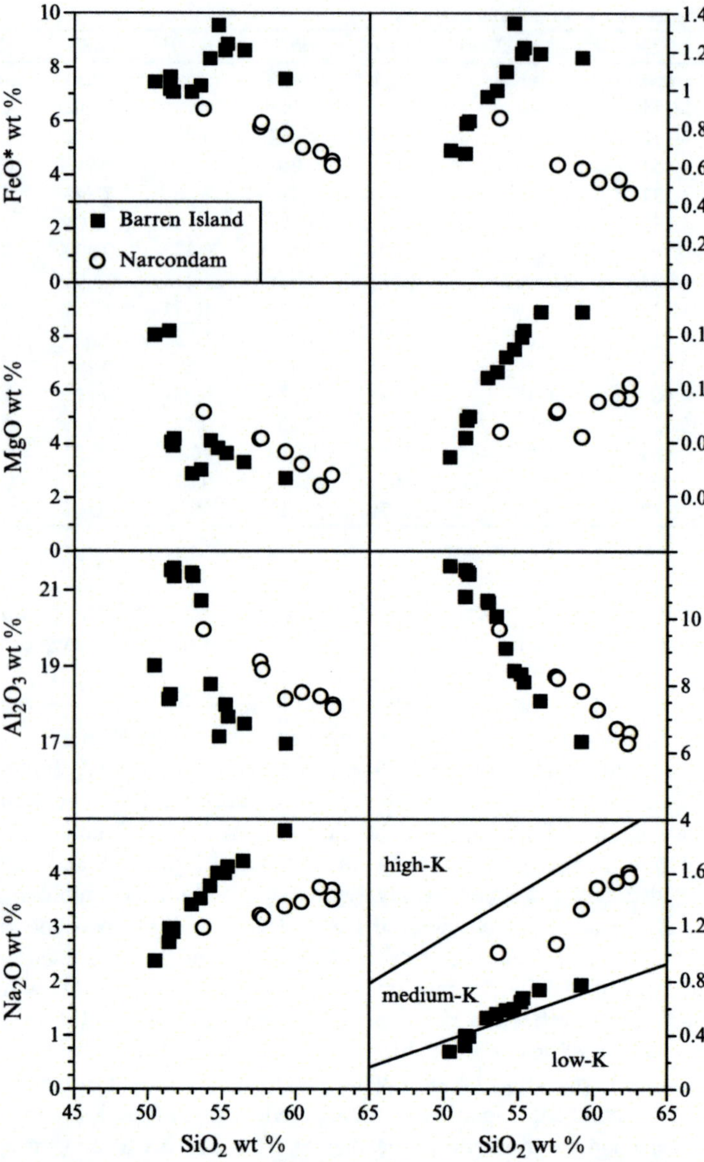

Fig. 11.5 Major elemental variations within Barren Island and Narcondam volcanic samples. Dividing lines in K_2O versus SiO_2 diagram are after Gill (1981)

observed on normalizated plots with progressively smaller positive Sr anomalies and expressed by strongly decreasing Sr/P (0.6–0.2) or Sr/Ce (25–9) with increased silica values. CaO contents decrease progressively from values of ~11.5 wt% in more primitive basalts to values ~7 wt% in andesites when plotted against SiO_2.

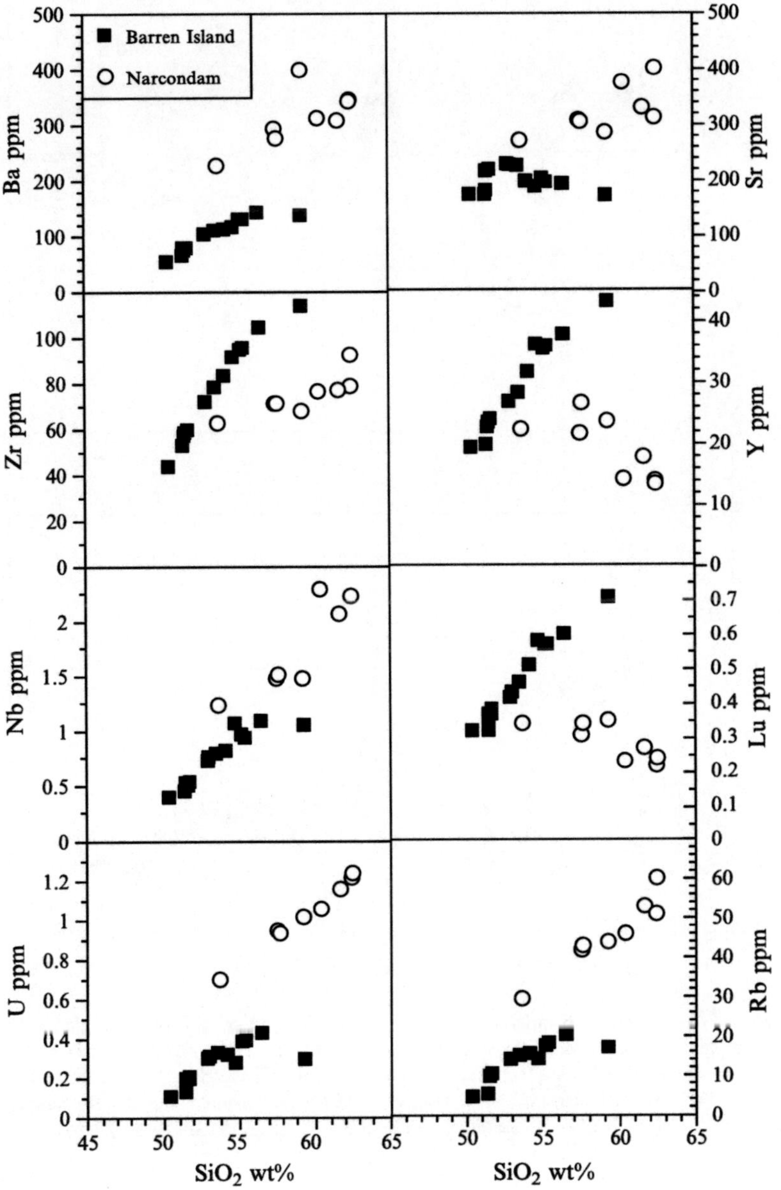

Fig. 11.6 Trace elemental variations within Barren Island and Narcondam volcanic samples

Al_2O_3 is lower in high Mg Barren samples, increasing in more evolved basalts before it decreases in andesites. This is emphasized if MgO content, instead of SiO_2, is used as the index of differentiation. And finally, a distinction between mostly prehistoric and historic Barren samples is evident on spiderdiagrams with

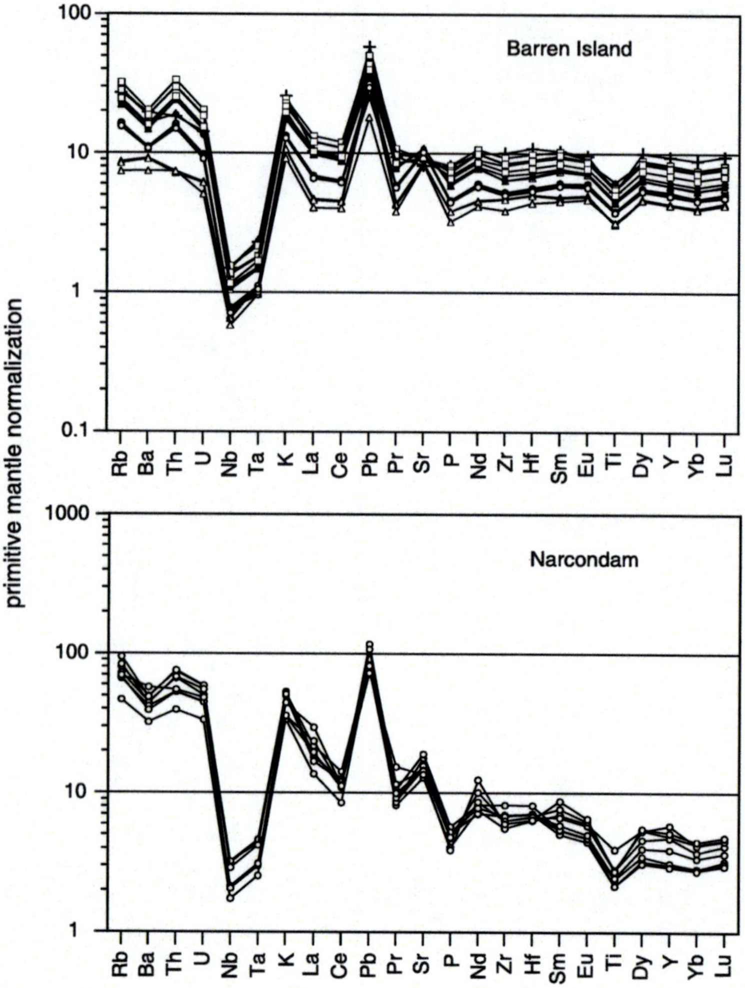

Fig. 11.7 Spiderdiagrams normalized to primitive mantle for Barren Island and Narcondam volcanic samples; normalization values for primitive mantle are taken from Sun and McDonough (1989). The symbols for Barren Island samples are as follows: *open triangle* = pre-historic samples, *solid circles* = 1787–1832 samples, *open squares* = 1991 samples, *open circles* = 1995 samples, the most silicic sample (P-1 of pre-historic age with 59.3 wt% SiO_2) is indicated by *cross*

regards to some elements such as Ba, Th, and Nb; ratios of Th relative to Ba and Nb are distinctly lower in pre-historic lavas than in most historic ones (Figs. 11.7 and 11.8); one historic lava was found to plot within the pre-historic ones and one pre-historic within historic ones.

Narcondam volcano lavas also generally define linear trends on element–element diagrams, although there is more scatter among these than within Barren Island lavas

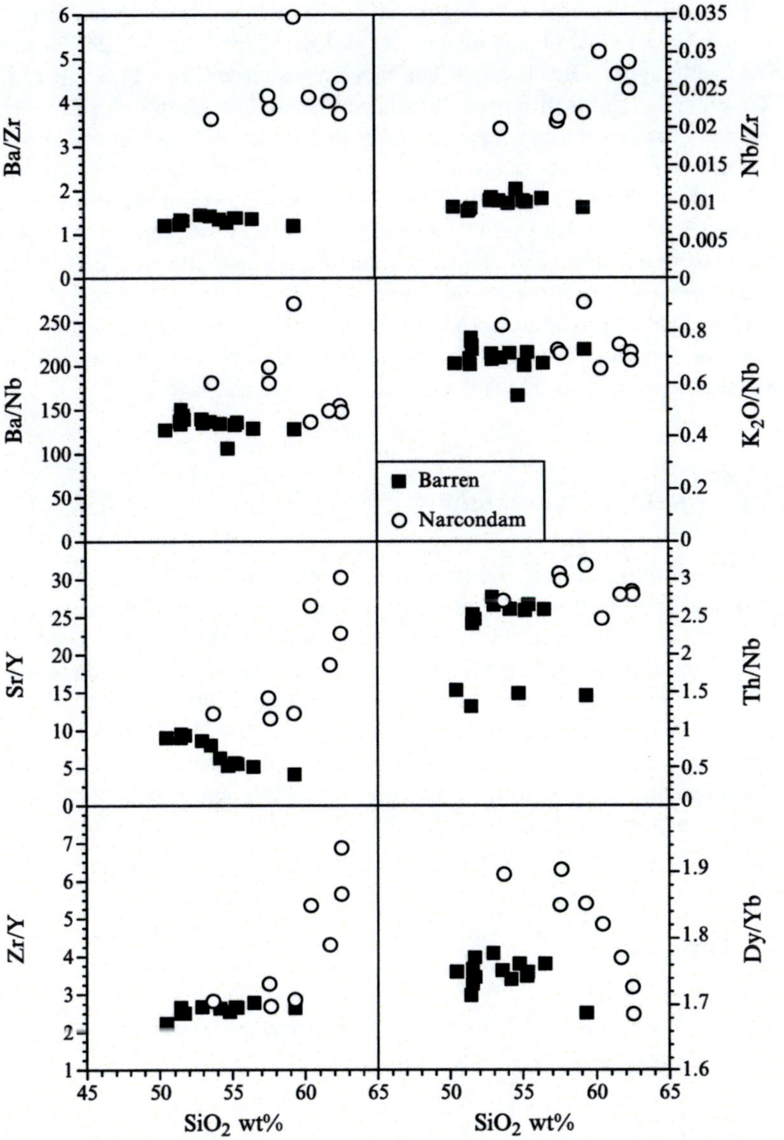

Fig. 11.8 Variations of incompatible trace element ratios among Barren Island and Narcondam volcanic samples

(Figs. 11.5 and 11.6). Maximum enrichments within the Narcondam sample suite (i.e., basaltic andesite to dacite) are also about twofold for the most incompatibly behaving elements like Rb, Th, U, Ta, and Nb. In contrast, some elemental trends within the suite are distinctly different compared to Barren Island lavas. Highlights include: (1) enrichments of Zr, Hf, and K are considerably less than Ta and Nb;

(2) HREE and Y decrease with higher SiO_2; (3) Sr increases despite a progressive decrease in CaO and Al_2O_3 and higher SiO_2 or lower MgO; and (4) LREEs generally increase with higher silica contents but data are scattered (Figs. 11.5 and 11.6).

Elemental variations of Barren Island and Narcondam samples lead to increased divergence between the two lava suites in many diagrams. Most commonly, the least evolved Narcondam lavas plot near Barren Island lavas near the "anchor point" from where the Narcondam trend diverges. However, there are elemental variations (e.g., Ba, U, Rb vs SiO_2) with a distinct offset between the two suites that is already evident in the most mafic Narcondam samples. Chemical offsets between suites are greatest with incompatible trace element ratios such as Ba/Zr, Nb/Zr, and U/Nb (Fig. 11.8). Enrichment trends of incompatible elements can be nearly parallel (e.g., Ta, Nb, Ba), have steeper slopes (e.g., U, Rb), or can have a shallower slopes (e.g., Zr, Hf) than trends defined by Barren Island lavas.

11.7.2 Radiogenic Isotopes

Barren Island and selected Narcondam isotope results (Table 11.4) indicate that Narcondam lavas are distinctly more radiogenic than Barren Island lavas resulting in fields that are offset (Fig. 11.9) from each other. For example, Narcondam lavas have $^{87}Sr/^{86}Sr$ ratios greater than 0.7051 while Barren Island lavas retain ratios lower than 0.7041. Pb isotope ratios also vary significantly with $^{206}Pb/^{204}Pb$ ratios of greater than 18.56 for Narcondam lavas and less than 18.31 for Barren lavas, respectively. In addition, smaller scale internal variations within Barren Island lavas are also present. Most notably, $^{208}Pb/^{204}Pb$, $^{206}Pb/^{204}Pb$, and $^{143}Nd/^{144}Nd$ ratios vary without corresponding changes in $^{207}Pb/^{204}Pb$ and $^{87}Sr/^{86}Sr$ (Fig. 11.9).

11.8 Discussion

11.8.1 Correlation of Chemical Signatures Between Volcanic Suites

Geochemical traits of Barren Island and Narcondam volcanoes are particularly interesting because Barren Island and Narcondam volcanoes share certain arc-like signatures, while also retaining characteristics that are distinct or become more distinctive with higher silica contents. Therefore, the two volcanoes provide information constraining when geochemical signatures are introduced and thus serve as an example for evaluating element fluxes in subduction related magmas.

Common features among Barren and Narcondam lavas include: (1) identical degrees of depletion of Ta and Nb and a depletion of Ti irrespectively of degree of magma evolution; (2) a lack of Zr and Hf depletion; and (3) identical normalized

Table 11.4 Isotopic ratios of Barren Island and Narcondam lavas

Sample	$^{87}Sr/^{86}Sr$	Error (2SE)	$^{143}Nd/^{144}Nd$	Error (2SE)	$^{206}Pb/^{204}Pb$	$^{207}Pb/^{204}Pb$	$^{208}Pb/^{204}Pb$
Barren Island							
Prehistoric:							
9-21-5-92	0.703884	0.000008	0.513111	0.000095	18.201	15.552	38.290
F1-1	0.703873	0.000010	0.512989	0.000010	18.134	15.551	38.194
1787–1832:							
X-L-1			0.512861	0.000011	18.265	15.562	38.367
L2-B-F2	0.703984	0.000014	0.512955	0.000010	18.309	15.562	38.411
1991:							
FG-91-L	0.704074	0.000011	0.512884	0.000011	18.246	15.558	38.359
8-9-2-92	0.704045	0.000011	0.512903	0.000010	18.237	15.558	38.337
1995:							
20-24-1-95	0.704027	0.000007	0.512903	0.000015	18.217	15.555	38.299
29-8-3-95	0.703978	0.000014	0.512898	0.000012	18.256	15.558	39.366
3C-8-3-95	0.704041	0.000007	0.512901	0.000011	18.273	15.563	38.369
1-11-5-95	0.703997	0.000007	0.512873	0.000010	18.237	15.564	38.318
17-6-10-95	0.704030	0.000008	0.512907	0.000013	18.063	15.557	38.098
Narcondam							
N87	0.705158	0.000011	0.512697	0.000009	18.565	15.644	38.861
8-22-2-87	0.705305	0.000008	0.512696	0.000010	18.617	15.655	38.936
Standards							
NBS987	0.710278	0.000009					
La Jolla			0.511849	0.000009			
NBS981					16.933	15.488	36.687
NBS981					16.931	15.486	36.682

trace element patterns for the most incompatible elements, Rb through U, with lower enrichments of Ba and U relative to Th and similar enrichments of Rb and Th (Fig. 11.7). These common features are also seen in diagrams involving K_2O/Nb, Ba/Nb, Ba/Th (Figs. 11.8 and 11.12). Exceptions are rare; one is the nearly horizontal normalized pattern from Rb to U of prehistoric Barren Island samples (Fig. 11.7), and another one is a Narcondam sample that has higher Ba (Figs. 11.6 and 11.7). In general, these similarities connect the two volcanoes and suggest these element patterns are ultimately derived from similar element fluxes from the subducting slab, similar mantle sources, and/or similar melt generation processes for primary mantle magmas (cf. Elliott 2003).

Contrary to similar geochemical features, are features that differentiate both lava suites. Among those are parameters with a pronounced offset between Narcondam and Barren Island lavas but with relative minor variations within each suite; these include Sr, Nd, and Pb isotopic ratios (Table 11.4, Fig. 11.9), Ba/Zr, and Nb/Zr (and analogue ratios involving Hf and Ta). Distinct offsets are also observed in single elements, however because of variability within each suite, it is the offset between sub-parallel trends as observed for element such as Rb, Ba, U, and Th when plotted against, e.g., SiO_2 (Fig. 11.6) that differentiates strongest the two lava suites.

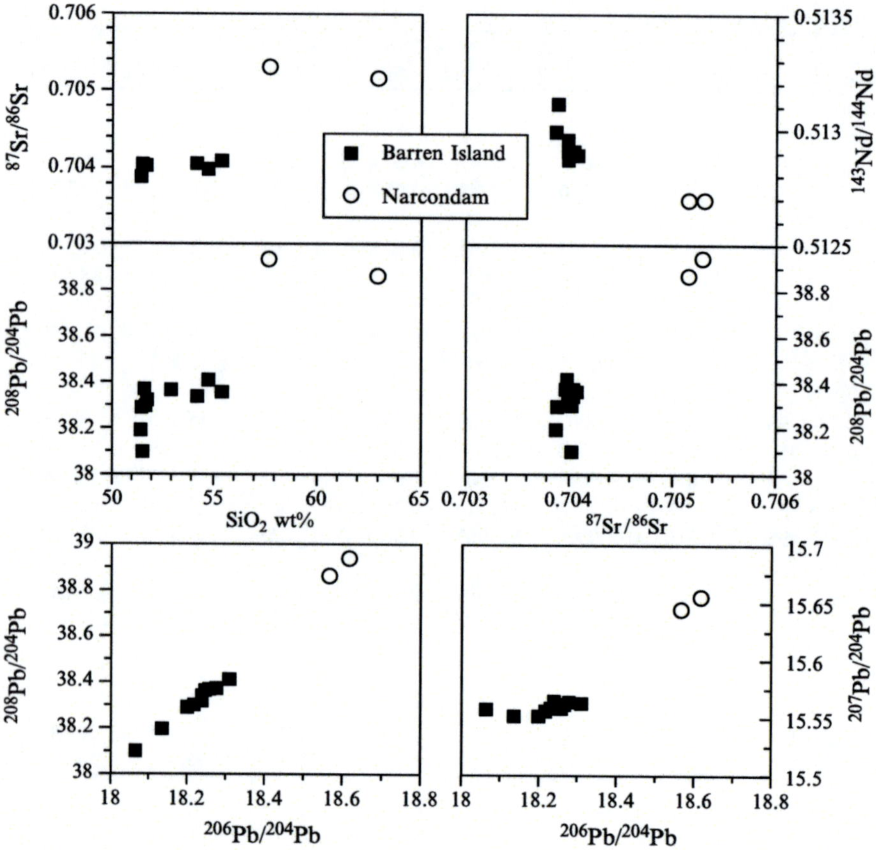

Fig. 11.9 Isotopic variations among Barren Island and Narcondam volcanic samples

Furthermore, there are a number of element and element ratios with systematics where values of Narcondam lavas at the mafic end are anchored within Barren Island lavas but diverge as lavas become more silicic (Figs. 11.6 and 11.8); the most notable are Sr/Y and Zr/Y and ratios with HREE. And lastly, MREE/HREE ratios (e.g., Dy/Yb) show the opposite behavior as the greatest differences are observed within mafic compositions between the Narcondam and Barren Island lavas but become more similar in more silicic lavas (Fig. 11.8). Narcondam lavas define a decreasing trend in e.g. Dy/Yb that is consistent with the influence of amphibole as a fractionating phase (Davidson et al. 2007) while Barren lavas retain similar values irrespective of silica contents.

As such, it is likely that shared geochemical traits seen in both Barren Island and Narcondam lavas, which do not change significantly with silica contents such as Ba/Th, Th/Nb or K/Nb, originate early during magma genesis, most likely in the source of magmas. We, however, argue that all geochemical differences are introduced during a later stage of differentiation, magma mixing, and contamination or a combination of all such processes.

To account for differences between Barren Island and Narcondam lavas, one could call upon changes in the involvement of (1) subducted sediments originating from the Bengal fan or (2) subducted altered oceanic crust. However, greater fluid addition, greater sediment melt addition, or other significant contributions from the subducting slab would ultimately modify Ba/Th, U/Th, Th/Nb, Ba/La, Nb/Ta and K/Ta ratios of primary melts. As shown above, except for pre-historic Barren Island samples, other Barren Island and Narcondam samples indicate nearly constant values in these ratios thus placing a limited role of variable sediment melt, fluid, or altered oceanic crust contributions as cause for the distinct geochemical differences between the two volcanic suites. (cf., Pearce et al. 1995; Patino et al. 2000; Elliott 2003; Leat et al. 2003). That these ratios remain nearly constant suggests that deep sources are not the principal cause for differences in geochemical signatures between Barren Island and Narcondam lavas. This, in turn, leaves processes and sources associated with the upper plate as explanation. Upper plate processes are particularly likely in cases where geochemical parameters correlate with increasing SiO_2 and/or correlate with petrographic evidence for open system processes such magma mixing and contamination. However, constant geochemical signatures shared by lavas at both volcanoes and the geochemical signatures of Barren Island basalts likely reflect deep subduction-related processes, such as fluid- and melt-driven element additions from the slab, and melting in the mantle wedge. Thus, the variations observed in lavas from these two Andaman arc volcanoes and the combination among parameters are remarkable and will be further explored in the future.

11.8.2 Evidence for Contamination by Felsic Material or Mixing with Silicic Magmas

Petrographic evidence offers important constraints for processes affecting Narcondam volcano lavas. Minerals are observed that would almost certainly not crystallize from melts similar to Narcondam lavas compositions. These minerals include quartz and zircon. Rather, such minerals are expected as late interstitial phases of Narcondam magmas if they reached high degrees of crystallinity. Quartz is typically not stable until a rhyodacitic (~70 wt% SiO_2) to rhyolitic (74 wt% SiO_2) melt composition is attained (cf., Barclay et al. 1998). Furthermore, quartz crystal sizes of the order of 200–800 μm make it unlikely that these are derived from a late crystallizing, interstitial melt. The presence of zircon is similarly problematic. Zircon crystallization depends on the amount of dissolved Zr, Al saturation, and temperature of the magma (Watson and Harrison 1983). Using observed bulk compositions, calculated zircon saturation would occur at ~670°C and ~700°C for 57 and 62.5 wt% silica, respectively (Watson and Harrison 1983). These temperatures are well below the solidus temperatures expected for either composition and thus indicates that zircon is an unlikely stable phase. Therefore, zircon must be inherited from either country rock or a more silicic magma. An additional phase that is not in equilibrium with host magma compositions is occasional olivine crystals which always have thin orthopyroxene reaction rims. Hornblendes, present as crystals with alteration rims or as pseudomorphs, also are unstable but this may not necessarily be due to compositional conditions (Streck 2008). However, hornblende acts as the main host for zircons which would suggest that both have a common source.

Disequilibrium textures and the presence of exotic mineral phases that are not stable in andesite/silicic andesite magma compositions offer unequivocal evidence that all Narcondam lavas containing these minerals have experienced silicic inputs. Furthermore, these silicic inputs may have occurred as a result of mixing with a silicic magma (e.g. rhyolite) as envisioned by Pal et al. (2007) or through addition of solid contaminants (country rock) during assimilation. Occurrence of euhedral, bipyramidal quartz crystals suggest the former while more radiogenic isotopic signatures of Narcondam lavas compared to Barren Island samples suggest the latter.

11.8.3 Implications of Hornblende as a Phenocryst Phase

A striking petrographic feature of Narcondam lavas is the abundance of both fresh hornblende and pseudomorphs after hornblende which are virtually absent in Barren Island lavas although Haldar et al. (2007) report that very rare hornblende phenocrysts occur in some prehistoric Barren Island lavas. Experimental phase equilibria studies have delineated the stability field of hornblende in basaltic to dacitic arc magmas to conditions typically deeper than 100 MPa (~3.5 km depth), cooler than ~875–900°C, and water saturation (e.g., Rutherford and Hill 1993; Rutherford and Devine 2003). Comparing the compositions of Narcondam hornblende phenocrysts with those of other natural and experimental samples indicates that hornblendes from silicic andesitic samples (i.e., ~62 wt% SiO_2) are very close in composition to natural and experimental hornblendes from silicic andesitic samples of Soufriere Hills volcano, Montserrat (Fig. 11.10). At Soufriere Hills, hornblende is stable at temperatures below 875°C and pressures of greater than 115 MPa (Barclay et al. 1998). In contrast, Narcondam hornblendes from basaltic andesitic inclusions, that are richer in Al_2O_3, are similar to those from Mt. St. Helens and require higher pressure conditions of 220 MPa (~7 km depth) and temperatures of less than 900°C (Rutherford and Hill 1993). This could indicate that hornblendes from Narcondam volcano are derived from a variety of depths. The compositional similarity of hornblendes from Soufriere Hills and Narcondam volcanoes suggests similar crystallization conditions of both systems that, in turn, constrains the depth of Narcondam hornblende crystallization. A pressure of above 100 MPa translates to a depth in excess of ~3.5 km. This places the shallowest possible Narcondam magma reservoir, where hornblendes grew at Narcondam

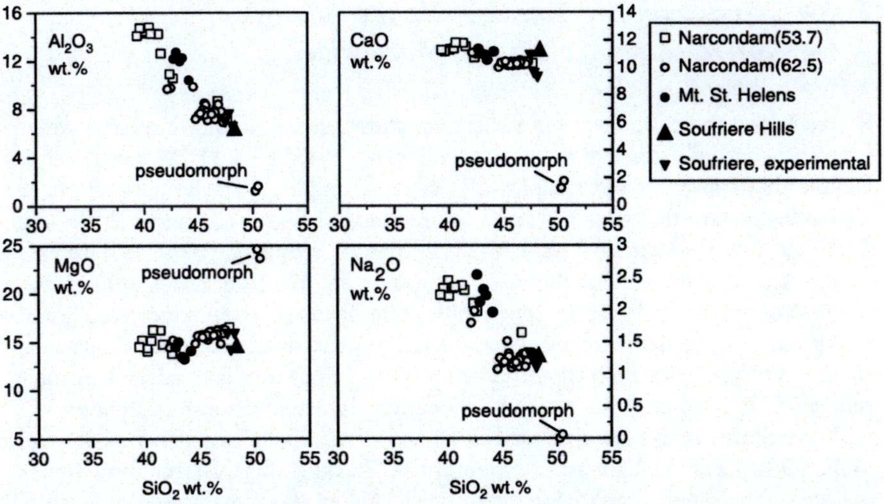

Fig. 11.10 Amphibole compositions in Narcondam lavas compared to natural and experimental amphibole of Soufriere Hills volcano (Barclay et al. 1998; Rutherford and Devine 2003) and Mount Saint Helens (Rutherford and Hill 1993). Pseudomorph indicates an orthopyroxene pseudomorph after amphibole (seen in Fig. 11.4b). *Numbers in parenthesis* after Narcondam are SiO_2 (wt%) of samples

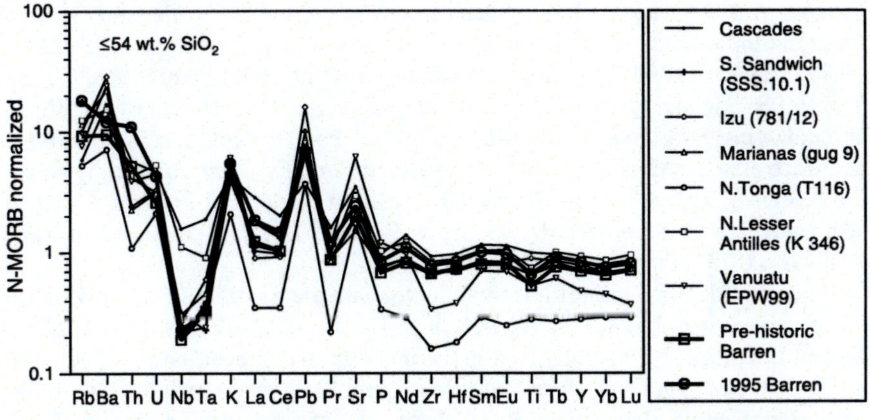

Fig. 11.11 Barren Island lavas compared to typical intra-oceanic arc tholeiites using select samples of Elliott (2003) and of a tholeiite from the Cascades (J. Streck, unpublished data 2007). N-MORB values are those of Sun and McDonough (1989)

volcano ~1–2 km below the seafloor. This further implies that the crust of the northern Andaman Sea underlying Narcondam volcano is evolved and not truly oceanic as the formation of calc-alkaline intermediate composition magmas (i.e., andesites and dacites) is commonly tied to continental or evolved arc crust and not to ridge-formed basaltic seafloor.

11.8.4 Andaman Arc Volcanoes in the Context of Volcanic Arc Systems Worldwide

Barren Island tholeiitic lavas have characteristics that are similar to other tholeiites from intra oceanic arcs like the South Sandwich arc (Fig. 11.11). Notable features include that Barren Island samples plot among samples with the greatest Nb, Ta anomalies for arc tholeiites but stand out because of high to extreme Th/Nb, high Th/U and low Ba/Th ratios, particularly in historic eruptions (Fig. 11.11). Shared characteristics with other arc tholeiites include overall flat patterns on 'spidergrams' (normalized to MORB or primitive mantle) that have relatively minor enrichments of the most incompatible elements compared to patterns of medium or high-K arc basalts but include the typically observed positive or negative anomalies in elements such as K, Pb, and Sr that are commonly enriched in subduction-related lavas.

The features above are emphasized when Barren Island samples are compared to the intra-oceanic island arc compilation by Elliott (2003). Barren tholeiites are extremely enriched compared to any other island arc lavas with respect to Th/Nb (1.5–2.5) (Fig. 11.12). However, some geochemical parameters are similar to typical island arc basalts but trend towards MORB (e.g., Pb/Ce of 0.2, U/Nb of ~0.35, and Nb/Ta of 10–14; cf. Fig. 9 in Elliott, 2003) while other parameters are virtually indistinguishable from MORB with Ba/Th ~60, U/La ~0.04, and $(La/Sm)_N$ = 0.9–1.3 (cf., Fig. 11.12). And lastly, some are more depleted than MORB (U/Th of ~0.17; Nb/Zr of ~0.01) (Fig. 11.13).

In regards to isotopes, Barren Island tholeiites are the least radiogenic lavas of the entire Indonesian arc system (Fig. 11.9) with, for example, values of $^{87}Sr/^{86}Sr$ of ~0.7040 and $^{207}Pb/^{204}Pb$ of ~15.55 (Luhr and Haldar 2006). While $^{87}Sr/^{86}Sr$ and $^{207}Pb/^{204}Pb$ ratios are uniform, Barren Island lavas are characterized by significant variations in $^{143}Nd/^{144}Nd$, $^{206}Pb/^{204}Pb$ and $^{208}Pb/^{204}Pb$. This combination of isotope and trace element variations is remarkable considering that the subducting slab at the Andaman Trench has one of thickest sediment piles (~3.5 km) and that these sediments are characterized by highly radiogenic isotopic compositions (e.g., $^{87}Sr/^{86}Sr = 0.731$; Plank and Langmuir 1998).

Andesitic/dacitic lavas of Narcondam volcano are medium K, arc lavas similar to those erupted along the arc in Sumatra and Java as well as in other places around the globe. Isotopically, and in comparison to compositions of Indonesian volcanic rocks in general, these lavas lie in the middle range with the least radiogenic lavas being Barren Island (cf., Figs. 14, 15 in Luhr and Haldar (2006) using a data compilation of Kimberly et al. 1998). In terms of trace elements (e.g. Ba/Zr), Narcondam lavas are comparable (Fig. 11.13) to typical basaltic andesites–dacites of island and continental arcs (e.g., Mt. St. Helens, Central America). Other parameters, mainly those unchanged from values observed among basaltic rocks of Barren Island, are unusual. Among those are extremely high Th/Nb (~2.7), very low, sub-MORB U/Th (~0.2) and very low, MORB-like Ba/Th (~60) (Figs. 11.12 and 11.13). One additional feature of Narcondam andesites/dacites is that lavas become progressively more adakite-like (Fig. 11.13) as Sr/Y ratios progressively increase with higher silica contents of the lavas

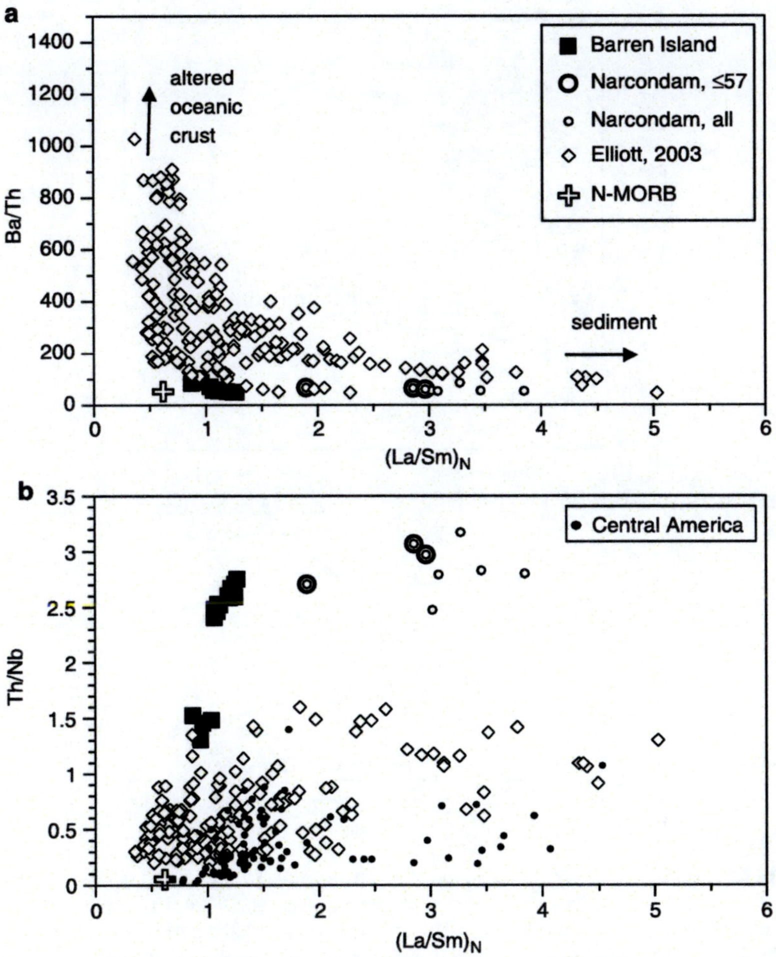

Fig. 11.12 Compositional ranges of lavas of Barren Island and Narcondam volcano superimposed on redrawn Figs. 4 and 8 of Elliott, (2003) with a global compilation of oceanic island arc lavas of <56 wt% SiO_2. Some Narcondam lavas range from 54 to 57 wt% SiO_2 and thus overlap or are close to silica range of lavas compiled by Elliott. The trend observed in Narcondam samples suggests involvement of a greater amount of sediment from the subducting plate than observed in Barren Island lavas. However, additional evidence suggests that sediments or continental crust additions from the overriding plate is more likely to have caused the distinct shift from Barren Island to Narcondam lavas

(cf. Defant and Drummond 1990). Thus, Narcondam andesites/dacites have interesting combinations of geochemical parameters that are commonly used to evaluate petrogenetic processes occurring at subduction zones. Some parameters are identical to those of basalts at neighboring Barren Island and those generally stand out in comparison to compilations of island arc basalts. Other Narcondam chemical parameters are, however, typical for island and continental arcs.

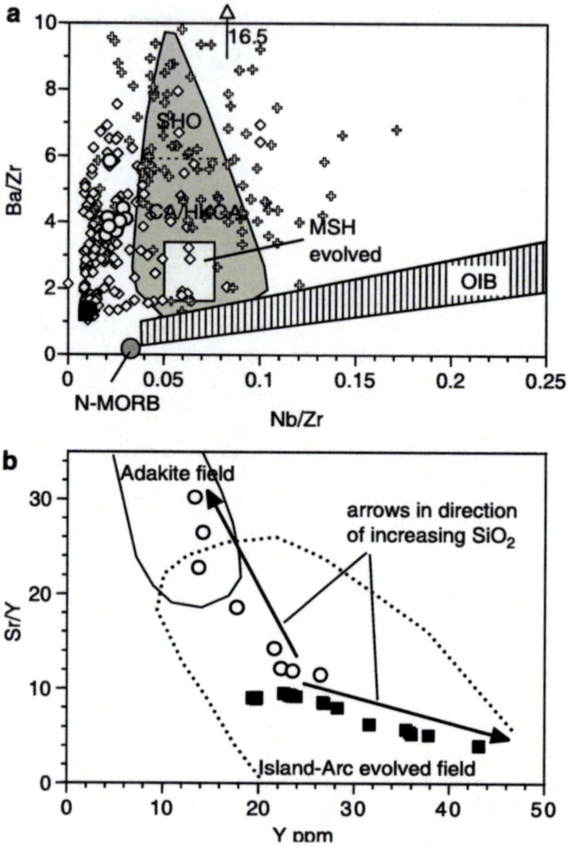

Fig. 11.13 (**a**) Barren Island (*solid squares*) and Narcondam (*open circles*) lava compositions compared to fields of Leeman et al. (2005) for the Cascades and OIB (OIB = ocean island basalt, SHO = shoshonite, CA = calc-alkaline, HKCA = high K calc-alkaline, MSH evolved = dacites of Mt. St. Helens), in comparison to N-MORB of Sun and McDonough (1989), and in comparison to data compilations by Elliott (2003) of intra-oceanic arcs (*open diamonds*) and of Central American lavas (<56 wt% SiO_2; *crosses*). Arrow indicates that 23 Central American samples and two intra-oceanic arc samples have Ba/Zr higher than 10 up to 16.5; (**b**) Barren Island and Narcondam lava compositions compared to fields by Defant and Drummond (1990)

11.9 Conclusions

The active Barren Island volcano and the Pleistocene-Holocene Narcondam volcano are the only two subaerially exposed arc volcanoes of the Andaman Sea and result from subduction of the Indian plate beneath the Burma plate. Lavas at Barren Island volcano range from basalt to andesite while lavas from the ~140 km distant Narcondam volcano range from andesite to silicic andesite/dacite. Lavas at both volcanoes indicate an unusual combination of trace elemental and isotopic characteristics. Similarities in the geochemistry of both lava suites include strong and comparable depletions in Nb and Ta, low MORB-like Nb/Zr, and nearly constant U/Th. Some of these similarities, such as high Th/Nb, stand out in comparison to other intra-oceanic arc systems. Shared geochemical features almost certainly originate early during magma genesis, most-likely in the source of magmas that do not change subsequently.

There are, however, distinct geochemical differences including elevated trace elemental abundances and elemental ratios in Narcondam lavas compared to Barren Island lavas. These include elevated Ba, Rb, and U concentrations and high Ba/Zr and Nb/Zr and significantly more radiogenic isotopic ratios of Narcondam lavas in comparison to Barren Island lavas. Additional geochemical parameters (e.g., Sr/Y, Zr/Y, Th/La, U/La, Ba/La) of Narcondam lavas positively correlate with increasing SiO_2 but are anchored at the mafic end within compositions observed at Barren Island volcano. Generally more-incompatible trace element enriched silicic Narcondam magmas are best explained by amphibole-dominated fractionation of a Barren Island-type basalt. Such a scenario would be consistent with an increase of Sr/Y correlated with increased silica and ubiquitous amphibole as major phenocryst phase in Narcondam samples. The shift in isotopic values from typical Barren Island lavas to those of Narcondam lavas is likely caused by crustal contamination resulting during magma residence in the upper plate within extended continental crust lithologies and/or sediments from the fan associated with the Irrawaddy Delta at the Myanmar continental margin, which extends to the more proximal position of Narcondam volcano.

Acknowledgments We like to thank the editors of this special publication for their invitation to contribute this paper. We also thank the two reviewers for constructive comments that helped to improve this manuscript.

Further more, we thank Tim Elliott for making his data compilations available.

References

Barclay J, Rutherford MJ, Carroll MR, Murphy MD, Devine JD, Gardner J, Sparks RSJ (1998) Experimental phase equilibria constraints on pre-eruptive storage conditions of the Soufriere Hills magma. Geophys Res Lett 25(18):3437–3440

Curray JR (2005) Tectonics and history of the Andaman Sea region. J Asian Earth Sci 25:187–232

Dasgupta S, Mukhopadhyay M (1993) Seismicity and plate deformation below the Andaman arc, northeastern Indian Ocean. Tectonophysics 225:529–542

Davidson J, Turner S, Handley H, Macpherson C, Dosseto A (2007) Amphibole sponge in arc crust? Geology 35:787–790

Defant MJ, Drummond MS (1990) Derivation of some modern arc magmas by partial melting of young subducted lithosphere. Nature 347:662–665

Duncan RA, Keller RA (2004) Radiometric ages for basement rocks from the Emperor Seamounts, ODP Leg 197. Geochem Geophys Geosyst 5: Art.No. Q08L03. doi: 10.1029/2004GC000704

Elliott T (2003) Tracers of the slab. In: Eiler J (ed) Inside the subduction factory. Geophys Monog 138:23–45

Gasparotto G, Spadafora E, Summa V, Tateo F (2000) Contribution of grain size and compositional data from the Bengal Fan sediment to the understanding of Toba volcanic event. Marine Geol 162:561–572

Gill JB (1981) Orogenic andesites and plate tectonics. Springer, Berlin, pp 1–391

Haldar D, Banerjee PK, Streck MJ, Mukherjee P (2007) Quaternary volcanism on the Barren and Narcondam Islands in the Andaman Sea: Arc magmatism within a rift tectonic environment. In: Ray J, Bhattacharyya C (eds) Igneous petrology: 21st century perspective. Allied Publishers Pvt. Ltd, New Delhi

Johnson DM, Hooper PR, Conrey RM (1999) XRF analysis of rocks and minerals for major and trace elements on single dilution Li-tetrahorate fused bead. Adv X-ray Anal 41:843–867

Khan PK, Chakraborty PP (2005) Two-phase opening of Andaman Sea: a new seismotectonic insight. Earth Plan Sci Lett 229:259–271

Kimberly P, Siebert L, Luhr JF, Simkin T (1998) Volcanoes of Indonesia, v. 1.0 (CD-ROM), Smithsonian Institution, Global Volcanism Program, Digital Information Series, GVP-1

Knaack C, Cornelius SB, Hooper PR (1994) Trace element analyses of rocks and minerals by ICP-MS. GeoAnalytical Lab, Washington State University, Washington DC

Koppers AAR (2002) ArArCALC – software for 40Ar/39Ar age calculations. Comput Geosci 28:605–619

Leat PT, Smellie JL, Millar IL, Larter RD (2003) Magmatism in the South Sandwich arc. In: Larter RD, Leat PT (eds) Intra-oceanic subduction systems: tectonic and magmatic processes. Geological Society of Special Publications 219, London, pp 285–313

Leeman WP, Lewis JF, Evarts RC, Conrey RM, Streck MJ (2005) Petrologic constraints on the thermal structure of the Cascades arc. J Volcanol Geotherm Res 140:67–105

Luhr JF, Haldar D (2006) Barren Island volcano (NE Indian Ocean): Island-arc high-alumina basalts produced by troctolite contamination. J Volcanol Geotherm Res 149:177–212

May WE, Watters RL (2004) Certificate of Analysis, Standard Reference Materials 997. National Institute of Standards and Technology (NIST), 1–2

Naslund HR (1989) Petrology of the Basistoppen sill, east Greenland: a calculated magma differentiation trend. J Petrol 30:299–319

Office of Naval Research (2009) The Andaman Sea, An Atlas of Oceanic Internal Solitary Waves, Global Ocean Associates, http://www.internalwaveatlas.com/Atlas_PDF/IWAtlas_Pg207_AndamanSea.PDF

Pal T, Mitra SK, Sengupta S, Katari A, Bandopadhyay A (2007) Dacite – andesite of Narcondam volcano in the Andaman Sea – an imprint of magma mixing in the inner arc of the Andaman–Java subduction system. J Volcanol Geotherm Res 168:93–113

Patino LC, Carr MJ, Feigenson MD (2000) Local and regional variations in Central American arc lavas controlled by variations in subducted sediment input. Contrib Mineral Petrol 138:265–283

Pearce JA, Baker PE, Harvey PK, Luff IW (1995) Geochemical evidence for subduction fluxes, mantle melting, and fractional crystallization beneath the South Sandwich Island arc. J Petrol 36:1073–1109

Plank T, Langmuir CH (1998) The chemical composition of subducting sediment and its consequences for the crust and mantle. Chem Geol 145:325–394

Raju KAK, Ramprasad PS, Rao PS, Rao BR, Varghese J (2004) New insights into the tectonic evolution of the Andaman basin, northeast Indian Ocean. Earth Planet Sci Lett 221:145–162

Ramos FC (1992) Isotope geology of the Grouse Creek Mountains, Box Elder County, Utah. MS thesis, UCLA, Los Angeles, CA, pp 1–92

Renne PR, Swisher CC, Deino AL, Karner DB, Owens TL, DePaolo DJ (1998) Intercalibration of standards, absolute ages and uncertainties in $^{40}Ar/^{39}Ar$ dating. Chem Geol 145:117–152

Rodolfo KS (1969) Bathymetry and marine geology of the Andaman Basin, and tectonic implications for Southeast Asia. Geol Soc Am Bull 80:1203–1230

Rutherford MJ, Devine JD (2003) Magmatic conditions and magma ascent as indicated by hornblende phase equilibria and reactions in the 1995–2002 Soufriére Hills magma. J Petrol 44:1433–1454

Rutherford MJ, Hill PM (1993) Magma ascent rates from amphibole breakdown: Experiments and the 1980-1986 Mount St Helens eruptions. J Geophys Res 98:19667–19685

Shanker R, Haldar D, Absar A, Chakraborty SC (2001) Pictorial monograph of the Barren Island Volcano: the lone active volcano in the Indian subcontinent. Geological Survey of India, Kolkata, p 87

Sun SS, McDonough WF (1989) Chemical and isotopic systematics of oceanic basalts: implications for mantle composition and processes. In: Saunders AD, Norry MJ (eds) Magmatism in the Ocean Basins. Geological Society of America Special Publication 42, pp 313–345

Stern RJ, Kohut E, Bloomer SH, Leybourne M, Fouch M, Vervoort J (2006) Subduction factory processes beneath the Guguan cross-chain, Mariana Arc: no role for sediments, are serpetinites important? Contrib Mineral Petrol 151:202–221

Streck MJ (2008) Mineral textures and zoning as evidence for open system processes. Rev Miner Geochem 69:595–622

Streck MJ, Grunder AL (1997) Compositional gradients and gaps in high-silica rhyolites of the Rattlesnake Tuff, Oregon. J Petrol 38:133–163

Thorne L et al (2005) The great Sumatra–Andaman earthquake of 26 December 2004. Science 308:1127–1133

Watson EB, Harrison TM (1983) Zircon saturation revisited: temperature and composition effects in a variety of crustal types. Earth Planet Sci Lett 70:295–304

Williams H, Turner FJ, Gilbert (1982) Petrography: an introduction to the study of rocks in thin sections. W.H. Freeman, New York, pp 1–626

Wolff JA, Ramos FC (2003) Pb isotope variations among Bandelier Tuff feldspars: no evidence for a long-lived silicic magma chamber. Geology 31:533–536

Extensional Volcanics

Chapter 12
Polybaric Evolution of the Volcanic Rocks at Gabal Nuqara, North Eastern Desert, Egypt

E.A. Khalaf, M. Khalaf, and F. Oraby

Abstract Field relations, petrography and chemical composition constrain old and young volcanic sequences for the formation of the Neoproterozoic-Early Cambrian Nuqara rock belts. These volcanics are a part of a post-subduction and extensional-related magmatic event in east Egypt. The two volcanic sequences differ in phenocryst mineralogy (hydrous vs anhydrous assemblages) and whole-rock major and trace element chemistry. Their eruptions were punctuated by occasional volcaniclastic deposits that generated fall, flow or reworked suites compositionally identical to the lava flows. The model proposed for their origin involves contrasting ascent paths and differentiation histories through crustal columns with different thermal and density gradients.

The old volcanic sequence covers a compositional range from basalt through andesite to rhyolite (51–73 wt% SiO_2) of the medium to high-K calc-alkaline series. They show lower contents of K_2O, Al_2O_3 and of several other incompatible trace element abundances and ratios than those of younger volcanics with comparable degree of evolution. The evolution of the whole old volcanic spectrum was governed mainly by crystal/melt fractionation of Ol, Cpx, Plag, Fe–Ti oxides and Ap in the intermediate varieties, and of Cpx, Plag, Fe–Ti oxides and Ap in the felsic varieties. At each stage of evolution, fractional crystallization was accompanied by variable degree of crustal contamination. The young volcanic sequence involves basaltic andesite and andesite rock suites and defines high-K tholeiitic series. Replenishment, tapping and subsequent fractional crystallization processes (RTF) within strongly zoned magma chamber led to the generation of tholeiitic melts. These young volcanics become enriched in LILEs and moderately compatible elements with time and this indicates that magma mixing become more important during the lifespan of the volcanic centers.

E.A. Khalaf (✉)
Geology Department, Cairo University, Cairo, Egypt
e-mail: ezz_khalaf@hotmail.com

M. Khalaf and F. Oraby
Nuclear Material Authority, Kottoymia, Egypt

J. Ray et al. (eds.), *Topics in Igneous Petrology*, DOI 10.1007/978-90-481-9600-5_12,

The source of the old and young volcanic magmas at Nuqara area was probably a metasomatised fertile asthenospheric mantle, where the long period of subduction (>300 Ma) in eastern Egypt prior to the onset of these volcanic magmatisms resulted in LILEs enrichment of the subcontinental mantle lithosphere. It is inferred that the old calc-alkaline melts came from a garnet-and amphibole-bearing mantle at deeper depth, modified by a subduction component, whereas the tholeiitic melts came from enriched phlogopite-bearing spinel lherzolite at shallower depth.

The contrasts in phase assemblages and differentiation trends reflect contrasting polybaric fractionation histories from two parental magmas. This polybaric evolution can account for the chemical discontinuity between the old and young volcanic magmas erupted on Nuqara area.

12.1 Introduction

Magmatic evolution is a complex interaction of processes governed by the physical properties, thermal budget, and composition of the magma and its environment. There is a general consensus that multisource, multi-component processes operate during the formation of island arcs and that this multiplicity is responsible for the genesis of the wide range in magma compositions, i.e. tholeiitic to potassic alkaline (Astis et al. 2000). However, the origin of the latter, and their relationship to the more voluminous calc-alkaline volcanic products, is still the subject of discussion. Many authors have documented the complex evolutionary history of magmas erupted on continental crust and have demonstrated that fractional crystallization, magma mixing and assimilation are important processes in the petrogenesis of these eruptives. In addition, evolution is polybaric and the compositions of eruptives reflect mixing of magmas which evolved via assimilation fractional crystallization (AFC) processes in separate, but interconnected chambers sited at different depths within the crust (Huijsmans et al. 1988). Polybaric origin of arc magmas in some volcanic arcs is supported by geophysical studies. These studies commonly identify two broad groups of magma-chambers beneath large stratovolcanoes; a crustal group at 5 to 20 km depth and a grouping of large chamber in the upper most mantle at the crust–mantle boundary (Balesta et al. 1977). The presence and variable thickness of continental crust has led other authors to stress the role of crustal contamination and low-pressure evolutionary processes to explain petrological and geochemical variability along converging plate margins (Esperanca et al. 1992). Moreover, the presence of regional tectonic frame not related to subduction can further influence the compositional features of the magmatic suites (Edwards et al. 1991).

Compositional variations of the magma may occur over short, medium and long time intervals (Gill 1981). Compositional variations occurring within a single eruption cycle (short term) have been discussed by Barton and Huijsmans (1986) for the post caldera lavas and reflect zoning of, or inhomogeneities in a magma chamber prior to eruption. Medium term variations are those which occur in a sequence of lava flows erupted over a period of several years to several thousands of years.

They are generally attributed to a variety of processes, including fractional crystallization, magma mixing, and assimilation. Within eruptive sequences, it is commonly found that the lavas become more differentiated with time (Luhr and Carmichael 1980) and the chemical variations are usually interpreted to indicate progressive crystal–liquid differentiation with or without the superimposed effect of magma mixing. Long-term variations may also occur over the life-time of some volcanic complexes and probably represent changes in the physical condition of magma generation and/or changes in the source rock towards more refractory compositions. Geochemical differences have been observed between different eruption centers and have been interpreted in the light of such long-term variations (Huijsmans et al. 1988).

Gabal Nuqara area is located near to the Red Sea coast at about 3 km southwest of Safaga city. It is bounded by latitude 26° 38 ′ and 26° 45′ N and longitudes 33° 51′ and 33° 56′ E. This area is rimmed on its flank by extensive outcrops of Neoproterozoic volcanics and volcaniclastics. In the same area, these volcanics are intensively dissected by epizonal granitic and gabbroic bodies as well as dyke swarms with sharp intrusive contact. The whole rock sequence has southwestward dips at moderate to high angle.

The purpose of this article is (i) to describe an exceptional association of highly contrasted volcanic types emplaced during Late Precambrian in a restricted area of North Eastern Desert (NED) (ii) to discuss the geochemistry of the volcanic suites that were erupted during and after the inferred culmination in crustal attenuation and (iii) to examine its petrogenesis in connection with the latest stage of the Pan African evolution.

12.2 Regional Geological Setting

Neoproterozoic closure of the Mozambique ocean sutured east and west Gondwana along the length of the East African Orogen (Stern 1994), stretching from Mozambique and Madagascar in the south, to the Arabian Peninsula and Eastern Desert of Egypt in the north (Kusky and Ramadan 2002). Nuqara area is located at the northern part of the Eastern Desert, where Safaga region marks the transition from collision ending at 615–600 Ma to extensional collapse occurred within the 600–575 Ma time span followed by transpressional tectonism along NW-trending strike-slip Najd faults and major shear zones until 530 Ma (Greiling et al. 1994). The Najd faults mark the closing stages of Late Precambrian crustal evolution in Afro-Arabia (Stern 1985). These faults also have been recognized in Egypt (Abu Zied 1984) and have affected Phanerozoic rocks (e.g. Duwi shear zone, Fig. 12.1a). The NW-SE directed shear zones common in the Central Eastern Desert (CED) represent the northernmost extension of the Najd fault system. These obvious zones of shears are not found to the north of the boundary between the North Eastern (NED) and Central Eastern Desert. El Gaby et al. (1996) mentioned that the Egyptian basement rocks have been affected from north to south by three shear zones; the dextral NE-SW Qena-Safaga, dextral Idfu-Marsa Alam and sinistral

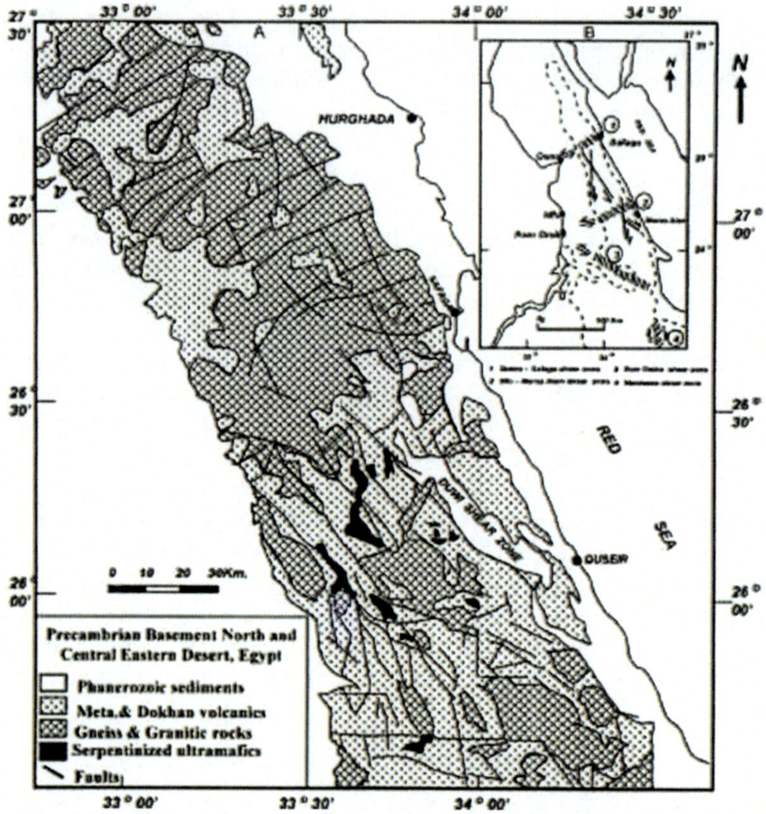

Fig. 12.1 (**a**) Generalized geological map of the Precambrian basement in the northern two thirds of the Central Eastern Desert and southern half of the north eastern Desert, Egypt (Mapped by Egyptian Geological Survey and Mining Authority). (**b**) Location of the shear zones outlined in the Precambrian belt in the Eastern Desert (After El Gaby et al. 1996)

NW–SE Kom Ombo shear zones respectivily (Fig. 12.1b). NW-trending structures including faults and elongated serpentinite bodies and granitic plutons are common over a wide region of the CED, documenting on a larger scale the structural elements observed by Abu Zied (1984) and Greene (1984). To the north, these are replaced by undeformed rocks (Dokhan volcanics and Hammamat facies sediments) and equidimensional plutons of the NED, indicating a lack of shearing in that region. Structural observations (Fritz et al. 1996) document a continuous decrease of shortening from the south to the north. The northern domain is characterized by abundant extensional structures that are overprinted by both reverse and strike-slip faulting further to the north. On the other hand, the southern domain is characterized by essentially compressional deformation (Greiling et al. 1994), which was subsequently overprinted by transpression. Clearly, the cause of the transition from areas affected by Najd movements to the south to areas affected by rifting to the north is main point of the Late Precambrian tectonic evolution of Afro-Arabia.

Nuqara volcanics extend in NW–SE trending structural lineaments. NW–SE – trending lineament – set corresponds to the junction between the Mesozoic (Duwi shear zone) and Precambrian terranes and appear to reflect long-lived planes of weakness in the basement. These structures have been repeatedly reactivated perhaps since the Late Precambrian. This basement is overlain by a series of continental and marine sediments of Phanerozoic age. NW–SE-striking Mesozoic graben structures formed during the northeastward drift of the NED and are filled with Cretaceous-Paleogene marine sediments.

12.3 Petrography of Lithofacies

Nuqara area comprises thick, essentially conformable massive volcanic sheets, hosted within and intercalated with pyroclastic and volcaniclastic deposits. Nuqara volcanics unconformably overlie a thick sequence of folded and faulted old basement. This old basement consists of Neoproterozoic metasedimentary and mafic–ultramafic rocks of debated ophiolitic origin, intruded by old granites (Fig. 12.1a). These volcanics crop out in large circular masses forming a 95 km^2 NW–SE trending belt, which extends from wadi Abu Asala in the north to wadi Nuqara in the south (Fig. 12.2). The maximum width of this belt exceeds 3 km. Nuqara volcanics are undeformed/slightly metamorphosed at low grade (greenschist facies) and have been affected by gentle tilting in fault-bound blocks. They have well preserved primary textures (e.g. spherulites, amygdules, welded shards, columnar joints, flow bands, bedding). There is also no indication of deep burial and the province has formed a long-lived region of positive topography. The volcanic rocks have erupted from fissures extending along NW–SE and NNE–SSW striking faults that appear to have largely controlled the distribution of these volcanics in the study area. Field photographs and relevant photomicrographs for those volcanic rocks are shown in Fig. 12.3(a–f). They have high relief with sharp peaks and steep slopes, with conspicuous columnar jointing (Fig. 12.3a). These rocks are intruded from west by younger granites and gabbros with sharp boundaries along which the granites send several apophyses of different thickness into these volcanics and take xenoliths of different shape and size from them. Juxtaposition of granites with the Nuqara volcanics suggests that at least 1 km of the overlying rocks (volcanics and/or volcaniclastics) have been eroded. These bedded volcanics are unconformably overlain eastward by Phanerozoic sediments (Fig. 12.3b). Mafic, microgranite and quartz–feldspar dykes traverse these volcanics as NE–SW trending parallel swarms particularly along their peripheries. Mafic dyke swarms are interpreted to be feeder dykes to lava flows and volcaniclastic units exposed at higher elevations on the flankes of these volcanics. In the course of a detailed stratigraphy, petrography and geochemistry, the Nuqara volcanics has been subdivided into old and young volcanic sequence (Fig. 12.4).

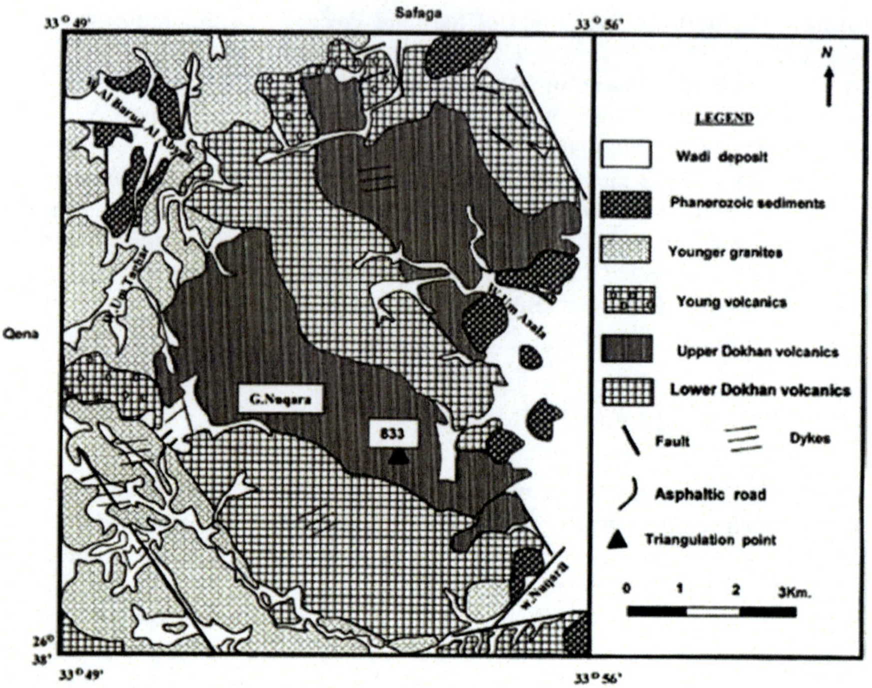

Fig. 12.2 Geological map of the Gabal Nuqara area

12.3.1 Old Volcanic Sequence

The old volcanic lineament is a sequence of remnant edifices that progressively young to the north and shows concomitant decrease in the degree of erosion with time. This sequence is a deeply eroded stratovolcano remnant, comprising a series of impressive bluffs and ridges about a central plateau at 833 m altitude. The old volcanic sequence comprises texturally and compositionally varied volcanic units that amount to 0.5–3 km in thickness, and are gently to moderately dipping (10°–30°). It is generally divided into two units: a lower mafic and an upper felsic volcanics. Thick package of pyroclastic and volcaniclastic deposits are common in both units. The mafic volcanics crops out in the form of three elongated parallel belts of NW-trending and are conformably overlain by the felsic volcanics, which appear as two intercalated alignment belts of the same trend. The sharp contact between mafic and felsic units is exposed. The radiometric age dating of mafic rocks at Nuqara area varies from 622 to 654 Ma (using Rb/Sr method), while age of the felsic rocks is 602 ± 12 Ma (Stern and Hedge 1985; Abdel Rahman and Doig 1987).

Fig. 12.3 **(a)** Nuqara volcanics show high rugged relief with conspicuous columnar jointing. **(b)** Nuqara volcanics are unconformably overlain by Phanerozoic sediments. **(c)** Clasts-supported breccias with chaotic fabric. **(d)** Photomicrograph showing plagioclase occurring subophitically with Clinopyroxene. Bar scale = 200 μm. **(e)** Field photograph showing presence of planar bedding in volcaniclastic rocks. **(f)** Field photograph of felsic volcanics showing presence of mafic xenoliths of irregular shape and size

12.3.1.1 Mafic Volcanics

Mafic volcanics represent the largest exposure of volcanic rocks with mafic to intermediate composition. These rocks are dark to light gray in color. Compared to the upper felsic volcanics, the mafic rocks form thin, extensive sheets with massive interiors and are highly propylitized particularly along the eastern border of the map nearby the Red Sea Coast. These volcanics start with volcanic breccias, thick sheets of andesitic lavas and intercalated volcaniclastic deposits.

The andesitic breccias are structureless, monolithological, relatively fine-medium grained, usually clast-supported rocks with a chaotic fabric (Fig. 12.3c).

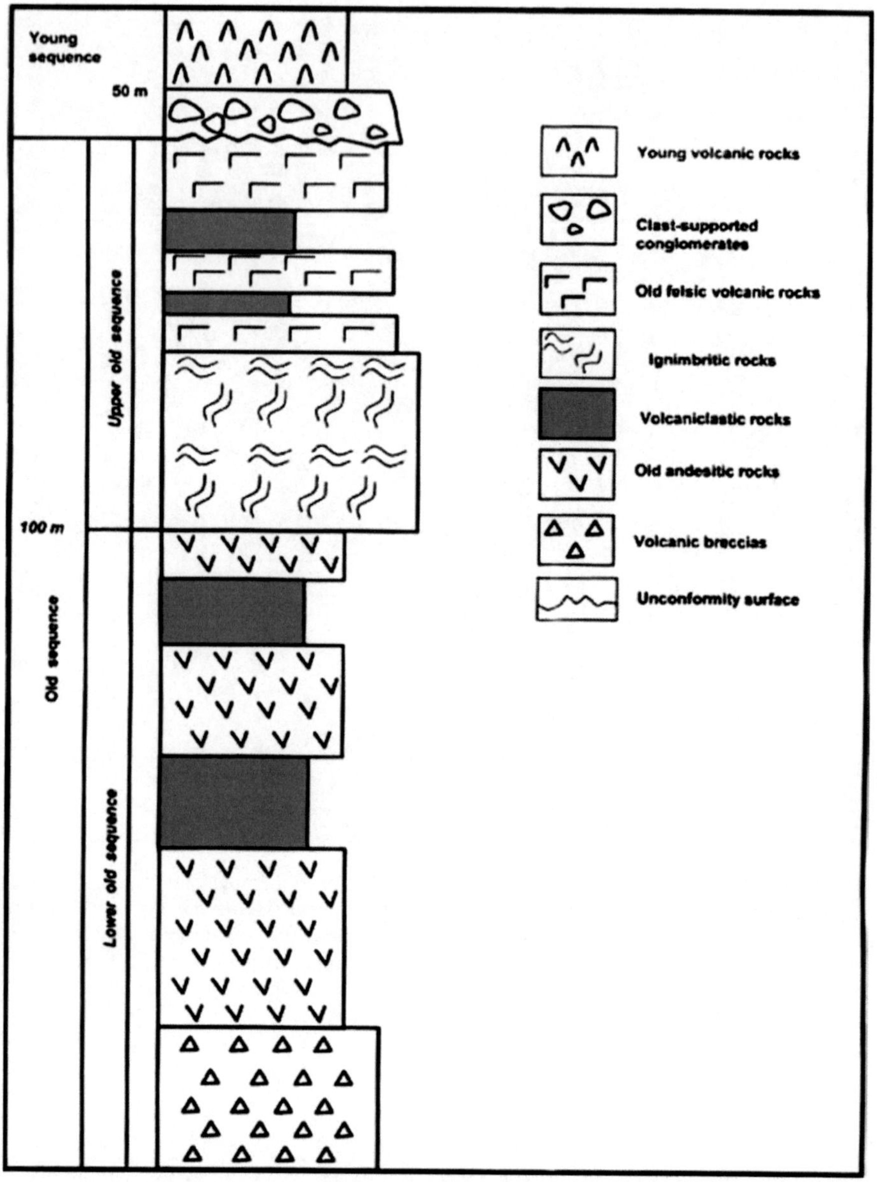

Fig. 12.4 The stratigraphic succession of Nuqara area

They constitute the base of the andesitic volcanics. Individual beds are not seen, owing to poor exposure. Underlying granites show strong hydrofracturing at the contact and has thin (cm-thick) pegmatitic apophyses into the breccias. At the contact, andesitic xenoliths or clasts from breccias are incorporated into granites.

These andesitic breccias consist of petrographically uniform granule to pebble andesitic clasts, usually less than 5 mm in size, with the largest up to ca.20 mm long. The clasts are isometric to elongate, and angular to subrounded in shape. They are uniformly dispersed within an aphanitic matrix of microcrystalline quartz with abundant hematite staining, with some broken plagioclase crystals and chlorite pseudomorphs after ferromagnesian minerals. The andesitic breccias probably formed due to quenching-induced fragmentation, with some influence of flow-related brecciation (Awdankiewicz et al. 2004). The latter process might have been of minor importance, as no aligned fabrics can be seen in the andesite breccias. Although the andesite breccias can be genetically considered as a type of hyaloclastite, they are characterized by relatively large proportion of hypocrystalline to crystalline, non-glassy clasts.

Andesitic rocks and their volcaniclastic deposits constitute about more than 60% of the total area of Gabal Nuqara. These volcanics crop out in the central part of the area under study. On a smaller scale, volcaniclastic rocks adjacent to the andesitic sheets show bending, folding and fracturing, and locally the andesites are discordant relative to the deformed bedding of the volcaniclastic rocks. The andesitic sheets generally display a monotonous internal structure and consist largely of massive rocks with predominant subvertical joints and prismatic to irregular joints near the margins. These lavas are dominanted by evenly porphyritic, with the largest phenocrysts ca. 2 mm long and coherent facies; textural variations are confined to bases and upper parts. Upper parts are amygdaloidal, while the basal parts are either brecciated, or comprise irregular-shaped or blocky clasts. Petrographically, these rocks are typified by plagioclase (An_{40}) and ferromagnesian minerals (augite and hornblende) set in a matrix of the same composition (Fig. 12.3d). Apatite, zircon and opaque are accessories, while chlorite, epidote, sericite and kaolinite are secondary minerals. Plagioclase phenocrysts show very well-developed oscillatory zoning, and textural features indicate disequilibrium crystallization, including some sieve-textured grains and grains showing resorption. Augite occurs in some andesitic samples as colorless twinned subhedral crystals. Hornblendes occur as subhedral crystals and are strongly pleochroic with the formula: α = pale-yellow, β = brownish-green, and γ = green. The phenocrysts are simply twinned, few are zoned and some exhibit corrosional embayments. The groundmass consists of equigranular plagioclase laths which are coated with alteration products. The volcaniclastic rocks occur as bedded to massive crystal and lithic-rich tuffs, lapillituffs, and as tuffaceous mudstone to crystal-rich breccia. The crystal fragments are mainly of plagioclase (An_{30-20}), quartz and chloritized hornblende. The rock fragments comprise basaltic andesites, proper andesites and reworked tuffs. These rocks are highly compact, hard, jointed and contain fragments ranging in size from ash to lapilli and even to boulders. Most intervals are less than 2 m thick. The laminae and thin beds are rich in feldspar crystals or blocky 2–6 mm, lithic clasts. Most of the rock units are overall planar bedded (Fig. 12.3e) with moderate to well sorted nature and resemble deposits from water-settled fall (Allen et al. 2008). However, some that show low-angle truncations of bedding may have been reworked by currents. Collectively, these volcaniclastic facies are interpreted to represent short time breaks between lava flows. Primary, or slightly reworked/resedimented pyroclastic units indicate that explosive mafic eruptions occurred intermittently throughout the effusive volcanism associated with Nuqara volcanics. Certainly, the slightly steeper dips and dominance of mafic-intermediate lavas are consistent with proximity to a volcanic cone. However, the preservation of well-bedded volcaniclastic intervals suggests that the setting was likely to be within a few kilometers of the summit close to the low-gradient ring plain (e.g., Schmincke 2004).

12.3.1.2 Felsic Volcanics

The upper volcanic sequence in Nuqara area is dominated by small- and moderate-volumes of effusive outpourings of felsic composition. Felsic volcanics occupy about ~40% of the total area of Gabal Nuqara. These rocks are formed of steeply sloping alternate sequence of blocky flow sheets and domes, with rubbly flow tops and bases and their associated pyroclastic deposits. They appear to be composite units that comprise groups of 10-m-thick; compositionally distinct feldspar-phyric lavas separated by thin ignimbrite and/or crystal-rich felsic pyroclastic facies. These felsic volcanics have a basal autobreccia or peperitic contact, overlain by the main coherent part with steeply dipping, and an upper compositionally mingled or amygdaloidal part. These rocks are highly weathered particularly along their contact with the mafic volcanics. They contain mafic xenoliths of different shape and size that range from 2.0 to 15 cm (Fig. 12.3f). The felsic volcanics are commonly porphyritic, massive, columnar jointed and white to pink in color with presence of sulphide disseminations.

Ignimbrite has a sharp and conformable contact with the rhyolitic lava dome. It is massive and at least 2.0–3.0 m thick as the upper contact is an erosion surface. Ignimbrite has eutaxitic texture defined by flattened pumice clasts (fiamme) in a matrix of glass shards that show varying degrees of plastic deformation (Fig. 12.5a). The volcanic clast population includes porphyritic rhyolite and wispy, lenticular, aphyric fiamme that are interpreted to be flattened, juvenile pumice lapilli (3–30%). Angular to subrounded andesitic clasts are also present and many are very coarse (up to 15 m across).The matrix comprises various proportions of lithic fragments and relict shards. In some ignimbritic units, ash lenses have been observed. These lenses are interpreted as circular accretionary lapilli on the basis of their fine-grained size and circular outline in three dimensions (Fig. 12.5b). The presence of such accretionary lapilli as proximal facies indicates a phreatomagmatic origin for these deposits (Cas and Wright 1987). The ignimbrite underwent hot welding compaction, as indicated by the subhorizontal compaction foliation defined by the fiamme, deformation of fiamme around the margins of lithic clasts, and plastic deformation of cuspate matrix shards. The presence of spherulitic and micropoikilitic texture within the fiamme resulted from high-temperature crystallization of a glassy precursor (Lofgren 1974). Rhyolitic lava is moderately widespread that outcrops to the northwest

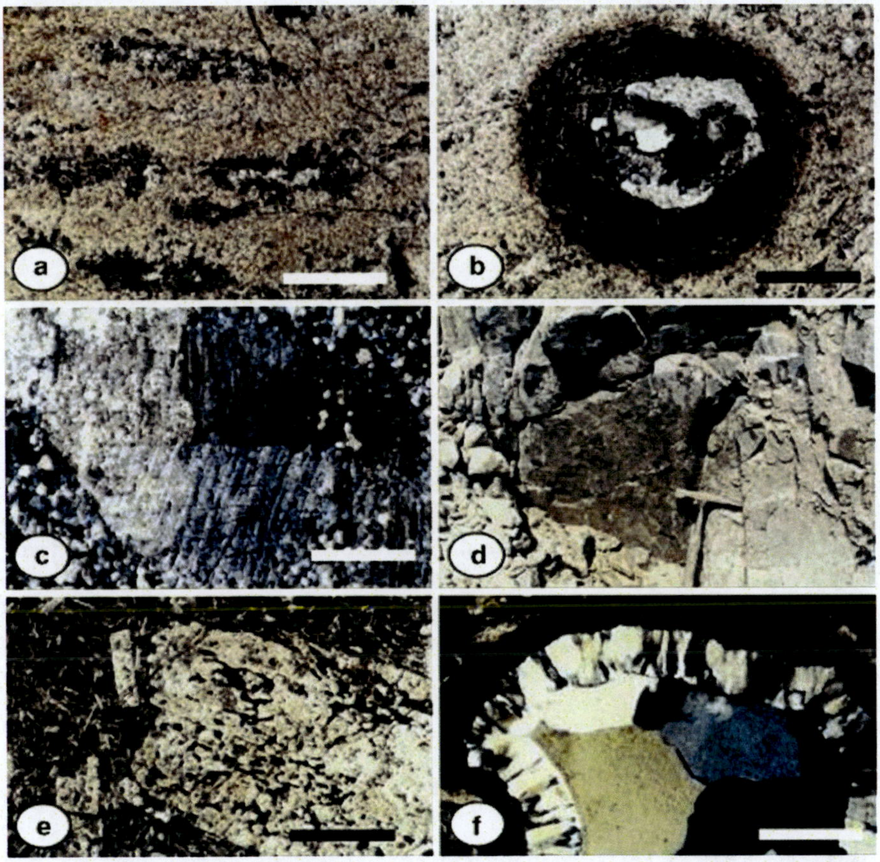

Fig. 12.5 (**a**) Sub-parallel pumice clasts forming fiamme fabric. Bar scale = 200 μm. (**b**) Oval shaped accretionary lapilli in ignimbritic rocks. Bar scale = 200μm. (**c**) Twinned potash feldspar megacrysts exhibit rod like perthitic and micrographic intergrowth. Bar scale = 200 μm. (**d**) Clasts-supported conglomerates show moderate sorting with no imbrication. (**e**) Large plagioclase phenocryst (An_{20}) contains black glass blobs embedded in the fine grained matrix; highly pigmented iron oxides. Bar scale = 200 μm. (**f**) Oval shaped quartz-rich amygdules in amygdaloi dal andesites. Bar scale = 200 μm

part of Nuqara area. It is a predominantly feldspar-phyric, 1.5–2.0 m thick with well-developed flow banding at the base. It lacks vitriclastic texture. This lava is crystal-rich (up to 40% crystals) with a predominantly spherulitic or granophyric-rich matrix. Megacrysts of perthitic K-feldspar, less common plagioclase (An_{15-20}), and corroded quartz are most abundant (up to 5 vol%) near the base (Fig. 12.5c). Fine-medium-grained, mafic volcanic clasts are sparsely scattered throughout (<2 vol%). The groundmass is also distinctive and comprises a finely microcrystalline quartz–feldspar mosaic that contains abundant, randomly oriented, 50–70 μm, feldspar laths and rare altered biotite.

The associated felsic pyroclastic deposits occur as conformable massive and bedded types. The latter rock types are typical of fall pyroclastic deposits. The individual beds may be internally structureless, cross-bedded, graded bedded or internally laminated. They range from breccia, vitric tuffs, to crystal-and lithic–vitric tuffs, and laminated, tuffaceous mudstone. These rocks are composed of pumice, crystals and lithic clasts in a matrix of granophyric, recrystallized and devitrified glass. Crystals, including sodic plagioclase, sanidine and rare euhedral zircon are commonly broken and enriched in the matrix relative to pumice. Minor accidental lithic clasts include andesite, pink rhyolite and rare tuffaceous mudstone with crystal-rich sandstone. The vitric- lithic-rich tuffs have lithofacies characteristics similar to those of primary subaerial or water-settled fallout. In comparison, the laminated, tuffaceous mudstone and crystal-rich sandstone intervals are probably composed of pyroclasts reworked by fluvial and/or lacustrine processes.

Clast-supported conglomerates occur as thin layers (~30 cm thick) unconformably overlying the lower volcanic sequence. These rocks are massive, ungraded, lithic-rich and has tuffaceous matrix. The clasts are granule to pebble in size, angular to subangular, unsorted and show no imbrication (Fig. 12.5d). The clast compositions involve mainly andesitic, rhyolitic and reworked volcaniclastic fragments that are similar to the main volcanic edifice of the underlying lower volcanic sequence. These volcanogenic conglomerates represent subaerial erosional unconformity locally, and indicate period of quiescence between volcanic episodes in the study area.

12.3.2 Young Volcanic Sequence

This sequence is well exposed in the northern and western part of the mapped area (see Fig. 12.2). Young volcanic sequence comprises small poorly exposed outcrops which straddle the rhyolite/ignimbrite boundary. These rocks are hard, massive, unfoliated, gray to dark gray and show well-developed joint pattern. This sequence conformably overlays old volcanic sequence with sharp erosional contact that is marked by clast-supported conglomerates (Khalaf 1986). The core of this sequence consists of massive and vesicular to amygdaloidal andesites. These andesites show a strong hydrothermal alteration and partial to total replacement of the primary volcanic phases with various secondary minerals, but retain their original crystal shapes and twinning habit. The old and young andesites differ in their phenocryst content and size. The old andesites are aphyric to weakly porphyritic, with the largest phenocrysts ca. 2.0 mm long. The young andesites however are strongly porphyritic, with up to 30 vol% phenocrysts, up to ca. 5.0 mm long phenocryst dimension and comprise glomeroporphyritic plagioclase ($An_{25–30}$), chlorite pseudomorphs, most likely after pyroxene and abundant titano-magnetite (Fig. 12.5e). Flow banding is locally defined by variable vesicle and/or phenocryst content. Apatite, sphene, zircon are accessories, whereas chlorite, carbonate, quartz and epidotite are secondary minerals. The amygdaloidal andesites are usually characterized by a strong flattening and alignment of vesicles that are filled with quartz, carbonate, chlorite and epidote (Fig. 12.5f).

The microcrystalline matrix consists of acicular feldspar laths, opaque minerals and alteration products. The acicular feldspar laths are usually arranged into fan-like aggregates and spherulites. Occasionally, some andesitic samples comprise six-sided clinopyroxene which form subophitic texture with plagioclase.

The young andesites are interpreted as plug and lava dome or short flows. The plugs were preferentially emplaced along the boundary of contrasting lithologies, and locally pierced their host rocks, resulting in lava effusion.

12.4 Geochemistry

12.4.1 Analytical Methods

Rocks were selected to characterize sequence geochemistry and geochemical stratigraphy. Major and trace element concentrations have been analyzed for 31 of the least altered volcanic rocks of which 21 are from old volcanics and rest ten are from young volcanic sequence (Table 12.1). Major elements were determined using the conventional wet chemistry techniques of Shapiro and Brannock (1962). SiO_2, TiO_2, Al_2O_3, and P_2O_5 were analyzed using spectrophotometric methods while Na_2O and K_2O were determined by flamephotometric technique. MgO, CaO and FeO(t) were measured by titration. The loss on ignition (L.O.I.) was determined gravimetrically. The trace elements were detected by XRF using Philips X-Unique II with automatic sample charge (PW 1510). Radioactive elements (U, Th) were determined by radioactive analysis. Calibration curves were prepared using international rock standards. All the analyses were performed at the laboratories of the Nuclear Material Authority (NMA), Kottoymia, Egypt.

12.4.2 Classification and Chemical Characteristics

The total alkali ($Na_2O + K_2O$) versus SiO_2 diagram (TAS) diagram (LeBas et al. 1986) shows that the investigated rocks of old sequence form a uniform volcanic suite with a wide compositional range of SiO_2 (51–73 wt%) ranging from mafic (basalt, basaltic andesite and andesite) to felsic (dacite and rhyolite) members. The rocks of young sequence have a restricted SiO_2 range (55–59 wt%) extending from basaltic trachyandesite to trachyandesite (Fig. 12.6a). On K_2O versus SiO_2 diagram (Fig. 12.6b), the old volcanic samples fall within medium to high-K field for mafic and felsic rocks respectively. The samples for young volcanic sequence fall within high-K field. Overall, the lowest K_2O contents are observed in the old volcanics and progressively younger stratigraphic groupings show a systematic shift to higher

Table 12.1 Chemical data of major and trace elements for Nuqara area, North Eastern Desert, Egypt

	Old volcanic sequence													
No	N1	N2	N3	N4	N5	N6	N7	N8	N9	N10	N11	N12	N13	N14
Type	B	BA	BA	BA	A	A	A	A	A	A	A	A	A	D
SiO2	51.48	53.63	54.92	56.61	57.87	57.91	58.64	58.82	61.86	62.18	62.18	62.78	63.24	64.04
TiO2	0.65	0.64	1.53	1.11	0.95	1.09	0.98	1.01	0.89	0.66	0.29	1.21	0.97	1.18
Al2O3	17.84	16.2	15.06	16.7	16.89	15.32	17.56	16.87	18.69	14.9	15.82	14.8	15.25	13.65
Fe2O3	8.39	8.39	7.99	6.93	7.15	7.99	5.81	6.38	3.8	5.59	4.39	5.38	6.39	4.99
CaO	8.4	7	8.4	4.26	4.4	5.6	4.01	6.32	4.01	6.4	5.6	4.2	4.2	4.2
MgO	4	4	4	3.85	2.93	4	2.9	2.5	1.22	3	4	3	3	2
Na2O	3.37	3.37	2.49	3.78	3.78	3.76	4.2	4.25	4.73	3.78	3.44	3.25	3.44	3.43
K2O	0.97	1.84	1.6	1.03	1.52	1.53	1.52	1.25	1.51	1.45	1.34	2.87	1.13	2.6
P2O5	1.2	1.28	1.38	0.41	0.43	1.28	0.47	0.46	0.42	1.39	1.39	0.72	1.04	1.52
L.O.I.	3.14	3.23	2.19	3.84	2.5	1.28	3.91	2.14	1.59	0.59	1.2	0.96	1.17	2.06
Total	99.44	99.58	99.56	98.52	98.42	99.76	100	100	98.72	99.94	99.65	99.17	99.83	99.67
CO	24	22	21	21	16	24	21	20	16	24	24	23	26	22
V	254	90	242	290	149	199	222	226	90	243	98	267	90	244
Cr	137	80	60	70	66	16	25	22	18	31	13	19	18	28
Ni	65	55	33	92	24	18	32	32	15	39	25	32	15	14
Cu	12	25	36	23	33	54	35	21	19	37	52	33	19	31
Zn	66	175	112	256	121	140	142	172	164	147	68	131	164	138
Pb	11	56	25	203	37	34	61	21	13	70	24	202	13	49
Ga	10	11	27	15	3	36	20	34	48	18	10	31	48	25
Rb	53	87	67	23	100	74	33	44	70	71	29	61	70	91
Sr	265	570	809	517	661	750	804	779	527	876	513	902	627	668
Ba	106	643	479	376	226	531	327	359	656	447	243	1,139	658	423
Zr	118	323	152	138	82	308	233	265	233	217	167	242	333	168
Y	13	2	7	6	4	11	7	11	15	14	16	18	15	16
Nb	13	<2	3	5	4	3	2	6	8	3	4	3	20	12
U	8	UDL	2	1	2	3	UDL	2	5	3	7	1	1	3
Th	10	4	2	5	2	2	4	5	3	2	10	3	3	3
Nb/Y	1	–	0.42	0.83	1	0.3	0.28	0.54	0.53	0.21	0.25	0.17	1.33	0.75
Y/Nb	1	–		2.33	1	3.67	3.5	1.83	1.88	4.67	4	6	0.75	1.33
K/Rb	288	418	108	540	106	106	555	342	554	250	557	567	195	344
Rb/Zr	0.27	0.45	0.44	0.17	1.2	0.34	0.14	0.17	0.3	0.32	0.17	0.25	0.2	0.54
Zr/Nb	24	–	51	27.6	20.5	73	116	44	29.1	72	41.75	81	15.9	84
Fe/Mg	2.99	2.09	1.99	2.43	2.44	1.99	3.05	2.55	3.11	1.86	1.09	1.79	2.13	2.49
Al2O3/TiO2	27	25	9.84	15	18	14	17.92	16.7	21	22.5	54.5	12.23	14.69	10.72
Zr/Y	24.8	59	22	23	21	20	33	24	15.5	15.5	10.43	13.44	15.33	10.5
Sr/Y	20	285	116	86	165	68	115	80	35	60	32	50	42	42
Mg number	0.23	0.36	0.37	0.33	0.36	0.37	0.26	0.32	0.27	0.39	0.52	0.3	0.36	0.32

Fe_2O_3: total iron as ferric oxides, L.O.I. = loss on ignition, major elements in wt%, trace elements D = dacite; R = rhyolite, Mg number = Mg/Mg + Fe (t)

						Young volcanic sequence										
N15	N16	N17	N18	N19	N20	B1	B2	B3	B4	B5	B6	B7	B8	B9	B10	B11
D	D	R	R	R	R	BA	BA	BA	BA	BA	BA	BA	BA	A	A	A
65.42	67.21	67.34	69.37	72.08	73.15	55.51	55.62	55.86	56	56.03	56.19	56.19	56.5	57.54	57.94	58.96
0.48	0.39	0.18	0.14	0.19	0.04	1.51	1.35	1.39	1.19	1.39	1.35	1.33	1.33	1.35	1.34	1.34
15.36	13.91	13.39	14.49	14.12	13.11	17.8	16.69	17.08	17.08	16.95	16.89	16.56	16.84	16.32	17.06	17.15
4.63	3.79	5.39	3.19	2.08	3.89	7.52	6.16	6.27	6.29	6.41	6.32	5.92	5.8	5.97	5.85	5.89
2.74	4.2	2.8	4.2	1.92	1.4	3.99	4.45	4.31	4.54	4.95	4.42	5.7	3.92	3.89	2.71	3.05
2.32	2	2	2	0.99	1	0.87	2.58	1.77	2.43	2.78	2.05	0.84	2.59	1.22	2.11	2.42
4.25	4.16	4.05	3.73	4.78	4.16	4.91	4.68	5.38	3.98	3.53	4.72	4.45	5.21	5.12	4.96	5.46
1.25	2.41	2.68	2.01	4.05	2.97	2.92	2.09	2.54	3.25	3.11	2.96	2.36	1.79	2.37	3.27	1.65
0.3	0.69	0.2	0.33	0.22	0.2	0.64	0.66	0.69	0.5	0.66	0.69	0.64	0.64	0.65	0.59	0.63
1.26	0.91	0.85	0.16	0.56	0.04	3.24	3.63	3.49	3.7	3.26	3.19	4.78	4.16	4.07	3.32	2.64
98.01	99.67	98.88	99.62	100.99	99.96	98.91	97.91	98.78	98.96	99.07	98.78	98.77	98.78	98.50	99.15	99.51
10	4	9	5	5	5	13	12	13	14	13	11	12	8	9	6	9
34	20	73	4	6	4	146	180	178	104	174	172	170	167	168	163	178
13	17	15	21	18	13	86	63	72	225	67	63	65	61	62	60	71
19	14	6	6	23	8	44	64	59	51	51	65	49	54	44	33	47
41	11	26	11	10	12	126	66	14	45	90	66	134	64	69	83	69
106	63	66	136	101	133	70	84	87	84	74	87	71	75	48	79	78
22	13	51	6	5	8	6	12	17	20	8	18	15	13	16	21	10
19	20	27	43	26	26	23	28	22	23	21	23	18	21	17	23	20
128	125	169	111	144	110	50	30	50	62	49	52	57	48	49	67	39
857	373	180	156	89	136	1,104	1,216	1,004	877	1,580	1,176	1,037	978	1,004	1,145	957
450	1,333	985	476	287	416	1,262	722	700	773	1,194	1,010	796	1,185	786	1,274	890
437	526	335	299	375	376	237	267	274	435	278	269	259	259	260	249	260
30	33	50	76	44	60	21	19	20	22	20	20	18	19	19	20	20
16	19	26	64	50	64	10	10	9	18	8	10	10	9	10	10	11
3	6	5	15	6	20	0	0	0	0	0	0	0	0	0	0	0
8	11	3	24	14	24	0	0	0	5	0	0	0	0	0	0	0
0.53	0.48	0.52	0.84	1.13	0.9	0.48	0.52	0.45	0.82	0.4	0.5	0.56	0.47	0.53	0.5	0.55
1.88	1.74	1.92	1.19	0.88	0.94	2.1	1.9	2.22	1.22	2.5	2	1.8	2.11	1.9	2	1.82
342	232	120	218	337	322	485	578	422	435	527	473	344	310	402	405	351
0.29	0.24	0.5	0.3	0.38	0.3	0.21	0.11	0.18	0.14	0.18	0.19	0.22	0.19	0.19	0.27	0.15
	28	12.9	4.7	7.5	6	23.7	26.7	30.44	24	34.8	26.9	25.9	28.8	26	24.9	23.6
1.99	0.69	2.69	1.59	2.1	3.89	8.46	2.22	3.36	2.42	2.14	2.9	6.92	2.08	4.63	2.59	2.25
32	35.66	74	35	80	3.27	11.79	12.36	12.29	14.35	12.19	12.51	12.45	12.66	12.08	12.73	12.8
14.57	15.9	6.7	3.93	8.5	6.3	11.29	14.05	13.7	19.77	13.9	13.45	14.39	14.39	13.68	12.45	13
28.57	11	3.6	2.05	2.02	2.27	53	64	50	40	79	59	58	51	53	57	48
0.37	0.35	0.3	0.42	0.36	0.23	0.12	0.33	0.25	0.31	0.34	0.28	0.14	0.34	0.19	0.3	0.33

in ppm, UDL = under detection limit, rock types: B = basalt; BA = basaltic andesite; A = andesite

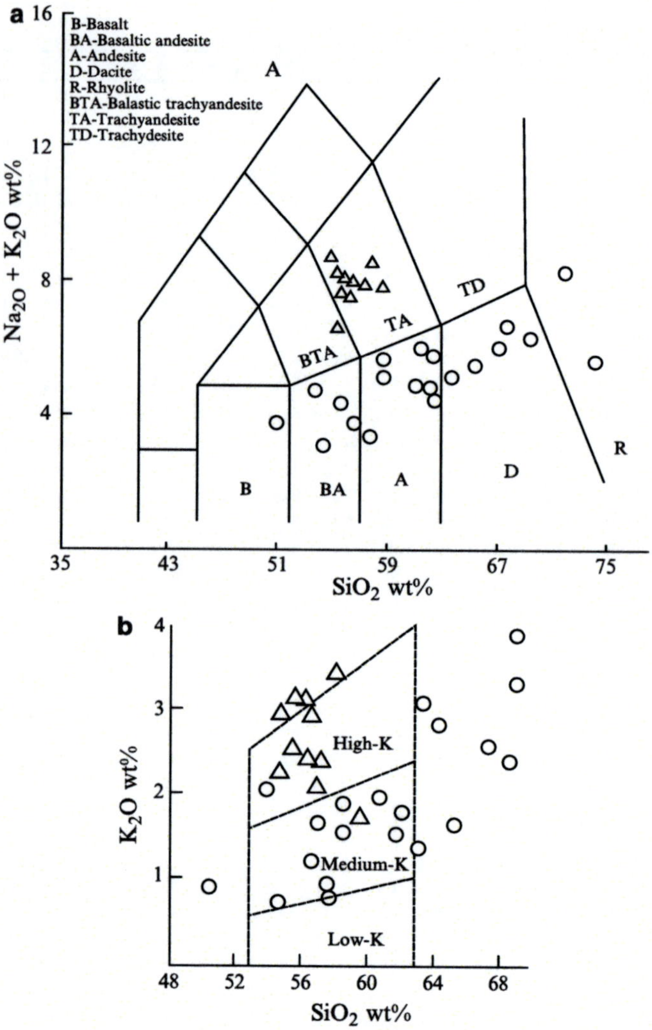

Fig. 12.6 Geochemical classification diagrams for Nuqara volcanics. (**a**) Total alkali versus silica diagram (Le Bas et al. 1986) (**b**) K_2O versus SiO_2 diagram (Gill 1981) and (**c**) Iron/magnesia versus SiO_2 diagram (Arculus 2003). Explanation of symbols: *unfilled circle*: old volcanic sequence and *unfilled triangle*: young volcanic sequence

K_2O contents at a given SiO_2 value. On the total iron/MgO ratio versus SiO_2 diagram (Fig. 12.6c), the mafic and felsic rocks of old sequence plot in medium to low-Fe field of calc-alkaline affinity, whereas the mafic rocks of young sequence plot in high-Fe field of tholeiitic affinity.

The old mafic volcanics (~51–63 wt% SiO_2) are silica-saturated. Their Mg numbers [(Mg# = Mg/Mg + $Fe_{(t)}$)] range from 0.52 to 0.23 (Table 12.1) suggesting that these rocks are not primary (mantle-derived) melts. These rocks are characterized

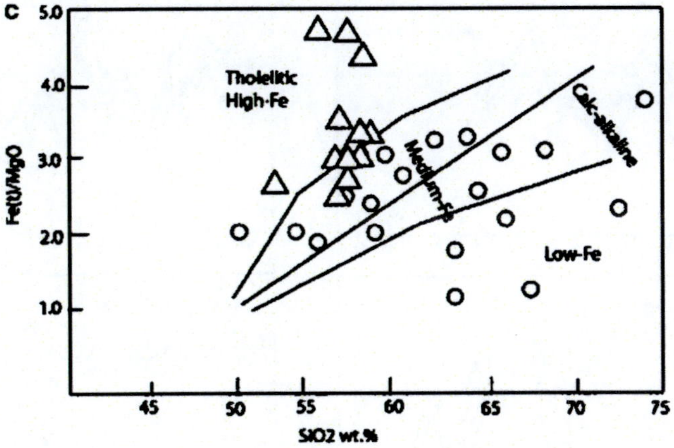

Fig. 12.6 (continued)

by rather high contents and wide range of Al_2O_3 (~14–18 wt%), TiO_2 (~0.29–1.53 wt%), P_2O_5 (~0.41–1.39 wt%), Rb (~23–100 ppm), Ba (~106–1,139 ppm), Sr (265–902 ppm), Zr (82–333 ppm) and Y (2–18 ppm). The old felsic volcanics (~64–73 wt% SiO_2) are marked by relatively high total alkalis (~6–9 wt%), Rb (91–169 ppm), Zr (168–526 ppm), Y (16–76 ppm) and Nb (12–64 ppm) and low Al_2O_3, MgO, CaO and Sr. The young mafic volcanics are chemically distinguishable from the old mafic volcanics. These rocks are enriched in Al_2O_3 (~16–18 wt%), TiO_2 (~1.19–1.50 wt%), total alkalies (~6.54–8 wt%), Sr (877–1,580 ppm), Ba (700–1,274 ppm), Zr (237–435 ppm), Y (18–22 ppm) and Nb (8–18 ppm) and low in Mg number (0.12–0.34).

In the primordial-mantle-normalized multi-element spiderdiagrams (Fig. 12.7), the studied old mafic volcanics show overall enrichment in the strongly incompatible large ion lithophile elements (LILEs) such as Rb, Ba, K; enrichment in U, Th, Pb; depletion in Nb relative to K and Th and marked depletion in high field strength elements (HFSEs) such as Zr, Ti and Y and compatible elements (like Cr, Ni, and Co; Fig. 12.7a and b). Additionally, the felsic rocks have negative anomalies for Sr, P and Ti (Fig. 12.7c). Such anomalies may be attributed to fractionation of plagio clase, apatite and Fe–Ti oxides respectively. Such profiles are typical of hydrous arc magmas in which plagioclase fractionation is suppressed by high H_2O content (Richard et al. 2006).The samples for young volcanic sequence are distinguished by more LILE- and HFSE-enrichment (Fig.12.7d). The youngest lavas generally have the highest values of LILE, HFSE and Ti and the oldest lavas have the lowest.

12.4.3 Major and Trace Element Variations

Silica variation diagrams of major and trace elements are illustrated in Fig. 12.8. The rocks of the entire old volcanic suite define correlated and continuous variation trends. Total iron, CaO, TiO_2 and P_2O_5 contents decrease with increasing SiO_2.

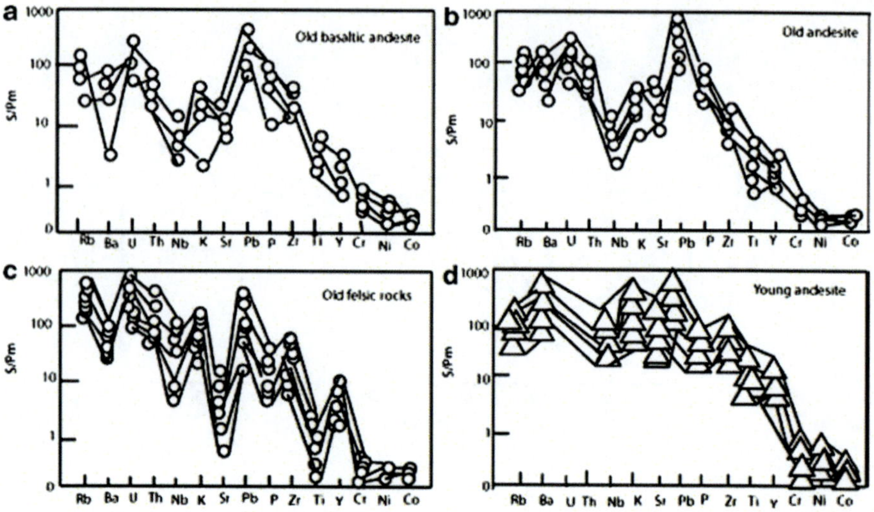

Fig. 12.7 Primitive mantle-normalized (Sun and Mc Donough 1989) elemental compatibility diagrams. Explanations of symbols are same as in Fig. 12.6

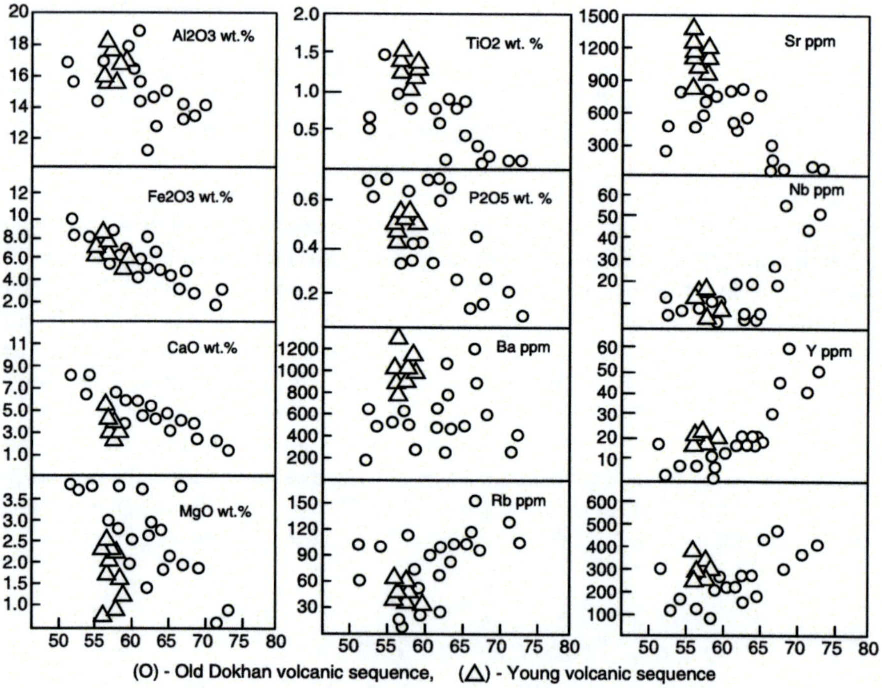

Fig. 12.8 Variation diagrams of some major element oxides and trace elements plotted against SiO_2 wt% along X-axes. Symbols are same as in Fig. 12.6. SiO_2 wt% along X-axes

Total alkali contents increase with increasing SiO_2 (see Fig. 12.6a). Al_2O_3 nearly increases with increasing silica during fractionation of mafic rocks, but decreases later during the crystallization of felsic rocks (Fig. 12.8). Rb, Nb, Y and Zr increase systematically with increasing SiO_2. Ba pattern shows much less coherent variation with respect to silica (Fig. 12.8). Sr shows behaviour similar to that of Al_2O_3 (Fig. 12.8). It increases up to 900 ppm over 51–63 wt% SiO_2 and gradually decreases to very low values (<100 ppm) implying plagioclase fractionation. Most of the major and trace elements of the young mafic rocks show no clear co-variations when plotted against SiO_2 suggesting possible role of assimilation/open system magmatic interaction.

12.5 Tectonic Environment

The entire volcanic sequences have major and trace element compositions similar to magma generated in orogenic and anorogenic settings. Trace elements, unlike many major elements, are often immobile during weathering and hydrothermal alteration. Therefore, the immobile trace elements can be used to constrain the tectonic setting for lavas. On the TiO_2 versus Zr diagram (Fig. 12.9), it is clearly shown that the old volcanic rocks plot in volcanic arc and within plate field. Accordingly, these volcanics have geochemical characteristics of both subduction-related and within-plate settings. This was previously noted by many authors and had led to a debate on the convergent margin versus within-plate geotectonic setting for these lavas in Nuqara area. The young mafic rocks plot in within-plate setting. Ressetar and Monard (1983) described the tectonic setting of the Dokhan volcanics which are similar to

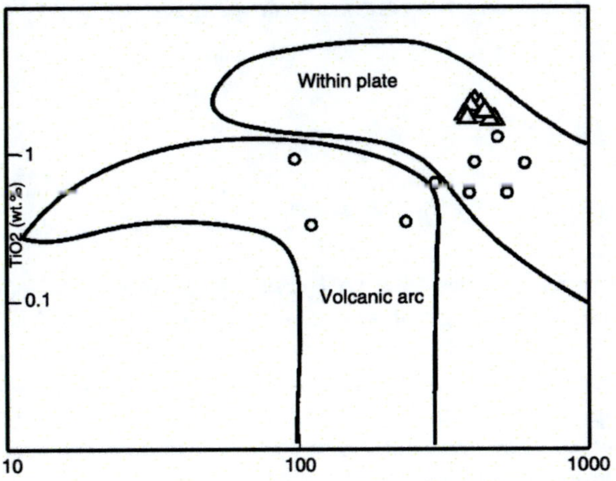

Fig. 12.9 Tectonic discrimination diagrams for Nuqara volcanics (TiO_2 vs Zr plot after Pearce 1980). The left hand closures for both the 'Volcanic Arc' and 'Within Plate Volcanics' are slightly extended towards TiO_2 axis. Symbols are same as in Fig. 12.6

the old volcanics in the Fatira–Safaga area as transitional between compressional and extensional settings. Stern and Hedge (1985) believed that fundamental transition in tectonic setting occurred about 600 Ma ago where the tectonic environment changed from compressional to extensional. Moreover, the Rb/Sr age of 620–560 Ma for the Qena–Safaga and Fatira Dokhan volcanics mark the 606–580 Ma ages of the post-tectonic granites (Fullagar and Greenberg 1978). Stern et al. (1984) regarded the Dokhan volcanics and the younger granites as manifestation of bimodal igneous activity characteristic of rifting. Therefore, the proposed setting of the studied old volcanics and Dokhan volcanics in general (i) agrees partially with the interpretation of Ressetar and Monard (1983) and (ii) agrees completely with the interpretation given by Ghoneim and Lebda (2001) and (iii) disagrees with the concept of bimodal rift-related magmatism given by Stern (1994) [that these volcanics are not orogenic but their emplacement follow the cessation of subduction in an extensional-related tectonic setting]. The eruption of high-K magmas with a typical arc trace element signature is recorded post-dating active subduction and occurs synchronous with uplift, extension or strike-slip motion (Sloman 1989).

Interpretation of the tectonic control based on tectonic discrimination diagrams must be used in conjunction with knowledge of the geology of the Nuqara area. The stratigraphy and physical volcanology of the study area mostly resemble that of continental extensional or hotspot-related environments such as the Yellowstone-Snake River plain region (Geist and Richards 1993). The sequence has the large volumes of reworked volcaniclastic deposits that are typically interstratified with volcanic rocks in arc terranes (Van Wagoner et al. 2002). The juxtaposition of silcic proximal and distal volcanic facies suggests that there were many small volcanic centers in the area. No caldera complexes have been identified. The occurrence of interbedded continental-facies volcaniclastic rocks throughout the sequence indicates that the area remained at or near the sea level throughout the volcanic activity and that the basin was continually subsiding.

12.6 Petrogenesis

12.6.1 Fractional Crystallization

Major/trace elements versus SiO_2 wt% (Fig. 12.8) exhibit two separate variation patterns for old and young volcanic sequence. On the Al_2O_3/TiO_2 versus TiO_2 (Fig. 12.10), the old volcanics define a curvi-linear continuum with the most primitive sample having the highest ratio and lowest TiO_2 content, while the young volcanics display another trend. According to Sun and Nesbitt (1978), the two volcanic sequences have two independent magma sources and each sequence has its own evolutionary path.

Many petrographic and geochemical observations provide fundamental constraints on the petrogenesis of the old volcanic sequence, including: (1) the closely overlapping profiles and temporal associations from mafic to felsic rocks suggest a broadly

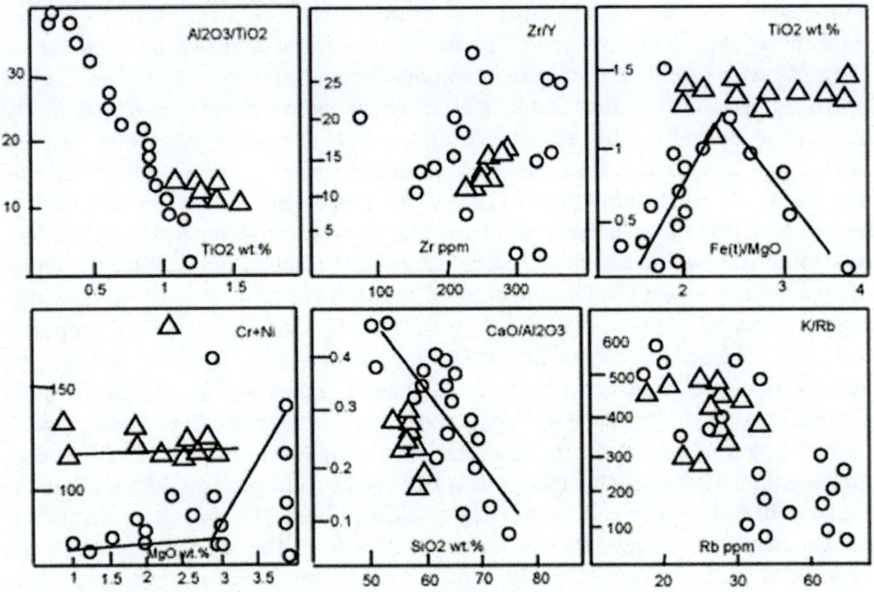

Fig. 12.10 Binary plots showing geochemical variations of the Nuqara volcanics. Al_2O_3/TiO_2 versus TiO_2 plot is after Sun and Nesbitt (1978). Other plots depicted are:(Cr + Ni) versus MgO, Zr/Y versus Zr, CaO/Al_2O_3 versus SiO_2, TiO_2 versus iron/magnesia and K/Rb versus Rb. Symbols are same as in Fig. 12.6

comagmatic relationship (having SiO_2: ~51–73 wt%); (2) the systematic changes in phenocryst assemblage with increasing SiO_2 from the mafic to the felsic rocks; (3) initial $^{87}Sr/^{86}Sr$ ratio does not vary over a large compositional range in Qattar–Dokhan and Nuqara area (Stern and Hedge 1985); (4) major and trace element abundances that vary along continuous trends of decreasing Al_2O_3, Fe_2O_3(total),MgO, CaO, P_2O_5 and Sr and increasing total alkalis, Rb, Ba, Nb, Y and Zr with increasing SiO_2; (5) overall decreasing range of K/Rb ratio with increasing Rb (Fig. 12.10). The above-mentioned characteristics support qualitatively that the old volcanics evolved predominantly through fractional crystallization of the phenocryst assemblages; which are Ol, Cpx, Plag, opaque and Ap in the mafic rocks and Cpx, Plag, opaque and Ap in the felsic varieties. The appreciable fractionation of olivine and pyroxene is marked by a slight increase in Zr/Y versus Zr (Fig. 12.10) which is in line with the higher incompatibility of Zr with respect to the fractionating pyroxene (Pearce and Norry 1979), decreasing of Cr + Ni versus MgO wt%, increasing Zr, Y, and Nb with increasing silica (see Fig. 12.8), and decreasing of CaO/Al_2O_3 versus SiO_2 wt% (Fig. 12.10). The drop in the CaO/Al_2O_3 ratio may indicate a decrease in the ratio of pyroxene to plagioclase in the melt. Plagioclase fractionation is indicated by the decrease in Sr versus SiO_2. Moreover, the decrease of Sr/Y values (285–2.02) for the mafic–felsic association with (overall) increasing SiO_2 is also attributed to plagioclase fractionation. Fe–Ti oxides and apatite fractionation is evidenced by the decrease of Ti/Zr with increasing SiO_2 (Table 12.1), nature of variation of $Fe_2O_{3(t)}/MgO$ ratio with respect to TiO_2 (Fig. 12.10) and decrease of P_2O_5 with increasing silica respectively (Fig. 12.8). Ti shows significant evidence of depletion by fractionation of Fe–Ti oxides and to a lesser extent amphibole.

Binary variation diagrams depict a consistency and overlapping plots among the major/trace elements in the young volcanics, which support its genesis by complex mixing and assimilation fractional crystallization processes, as was described and modeled by O'Hara and Mathews (1981), Cox (1988) and Voshage et al. (1990) as replenishment/refilling, tapping and fractional crystallization processes (RTF). The RTF model is consistent with the cyclic nature of volcanism and with the pattern of change noticed upward in the sequence of having more primitive magma compositions (i.e. low SiO_2, and low FeO/MgO) (O'Hara 1977). Hogg et al. (1989) suggested that the RTF process is most efficient in small magma reservoirs, often located in a densely fractured continental crust. Small magma reservoirs are indicated in the Nuqara area by the presence of thin, low-volume young mafic lava flows. This is further supported by appearance of a kink in the Cr + Ni versus MgO wt% variation diagram (Fig. 12.10). Such a mechanism takes place in magma chambers where the recharge of basic melts and the tapping of differentiated melt from this chamber is accompanied and followed by fractional crystallization. The relative high Al_2O_3 and Sr in these volcanics as compared with the old volcanics argue against extensive fractionation of plagioclase. The constancy of TiO_2 with increasing $Fe_2O_{3\,(t)}$/MgO ratio in the young volcanics indicates fractionation of Fe–Ti oxides is not involved. They show a moderate increase of Zr/Y from the basaltic andesite to the andesite, which emphasizes hornblende fractionation; to which Zr is incompatible with a K_D value of 0.5, whereas Y is compatible with a K_D of 2.5 (Pearce and Norry 1979). The hornblende fractionation is also evidenced by no clear correlation of HFSEs with increasing SiO_2. So, the andesites evolved via assimilation fractional crystallization from basaltic andesites by the separation of Cpx, Hb, Plag, opaque and Ap.

The two evolutionary fractionation paths for the old and young volcanics reflect a difference in redox conditions. The old calc-alkaline volcanic suite formed at greater depth under high P_{H2O} and high f_{O2} which are favourable condition for amphibole stability (Romick et al. 1992), whereas the young tholeiitic volcanic suite occurred at shallow crustal depth under low P_{H2O} and low f_{O2}.

12.6.2 Quantitative Evolution Model for the Nuqara Volcanics

To quantify the effect of fractional crystallization, conventional major element least-squares mixing calculations devised by Le Maitre (1981) with the assumed chemical compositions of the fractionated mineral phases after Deer et al. (1980) were carried out (Table 12.2). The multi-step calculations for the old volcanic sequence involve the most basic sample (N.1) to basaltic andesite (N.4), basaltic andesite (N.4) to andesite (N.11), andesite (N.11) to dacite (N.16), and dacite (N.16) to rhyolite (N.19). A reasonable model can be achieved by about 22%, 31%, 35%, and 12% crystallization of Cpx, Plag, Op. and Ap. with 78%, 47%, 12% and zero%, residual liquids

Table 12.2 Results of the least-square mass balance calculations and trace element modeling

	Using the equations of Bryan et al. (1969)								
Fc	Percent	F	R2	Traces	P	D	CD	R	Bulk D
Poly-step calculations									
				B (N,1) → BA (N,4)					
SiO2 range : 51.48–56.61									
				Co	24	21	23	−2	1.212
				V	254	290	161	129	2.8466
				Cr	137	70	15	55	10.5081
				Ni	62	92	27	65	4.3306
				Rb	53	23	67	−44	0.0455
				Sr	265	517	550	−33	1.1674
CPX	24.68	0.78	0.23	Ba	106	376	131	245	0.1456
Plag	63.02			Zr	118	138	408	−270	0.0552
OP	9.7			Y	13	6	16	−10	0.2543
AP	2.6			Nb	13	5	17	−12	0.0396
				BA (N,4) → A (N,11)					
SiO2 range : 56.61–62.18									
				Co	21	24	7	17	2.4574
				V	290	98	163	−65	1.7624
				Cr	70	13	0	17	9.2422
				Ni	92	25	4	21	5.0842
				Rb	23	29	47	−18	0.0461
				Sr	517	513	455	58	1.1679
CPX	24.96	0.47	1.2	Ba	376	243	337	−94	1.1456
Plag	68.6			Zr	138	167	283	−116	0.0557
OP	5.48			Y	6	16	11	5	0.256
AP	0.97			Nb	5	4	11	−7	0.0227
				A (N,11) → D (N,16)					
SiO2 range : 62.18–67.21									
				Co	24	4	6	−2	1.6779
				V	98	20	54	−34	1.2832
				Cr	13	17	0	17	14.0425
				Ni	25	14	0	14	6.9262
				Rb	29	125	217	−92	0.0492
				Sr	513	373	581	−208	0.9414
CPX	43.98	0.12		Ba	243	1,333	1,570	−237	0.1159
Plag	50.09			Zr	167	526	1,252	−726	0.0496
OP	2.63			Y	16	33	80	−47	0.2404
AP	3.33			Nb	4	19	33	−14	0.0116
				D (N,16) → R (N,19)					
SiO2 range : 67.21–72.06									
				Co	4	5	5	0	0.927
				V	20	4	39	−35	0.6865
				Cr	17	18	0	18	13.2796
				Ni	14	23	0	23	5.8563

(continued)

Table 12.2 (continued)

	Using the equations of Bryan et al. (1969)								
Fc	Percent	F	R2	Traces	P	D	CD	R	Bulk D
Poly-step calculations									
				Rb	125	144	937	−793	0.0541
				Sr	373	89	289	−200	1.1202
CPX	36.92	0	0.27	Ba	1,333	287	986	−699	1.1389
Plag	60.15			Zr	526	375	973	−598	0.7009
OP	2.32			Y	33	44	131	−87	0.3521
AP	0.6			Nb	19	50	86	−36	0.2982

(daughters) respectively. All the sums of the squares of the residuals (R^2) are <1.5 (0.75, 0.23, 1.2 and 0.27 respectively), which indicate a good fit (Luhr and Carmichael 1980). Mono-step calculations for the young volcanic sequence show that the basaltic andesites/andesites (B.10 and B.11) could be produced from the least evolved basaltic andesite (B.4) from ~39% to 30% crystallization of Cpx, Hb, Plag, Op and Ap phases in an average proportions of 46%, 18%, 32%, 4%, and 0.64% respectively.

In order to test the validity of the major element modeling, trace element modeling were also attempted, applying the same fractionating phases and proportions Table 12.2). For the old volcanic sequence, the calculated daughter (CD) has generally high LILEs and HFSEs. The high HFSE contents may be due to the higher modal contents of hornblende, while the high contents of LILE reflect higher amount of plagioclase. Regarding the young volcanic sequence, the CD has lower Y contents, which may be due to the higher amount of hornblende fractionation. So, the general discrepancies between the observed (D) and calculated (CD) daughters in two sequences and high values of R^2 especially for young sequence could thus occur by such processes as small degrees of assimilation or periodic magma replenishment. Some mobile elements such as Rb, Sr and Ba are sensitive to post-eruptive alteration and to loss by volatile-rich phases (e.g. Clague and Frey 1982; O' Hara and Mathews 1981).

12.6.3 Crustal Contamination/Magma Mixing

It is important to assess first whether or not the studied volcanics have undergone crustal contamination. This is inferred from the limited distribution of the mafic volcanics and most of the rock varieties are porphyritic suggesting that they paused in crustal magma chambers prior to eruption and would have had adequate chance to interact with continental crust. For this aim, evidences from trace element ratios are considered. The LILEs (such as Rb, Ba, and K) and Zr are incompatible with respect to pyroxene, amphibole and plagioclase in the investigated mafic volcanics (Green 1980). Davidson et al. (1988) argue that ratios, such as K/Rb and Rb/Zr do not significantly change by simple fractional crystallization, whereas variation in these ratios are preferably related to crustal contamination, by assimilation fractional crystallization processes. In the entire volcanic sequences, K/Rb and Rb/Zr versus SiO_2 (Fig. 12.11a and b) show an apparent variations and wide scatter of plots

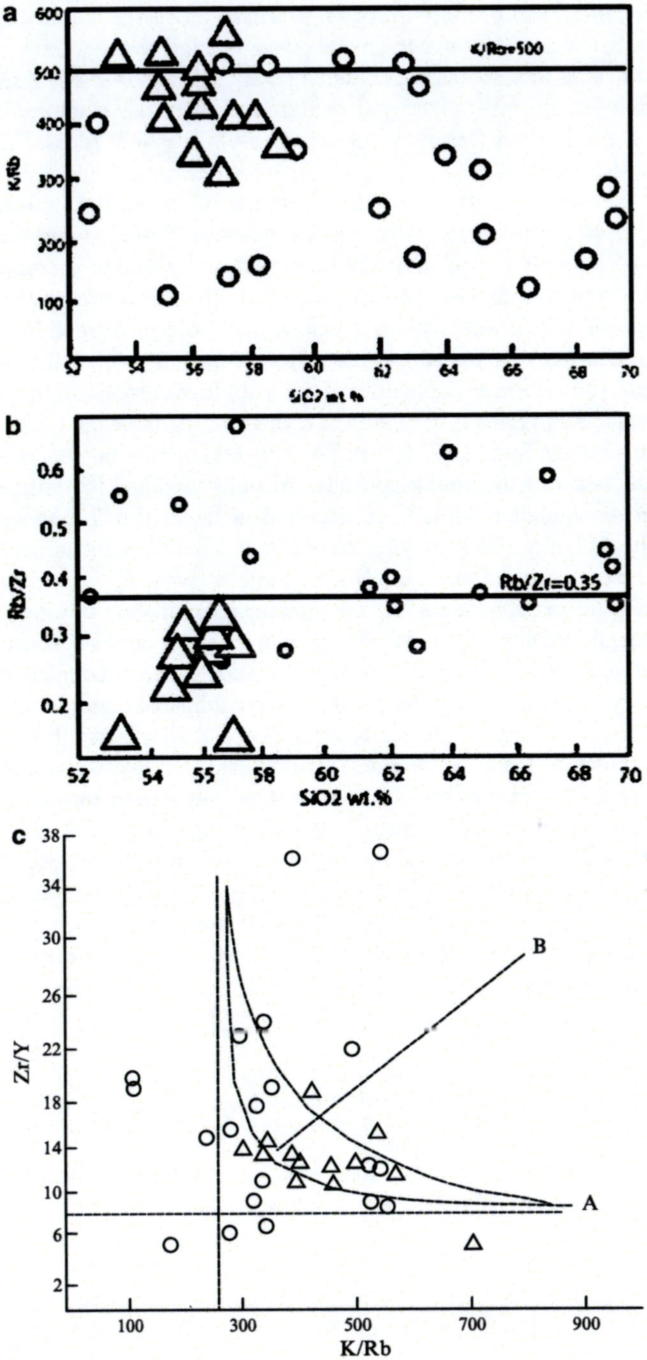

Fig. 12.11 K/Rb versus SiO_2 and Rb/Zr versus SiO_2 diagrams for Nuqara volcanics (for (**a**) and (**b**) respectively). (**c**) Zr/Y versus K/Rb relation showing magma mixing trend (trend A) and fractional crystallization trend (trend B). Symbols are same as in Fig. 12.6

indicating open-system evolution processes. These processes are further supported by Nb/Y ratio, which is sensitive to crustal contamination of mafic melts and subduction-related mantle metasomatism; contamination leads to lower Nb/Y ratio (Wilson 1989). Moreover, the role of crustal contamination in the evolution of Dokhan volcanics in the Eastern Desert is confirmed by Wilde and Youssef (2000), who recognized ~680 Ma old ziron xenocrysts in the andesites at Gebel Dokhan area.

Crustal contamination may also have occurred as a result of mixing between mantle melts and partial melts of the continental crust (felsic magma, Van Wagoner et al. 2002). The mixing mechanism is supported by hyperbolic incompatible Zr/Y versus K/Rb distributions (see Langmuir et al. 1978) and confirms that mixing is indeed operative in the diversification of young mafic volcanics (trend A, Fig. 12.11c). However, some plots of old volcanics follow trend (B) which indicates crystal fractionation. The incoherent behaviour of compatible elements with increasing MgO for young sequence implies assimilation of country rocks following fractional crystallization (Huppert and Sparks 1985; Friz-Töpfer 1991) or mixing processes. Owing to the small thickness of the young volcanics in study area and the implied low heat capacity any assimilation must have occurred at depth (Pudlo and Franz 1994). However, the difficulty arises in trying to determine a suitable contaminant, especially given that the basement rocks for the Nuqara terrane have not yet been characterized. The only real clue to the nature of the surrounding rocks in the study area is the felsic rocks. Although mobility of Rb may be a factor, on a plot of Nb/Y versus Rb/Y (not shown), the young mafic volcanics display variation in Rb/Y compared to Nb/Y, a trend contrary to mantle melts, but similar to syncollisional granites (Pearce et al. 1984) and the felsic rocks of the study area. Cox and Hawkesworth (1985) studied continental tholeiites from the Deccan Traps and showed two sequences with good isotopic evidence for crustal contamination that have trace element variations similar to those of the young mafic volcanics. These mafic volcanics show a wide range in some incompatible element ratios, and are associated with large volumes of felsic rocks indicative of crustal anatexis. Thus, the mafic magmas probably resided within or at the base of the continental crust for long periods of time, providing sufficient heat to produce crustal anatexis, and therefore had ample opportunity to become contaminated.

12.7 Nature of the Parnet Magma

None of the basaltic lavas for the two volcanic sequences are primary mantle melts, judging from their low Mg numbers and trace element chemistry (Table 12.1). These volcanics are far from the expected composition of melts in equilibrium with mantle peridotites (Cox 1980). These characteristics suggest that they underwent crystal fractionation from parental magmas either in magma chambers or en route to the surface. The calc-alkaline basalt (N.1) and tholeiitic basalt (B.4) are assumed to be the parental magmas for the old and young volcanic sequences respectively. Their Ni and Cr contents and low Mg numbers, much lower than primary mantle melts (>250 ppm Ni, >1,000 ppm Cr, >0.65 Mg number, Wilson 1989) suggested that they have suffered moderate to extensive olivine and pyroxene fractionation.

The low Y content for the old parent basalt (N.1) is consistent with mantle melting of a garnet-bearing peridotite. Modal melting model (assuming batch melting) shows that melting of about 5–10% of a variable two times mantle enriched garnet lherzolite (53% olivine, 30% clinopyroxene, 10% orthopyroxene, 5% garnet, 2% amphibole) can generate its trace element patterns (Table 12.3). Similar mantle melting conditions were reported from volcanic rocks at Gabal Dokhan area (Stern and Gottfried 1986; Khalaf 1994) and continental margins in general (Tatsumi et al. 1992) and also for within-plate basalts (Klewin and Shirey 1992). However, the pronounced negative Nb anomaly in the calc-alkaline basalt implies some modification by subduction slab-derived magma batches or fluids, because such anomalies are characteristic of continental margin settings (Saunders et al. 1988). They were also described from continental tholeiites where the magma genesis is influenced by a subduction component in the mantle (Fitton et al. 1988).

The following arguments are used to put constraints on the source of young volcanic sequence. First, its relatively high Y content indicates the absence of a residual garnet from the source area at the magmas (Pearce and Norry 1979). Second, investigations on mantle xenoliths from Cenozoic basalts associated with Red Sea rifting helped to select the suitable mineralogy of the mantle source (Stein et al. 1993). A metasomatized lithospheric mantle source of phlogopite-bearing spinel lherzolite is selected (65% olivine, 20% orthopyroxene, 10% clinopyroxene, 4% spinel, 1% phlogopite). Table 12.2 shows that the trace element compositions of the least evolved sample (B.4) fits that of the 1–5% partial melt.

The deviations in the two sequences are attributed to uncertainties in the selection of distribution coefficients and the composition of the source. Accepted criteria for considering primary basaltic magma according to Clague and Frey (1982) are Mg number >65 and high Ni contents >235 ppm. None of the investigated mafic rocks fulfils this requirement. A moderate amount of olivine and pyroxene fractionation of the primary magma in the magma chamber before eruption (not encountered in the suite) could have brought about the observed drop of Ni and Cr from 272 to 65 and 232 to 137 respectively for old and 183 to 65 and 364 to 225 respectively for young sequence. Such low-pressure fractionation is most probably the reason for the absence of primary lava in the studied volcanics. To summarize, the primitive melts that gave rise to the investigated volcanic suites were derived from possibly a garnet peridotite for old sequence and phlogopite-bearing spinel lherzolite mantle source for young sequence by a low (5–10%) modal batch partial melting.

The high incompatible trace element compositions of all the Nuqara volcanics require that the mantle sources for both older and younger magmas were affected by metasomatism. The similar incompatible trace element patterns for the two sequences points to an identical style of metasomatic enrichment for the two sources (though differences in incompatible element abundances, together with some variations of incompatible element ratios, suggest different degrees of metasomatic modification). Thus, the transition from early calc-alkaline magma to late erupted tholeiitic magma is most likely due to different degrees of metasomatism of sources

Table 12.3 The input data and the results of the modal batch partial melting of the least evolved Nuquara volcanics

			Mineral – melt distribution coefficients		Old volcanic sequence						Young volcanic sequence								
Element	Ol	Opx	Cpx	Hb	Gt	Sp	Phlog	D	Cs	F = 1	F = 5	F = 10	SB25	D	Cs	F = 1	F = 5	F = 10	B5
K	0.0068	0.014	0.038	0.96	0.015	–	1.5	0.03635	405	8,807	4,791	3,052	8,049	0.026	1,008	28,204	13,494	8,169	26,968
Rb	0.0098	0.022	0.031	0.29	0.042	0.01	1.7	0.02459	1.72	50	25	14	53	0.031	3.44	85	43	27	62
Ba	0.0099	0.013	0.001	0.42	0.023	0.01	1.5	0.0154	7.56	298	117	66	106	0.025	30.24	870	410	247	773
Sr	0.014	0.04	0.06	0.46	0.012	0.01	0.081	0.03922	23	471	264	170	265	0.024	34.5	918	474	284	877
U	0.00002	0.015	0.013	0.1	–	–	0.011	0.01506	0.081	3	1	0.74	8	0.004	-----	-----	-----	-----	-----
Th	0.000052	0.0056	0.014	0.5	–	–	0.12	0.019	0.288	10	4	2	10	0.004	0.096	7	2	1	5
Ti	0.02	0.1	0.4	1.5	0.3	0.048	0.98	0.1856	1,200	6,194	5,302	4,494	4,059	0.085	1,200	12,837	9,235	6,837	7,432
Zr	0.012	0.18	0.12	0.5	0.65	0.14	0.6	0.09686	11	104	83	59	118	0.067	22	288	194	137	435
Nb	0.01	0.15	0.005	0.8	0.02	0.02	0.14	0.0388	0.62	13	7	5	13	0.039	1.24	42	14	9	18
Y	0.01	0.18	0.9	1	9	–	0.03	0.8933	9.74	11	11	11	13	0.133	4.55	32	26	21	22
Ni	16.3	5	2.75	6.8	–	–	1.3	7.993	2,080	263	272	281	65	11.883	2,080	177	183	193	51
Cr	1.92	10	11.5	12.5	1.345	–	–	11.851	2,625	224	232	243	137	7.532	2,625	352	364	382	225
Co	0.17	1.2	3.2	2	0.955	0.57	8.3	3.923	470	121	124	129	24	0.776	470	604	597	589	14
V	0.06	0.6	1.2	3.4	–	–	–	0.5653	164	288	279	269	251	0.279	164	573	521	467	104
P	0.043	0.03	0.0044	–	–	–	–	0.0266	180	4,954	2,391	1,,452	5,237	0.034	180	4,087	2,177	1,375	2,182

D is the bulk distribution coefficient calculated for the source. Amphibole-bearing garnet lherzolite and phlogopite bearing spinel lherzolite are used for old and young volcanic sequence respectively. Minerals used in the calculations for old and young sequence are in the following proportions: Ol : Opx: Cpx: Gt: amph = 53: 10: 30: 5: 2 (for amphibole-bearing garnet lherzolite); OL: Opx: Cpx: Sp: Phlog = 65: 20: 10: 4: 1 (for phlogopite-bearing spinel lherzolite). Cs is the composition of the source. Distribution coefficients are from the compilation of the Rollinson (1993) and from the Geochemical Earth Reference Model (GERM). Ni, Cr and Co are similar to the primitive mantle compositions (Sun and McDonough 1989). LILE, U and Th are taken as three times the chondrite whereas Zr, Y and Ti are taken as two times the chondrite that were tapped at different times during the evolutionary history of the Nuqara area. These two contrasting magma series which are spatially and temporally associated indicate that chemical heterogeneities produced in the mantle via metasomatism, can exist for substantial periods of time after cessation of subduction (Rogers et al. 1987; Sloman 1989). It is likely that the long period of subduction (i.e. between 900 and 620 Ma; Stern 1994) in eastern Egypt (prior to the onset of the Dokhan volcanism and Nuqara volcanics in particular) has resulted possibly in LILE metasomatic enrichment of the subcontinental mantle lithosphere.

Normal Arabo-Nubian lithosphere is 100–120 km thick (Mooney et al. 1985). The upper lithosphere (or mechanical boundary layer) is presumed to comprise spinel-bearing (<80 km) and garnet-bearing (>80 km) lithospheric mantle. So, melting conditions were attained deeper in the asthenospheric source for old magmas, followed by relatively shallow-melting of mantle producing young volcanic sequence. These estimates, together with systematic differences in trace element compositions, indicate that the upper mantle beneath NED is characterized by variable degree of metasomatism with depth condition.

12.8 Scenario for the Nuqara Volcanics

The greatest volume of Nuqara volcanics is of calc-alkaline mafic to felsic composition and to less extent tholeiitic mafic volcanics. These two contrasting magma types were erupted over a long time period and their mantle source regions coexisted, in a single tectonic environment. The observed compositional range in some regions have been explained by polybaric fractional crystallization (Smedley 1986); therefore, a multi-stage evolutionary scenario for the whole magmatic complex of Nuqara area has been postulated. After 5% melting of an enriched, garnet-and amphibole-bearing mantle, moderate degrees of olivine and pyroxene fractionation generated a magma composition similar to the calc-alkaline basalt, probably at high pressure (i.e., at mantle–crust boundary) during magma ascent through the crust. Different degrees of further fractionation of olivine, clinopyroxene and plagioclase led to basaltic andesitic magmas. Macdonald (1975) suggests that some of the basic rocks have undergone high pressure crystallization of clinopyroxene (>13 kbar) or of olivine + clinopyroxene + plagioclase (>9 kbar). Fractional crystallization continued but the relative proportions and types of phases changed in response to total pressure and P_{H2O}. Crystallization at moderate pressure can occur if the crust acts as a density filter and the mafic rocks are held in a crustal magma reservoir. Thus at mid-crustal depth under moderate P_{H2O}, fractionation was dominated by Cpx, Plag and Fe–Ti oxides with apatite and produced andesitic magmas. The presence of amphibole in this step indicates that andesitic magmas were relatively hydrous, with elevated water contents (>4 wt%) and pressures of ~10 kbar with high temperature (<1,050°C), by comparison with similar rocks (Romick et al. 1992). However, felsic magmas ascended to shallower, upper-crustal magma reservoirs and the reduced total pressures permitted the crystallization of large proportions of plagioclase. The relatively small crystal size of the felsic rocks and their granophyric textures in some intrusions indicate relatively shallow depths of crystallization (Allen et al. 2008). Finally, magma mixing between mafic and felsic magma-types occurred, as documented by the occurrence of mafic enclaves in the felsic rocks. This process was probably due to rapid magma ascent from the magma chamber at depths of about 25–30 km to a very high crustal level (<10 km; Pudlo and Franz 1994). A depth of the mantle–crust boundary at 25–30 km and the presence of calc-alkaline magma chambers at plate margin settings were also reported by Blundy and Shimzu (1991).

The tholeiitic young volcanic sequence is relatively reduced and water poor (Cpx + plag + Fe–Ti oxides + apatite) in comparison with the old calc-alkaline sequence. The genesis of the tholeiitic mafic volcanics is very different. Firstly, mantle melts ascended into the crust or the crust–mantle boundary until they reached an isostatic equilibrium and formed the magma chamber consisting of the tholeiitic mafics. During the magma ascent and emplacement, crustal material contaminated this magma. In the magma chamber, fractional crystallization, magma recharge and the tapping of differentiated magma occurred. The tapped magma probably stagnated at a higher crustal level, forming another magma chamber. A poorly constrained depth estimate of 1–2 kbar for evolution of the tholeiitic mafics is in agreement with published phase equilibria for magmas of similar composition (Moore and Carmichael 1998). The occurrence of clinopyroxene intergranular between plagioclase laths implies crystallization of plagioclase followed by clinopyroxene as it is typical of quartz tholeiites at low P_{H2O} at depth less than 15 km (Green and Ringwood 1967). Assuming that tholeiitic magmas ascended through relatively dense crust to shallow levels, they may result from degassing of water-rich parental basalt similar to that of the calc-alkaline lavas. There, further assimilation of country rocks and fractional crystallization processes led to more differentiated melt composition (Harms 1989) which formed the magmas for the plutonic rocks of study area. RTF processes within the tholeiitic magma chamber also caused magma recharge in the overlying plutonic magma. In this instance, the magma chamber had presumably different compositions in different parts (zoned magma chamber). Recharge and tapping processes in the lower part of the chamber would then result in interaction with the upper part of the chamber. In any case, the scenario of magma tapping from a more basic magma body and its recharge into the more differentiated plutonic magma is still the same (Pudlo and Franz 1994).

The two volcanic sequences were not observed in the granitic core of the Nuqara pluton and it is therefore probable that the granitic melts intruded after the emplacement of calc-alkaline and tholeiitic volcanics. A small gabbro stock in the study area is genetically related to the tholeiitic volcanics and is possibly a part of the postulated tholeiitic magma. These gabbro bodies send off-shoots into the old volcanic sequence, while are intruded by the younger granites. The post-magmatic infiltration of a fluid phase in some of the calc-alkaline and tholeiitic volcanics led to high Ba and Sr concentrations in these rocks.

12.9 Implications of Volcanism in Nuqara Area

During the Pan African orogeny, the Nuqara area formed a part of the Arabo-Nubian Shield. Intensive post-orogenic magmatism took place at the subduction–extension transition in/on consolidated folded blocks which suffered intensive deformation. Thus, the supracrustal setting for magma ascent and eruption/emplacement of the early andesitic and late andesitic phases (plugs and flows) was that of solid blocks under a extensional regime in the North Eastern Desert (NED). In the study area, the unconsolidated conglomerates (sandwiched between the old and young volcanic sequence) were marking the weak zone and these were encountered by andesitic magma during their ascent (see Einsele 1982). The old andesite lavas were extruded into the folded basement rocks along faults and moved on laterally, forming the lava flows (Fig. 12.12). These lavas were followed by ignimbrites and rhyolitic lavas with their interbedded pyroclastic deposits. The final stage of the andesitic magmatism included the formation of intrusive veins and plugs along the ignimbrite/rhyolite boundary and local effusion of short-lived lava flows, where the ignimbrite seal was pierced by the andesitic magma.

12.10 Petrogenetic Model: Regional Implications

Nuqara volcanics are one of the northernmost fragments of the Arabian–Nubian Shield (ANS). The ANS extends over large areas in NE-Africa and Arabia and is considered to be the product of one of the most intensive episodes of continental crust formation in Earth history (Stein and Hofmann 1994). The geological history of the northern ANS has been divided into four main phases (Bentor 1985). Phase I (~900–870 Ma) was characterized by the formation of a several kilometer thick sequence of tholeiitic basalts, which probably erupted in an oceanic environment and formed an oceanic plateau (Stein 1999). Phase II (~870–650 Ma) was characterized by mainly calc-alkaline, island arc magmatism, followed by extensive metamorphism. The subduction processes (~900–600 Ma) in the crustal evolution of the ANS were marked by the emplacement of calc-alkaline, collisional, I-type granitoids (phase III) and correspond to following time-spans: 640–610 Ma in Saudi Arabia (Fleck et al. 1980); 670–610 Ma in Egypt (Stern and Hedge 1985) and 615 Ma in Jordan (Mc Court 1990). Phase IV, which extended into the Cambrian (~630–600 Ma), was characterized by intrusion and extrusion of alkaline magmas. Outcrops of phase IV alkaline magmas form a small part of the ANS volume, yet they are scattered over the entire shield (Bentor 1985; Black and Liegeois 1993). In the northeastern ANS, alkaline magmas were generated during the transition from orogenic activity to intra-plate stable conditions. In Egypt, this period (~600–530 Ma) was also characterized by 610–550 Ma Dokhan volcanics and the 600–585 Ma molasses-type Hammamat sediments, dike swarms and A-type granites (Stern and Hedge 1985). The onset of stable platform conditions in this region is marked

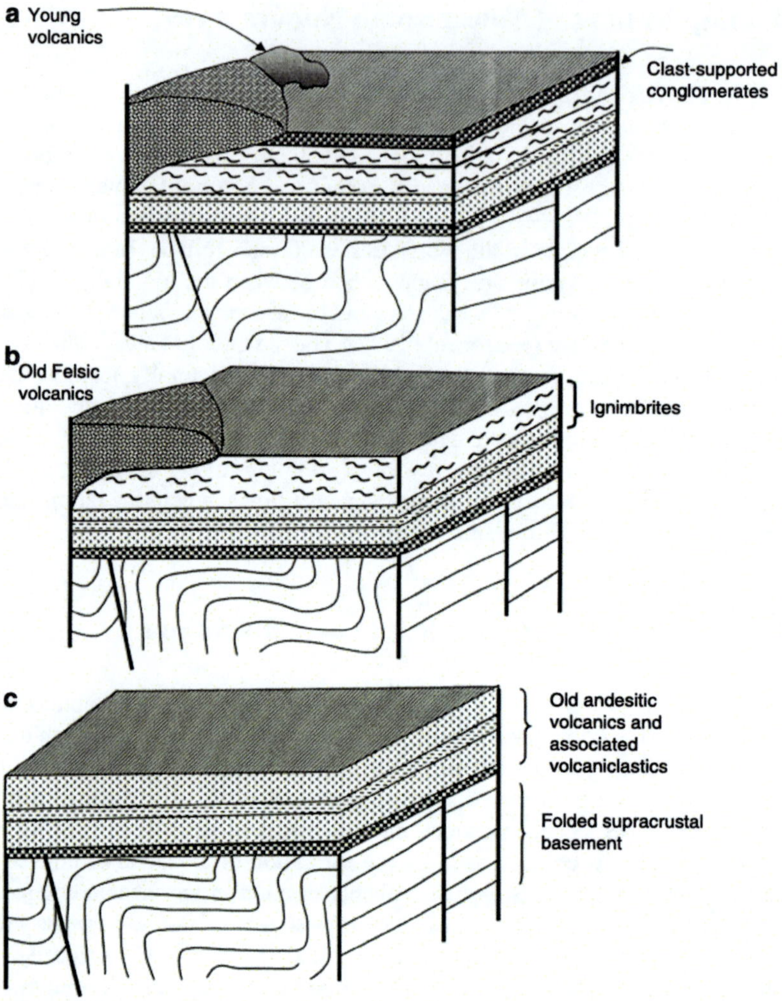

Fig. 12.12 Simplified model of the magma emplacement in Nuqara block. (**a**) Old andesitic stage. Folded supracrustal basement is overlain by an old andesitic sequence. (**b**) Felsic magma rises through the basement along faults and intrudes laterally forming ignimbrite and rhyolitic lavas. (**c**) Nuqara rock complex develops as magma is supplied from depth. Locally the andesitic magma is extruded along the ignimbrite/rhyolite boundary in the sequence above and finally extrudes at the surface forming the so called Young andesites

by the deposition of Early Cambrian sandstone (e.g. Parnes 1971). Intra-plate sedimentation and recurring cycles of alkaline magmatism characterized this region through most of the Phanerozoic (e.g. Garkunkel 1989).

Based on the low and relatively uniform $^{87}Sr/^{86}Sr$ and $\epsilon_{Nd\ initial}$ values of magmas from the various stratigraphic stages in the evolution of the ANS, and on the distinction from the mid-ocean ridge basalt (MORB)-type Gebel Gerf ophiolite,

Stein and Goldstein (1996) proposed a lithospheric mantle source for these magmas. Stein et al. (1997) suggested a chromatographic model for the transport of trace elements in the mantle wedge (above the ANS subduction zone). According to this model, the upper zones in the chromatographic column will be enriched in the incompatible and mobile elements such as Rb, Pb and LREE. The moderate enrichment of the two volcanic sequences in these elements is consistent with their derivation from such an enriched part of the lithosphere.

Nuqara area comprises two independent volcanic sequences which are previously mapped as Dokhan volcanics. The field, petrographic and geochemical data differentiate these volcanics into old calc-alkaline and young tholeiitic volcanic sequence. The data showing comparison between the volcanic rocks in Nuqara area with the same volcanic sequences described in the Eastern Desert of Egypt (El Ramly 1972; Abu Zied 1984, 1985; Khalaf 1986) and in the Saudi Arabia (Greenwood et al. 1976, 1980; Delfour 1979) are given in Table 12.4. From this table, it is obvious that the old volcanic sequence would then equated with the Dokhan volcanics, Hamrewein lower volcanic and Esh volcanic sequence. Equivalent rock units in Saudi Arabia include the Hulayfah and Shammer Group which is widespread in the northern part of the Arabian Shield with a thickness of 1,000–1,400 m (Baubron et al. 1976; Hadley and Schmidt 1980). It consists of undeformed volcanic rocks interbedded with coarse continental clastics, which indicate an alluvial to lacustrine environment and intruded by alkali granite at about 600–571 Ma (Baubron et al. 1976). In NE Sudan, the Amaki Series is Neoproterozoic undeformed volcanic rocks and molasses-type sediments, which have a composition similar to the Dokhan volcanics and the Hammamat Group in Egypt (Almond et al. 1984). On the other hand, the young volcanic sequence is possible equivalent to the Hamrewein Upper volcanic sequence, Bali volcanic sequence and Jibalah Group in Saudi Arabia. Abu Zied (1984) determined the age of the Hamrewein Upper volcanics to be about 514 Ma. The age dating of this sequence lies within the range of K/Ar minimum ages, 520–500 Ma of the andesitic flows of the Jibalah Group (Deflour 1979). These volcanic eruptions coincided with the movement along the Najd fault zones of Saudi Arabia (Delfour 1979; Greenwood et al. 1980). Baubron et al. (1976) and Delfour (1979) reached the same conclusion on the mantle-derived origin of the andesitic Jibalah Group that confirm with the origin of the Young volcanics at study area. Black et al. (1985) showed that large-scale strike-slip movements and reactivation of transcurrent faults play a major role during the emplacement of anorogenic alkaline felsic magmas. Recently, some authors recorded new occurrences of calc-alkaline Dokhan volcanics younger than the younger granites in the Eastern Desert (Nasr and El-Sherbeni 1998; Nasr et al. 2002), but there is no age dating performed on such volcanics.

Similar magmatic sequences and whole-rock chemistry relate most of the post-orogenic tholeiitic magmatism of the northeastern ANS to a common petrogenetic process. These rocks suggest the presence of large unexposed mafic bodies, all derived from the sub-continental mantle beneath the ANS. Despite its relatively small areal expression, post-orogenic tholeiitic magmatism in the northeastern ANS marks a significant lithospheric melting event during the maturation stages of

Table 12.4 Stratigraphical correlation of the basement rocks of Nuqara area with the accepted classification at the Eastern Desert of Egypt and northern Saudi Arabia

Sequence units in the basement rocks of Egypt taken from El Shazly 1977	Stratigraphic rock sequences in wadi Hamrawein area, north Quseir (Abu Zied 1984)	Stratigraphic succession in the wadi Esh-wadi Bali, Eash El Mellaha Range, Eastern Desert, Egypt (Khalaf 1986)	Stratigraphic volcanic sequences in Nuqara area, northern Eastern Desert, Egypt	Possible equivalent rock sequences at Saudia Arabia (Greenwood et al. 1980; Delfour 1979)
Post-tectonic younger granitoids (590 million years)	Upper volcanic sequence (minimum age, 514 million years) and Hammamat facies sediments of the fault-bounded syncline of wadi Hamrawein	Bali volcanic sequence and associated volcaniclastic sediments.	Young volcanic sequence	Contiunation of Najd (500 million years)
				Jibalah Groups: (520–500 million years) Clastic sediments, alkali basalts and andesitic flows. Najd faulting (580–500 million years)
Post Hammamat felsite (El-Ramly 1972)				NW-trending left lateral wrench faulting. All the north and northeast structural elements of the older rocks related to the NW-trend in the Najd fault zone.

Late-orogenic plutonites (585?–485 million years)				Bishah Orogeny: Folds, faults and greenschist metamorphism.
	Hammamat sediments of wadi Queih			
Post-geosynclinal (Hammamat) sediments				Murdama Group (570–550 million years) Conglomerate, greywacke, minor andesite and rhyolite.
Major unconformity				Yavikh Orogeny:
Emerging geosynclinal Dokhan volcanics (660–602 ± 63 million years (Stern et al. 1984).	Abu Hamra volcaniclastic sediments and lower volcanic sequence (minimum age, 565–520 million years)	Esh volcanic sequence and associated Hammamat facies sediments	Old volcanic sequence (lower mafic and upper felsic rocks)	Folds and faults, N trends and greenschist metamorphism Hulayfah or Halaban Group (650–600 million years) intermediate with shallow water sediments (mature island arc assemblage formed in an active continental margin

the newly formed continental lithosphere. As suggested by Turner et al. (1992) for post-orogenic 490 Ma A-type granites from South Australia, these tholeiitic volcanics represent a considerable volumetric addition of mantle material to the continental crust during the final phase of a large orogenic event.

12.11 Summary and Conclusions

1. On the basis of field, petrographic and geochemical data, Nuqara area is stratigraphically subdivided into two volcanic suites: an older volcanic suite of mafic to felsic composition (consisting of basalt, basaltic andesite, andesite, dacite and rhyolite) and a younger volcanic suite comprising basaltic andesite and andesite. Volcaniclastic deposits separate these two suites. Change of styles of volcanic activities in this area is apparent.
2. The old volcanic suite defines medium to high-K calc-alkaline series that is oxidized, hydrous and formed under high f_{O2}. The young volcanic suite demonstrates high K-tholeiitic series that is less oxidized, anhydrous and formed under low f_{O2}. These two suites have different styles of magmatic evolution.
3. The both suites are enriched in LILEs (Rb, Ba, K, Th, Pb) relative to HFSEs (Nb, Zr, Ti, Y). The high LILE/HFE ratios are suggestive of subduction component from mantle material which constituted the mantle wedge during previous subduction events in eastern Egypt.
4. The old volcanic suite evolved through crystal/melt fractionation of Ol, Cpx, Plag, Fe–Ti oxides and Ap in the mafic varieties, and Cpx, Plag, Fe–Ti oxides and Ap in the felsic varieties. Replenishment, tapping, mixing and subsequent fractional crystallization process within strongly zoned magma chamber led to the generation of young volcanic suite. The influx of new batches of primitive magma took place periodically and that mixing of magmas (young and old magma) accounts for chemical zonation.
5. Crustal contamination plays a role in the genesis of Nuqara magmatism, since the basement lithologies beneath the NED region are markedly different in composition.
6. The chemical variations within these volcanic suites as described above require relatively high eruption rates. There is ample evidence from studies of lavas from other localities (e.g., Wyers and Barton 1986,1989) and from theoretical work (e.g., Huppert and Sparks 1985) that influx of hot, primitive magma (like young suite) into a chamber which contains evolved magma (like old suite).
7. Calc-alkaline and tholeiitic magma chambers existed in NED during Neoproterozoic Pan-African: the depth of the calc-alkaline magma chamber generating the old suite was 25–30 km, whereas the magma chamber for the tholeiitic suite was less than 15 km.
8. Differences between the calc-alkaline and the tholeiitic suites are due to variable source characteristics: calc-alkaline melts generated from a garnet-and amphibole-bearing, enriched mantle (modified by the release of magma batches or fluids from a subducted slab) whereas the tholeiitic melts generated from phlogopite-bearing spinel lherzolite.

9. The close spatial stratigraphic relations between two magma types clearly indicate that there were no variations in tectonic conditions (emplaced in an extensional environment) with changes in magma types.
10. Polybaric evolution can account for the chemical discontinuity between the old and young volcanic magmas erupted at Nuqara area.

References

Abdel-Rahman AM, Doig R (1987) The Rb-Sr geochronological evolution of the Ras Gharib segment of the northern Nubian Shield. J Geol Soc London 144:577–586

Abu Zied HT (1984) Geology of the wadi Hamrawein, Red Sea hills, Eastern Desert, Egypt. Ph.D. thesis, South Carolina University, 206p

Abu Zied HT (1985) Wadi Hamrawein Upper volcanic sequence: a possible Late Proterozoic-Early Phanerozoic volcanic activity younger than Dokhan volcanics, Eastern Desert, Egypt. Geological Society of Egypt (abstract), 17–18

Allen SR, Mcphie J, Ferris G, Simpson C (2008) Evolution and architecture of a large felsic igneous province in western Laurentia: the 1.6 Ga Gawler Range volcanics, south Australia. J Volcanol Geotherm Res 172:132–147

Almond DC, Ahmed F, Dawoud AS (1984) Tectonic, metamorphic and magmatic styles in the northern Red Sea hills of Sudan. Bull Fac Sci, King Abdulaziz University, Jeddah, 6:449–458

Arculus RJ (2003) Use and Abuse of the terms calc-alkaline and calcalkalic. J Petrol 44:929–935

Astis AG, Peccerillo A, Kempton PD, Volpe L, Wu TW (2000) Transition from calc-alkaline to potassium-rich magmatism in subduction environments: geochemical and Sr, Nd, Pb isotopic constraints from the island of Vulcano (Aeolian arc). Contrib Mineral Petrol 139:684–703

Awdankiewicz M, Breitkreuz C, Carloehling B (2004) Emplacement textures in Late Palaeozoic andesite sills of the Flechtingen-Roblau Block, north of Magdeburg, Germany. Geol Soc London 234:51–66

Balesta AT, Farberov AI, Smirnov VS, Tarakanovsky AAJ, Zubin MI (1977) Deep crustal structure of the Kamchatkan volcanic regions. Bull Volcanol 40:260–266

Barton M, Huijsmans JP (1986) Post-caldera dacites from the Santorini volcanic complex, Aegean Sea, Greece: an example of the eruption of lavas of near constant composition over a 2200 year period. J Petrol 94:472–495

Baubron JC, Delfour J, Vialette Y (1976) Geochemical measurements (Rb/Sr & K/Ar) on rocks of the Arabian Shield, Saudi Arabia. Bur Res Geol Min open file rept 76, 152p

Bentor JK (1985) The crustal evolution of the Arabo-Nubian Massif with special reference to the Sinai Peninsula. Precamb Res 208.1–74

Black R, Liegeois JP (1993) Cratons, mobile belts, alkaline rocks and continental lithospheric mantle; the Pan-African testimony. J Geol London 150:89–98

Black R, Lameyre J, Bonin B (1985) The strucural setting of alkaline complexes. J African Earth Sci 3:5–16

Blundy JD, Shimizu N (1991) Trace element evidence for plagioclase recycling in calc-alkaline magmas. Earth Planet Sci Lett 102:178–197

Bryan WB, Finger LW, Chayes F (1969) Estimating proportions in petrographic mixing equations by least-squares approximation. Science 163:926–927

Cas RAF, Wright JV (1987) Volcanic successions – modern and ancient: geological approach to processes, products and successions. Allen & Unwin, London, p 529

Clague DA, Frey FA (1982) Petrology and trace element geochemistry of the Honolulu volcanics, Qahu: implication of the oceanic mantle below Hawaii. J Petrol 23:447–504

Cox KG (1980) A model for flood basalt volcanism. J Petrol 21:629–650

Cox KG (1988) Numerical modeling of a randomized RTF magma chamber: a comparison with continental flood basalt sequences. J Petrol 29:681–697

Cox KG, Hawkesworth CJ (1985) Geochemical stratigraphy of the Deccan Traps at Mahabaleshwar, western Ghats, India: implication for open system magmatic processes. J Petrol 26:355–377
Davidson JP, Ferguson KM, Colucci MT, Dungan MA (1988) Crust-magma interactions and the evolution of arc magmas: the San Pedro-Pellado volcanic complex, southern Chilean Andes. Geology 15:443–446
Deer WA, Howie RA, Zussman J (1980) An introduction to the rock-forming minerals. Longman, London, 528p
Delfour J (1979) Upper Proterozoic volcanic activity in the northern Arabian Shield, Saudi Arabia. In Al Shanti AMS (eds) Evolution and mineralization of the Arabo-Nubian Shield. Institute of Applied Geology, King Abdul Aziz University, Jeddah, Bulletin 3, pp 59–75
Edwards CMH, Menzies MA, Thirlwall MF (1991) Evidence from Muriah, Indonesia for the interplay of supra-subduction zone and intraplate processes in the genesis of potassic alkaline magmas. J Petrol 32:555–592
Einsele G (1982) Mechanism of sill intrusion into soft sediment and expulsion of pore water. Deep Drill Proj 64:1169–1176
El Gaby S, List FK, Tehrani R (1996) Ancient fracture systems and Red Sea tectonics. Geology of the Arab World. Cairo University, Egypt, pp 143–160
El Ramly MF (1972) A new geological map for the basement rocks in the Eastern and Southwestern Desert of Egypt. Ann Geol Surv Egypt 2:1–18
El Shazly EM (1977) The geology of the Egyptian region. In: Nairn AM, Stehli FG (eds) Ocean basins and margins, vol 44. Plenum, New York, pp 379–444
Esperanca S, Crisci GM, DeRosa R, Mazzuoli R (1992) The role of the crust in the magmatic evolution of the island of Lipari (Aeolian Islands, Italy). Contrib Mineral Petrol 112:450–462
Fitton JG, James D, Kempton PD, Ormerod DS, Leeman WP (1988) The role of lithospheric mantle in the generation of Late Cenozoic basic magmas in the western United States. J Petrol, Special Litosphere issue, 331–349.
Fleck RJ, Greenwood WR, Hadley DG, Anderson RF, Schmidt DL (1980) Rubidium-Strontium geochronology and plate tectonic evolution of the southern part of the Arabian Shield. U S Geol Surv 1131:38p
Fritz H, Wallbrecher E, Khudeir AA, El Ela F, Dallmeyer DR (1996) Formation of Neoproterozoic metamorphic core complexes during oblique convergence. J Afr Earth Sci 23:311–329
Friz-Töpfer A (1991) Geochemical characterization of Pan-African dike swarms in southern Sinai: from continental margin to intraplate magmatism. Precamb Res 49:281–300
Fullagar PD, Greenberg JK (1978) Egyptian younger granites: a single period of plutonism. I A G Res Ser 4:62–63
Garkunkel Z (1989) Tectonic setting of Phanerozoic magmatism in Israel. Israel J Earth Sci 38:51–74
Geist D, Richards M (1993) Origin of the Columbia plateau and Snake River Plain: deflection of the Yellowstone plume. Geology 21:789–791
Ghoneim MF, Lebda EM (2001) Petrogenetic diversity of Dokhan volcanics in relation with older and younger granites, Eastern Desert, Egypt. In: 2nd international conference on the geology of Africa, Assiut University,11, pp 167–181
Gill JB (1981) Orogenic andesites and plate tectonics. Springer, Berlin, Heidelberg, New York, p 390
Green TH (1980) Island arc and continent-building magmatism: a review of petrogenetic models based on experimental petrology and geochemistry. Tectonophysics 63:367–385
Green DH, Ringwood AE (1967) The genesis of basaltic magmas. Contrib Mineral Petrol 15:103–190
Greene DC (1984) Structural geology of the Quseir area, Red Sea coast, Egypt. Department of Geology and Geography, University of Massachusetts, Amherst, contribution no.52
Greenwood WR, Hadley DG, Anderson RE, Fleck RJ, Schmidt DL (1976) Late Proterozoic cratonization in southwestern Saudi Arabia. Phil Trans R Soc London 280:517–527
Greenwood WR, Anderson RE, Fleck RJ, Robert RJ (1980) Precambrian geologic history and plate tectonic evolution of the Arabian shield. Dir Miner Res Tech Rep 24:35p
Greiling RO, Abdeen MM, Dardir AA, El Akhal H, El Ramly MF, Kamal ElDin GM, Osman AF, Rashwan AA, Rice AN, Sadek MF (1994) A structural synthesis of the Proterozoic Arabian-Nubian Shield in Egypt. Geol Rundsch 83:484–501

Hadley DG, Schmidt DL (1980) Sedimentary rocks and basins of the Arabian shield and their evolution. Institute of Applied Geology, Jeddah Bulletin 3: 25–50

Harms U (1989) Krustenentwicklung in Nordost-Afrika: geochemische und isotopengeochemische untersuchungen von Granitoiden aus Südägypten und Nordsudan. Berl Geowiss Abh (A) 108, 152p

Hogg AJ, Fawcett JJ, Gittins J, Gorton MP (1989) Cyclical variation in composition in continental tholeiites of East Greenland. Canad J Earth Sci 26:534–543

Huijsmans JPP, Barton M, Salters VJM (1988) Geochemistry and evolution of the calc-alkaline volcanic complex of Santorini, Aegean Sea, Greece. J Volcanol Geotherm Res 34:283–306

Huppert HE, Sparks RSJ (1985) Cooling and contamination of mafic and ultramafic magmas during ascent through continental crust. Earth Planet Sci Lett 74:371–386

Khalaf EA (1986) The volcanics and volcaniclastic sediments of wadi Esh-wadi Bali District, Esh El Mellaha Range, Eastern Desert, Egypt. M.Sc. thesis, Cairo University, 147p

Khalaf EA (1994) Petrography,geochemistry and petrogenesis of volcanics and associated rocks of Gabal Dokhan area, northeastern Desert, Egypt. Ph.D. thesis, Cairo University, 172p

Klewin WK, Shirey BS (1992) The igneous petrology and magmatic evolution of the Midcontinent rift system. Tectonophysics 213:33–40

Kusky TM, Ramadan TM (2002) Structural controls on Neoproterozoic mineralization in the south Eastern Desert, Egypt: an integrated field,landsat TM, and SIR-C/XSAR approach. J Afr Earth Sci 35:107–121

Langmuir CH, Vocke RD, Hanson GN, Hart SR (1978) A general mixing equation with applications to Icelandic basalts. Earth Planet Sci Lett 37:380–392

LeBas MJ, LeMaitre RW, Streckeisen A, Zanettin BA (1986) Chemical classification of volcanic rocks based on the total alkali-silica diagram. J Petrol 27:745–750

LeMaitre RW (1981) GENMIX – a generalized petrological mixing model program. Comput Geosci 7:229–247

Lofgren G (1974) An experimental study of Plagioclase crystal morphology: isothermal crystallization. Am J Sci 274:243–273

Luhr JF, Carmichael ISE (1980) The Colima volcanic complex, Mexico. Contrib Mineral Petrol 71:343–372

Macdonald R (1975) Petrochemistry of the early Carboniferous (Dinantian) lavas of Scotland. J Geol 11:269–314

McCourt WJ (1990) The geochemistry and tectonic setting of the granitic rocks and associated rocks in the Aqaba and Araba complex, southwest Jordan. Nat Res Auth Bull 10:1–96

Mooney WD, Gettings ME, Blank HR, Healy JW (1985) Saudi Arabian seismic deep refraction: a travel-time interpretation of crust and upper mantle structure. Tectonophysics 111:173–246

Moore G, Carmichael ISE (1998) The hydrous phase equilibria to 3 kbar of an andesite and basaltic andesite from western Mexico: constraints on water content and conditions of phenocryst growth. Contrib Mineral Petrol 130:304–319

Nasr BB, El Sherbeni H (1998) New occurrences of Dokhan type volcanics, south Eastern Desert. Egypt. Annal Geological Survey, Egypt, XXI, pp 23–39

Nasr BB, El Sherbeni H, Masoud MS (2002) Geology of wadi Al Ghuza area: Molasse sediments and dyke swarms, north Eastern Desert, Egypt. Ann Geol Surv, Egypt XXV:1–9

ÓHara MJ (1977) Geochemical evolution during fractional crystallization of a periodically refilled magma chamber. Nature 266:503–507

ÓHara MJ, Mathews RE (1981) Geochemical evolution in an advancing, periodically replenished, periodically tapped, continuously fractionated magma chamber. J Geol Soc London 138: 237–277

Parnes A (1971) Late Lower Cambrian trilobites from the Timna and Har'Amram, southern Negev, Israel. Israel J Earth Sci 20:179–205

Pearce JA (1980) Geochemical evidence for the genesis and eruptive setting of lavas from Tethyan ophiolites. In: Panayiotou A (ed) Ophiolites. Geological Survey Department, Cyprus, pp 261–272

Pearce JA, Norry MJ (1979) Petrogenetic implications of Ti,Zr,Y and Nb variations in volcanic rocks. Contrib Mineral Petrol 69:33–47

Pearce JA, Harris NBW, Tindle AG (1984) Trace element discrimination diagrams for the tectonic interpretation of granitic rocks. J Petrol 25:956–983

Pudlo D, Franz G (1994) Dike rock generation and magma interactions in the Bir Safsaf igneous complex, south-west Egypt: implications for the Pan-African evolution in north-east Africa. Geol Rundsch 83:523–536

Ressetar R, Monard JR (1983) Chemical composition and tectonic setting of the Dokhan volcanic formation, Eastern Desert,Egypt. J Afr Earth Sci 1:103–112

Richard JP, Uilrich T, Kerrich R (2006) The Late Miocene-Quaternary Antofulla volcanic complex, southern Puna, NW Argentina: protracted history, diverse petrology, and economic potential. J Volcanol Geotherm Res 152:197–239

Rogers NW, Hawkeswork CJ, Mattey DP, Harmon RS (1987) Sediment subduction and the source of potassium in orogenic leucitites. Geol 15:451–453

Rollinson H (1993) Using geochemical data: evaluation, presentation, interpretation. Longman, UK, p 352

Romick JD, Kay SM, Kay RW (1992) The influence of amphibole fractionation on the evolution of calc-alkaline andesite and dacite tephra from the central Aleutians, Alaska. Contrib Mineral Petrol 5:101–118

Saunders AD, Norry MJ, Tarney J (1988) Origin of MORB and chemically depleted mantle reservoirs: trace element constraints. J Petrol, Special Litospheric Issue, 415–445

Schmincke HU (2004) Volcanism. Springer, Berlin, Heidelberg, p 324

Shapiro L, Brannock WW (1962) Rapid analysis of silicate, Carbonate and phosphate rocks. US Geol Surv Bull 1144A:56p

Sloman LE (1989) Traissic shoshonites from the dolomites, north Italy, alkaline arc rocks in a strike-slip setting. J Geophys Res 94:4655–4666

Smedley PL (1986) The relationship between calc-alkaline volcanism and within plate continental rift volcanism: evidence from Scottish Palaeozoic lavas. Earth Planet Sci Lett 76:113–128

Stein M (1999) The Pan-African plume event and growth of Gondwana. In: Ninth Annual V.M. Goldschmidt conference, Harvard University Press, Cambridge, MA, 284p

Stein M, Goldstein SL (1996) From plume head to continental lithosphere in the Arabian-Nubian shield. Nature 382:773–778

Stein M, Hofmann AW (1994) Mantle plumes and episodic crustal growth. Nature 372:63–68

Stein M, Garfunkel Z, Jagoutz E (1993) Chronothermometry of peridotitic and pyroxenitic xenoliths: implications for the thermal evolution of the Arabian lithosphere. Geochimica Cosmochim Acta 57:1325–1337

Stein M, Navon O, Kessel R (1997) Chromatographic metasomatism of the Arabian-Nubian lithosphere. Earth Planet Sci Lett 152:75–91

Stern RJ (1985) The Najd fault system, Saudi Arabia and Egypt: a Late Precambrian rift-related transform system. Tectonics 4:497–511

Stern RJ (1994) Arc assembly and continental collision in the Neoproterozoic East African orogen: implications for the consolidation of Gondwanland. Ann Rev Earth Planet Sci 22:319–351

Stern RJ, Gottfried D (1986) Petrogenesis of a Late Precambrian (575-600 Ma) bimodal suite in North East Africa. Contrib Mineral Petrol 92:492–501

Stern RJ, Hedge CE (1985) Geochronologic and isotopic constraints on Late Precambrian crustal evolution in the Eastern Desert, Egypt. Am J Sci 285:97–127

Stern RJ, Gottfried D, Hedge CE (1984) Late Precambrian rifting and crustal evolution in the northeast Desert of Egypt. Geology 12:168–172

Sun SS, McDonough WF (1989) Chemical and isotopic systematics of oceanic basalts: implications for mantle composition and processes. In: Saunders AD, Norry MJ (eds) Magmatism in the ocean basins. Geological Society, London, Special Publications 42, London, pp 313–345

Sun SS, Nesbitt RW (1978) Petrogenesis of Archean ultrabasic and basic volcanic rocks: evidence from rare earth elements. Contrib Mineral Petrol 65:301–325

Tatsumi Y, Murasaki M, Nohda S (1992) Across-arc variations of lava chemistry in the Izu-Bonin Arc: identification of subduction components. J Volcanol Geotherm Res 49:179–190

Turner SP, Foden JP, Morrison RS (1992) Derivation of some A-type magmas by fractionation of basaltic magma: an example from the Padthaway Ridge, South Australia. Lithos 28:151–179

Van Wagoner NA, Leybourne MI, Dadd KA, Baldwin DK (2002) Late Silurian bimodal volcanism of southwestern New Brunswick, Canada: products of continental extension. GSA Bull 114:400–418
Voshage H, Hofmann AW, Mazzuchelli M, Rivalenti G, Sinigoi S, Raczek I, Demarchi G (1990) Isotopic evidence from the Ivrea Zone for a hybrid lower crust formed by magmatic underplating. Nature 347:731–736
Wilde SA, Youssef K (2000) Significance of SHRIMP U-Pb dating of Imperial porphyry and associated Dokhan volcanics, Gabal Dokhan, north Eastern Desert, Egypt. J Afr Earth Sci 31:403–413
Wilson M (1989) Igneous petrogenesis: a global tectonic approach. Unwin Hyman, London, 466p
Wyers GP, Barton M (1986) Petrology and evolution of transitional alkaline-subalkaline lavas from Patmos, Dodecanesos, Greece: evidence for fractional crystallization, magma mixing and assimilation. Contrib Mineral Petrol 93:297–311
Wyers GP, Barton M (1989) Polybaric evolution of calc-alkaline magmas from Nisyros, southeastern Hellenic Arc, Greece. J Petrol 30:1–37

Ophiolites

Chapter 13
Textural Fingerprints of Magmatic, Metamorphic and Sedimentary Rocks Associated with the Naga Hills Ophiolite, Northeast India

N.C. Ghose and Fareeduddin

Abstract The Naga Hills Ophiolite (NHO) in the northern sector of Indo-Myanmar Ranges (IMR) constitutes a long (200 km), narrow (2–15 km wide), NNE-SSW trending, westerly convex sigmoidal belt of Maestrichtian–Paleocene age. It comprises peridotite tectonite, serpentinite, garnet lherzolite xenolith, layered sequence of ultramafic–mafic cumulates, plagiogranite, minor dolerite dykes, basalts, pyroclastic rocks, and pelagic sediments containing radiolaria. The litho-assemblages are highly tectonised and dismembered and occur as imbricated sheets. Presence of a suite of low-grade prehnite-clinochlore schist and greenschist, and high-pressure rocks, viz., glaucophane schist and metachert containing glaucophane and eclogite, gives evidence of their development in convergent lithospheric plate junctions. The rocks have undergone three phases of deformations correlatable with the Himalayan orogeny accompanied by closure of the Tethyan Sea.

Petrographic study of magmatic and metamorphic rocks from the ophiolite suite reveals several new textural and mineralogical features that are reported for the first time in this work. Notable amongst these are quenched and vitrophyric textures in basalts, interaction of pelitic xenolith with basalt, and presence of a wide range of volcaniclastic rocks. Occurrences of perovskite in crystal–lithic tuff and welded tuff suggest their derivation at high temperature. The quenched texture in basalt characterized by elongated, radiating and curved crystals of hollow core plagioclase filled with clinopyroxene indicates rapid cooling in aqueous environment. Veins of plagiogranite in such basalts strongly support their derivation due to partial melting at high temperature. These features together with other supporting evidence like chemistry (from published works) of basalts reflect close similarity with those

N.C. Ghose (✉)
Formerly Patna University, G/608, Raheja Residency, Koramangala, 3rd Block, Bangalore 560034, India
e-mail: ghosenc2008@gmail.com

Fareeduddin
Geological Survey of India, AMSE Wing, Kumaraswamy Layout, Bangalore 560078, India
e-mail: fareedromani@hotmail.com

J. Ray et al. (eds.), *Topics in Igneous Petrology*, DOI 10.1007/978-90-481-9600-5_13,

formed at the present day mid-ocean ridge (MOR) of divergent plate boundaries. Absence of olivine in basalts and presence of K-feldspar in pyroclastics (lithic tuff) indicates pronounced fractionation of lava prior to eruption.

Micro-structural features like 'mica-fish' in glaucophane schist and 'S-C mylonite' in garnet lherzolite xenolith support emplacement of ophiolite in ductile shear zone, contradicting advocates of NHO representing an overthrusted klippe. This lends strong evidence to the idea that the IMR defines the eastern suture of the Indian plate and its emplacement is due to the result of eastward convergence of the Indian plate colliding with the Eurasian plate (Sino-Myanmar microplate).

13.1 Introduction

Occurrence of ophiolites is known from northern part of the Indian plate bordering Tibet and eastern part in the Indo-Myanmar Ranges (IMR). Such development of thin detached outcrops of ophiolite or a geosuture, representing oceanic crust of Upper Cretaceous-Eocene age, demarcates a zone of collision between Laurasia and Gondwana supercontinents. The Gondwana crustal blocks of India and Arabia first underthrust the Eurasian margin as the Tethyan Ocean was closed during the Late Mesozoic and Cenozoic, and subsequently the continental blocks themselves collided with Eurasia (Stoneley 1974).

The extensively studied Indus ophiolite (Upper Cretaceous-Paleocene) in the northwest Himalaya (Gansser 1964; Srikantia and Razdan 1980; Srikantia 1986; Desai et al. 1986; Singh et al. 1989; Virdi 1989; De Sigoyer et al. 1997; Mukherjee and Sachan 2001), occurring as a linear discontinuous belt of mafic-ultramafic rocks and high-P metamorphites, has been emplaced in the Dras Volcanic Formation of Sangeluma Group (Lower Cretaceous-Miocene). Passing eastward the suture takes a southward swing along the syntaxial bend of Himalaya in the northeast, and continues further through the Dibang valley of Arunachal Pradesh (Ghosh and Ray 2003; Ghosh et al. 2007), and enters the IMR. The suture then passes through Andaman–Nicobar islands (Vohra et al. 1989) reaching Indonesian trench, along which the Indo-Australian plate is still underthrusting a continental margin. Geophysical data support underthrusting of the Indian plate beneath the Eurasian plate both along the northern and eastern margins generating intermediate to shallow focus earthquakes (Verma 1985), land slides and terrace uplifts (Ghose and Agrawal 1989)

The Indo-Myanmar Ranges (IMR, also referred to as Assam–Arakan basin), constitute a part of the Cenozoic fold mountain belt at the eastern margin of the Indian plate, the history of which begins from the Upper Cretaceous onward (Brunnschweiler 1966). This northward trending belt of about 1,200 km in length comprises three segments which from south to north are Arakan Yoma, Chin Hills (both in Myanmar) and Naga Hills (between the states of Manipur and Nagaland within Indian territory) (Fig. 13.1) and is marked by a linear zone of mafic-ultramafic rock all along its central part showing analogy with the Alpine peridotite. The NHO

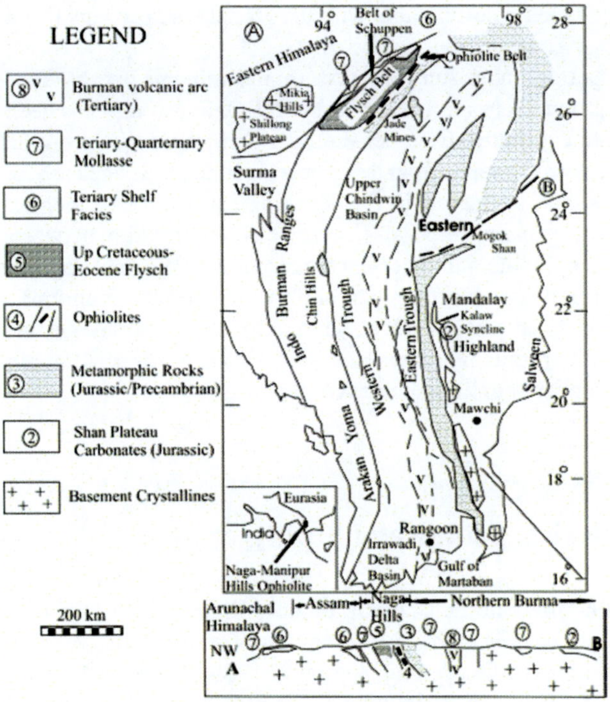

Fig. 13.1 Simplified structural layout of Indo-Myanmar Ranges (After Mitchell and McKerrow 1975; Agrawal and Ghose 1986). Geological section along A–B shows tectonic units across India–Myanmar

onsists of a suite of mafic and ultramafic rocks and pelagic sediments trending NNE–SSW direction following the regional tectonic framework. Occurrence of high-P/low-T assemblages of blueschists (glaucophane schist and metachert containing glaucophane) and eclogite (Ghose and Singh 1980; Ghose et al. 1984), endorses a tectonic setting of collision zone that developed between the Indian plate and the Sino-Myanmar continental plate.

Almost a complete litho-assemblage of ophiolite as typically observed in ideal section (e.g. Semail Mountains in Oman) is encountered in the Naga Hills (Anonymous 1978; Agrawal and Kacker 1980; Chattopadhyay et al. 1983; Agrawal and Ghose 1986; Ghose et al. 1986; Mitra et al. 1986). The litho-assemblages are highly dismembered and emplaced as imbricated sheets like any other Tethyan counterparts. A complementary island arc with bi-modal volcanism has developed along the volcanic line of Central Myanmar Lowland (Mitchell and Mckerrow 1975). The dyke swarms that characterize many ophiolites is absent in the NHO, possibly due to high level crystallization of gabbro that comprises the layered sequence immediately underlying the oceanic crust (Layer 3A,B). The rocks have undergone polyphase tectonism, metamorphism, magmatism and metallogenic events that are broadly correlatable with the Himalayan orogeny during the closure of the southeast Tethyan Sea.

Based on petrographic study on texture and mineral paragenesis of magmatic and metamorphic rocks associated with the NHO, an attempt has been made in this paper to evaluate: (i) difference between the peridotites that are subjected to high degree of tectonism (Layer 4) from those developed in the layered sequence (Layer 3), (ii) transformation of minerals in the development of layered magmatic sequence, (iii) sequence of volcanic eruptions (Layer 2), (iv) correlation between basalts associated with the ophiolite and those encountered in the present day ocean, (v) mode of fragmentation of pyroclastic rocks in phreatomagmatic eruptions, (vi) mineralogical changes of mafic rocks and oceanic sediments in burial metamorphism, (vii) development of chromite in ultramafics, and (viii) mechanism of ophiolite emplacement based on microstructural evidence.

13.2 Geological Setting and Stratigraphy

The Naga hills orogenic belt can be broadly divided into four distinct tectonostratigraphic units (Fig. 13.1) (Ghose and Singh 1981; Bhattacharyya and Venkataramana 1986; Ghose and Agrawal, 1989, Table 13.1, Fig. 13.2). These are: (i) Belt of Schuppen: a sedimentary nappe of Eocene-Upper Miocene age separated by Disang thrust, along which the Naga Hills have relatively moved towards northwest (Evans 1964; Dasgupta 1977), (ii) Disang: a thick pile of clastic flyschoid sediments dominantly argillaceous of Upper Cretaceous-Eocene age derived from the Proterozoic hinterland of the Indian shield and Myanmar (Chattopadhyay et al. 1983), (iii) Ophiolite: a suite of magmatic, metamorphic and marine sediments of Maestrichtian–Palaeocene age based on macro- and micro-fossil evidences (Bender 1983; Acharyya et al. 1986; Chungkham and Jafar 1998) and (iv) Nimi/Naga Metamorphics: a low-grade pelitic-psammitic and carbonate sequence of Mesozoic/Proterozoic (?) age, occupying the eastern fringes of India and Myanmar. The Nimi Formation forms a sedimentary cover sequence of accretionary prism (Agrawal 1985; Agrawal and Ghose 1986), overlying the Naga Metamorphics possibly representing the basement (Acharyya et al. 1986), and (v) a younger sedimentary sequence of ophiolite derived rocks unconformably overlying the ophiolite as a carapace (Jopi/Phokphur Formation) (Table 13.1). Presence of plant fossils and carbonaceous matter suggest a humid climate and paralic depositional environment for cover sediments. It is least deformed and attributed to be Eocene (Acharyya et al. 1986) or Oligocene (Agrawal and Ghose 1986). A generalized stratigraphic sequence of the ophiolite belt is given in Table 13.1.

The ophiolite has a tectonic contact on its east overridden by the metamorphic complex (Fig. 13.2). The western boundary of the ophiolite also shows a thrust contact with the Disang flysch. The ophiolite is emplaced both onto and into the

Table 13.1 Simplified stratigraphy of the ophiolite belt of Naga Hills, Northeast India (After Agrawal 1985; Agrawal and Ghose 1986; Acharyya et al. 1986)

Age	Formation	Lithostratigraphy
Late Eocene–Oligocene	Jopi/Phokphur Formation	Tuffaceous shale, sandstone, greywacke, grit and conglomerate. Minor limestone and carbonaceous matter.
Unconformity		
Maestrichtian–Paleocene	Ophiolite	Marine sediments (slate, phyllite, greywacke, iron-rich sediments, chert and limestone containing radiolaria), Plagiogranite, Volcanics (basalt with quenched texture, trachy basalt, plagioclase phyric basalt and spilite), Volcaniclastics, Metabasics (prehnite–clinochlore schist, greenschist, glaucophane schist and eclogite), Minor dolerite, Cumulate magmatic rocks (peridotite, pyroxenite, gabbroids and anorthosite), and Peridotite tectonite/Serpentinite with Garnet lherzolite mantle xenolith. Deposits of podiform chromite and nickeliferous magnetite, and minor mineralization of Cu–Mo sulphides associated with late felsic intrusives.
Thrust		
Upper Cretaceous–Upper Eocene	Disang Flysch	Dominantly shale, siltstone, passing into graphite slate, phyllite, schist and sandstone (sub-greywacke to greywacke), calcareous shale and minor limestone. In Manipur it is olistostromal in facies.
Thrust		
Mesozoic(?)	Nimi Formation/Naga Metamorphics	Feebly metamorphosed limestone, phyllite, quartzite and quartz-sericite schist in the upper part (called Nimi Formation), overlying a sequence of mica schist, granitoid gneiss, feldspathic metagreywacke, etc. called Naga Metamorphics. Tectonic slices of ophiolite are also present of variable dimensions.

Disang sediments (Ghose and Agrawal 1989), and are intermixed and interbeded (Bhattacharjee 1991). The western contact of the NHO is marked by the presence of high-pressure rocks, viz., glaucophane schist and eclogite, indicating their involvement in a convergent plate boundary (Ghose et al. 1984, 1986). The high-pressure litho-assemblages are confined to the northwest margin of the NHO in close association with the Disang, occurring as lenses in the mafic volcanics and rarely with ultramafics. Another feature of petrological significance is the occurrence of garnet lherzolite xenolith in association with peridotite tectonite possibly derived within the mantle (Agrawal and Ghose 1986; Vidyadharan et al. 1986).

The major structural fabric noted includes three types of planar structures with concomitant lineations. Three phases of fold movements have been recognized, out of which the first fold is flexturally folded that is developed in all scales, and forms the regional fold (Ghose and Singh 1981; Ghose and Agrawal 1989).

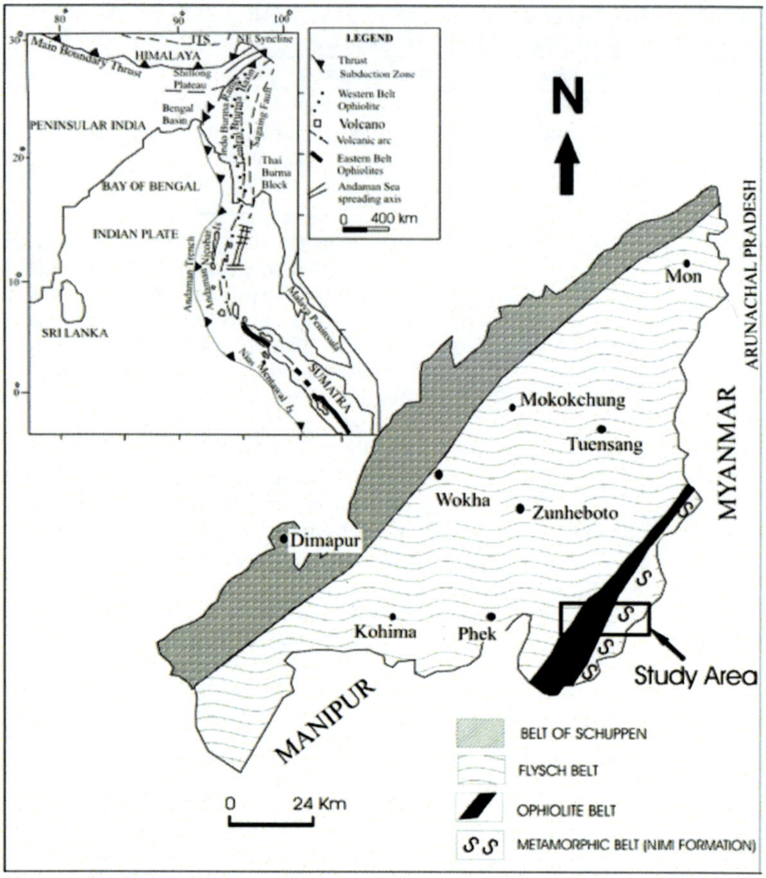

Fig. 13.2 Simplified geological map and tectonic divisions of Nagaland (After Ghose and Agrawal 1989). Inset showing the two parallel ophiolite belts– **western** (*dots*) and **eastern** (*bold dash*), in the Indo-Myanmar Range (After Acharyya 2007). ITS– Indus–Tsangpo suture

Although F_2 and F_3 folds are noted mostly on mesoscopic scale, however, their effects are observed on the regional first folds. Two phases of metamorphism are broadly coeval with the first and second fold movements, and correlatable with the Himalayan orogeny. The third fold movement may be related to the Miocene–Quaternary tectonic episode.

Numerous occurrences of metallic and non-metallic mineralization have been reported from the NHO. However, economically viable deposits are noted by the presence of podiform chromite in association with peridotite tectonite in Manipur, and stratabound-nickeliferous magnetite between ultramafic cumulate and younger cover sediments (Jopi/Phokphur Formation) in the state of Nagaland (Ghose and Shrivastava 1986; Agrawal and Ghose 1989; Venkataramana et al. 1986; Singh 2008).

13.3 Petrography

13.3.1 Meta-ultramafics or Peridotite Tectonite

The peridotites occur as highly deformed or sheared bodies, varying in mineralogical composition from dunite, harzburgite, lherzolite to wherlite (minor). These are variably altered to serpentinite. The olivine shows preferred orientation and in a matrix of serpentine, develop strong foliation (Fig. 13.3a). The porphyroclastic peridotites usually have two modes of olivine, the larger one (porphyroclast >2 mm) is surrounded by smaller grains of olivine (neoblasts <1 mm) and chrome spinel (Fig. 13.3c, d). The olivine porphyroclasts are elongated, tapered or curved (Fig. 13.3d). Olivine is also found as finely granulated grains in serpentine matrix sandwiched between elongated crystals of olivine (Fig. 13.3e). Presence of strain shadow (Fig. 13.3b) and undulose extinction in olivine, and kink bands of cleavage traces and bent twin lamellaes in pyroxenes (Agrawal and Ghose 1986) are common in meta-ultramafics. These meta-ultramafics constitute possibly a part of the upper mantle.

The tectonites are essentially spinel-bearing, and free from any plagioclase, suggesting derivation from deeper source in the mantle beyond the stability limit of plagioclase. The spinels are usually octahedral in shape (Fig. 13.3f), sometime lensoid or equidimensional. Abundance of spinel is greater in tectonites than those associated with the peridotites formed in layered sequence. Alumina-rich chromite occurring as pods or lenses within the tectonites substantiates deep mantle origin (Ghose and Shrivastava 1986). Based on single clinopyroxene and olivine-spinel geobarometric studies of peridotite from Ukhrul in Manipur, equilibrium pressure between 19–25 kbar has been estimated (Maibam and Foley 2008), corresponding to 60–75 km depth.

13.3.2 Garnet Lherzolite

The metaultramafics in the northern part of the NHO are associated with lenses of garnet lherzolite xenolith – a mantle protolith. The garnet is colourless, almost free from cracks and alteration (Fig. 13.4a), and contains inclusions of olivine, orthopyroxene (enstatite) (Fig. 13.4a, c) and spinel. Olivine is granular and highly fractured. Large prisms of orthopyroxene are marked by streaks of opaque aligned parallel to the cleavage traces and in turn are traversed by diagonal fractures (Fig. 13.4d). Chrome spinels occur as octahedral, subhedral or skeletal form and contain inclusion of olivine.

Presence of a micro-shear plane and step-faulting across it (Fig. 13.4b), supports emplacement of garnet lherzolite in ductile environment. Olivines are brecciated and pulverized along the shear plane (Fig. 13.4b, d), and occur at an angle with the host olivine. A thin layer of chromite is dragged along the fault in the shear plane

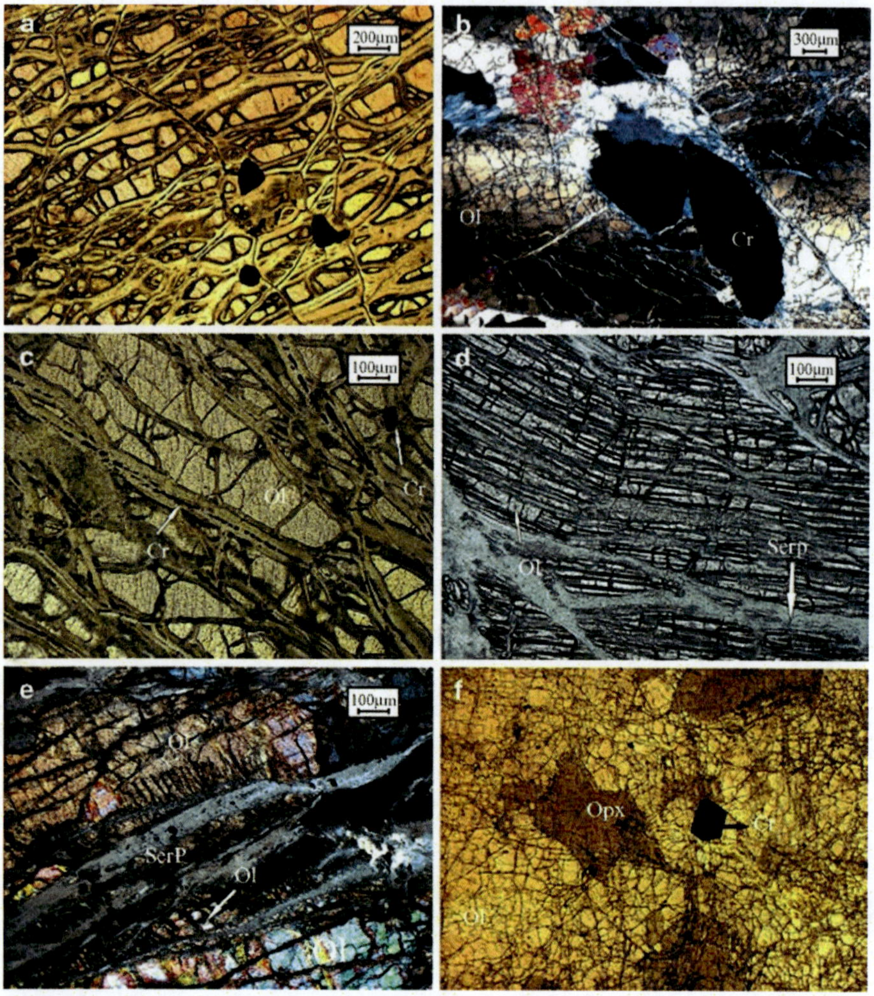

Fig. 13.3 Peridotite Tectonite: (**a**) Olivine entrapped in chords of serpentine in dunite showing preferred orientation. (**b**) Olivine showing strain shadow (across ol) in dunite and traversed by fine strings of serpentine. (**c**) Olivine porphyroclasts in dunite. Chords of serpentine containing trail of fine granular chrome spinels. (**d**) Porphyroclastic sheared dunite with bent olivine crystals intricately veined by serpentine. (**e**) Fine granular and crushed olivine grains trapped in serpentine sandwiched between stretched olivines. (**f**) Harzburgite containing olivine, large plates of orthopyroxene and octahedral chrome spinel. Abbreviations: ol – olivine, cr – chromite, serp – serpentine, opx – orthopyroxene

(Fig. 13.4b). Post-crystalline carbonate veins cutting across both olivine and shear plane indicating efficacy of volatiles in the late stage geological events.

Where olivine is altered to serpentine, corona around garnet forms symplectic intergrowth with chlorite. Serpentinization of olivine cause release of iron both

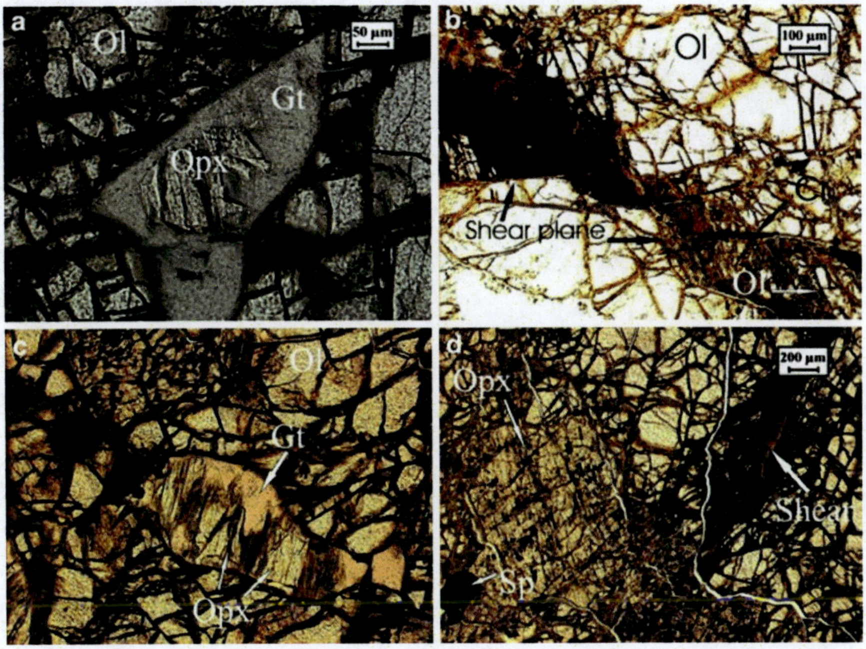

Fig. 13.4 Garnet Lherzolite Xenolith: (**a**) Colourless unaltered garnet (gt) containing inclusions of orthopyroxene (opx) surrounded by olivine in garnet lherzolite. (**b**) Step-faulting across a micro-shear plane (*dark shade*) in garnet lherzolite. The shear zone contains crushed and granular olivines giving evidence of ductile deformation. A vein of chromitite is dragged along the shear plane. (**c**) Colourless lenticular garnet contains inclusions of orthopyroxene (enstatite) in garnet lherzolite. (**d**) Highly fractured olivine, prismatic orthopyroxene and spinel in garnet lherzolite. A micro-shear plane (on *right, darker shade*) is filled with pulverised olivine. The orthopyroxene shows schiller structure due to the presence of magnetite along cleavage traces. Late carbonate veins (*thin white*) cutting across olivine, orthopyroxene and the micro-shear. Abbreviations: ol – olivine, grt – garnet, opx – orthopyroxene, sp – spinel, cr – chromite

within and at the outer margin of serpentine. Slender crystals of orthopyroxene are noted along the striations of large plate of clinopyroxene as exsolution lamellae.

13.3.3 Serpentinite

Three principal types of serpentinite have been distinguished viz., massive, sheared and fibrous. Besides the two principal constituents, viz., olivine and pyroxene, contributing largely to the formation of serpentine due to hydration (Ghose and Agrawal 1989), other minor components like chromite, magnetite, talc, tremolite, picotite, garnet and brucite are also present. Brucite is present in low temperature hydrothermal veins. Folding in serpentinite shows concentration of spinel along the

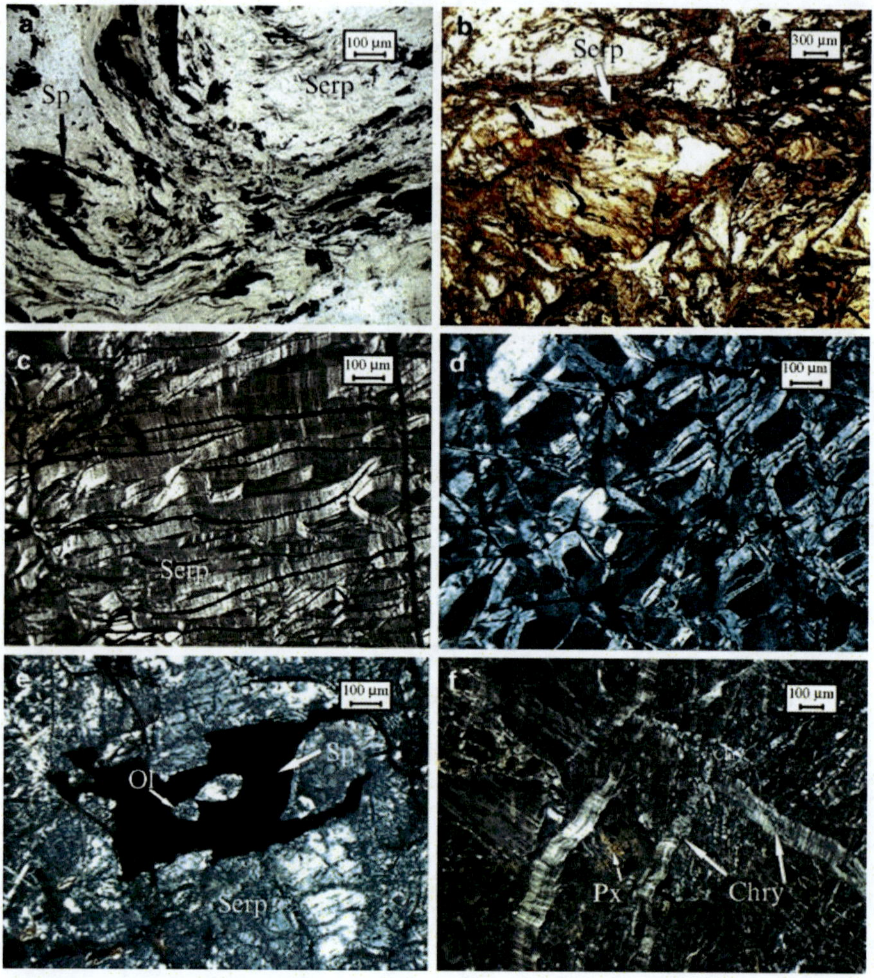

Fig. 13.5 Serpentinite: (**a**) Folding in serpentinite showing concentration of spinel along fold axis and crenulation cleavage. (**b**) Mylonitic tectonic breccia in serpentinite. (**c**) Ribbon texture in serpentinite. (**d**) Reticulate texture in serpentinite. (**e**) Serpentinite showing mat texture. Skeletal magnetite containing inclusions of serpentinised olivine. (**f**) Veins of chrysotile in serpentinite containing relict pyroxene. Abbreviations: sp – spinel, serp – serpentine, ol – olivine, px – pyroxene, chry – chrysotile

fold axis and crenulation cleavages (Fig. 13.5a). Highly sheared and foliated variety contains talc and tremolite, and in extreme case develops mylonitic tectonic breccia (Fig. 13.5b). Garnets in sheared serpentinite, associated with chrysotile, chlorite and relict pyroxene, occur as small lobate crystals. This may represent either a recrystallised residual mineral that formed at higher temperature, or a garnet-bearing lherzolite involved in low-grade metamorphism and hydrothermal alteration.

Mesh (honeycomb) texture is common, consisting of primary, secondary and tertiary cords of serpentine enclosing isolated grains of olivine and pyroxene (Fig. 13.3a, c). Core of cords at times entrap trail of fine granules of chrome spinel (Fig. 13.3c). Sometime serpentine exhibits ribbon texture composed of parallel and undulating bands (Fig. 13.5c) and reticulate texture with cross fibres (Fig. 13.5d). The bladed-mat texture contains inter-felted mass of fine-to-medium size blades of serpentine (Fig. 13.5e), cemented in a fine grained mass of serpentine and spinel. Olivines in such serpentinite are entrapped in skeletal magnetite grain (Fig. 13.5e). X-ray diffraction studies of serpentinite confirm that these are mostly composed of antigorite with minor lizardite and chrysotile (Fig. 13.5f) (Ghose and Agrawal 1989).

13.3.4 *Layered Magmatic Rocks*

13.3.4.1 Ultramafics

The layered magmatic sequence in the Naga hills ophiolite may be classified into peridotite (dunite, harzburgite, lherzolite and minor wherlite), pyroxenite (olivine websterite, websterite, olivine clinopyroxenite and clinopyroxenite) and gabbroid (olivine gabbronorite, olivine gabbro, norite, gabbro, leucogabbro, hornblende gabbro and anorthosite). Late intrusion of plagiogranite cuts across the entire magmatic sequence. All these lithological units are disposed as tectonic slices. However, based on mutual field relationship and mineral paragenesis, a magmatic sequence may be established beginning with peridotite cumulate at the base succeeded by pyroxenite, gabbro and anorthosite (Agrawal and Ghose 1986), similar to those encountered in the Precambrian layered mafic intrusions. The peridotite of the cumulate sequence forms host to the stratabound nickeliferous magnetite deposit overlain by ophiolite derived younger cover sediments (Jopi/Phokphur Formation) in Nagaland.

The peridotite cumulate shows protogranular texture. The crystals are coarse-to-fine granular, tabular and smoothly curved (Fig. 13.6a, b). Absence of any strain effect, or foliation and lineation characterize these peridotites (Fig. 13.6a, b, e). Spinels are subhedral to anhedral, skeletal and less frequently octahedral. Cracks in olivine and grain boundaries are commonly filled with spinel, however, their distribution is uneven. Grain boundaries between olivine and larger grains of pyroxenes and spinels are often filled with serpentine as a result of later alteration (Fig. 13.6a). Orthopyroxene contains inclusions of olivine and spinel. Poikilitic clinopyroxene contains inclusions of olivine and orthopyroxene in lherzolite (Fig. 13.6c). Polysynthetic twinning in clinopyroxene (Fig. 13.6f) is observed in pyroxenite.

Olivine of harzburgite contains amorphous/non-crystalline (?) greyish core (Fig. 13.6b). Enlarged view of the core reveals fine streaks of black opaque mineral traversing diagonally across it within the crystal outline of host olivine. A thin yellowish-white layer consisting of fibrous chrysotile (?) separates the core from the dark iron-rich border of olivine (Fig. 13.6b), which alters to reddish brown hematite.

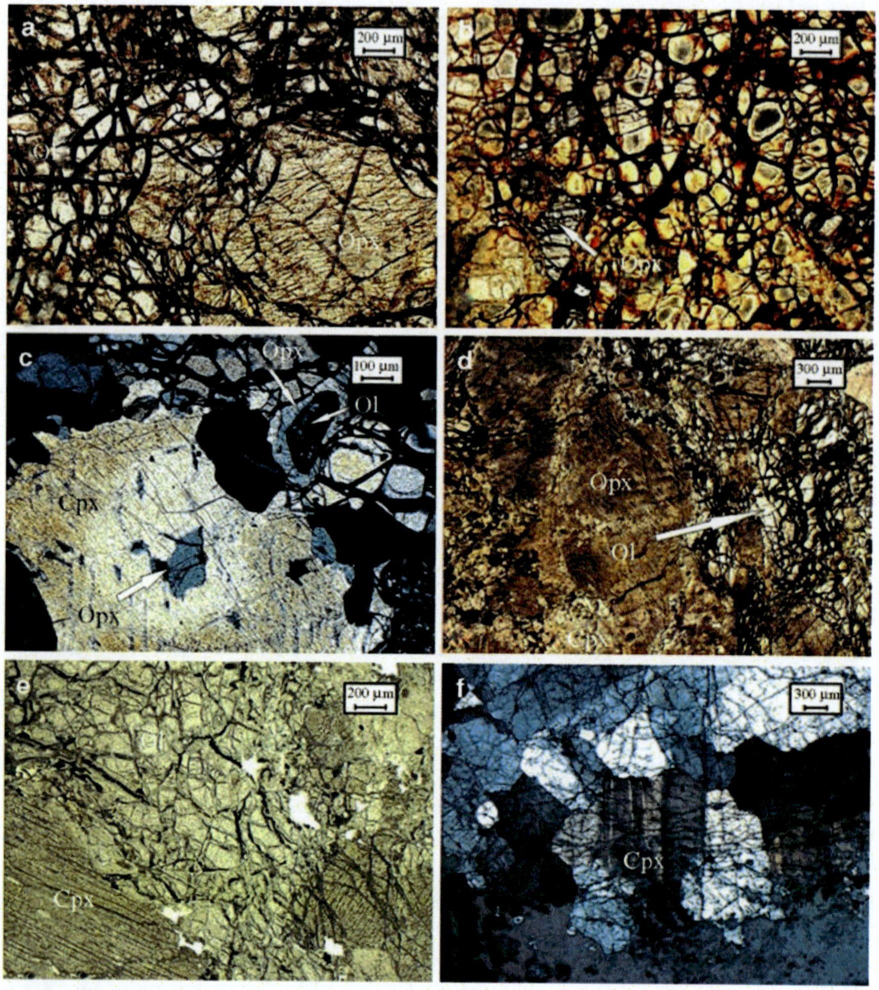

Fig. 13.6 Ultramafic Cumulates: **(a)** Coarse protogranular harzburgite containing orthopyroxene with inclusions of olivine. **(b)** Two modes of olivine of variable dimension, a fresh looking and the other containing dark glassy core (dominant) traversed by fine streaks of iron oxide (on closer view) in harzburgite. **(c)** Poikilitic clinopyroxene in lherzolite containing inclusions of orthopyroxene and olivine. Exsolution lamellaes of orthopyroxene cutting across the cleavage traces of clinopyroxene. **(d)** Long tapered prismatic crystals of orthopyroxene and granular olivine occupying the interspaces in olivine websterite. Post-crystalline fractures in olivine are filled by opaque minerals. **(e)** Coarse protogranular wherlite containing large plates of clinopyroxene and olivine. **(f)** Polysynthetic twinning of clinopyroxene in clinopyroxenite. Abbreviations: ol – olivine, opx – orthopyroxene, cpx – clinopyroxene

Uniform presence of such feature in olivine grains throughout the rock points to its formation possibly under extreme thermal conditions. Wherlite is essentially composed of olivine and large prismatic clinopyroxene (Fig. 13.6e). Secondary products of serpentine and iron oxides are developed along grain boundaries.

13.3.4.2 Gabbroids

The clinopyroxene in olivine gabbro contains inclusions of both olivine and orthopyroxene (Fig. 13.7a). Exsolution lamellaes of orthopyroxene along striations are common in such clinopyroxene (Fig. 13.7a, b). Large plate of clinopyroxene is surrounded by orthopyroxene and plagioclase apparently rendering a porphyritic texture to olivine gabbro (Fig. 13.7d). Recrystallization of olivine to chlorite (Fig. 13.7a) is connected by a feeder channel originating outside the host pyroxene. The clinopyroxene shows alteration to green amphibole both within and outer margin of the grain boundary (Fig. 13.7d). The clinopyroxene of gabbronorite occurs as inclusion in plagioclase (top right), favouring crystallization of the former to have preceded the latter (Fig. 13.7c). However, interfingering relationship between the plagioclase and clinopyroxene, and inclusion of the former in the latter (lower half), suggest that crystallization of plagioclase continued beyond clinopyroxene (Fig. 13.7c). The anorthosite is composed of large plates of plagioclase showing profuse growth of chlorite and epidote due to late recrystallization.

13.3.5 Plagiogranite

The plagiogranites varies in mineralogical composition from diorite, tonalite and trondhjemite (Agrawal and Ghose 1986). Intrusive relationship of plagiogranite with basalt/spilite lends support of being youngest amongst magmatic rock. These are coarse grained and show hypidiomorphic to allotriomorphic granular texture, but tend to be porphyritic, panidiomorphic granular, or gneissic in parts under the microscope in the same section. Laths of plagioclase are surrounded by quartz, brown chlorite, muscovite rimmed by biotite, epidote and opaque. Long anhedral and small granular quartz showing sutured border dominate over plagioclase. Potash feldspar is absent. Plagioclase grains are feebly altered to sericite or a mixture of quartz, calcite and epidote. These are frequently zoned and show simple or polysynthetic twinning. Greenish brown amphiboles occur in clusters (Fig. 13.7e). Breakdown of brown amphibole to green chlorite, opaque and epidote suggests recrystallization. Carbonate veins cut across both the plagioclase and amphibole as late hydrothermal product.

13.3.6 Dolerite Dykes

The dolerite dykes are rare and occur as intrusives in ultramafic cumulates. These are medium- to fine-grained showing sub-ophitic texture, and consist of labradorite, augite, magnetite and secondary chlorite. Fractures and vesicles are filled by carbonate (Fig. 13.7f). The groundmass is composed of reddish brown glass, microlaths of plagioclase and pyroxene. Prismatic or sub-rounded grains of clinopyroxenes

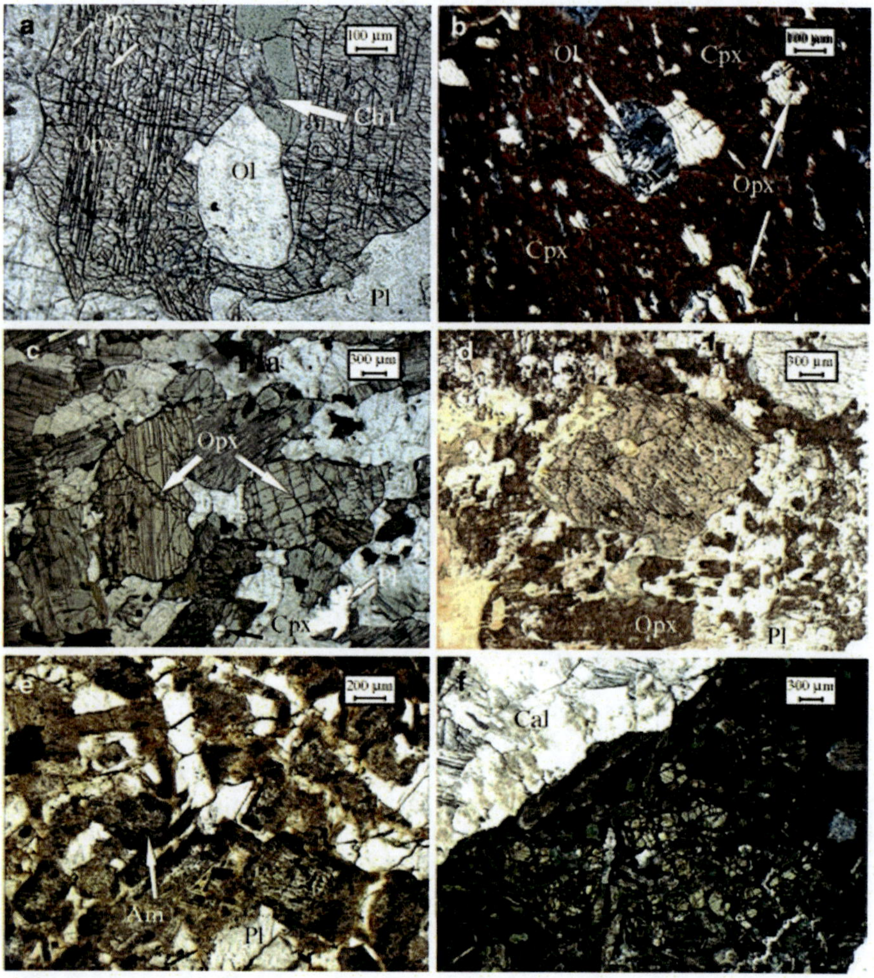

Fig. 13.7 Gabbroids, Plagiogranite and Dolerite: (**a**) Poikilitic clinopyroxene in olivine gabbro containing inclusions of olivine (*white*) and chlorite (*grey*). Exsolution lamellaes of orthopyroxene are seen as small oblong white blebs along striations. (**b**) Large plate of clinopyroxene (*dark, under crossed Nicols*) containing exsolution lamellae of orthopyroxene (*white*) and inclusion of olivine (*grey*) in olivine gabbro. (**c**) Gabbronorite containing pleochroic green orthopyroxene, clinopyroxene, plagioclase and spinel. (**d**) Large plate of clinopyroxene with exsolution lamellaes of orthopyroxene is surrounded by short prisms of orthopyroxene (*brown*) and plagioclase giving porphyritic appearance to olivine gabbro. (**e**) Panidiomorphic granular texture in plagiogranite (tonalite) containing zoned plagioclase, quartz and cluster of chloritised ampliboles. (**f**) Olivine dolerite showing ophitic texture in contact with a vein of calcite. Abbreviations: ol – olivine, pl – plagioclase, opx – orthopyroxene, cpx – clinopyroxene, cal – calcite, am – amphibole, chl – chlorite

show low-grade metamorphism to fine needles of radiating tremolite and iron oxides. Plagioclases are short rectangular, untwinned and altered. Small granular opaques are scattered throughout the rock.

13.3.7 Volcanics

13.3.7.1 Basalt

The basalt constitutes one of the dominant components of ophiolite next to peridotite/serpentinite in order of abundance. Presence of minor amounts of andesite, spilite, trachyte, keratophyre, hyaloclastite and volcaniclastics is also reported (Agrawal and Ghose 1986). In the present study some new textural variants of basalt have been identified along with pyroclastic rocks for the first time from the central part of NHO. The volcanics are widespread, those exposed along the boundary thrusts of ophiolite with the Disang in west and the Nimi Formation in east, form a tectonic mixture, and are highly brecciated, fractured, silicified and schistose.

The volcanics are interlayered, intercalated and intermixed with oceanic sediments, viz., chert and limestone. Multiple flows differing in degree of crystallinity, colour, glass content and vesicularity are observed. A thickness of over 800 m has been inferred for the volcanics in the central part of ophiolite (Agrawal and Ghose 1986). Four distinctive types of mafic flows have been identified, viz., (a) basalts with quenched texture (Fig. 13.8) characterized by intergranular (Fig. 13.8c) and (b) vitrophyric textures (Fig. 13.9c, d), (c) trachy basalt (Fig. 13.9a, b), and (d) plagioclase phyric basalt (top left, Fig. 13.9a) containing megacryst of plagioclase (Fig. 13.9e, f). Basalts are free from olivine or orthopyroxene suggesting pronounced fractionation prior to eruption. Contact between the successive flows is marked by chilled margin, thin line of sinuous siliceous material (Fig. 13.9a), or a layer of pyroclastics/hyaloclastite (top left, Fig. 13.8d).

The quenched texture in basalt is characterized by elongated, parallel or radiating crystals of hollow-core plagioclase and granular pyroxene (Fig. 13.8b, c). The hollow-core of the plagioclase is filled by granular brown pyroxene and green glass (Fig. 13.8c). Pyroxenes are also trapped in between laths of plagioclase producing intergranular texture. Some plagioclases of such flow show dog-tooth structure (Fig. 13.8d) and are curved (Fig. 13.8d, e). Late carbonate vein contains euhedral crystal of dark brown amphibole (kaersutite?) rimmed by thin chlorite with a core of clinopyroxene (Fig. 13.8f). Presence of plagiogranite (tonalite) vein showing intrusive relationship with quenched basalt supports partial fusion of oceanic crust at high temperature.

Another quenched texture in basalt is noted with the development of vitrophyric texture in which phenocrysts of pyroxene and hollow-core plagioclase are set in a

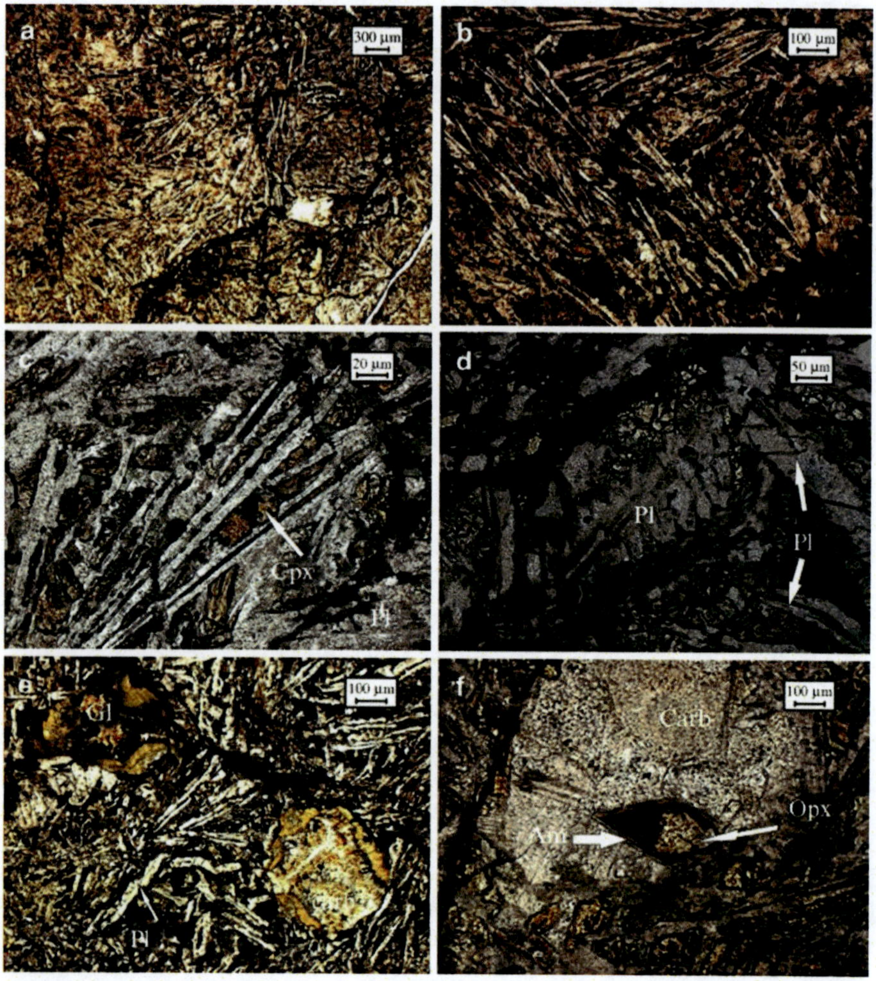

Fig. 13.8 Basalt: (**a**) Spilite showing variolitic texture. (**b**) Quenched texture in basalt, containing long, parallel and radiate plagioclase and granular pyroxene (*grey*). (**c**) Enlarged view of quenched texture showing hollow core plagioclase filled by granular pyroxene and green glass. (**d**) Basalt with quenched texture containing curved and dog-tooth plagioclase in contact with hyaloclastite (*top left corner*). (**e**) Basalt with quenched texture containing hollow core and curved plagioclase and pyroxene in association with clasts of glass and carbonate developed in vesicular cavity. (**f**) Euhedral brown amphibole (am, kaersutite?) with a core of pyroxene (cpx) in carbonate clast of quenched basalt by late hydrothermal activity. Abbreviation: pl – plagioclase, cpx – clinopyroxene, gl – glass, carb – carbonate, am – amphibole

groundmass essentially composed of glass, with minor amount of spinel and sphene (Fig. 13.9c, d). Resorption of clinopyroxene and their radial alignment (Fig. 13.9d), together with hollow-core plagioclase give evidence of a turbulent glass-rich magma, pointing to rapid cooling at very high temperature. The rock shows ocellar

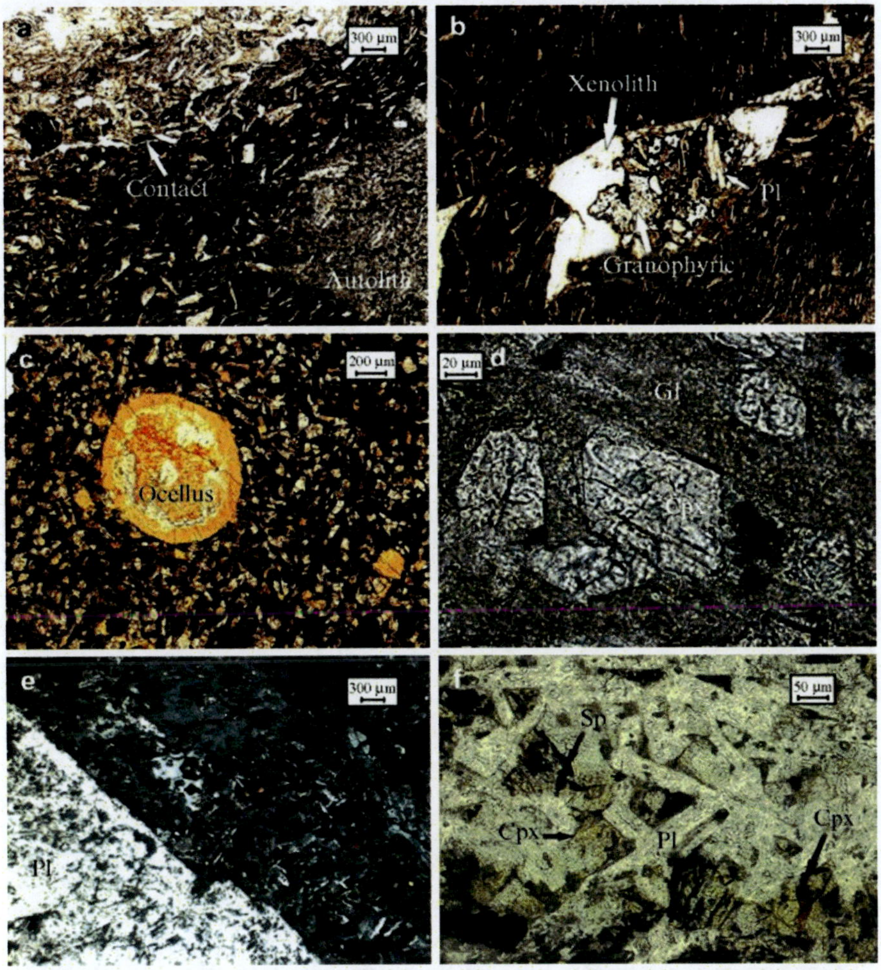

Fig. 13.9 Basalt: (**a**) Trachy basalt (basal) is separated from plagioclase phyric basalt (*top*) by a thin line of sinuous silicious material. The flow directions of the two lavas can be distinguished by the orientation of feldspars. A rectangular xenolith of the basal flow may be seen above the contact (*on left*). An autolith is seen within the basal flow. (**b**) Trachy basalt containing lensoid xenolith (*white*) involved in assimilation resulting in granophyric texture (*left*) and development of plagioclase laths (*on right*). (**c**) Basalt showing vitrophyric texture containing ocellus. The phenocrysts of clinopyroxene and less abundant hollow-core plagioclase are set in a matrix largely composed of glass with minor spinel and sphene. (**d**) Close view of vitrophyric basalt showing radial phenocrysts of pyroxene set in a groundmass of glass. (**e**) Basalt with plagioclase megacryst (*white*) is flanked by fine groundmass (*dark*) consisting of plagioclase, pyroxene and spinel. (**f**) Sub-ophitic texture in basalt containing plagioclase megacryst. Abbreviations: pl – plagioclase, gl–glass, cpx – clinopyroxene, sp–sphene

structure (Fig. 13.9c). The rim of a circular ocellus is composed of an orange coloured phase (?), while the core is a mixture of deep orange and pale yellowish white crystalline material (?). Presence of lithic fragment in the rock may represent the source material/cognate lapillus.

The flow texture of trachy basalt is exhibited by microcrystalline laths of plagioclse (Fig. 13.9a, b). This is overlain by a younger flow of plagioclase phyric basalt containing a xenolith (a rectangular clast above the line of contact on left) of older (basalt) flow at the contact (Fig. 13.9a). A closely similar rock with trachytic to pilotaxitic texture, low-Mg basalt (Mg# <50), has been reported by Venkataramana and Datta (1987). The rock consists of plagioclase (An_{30-50}), pigeonite (2V = 20°) and augite in a groundmass containing fine-grained K-feldspar (?), plagioclase and green palagonite glass, showing chemical affinity with alkali basalt containing normative nepheline.

The basalt with megacryst of plagioclase showing sub-ophitic texture is composed of very large phenocryst of plagioclase (white, Fig. 13.9e) set in a fine groundmass (dark) of plagioclase and pyroxene (Fig. 13.9f). Cubic crystals of magnetite are associated with chlorite in the groundmass. The megacryst contains inclusion of older twinned plagioclase. Pyroxene shows alteration to radiating crystals of tremolite. The vesicles are filled with chlorite and rimmed by yellowish green glass.

Basalts with typical pillow structure show a chilled margin and massive core, displaying variolitic (Fig. 13.8a, a type of quenched texture), sub-ophiotic or intersertal textures (Ghose and Singh 1980; Singh and Ghose 1981; Agrawal and Ghose 1986; Singh et al. 2008). Spilite contains plagioclase (An_{10-30}), chlorite with relict clinopyroxene and spinel. A host of secondary minerals like calcite, epidote, quartz, rhodochrosite and sphene are also present as altered products of primary minerals.

13.3.7.2 Volcaniclastics

Presence of a wide range of volcaniclastic rocks is known from the NHO, viz., ash (Fig. 13.11b), tuff (Fig. 13.10a), bomb, agglomerate, scoria, pyroclastic breccia (Fig. 13.11f) and glass (Fig. 13.10f) from the central part of the ophiolite belt (Agrawal and Ghose 1986). In addition, vitric tuff (Fig. 13.11c, d) with xenocryst of clinopyroxene entrapped in fine particle, crystal–vitric tuff (Fig. 13.10b), crystal tuff (Fig. 13.11e), lithic tuff with clast of rhyolite showing granophyric texture (Fig. 13.10c), crystal–lithic tuff (Fig. 13.11a), welded tuff (Fig. 13.10e) or ignimbrite with deformed glass shards (Fig. 13.10d), and interlayered chert and glass have been identified in this study. The crystal tuffs usually contain clasts of white plagioclase (Fig. 13.10e) or pink potash feldspar (sanidine?, Fig. 13.10b). However, the one containing lithic (basalt-central part) fragment (Fig. 13.11a) is associated with perovskite (?) as well (Fig. 13.11e). Presence of xenocryst of clinopyroxene trapped in fine particles of vitric tuff and potash feldspar in crystal vitric tuff (Fig. 13.10b) indicates these are derived from a well fractionated magma.

The pyroclasts like glass, ash and volcanic breccia are chloritised and associated with carbonate, quartz and iron oxide. The field evidences noted by the presence of scarp faces, repeated sequence of vesicular and amygdaloidal basalts, and pyroclastic deposits support upwelling of lavas to have taken place through several eruptive centres (Agrawal and Ghose 1986).

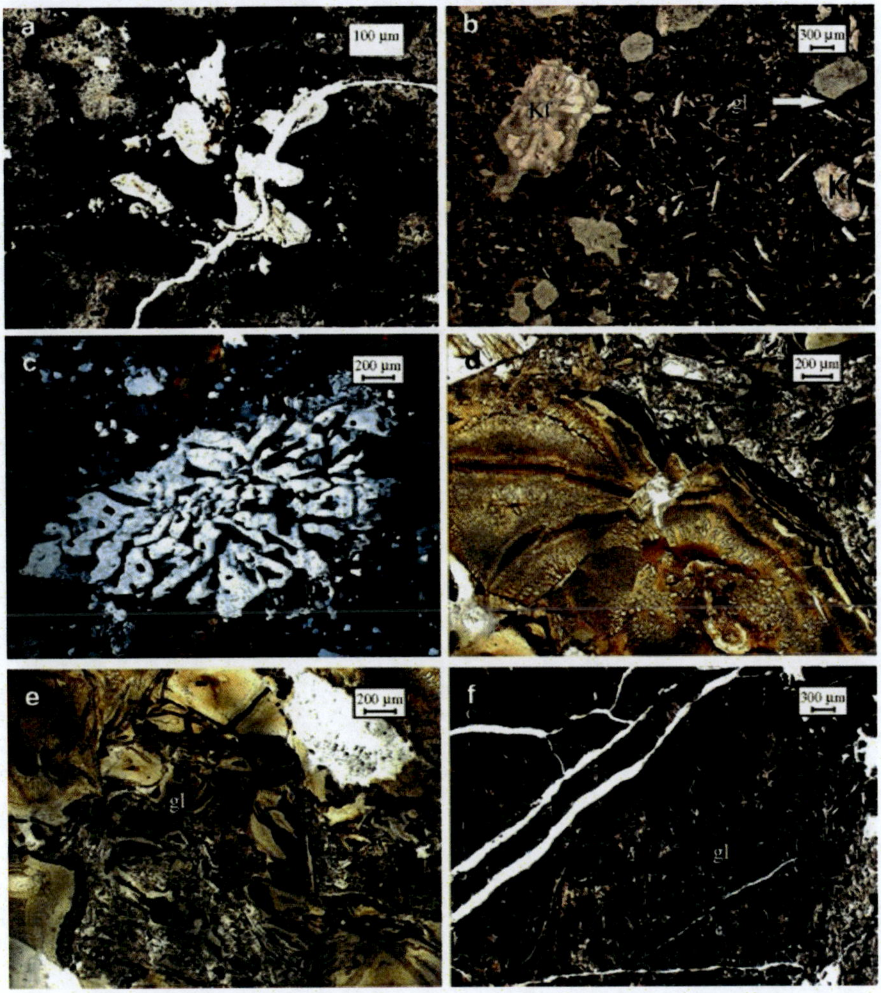

Fig. 13.10 **Volcaniclastics:** (**a**) Fine lapilli tuff showing a silica vein passing through lapillies (**b**) Crystal-vitric tuff (rhyolite tuff) containing large plates and microlites of pink potash feldspar (sanidine?). Larger phenocrysts are pseudomorphed by glass or rimmed by thin secondary silica. (**c**) Lithic tuff containing clast of rhyolite showing granophyric texture. Angular and broken clasts of variable size are cemented in a fine grained tuffaceous matrix. (**d**) Cuspate glass shard in crystal–vitric tuff showing post-consolidation faulting and folding around the rim due to late tectonic movement. *Grooved channels* and *white vesicles* represent escape routes of volatiles. (**e**) Welded crystal (*white*) tuff or ignimbrite. (**f**) Volcanic glass is traversed by quartz veins of late hydrothermal origin. Abbreviations: gl – glass, kf – potash feldspar

13.3.8 Metamorphics

The NHO is associated with low-to medium-grade metamorphic rocks composed of mafic volcanics and oceanic sediments. The low-grade mineral paragenesis

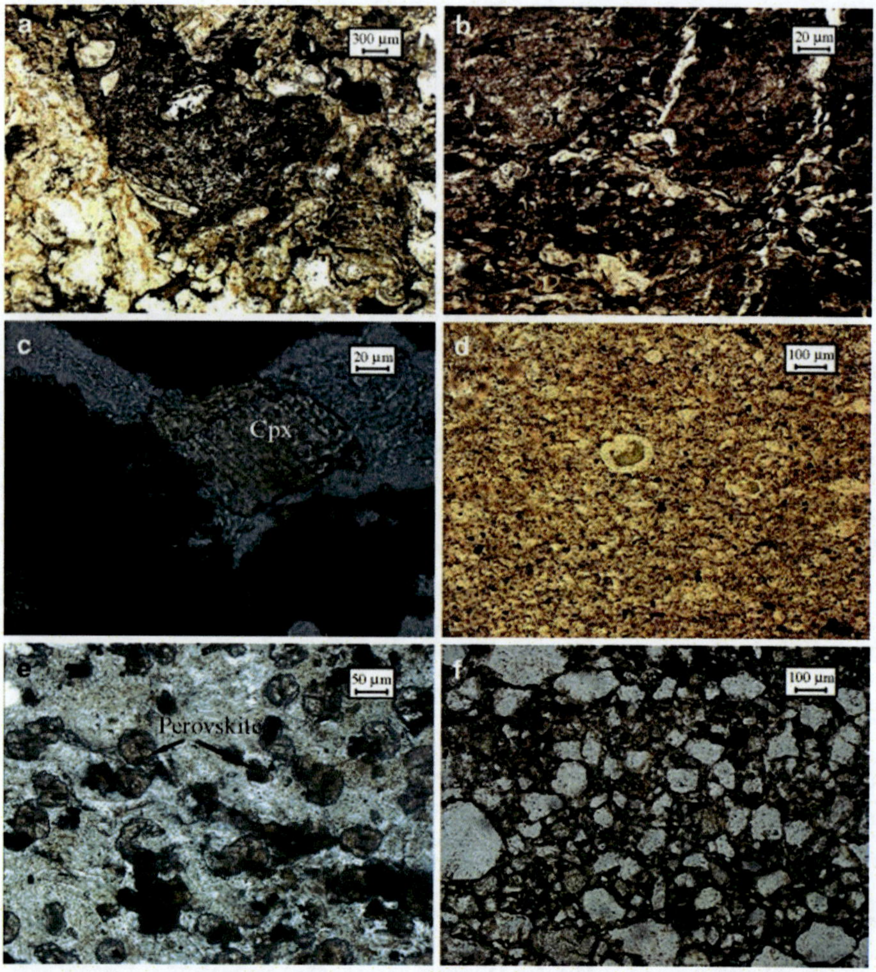

Fig. 13.11 Volcaniclastics: (**a**) Crystal (white plagioclase)-lithic tuff containing fragment of basalt at the center with fine microlites of plagioclase. (**b**) Fine ash tuff affected by late tectonism. (**c**) Vitric tuff containing xenocryst of clinopyroxene embedded in fine tuff (particles). (**d**) Vitric tuff showing devitrified glass filled with secondary mineral. (**e**) Sub-rounded crystals of brown perovskite (cauliflower structure, high relief and birefringent) are set in a recrystallised tuffaceous matrix of crystal tuff. (**f**) Angular and broken clasts and rock fragments of variable dimension are set in a fine grained matrix in volcanic breccia. Abbreviation: cpx – clinopyroxene

is represented by prehnite-clinochlore schist (Fig. 13.12a) and greenschists (albite-epidote-actinolite schist, Fig. 13.12b). The mafic volcanics are involved in subduction related crustal deformation and metamorphism producing high-pressure mineral paragenesis leading to formation of blueschist (glaucophane schist, Fig. 13.12c, d, f) and eclogite of group-C (Coleman 1977) (Fig. 13.12e). Schematic

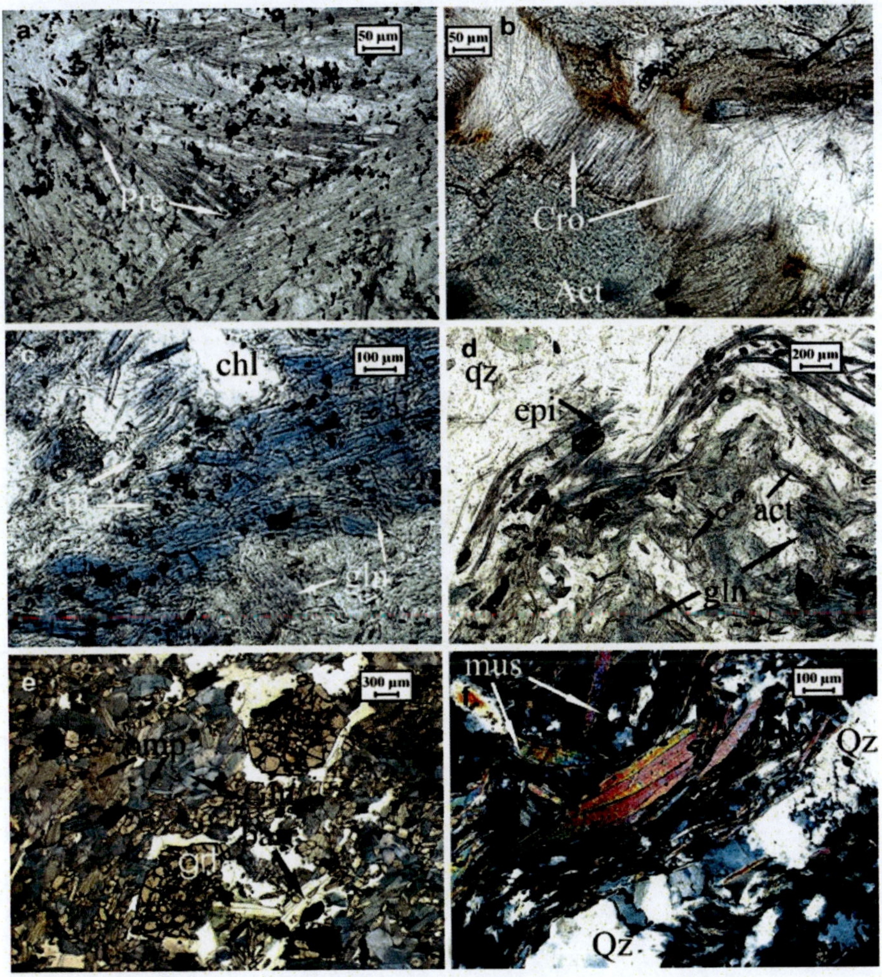

Fig. 13.12 Metamorphics: (**a**) Prehnite-(bladed, sheaf structure) clinochlore (*tiny dark mineral*) schist containing minor amounts of clinozoisite and opaque. (**b**) Coarse grained actinolitite (with minor amounts of quartz, albite, epidote and opaque) showing development of fibrous crocidolite (*blue*) between green actinolites. (**c**) Metachert containing albite, actinolite, epidote, carbonate, chlorite, opaque, glaucophane and quartz. A glaucophane–rich band with epidote (*darker shade*) occupying diagonally across the central part. (**d**) Warping in glaucophane schist containing actinolite, epidote and quartz. (**e**) Eclogite containing barroisite and glaucophane besides garnet and omphacite, quartz, albite, epidote, actinolite and rutile. (**f**) 'Mica fish' structure in glaucophane schist developing S-C mylonite. Abbreviations: pre – prehnite, act – actinolite, cro – crocidolite, gln – glaucophane, qz – quartz, epi – epidote, grt – garnet, omp – omphacite, chl – chlorite, mus – white mica (phengite)

sections of blueschist occurrences in the central part of NHO are given by Agrawal and Ghose (1986). The oceanic sediments that are subjected to low-grade metamorphism are converted into slate, phyllite, meta-greywacke, quartzite, crystalline limestone and metachert. The last one often contains radiolaria (Fig. 13.13).

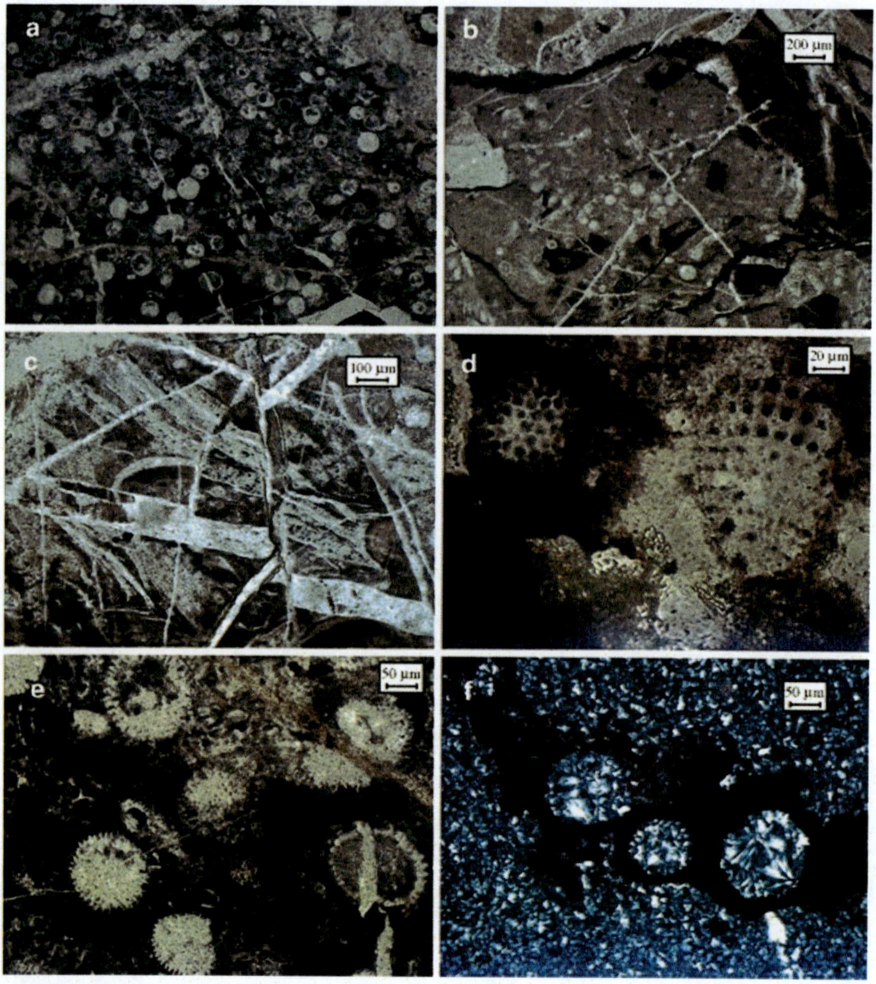

Fig. 13.13 Chert: (**a**) Red chert with radiolaria is traversed by two sets of silica veins, the longitudinal one showing displacement. (**b**) Brecciated chert containing radiolaria. Fragments of different composition and orientation could be seen from the variation of net working of silica veins. (**c**) Brecciated chert involved in step-faulting and dragging. Fault planes are traversed by recrystallised silica veins. (**d**) A colony of radiolaria in red chert. (**e**) Close view of radiolaria in red chert. (**f**) A view of radiolaria in red chert under crossed Nicols showing sectoral twinning

Phase transition of mafic volcanics at low temperature burial metamorphism has given rise to prehnite-clinochlore schist (Fig. 13.12a). Hydrothermal experiments of this critical transformation is established at temperatures below about 280°C in pressure range from 5 to 8 kb (Heinrichsen and Schurmann 1972) represented by following reaction:

$$7 \text{ anorthite} + 5 \text{ diopside} + 10H_2O \rightarrow 6 \text{ prehnite} + 1 \text{ clinochlore} + 3 \text{ quartz}$$

Development of blueschist and eclogite is controlled by deep burial. High-pressure mineral assemblages, e.g., glaucophane schist and eclogite have almost identical mineral association with ubiquitous presence of glaucophane and barroisite in both. The eclogite is characterized by presence of garnet and omphacite (Ghose et al. 1987). Omphacite is also present in glaucophane schist, and never in association with garnet (Agrawal and Ghose 1986). The mafic rock undergoes a complex recrystallization process in subducted lithospheric slab controlled by physical and chemical kinematics in presence of excess fluid. A close relationship between transformation of eclogite to greenschist with respect to texture, mineralogy and degree of deformation has been traced to a limited extent under microscopic studies. Thus, the parent mafic composition (medium to coarse grained, sub-ophitic, tholeiite) at the leading edge of down going slab is converted into eclogite on prolonged subduction near the crust–mantle inter-junction giving rise to garnet–omphacite–rutile–quartz–albite–glaucophane–barroisite–phengite–epidote with granoblastic texture (Fig. 13.12e).

Decoupling of the slab induced by tectonism and/or by thermo-chemical reactions like dehydration (decomposition) of antigorite in the subducted serpentinite slab (Bucher et al. 2005) possibly triggered exhumation. The exhumed mafic rock is transformed into blueschists containing garnet (highly sheared), omphacite (fine grained, schistose), massive glaucophanite (~ 80% glaucophane, nematoblastic) and greenschist (sheared with cataclasis and mortar texture) at high level emplacement in the upper crust. Overprinting of the greenschist facies metamorphism on high-pressure assemblages in the NHO is marked by two periods of fabric lineations in the metamorphites and associated oceanic sediments (Figs. 13.11b, 13.12d, 13.13a–c) that monitors the history of recrystallization. Ubiquitous presence of calcic-amphibole, besides albite, quartz, epidote, rutile, phengite/white mica, biotite and carbonate and other minor constituents, suggests change in paragenetic sequence of minerals caused by increased activity of volatiles during exhumation. Effects of cataclasis, granulation, shearing and mylonitization give indications that the rocks were subjected to extreme pressure during uplift.

Mineralogical transformations noted in the formation of blueschists are - pseudomorph of barroisite over omphacite, and carbonate over albite; alteration of omphacite/barroisite to glaucophane; inclusion of glaucophane in calcic-amphibole, and zoisite in albite; cross-cutting relationship between glaucophane with albite, and phengite and carbonate; and rim of glaucophane at the crystal margin of sodi-calcic-amphibole. It is apparent that glaucophane is formed replacing an earlier assemblage essentially composed of chlorite + zoisite + albite in a prograde reaction. These features clearly demonstrate that the sodic amphibole is likely to have developed both as prograde and retrograde reactions. Physical conditions in late formation of phengite in paragenetic sequence of minerals is controlled by high fluid pressure and low temperature.

Presence of glaucophane- and barroisite-bearing eclogite in the NHO is analogous to those occurring in the Upper Khagan nappe of northwest Himalaya in north Pakistan (Lombardo et al. 2000). The eclogites in both the areas generated shortly after continental collision, and the history of its exhumation that followed along a cooling path in the latter (Treloar 1995, 1997) is repeated in the former, indicating that the mechanism of emplacement in the northern and eastern margins of the Indian plate is identical.

13.3.9 Radiolarian Chert

The marine sediments associated with the NHO comprise chert, iron-rich sediments, limestone, shale/phyllite, quartzite and greywacke. Some cherts containing marine planktonic micro-organisms have been incorporated in this study. The cherts are grey, green or red in colour, the last one is usually rich in microfossils (Fig. 13.13a, b). These are composed of fine crypto-crystalline quartz and chalcedony. Presence of multiple veins of silica cross-cutting each other in different directions, brecciation (Fig. 13.13b), micro-faulting (Fig. 13.13c), step-faulting and dragging (Fig. 13.13c) in cherts suggest their involvement in tectonic movements. The fracture planes are traversed by silica veins showing recrytallization. The fossil assemblages indicate Maestrichtian age (Chattopadhyay et al. 1983; Acharyya et al. 1986; Chungkham and Jafar 1998).

13.3.10 Chromite Mineralization

Occurrences of pods and lenses of chromitite are reported in association with peridotite (Ghose and Shrivastava 1986; Venkataramana and Bhattacharyya 1989; Agrawal and Ghose 1989; Singh 2008). Workable deposits of chromitite have been encountered at Sirohi and Moreh in the Manipur state. Four types of chromitite ore have been distinguished, viz., massive, granular, disseminated and nodular. High degree of fracturing and brecciation indicate post-crystallization deformation. Nucleation of chromite in olivine (Fig. 13.14) suggests subsolidus replacement at high temperature by diffusion (Ghose and Shrivastava 1986). The massive ores are rich in iron, whereas the nodular types are rich in Al^{3+}. The chromitites are reported to contain traces of gold and platinum, and characterized by enrichment of IPGE (Ir, Os, Ru) with depletion of PPGE (Rh,Pd,Pt), and low Pt + Pd/Os + Ir + Ru ratios indicating its early separation from magma (Singh 2008).

13.4 Discussion

13.4.1 Lava Flows

One of the significant contributions of this study is the nomenclature of some mafic volcanic and volcaniclastic rocks associated with the NHO based upon detailed petrography. An attempt is made to establish the sequence of volcanic eruptions on textural and mineralogical criteria and mutual relationship as revealed under microscope in a dismembered ophiolite suite where relative stratigraphy is a misnomer. The basalt showing quenched texture may be considered as the earliest eruption.

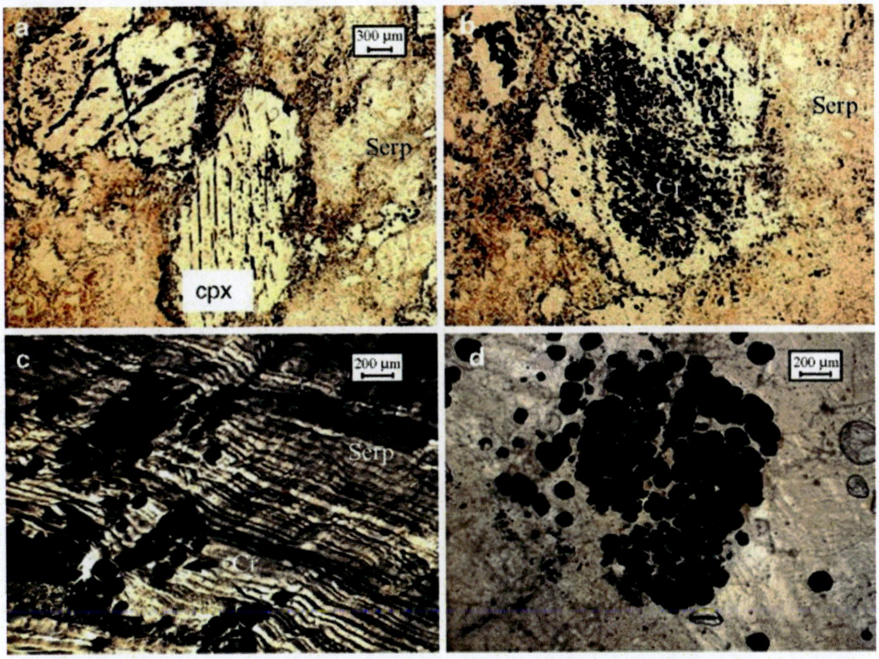

Fig. 13.14 Chromite Mineralisation: (**a**) Replacement of pyroxene by chrome spinel along the cleavage and grain boundary in a serpentinised peridotite. (**b**) Core of olivine is replaced by chromite in serpentinised peridotite. (**c**) Clustering of octahedral chromite grains in serpentinite across the foliation. (**d**) Enlarged view of concentration of octahedral chromite grains in serpentinite possibly leading to the formation of nodular chromite. Abbreviations: cr – chromite, serp – serpentine, px – pxroxene, ol – olivine

Presence of parallel, elongated and radiate plagioclase with hollow-core and curved crystals demonstrate lava to have congealed rapidly under water. Consequently the rock got granulated and shattered into small angular fragments that are closely associated and called hyaloclastite (Fig. 13.8d). A closely similar eruption at very high temperature is vitrophyric basalt containing ocellus. Such texture found in ophiolite has also been recorded in quenched MORB. This cooling pattern is in contrast to that of basalts erupted subaerially and cooling slowly (Smith and Perfit 2007). A temperature difference of about 1,200°C is estimated between eruption and quenching by these authors.

The flows with quenched texture is closely followed by trachy basalt containing autolith (Fig. 13.9a), and is succeeded by plagioclase phyric basalt. Contact between these two successive flows is marked by dark chilled margin and appearance of a thin sinuous siliceous material in between (Fig. 13.9a), indicating the eruption of the latter in quick succession. The flow direction of the two lavas can be distinguished by the orientation of feldspars possibly influenced by a change in tectonic conditions. The trachy basalt with aligned feldspar laths (Fig. 13.9a, b, sub-trachytic texture) is also developed in divergent plate boundary as a result of eruptive dynamics and the flow of lava (Smith and Perfit 2007, p. 8).

The trachy basalt containing a lensoid xenolith (white) shows a sharp contact and interaction with the melt demonstrating assimilation (Fig. 13.9b). This has resulted in granophyric texture (quartz-feldspar symplectite, on left) and formation of large plagioclase laths across the orientation of the xenolith (on right). The xenolith under crossed Nicols is strongly birefringent due to the presence of micaceous mineral supporting pelitic nature of the source.

The basalt with plagioclase megacryst is possibly the youngest eruptive phase representing a derivative of the parent tholeiitic magma. It contains considerable amount of vesicles filled inside by secondary chlorite rimmed with glass implying rapid cooling. The spilites are formed due to interaction of basalt with sea water that has undergone low grade metamorphism and metasomatism.

The mafic lava flows showing quenched texture identified first time in this study could not be supported by chemical data which need detailed investigation in future. The inference drawn from the published chemical data of basalts from the NHO reveals that these are essentially tholeiitic in composition (Ghose et al. 1986; Venkataramana et al. 1986; Sengupta et al. 1989; Srikanth et al. 2004). These are characterized by a flat REE pattern, absence of Eu-anomaly, and negative anomalies of chondrite normalized incompatible elements viz., Rb, Th, Nb, K, Sr and Zr (Srikanth et al. 2004). Negative anomalies of LILE (Rb, K and Sr) and HFSE (Th, Nb and Zr) strongly support limited crustal contamination or alteration. Chemistry of basalts thus shows a strong affinity with MORB in support of petrographic evidence. These basalts were derived by high degree of partial melting of a relatively non-depleted mantle source giving rise to a composition close to the present day MORB. The low-Mg type trachy basalt (Mg# 23–49) that is distinguished in the NHO, shows chemical affinity with alkali basalt, and has been related to seamount/OIB (Venkataramana and Datta 1987; Sengupta et al. 1989).

13.4.2 Pyroclastic Rocks

A wide range of pyroclastic deposits ranging from agglomerate to fine-ash, scoria, breccia with admixture of crystal, glass and lithic fragments are encountered in the ophiolite belt. The pyroclastic deposits give evidence of at least two periods of diastrophic movements as noted by their involvement in folding and micro-faulting. The vitric tuff and ash flows for their fine granularity (<2 mm) as products of explosive eruption of poorly crystallized magma are deposited far away from the vent. The coarser and denser crystals and lithic fragments are left behind closer to the vent, resulting in crystal enriched ignimbrites. Less vesiculation of cognate fragments in glassy clast reflects phreatomagmatic origin (Cas and Wright 1987).

The crystal tuffs show association of fragments of diverse origin ranging from plagioclase, potash feldspar (sanidine?), perovskite, welded ash-flow to lithic clast (basalt). The single crystal fragments in tuffaceous matrix possibly indicate complete fractionation of magma prior to eruption. Because of high viscosity of crystal-laden melt these were probably ejected close to the volcanic vent. The crystal tuffs are also considered as pyroclastic flow deposits (Cas and Wright 1987). Lithic fragments are dominantly rhyolitic in composition with frequent presence of clasts of potash feldspar and graphic texture. Fractionation in a high-level magma chamber is possibly responsible for bringing out these volcanic products.

The glassy rocks constitute only a very small proportion of these volcanic assemblage, however, their presence is of considerable significance in petrogenesis in deciding composition, temperature and viscosity of the melt. The volcanic glasses are associated with lithic/crystal tuff with single crystal fragments (plagioclase or pink potash feldspar) and are inter-layered with chert. Adjacent to crystal or lithic fragments, and where welding occurred (Fig. 13.10e), the glasses are associated with shard. The winged butterfly-shaped cuspate shard (Fig. 13.10d) and other such configurations noted with inimitable interior design of vesicles and channels, bear signatures through which the volatiles had escaped. Thus it is evident that the volatile species (C–H–S etc.) played a major role on the physical and chemical parameters in the formation of volcanic and pyroclastic rocks.

13.4.3 Eastern Suture of the Indian Plate

After examining the litho-assemblages of a cross-section of magmatic and metamorphic rocks associated with the NHO, it would be appropriate to check the existence of suture or collision boundary at the eastern Indian plate margin. The validity of Naga–Manipur Hills ophiolite belt as the easternmost suture of the Indian plate of late has been questioned (Sengupta et al. 1990; Acharyya 2007). According to these authors, two parallel ophiolite belts are in existence (Fig. 13.2, inset), one occurring at the Naga-Manipur Hills is called the *western ophiolite belt*, known for their close spatial relationship to a zone of negative gravity anomaly while the other one shows gravity high in the Central Myanmar Basin and is called the *eastern ophiolite belt*. The former is extensively studied, whereas the latter is poorly preserved, ill exposed and been sparsely studied. The 'eastern ophiolite belt' is said to have been overthrusted westward as flat-lying nappes and juxtaposed with the flyschoid sediments now constituting the Naga–Manipur Hills ophiolite (Acharyya 2007). The geological and petrological account put forward in the preceding pages constrain the development of the NHO as a fragmented nappe. The presence of ductile shear in the NHO (western ophiolite belt) rejects the contention of its formation by transmigration.

The contentious issue of 'transmigration model' is in the interpretation of negative gravity anomaly across the region between Assam and Myanmar that includes a thin wedge (2–15 km thick) of the Naga–Manipur Hills ophiolite (Mukhopadhyay and Dasgupta 1988). This is considered to be atypical of ophiolites in sutures (Acharyya 2007). Such inconsistency is related to flexure of the crust and mantle as a result of collision between two plates which can cause large negative isostatic anomalies of the order of few hundred milligals (Watts and Talwani 1974). It is further advocated that large negative isostatic anomaly as observed in Myanmar is related to flexure of the lithosphere and focal mechanism solutions for earthquakes taking place in the area (Verma 1985, p. 167).

13.5 Conclusion

From the foregoing discussion it may be concluded that some basalts from the NHO bear petrological fingerprints as a product of mid-ocean spreading center, formed due to rifting and opening of the southern Tethys. Collision of the Indian plate with the Eurasian plate (Sino-Myanmar highland) commenced since Late Cretaceous and is still continuing. The oceanic lithosphere was uplifted, accreted and exhumed into active continental plate margin in the IMR as a result of obduction. Occurrence of high-P assemblages (eclogite and blueschist) at the western tectonic margin of the ophiolite defines the existence of a suture. Microscopic observations of S–C mylonite in garnet lherzolite and 'mica-fish' in glaucophane schist substantiate that the shear zone along which the ophiolite was emplaced has been deformed by ductile deformation.

Acknowledgments This paper has been richly benefited from the constructive reviews of Prof. R.K. Srivastava and two anonymous reviewers of the early text. The authors take the responsibility of the interpretations based essentially on microscopic observations. Thanks are also due to the Geological Survey of India for providing facilities to undertake photomicrographs.

References

Acharyya SK (2007) Collision emplacement history of the Naga-Andaman ophiolites and the position of the eastern Indian suture. J Asian Earth Sci 29:229–242

Acharyya SK, Roy DK, Mitra ND (1986) Stratigraphy and palaeontology of the Naga Hills ophiolite belt. Memoir Geol Surv India 119:64–79

Agrawal OP (1985) Geology and geochemistry of the mafic-ultramafic complex of Indo-Burman ranges between Meluri and Awankhoo, Phek district, Nagaland, India. Unpublished Ph.D. thesis, Patna University, Patna

Agrawal OP, Ghose NC (1986) Geology and stratigraphy of the Naga Hills ophiolite between Meluri and Awankhoo, Phek district, Nagaland, India. In: Ghose NC, Varadarajan S (eds) Ophiolites and Indian plate margin. Sumna Publications, Patna, pp 163–195

Agrawal OP, Ghose NC (1989) Mineral resources in the ophiolite belt of Nagaland, N E India. In: Ghose NC (ed) Phanerozoic ophiolites of India. Sumna Publications, Patna, pp 245–280

Agrawal OP, Kacker RN (1980) Nagaland ophiolites, India – a subduction zone ophiolite complex in Tethyan orogenic belt. In: International Ophiolite Symposium, Cyprus, pp 454–461

Anonymous (1978) Status of geological work and inventory of mineral discoveries in Nagaland. Misc Pub 1, Directorate of Geology and Mining, Nagaland: 38p

Bender F (1983) Geology of Burma. Gebruder Borntraeger, Berlin-Stuttgart, p 293

Bhattacharjee CC (1991) The ophiolite of northeast India – a subduction zone ophiolite complex in the Indo-Burman orogenic belt. Tectonophysics 191:213–233

Bhattacharyya S, Venkataramana P (1986) Structure and metamorphism of Nagaland ophiolite. Memoir Geol Surv India 119:28–32

Brunnschweiler RO (1966) On the geology of Indo-Burman ranges. J Geol Soc Austr 13:137–194

Bucher K, Fazes F, de Capitani C, Grapes R (2005) Blueschists, eclogites, and decompression assemblages of the Zermatt-Saas ophiolite: high-pressure metamorphism of subducted Tethys lithosphere. Am Mineral 90:821–835

Cas RAF, Wright JV (1987) Volcanic successions – modern and ancient. Allen & Unwin, London, p 361

Chattopadhyay B, Venkataramana P, Roy DK, Bhattacharyya S, Ghosh S (1983) Geology of Naga Hills ophiolites. Geol Surv India Record 112(pt. 2):59–115

Chungkham P, Jafar SA (1998) Late Cretaceous (Santonian-Maestrichtian) integrated Coccolith-Globotruncanid biostratigraphy of pelagic limestone from the accretionary prism of Manipur, northeast India. Micropalaeontology 44:68–83

Coleman RG (1977) Ophiolites: ancient oceanic lithosphere. Springer, Berlin, p 229

Dasgupta AB (1977) Geology of Assam–Arakan region. Q J Geol Min Met Soc India 49:1–54

Desai AG, Pardeshi RG, Karmalkar NR, Powar KB (1986) Petrology of the ultramafic rocks of the Indus ophiolite belt, Ladakh himalaya, India. In: Ghose NC, Varadarajan S (eds) Ophiolites and Indian plate margin. Sumna Publications, Patna, pp 63–73

De Sigoyer J, Guillot S, Lardeaux JM, Mascle G (1997) Glaucophane-bearing eclogites in the Tso-Morari dome (eastern Ladakh, NW Himalaya). Eur J Miner 9:1073–1083

Evans P (1964) The tectonic framework of Assam. J Geol Soc India 5:80–96

Gansser A (1964) Geology of the Himalayas. Wiley Interscience, London, p 289

Ghose NC, Agrawal OP (1989) Geological framework of the central part of Naga Hills. In: Ghose NC (ed) Phanerozoic ophiolites of India and associated mineral resources. Sumna Publications, Patna, pp 165–188

Ghosh B, Ray J (2003) Mineral chemistry of ophiolitic rocks of Mayodia-Hunli area of Dibang valley district, Arunachal Pradesh, Northeastern India. Memoir Geol Soc India 52:447–471

Ghose NC, Shrivastava MP (1986) Podiform chromite of Naga Hills ophiolite, N.E. India. In: Petrascheck W and 5 others (eds) Chromites. Theophrastus, Athens, pp 263–284

Ghose NC, Singh RN (1980) Occurrence of blueschist facies in the ophiolite belt of Naga Hills, east of Kiphire, N.E. India. Geol Rundschau 69:41–43

Ghose NC, Singh RN (1981) Structure of the Naga Hills ophiolites and associated sedimentary rock in the Tuensang district of Nagaland, N.E. India. Ofioliti 6:237–254

Ghose NC, Agrawal OP, Windley BF (1984) Geochemistry of the blueschist-eclogite association in the ophiolite belt of Nagaland, India. In: Seminar on Cenozoic crustal evolution on the Indian plate margin, Abstracts, Patna, pp 27–30

Ghose NC, Agrawal OP, Singh RN (1986) Geochemistry of the ophiolite belt of Nagaland, N.E. India. In: Ghose NC, Varadarajan S (eds) Ophiolites and Indian plate margin. Sumna Publications, Patna, pp 241–291

Ghose NC, Agrawal OP, Srivastava SC (1987) Metamorphism of the ophiolite belt of Nagaland, N E India. In: Goirala VK (ed) Tertiary orogeny. Khandelwal Offset, Varanasi, pp 189–213

Ghosh B, Mahoney J, Ray J (2007) Mayodia ophiolites of Arunachal Pradesh, Northeastern Himalaya. J Geol Soc India 70:395–604

Heinrichsen Th, Schurmann K (1972) Mineral reactions in low- grade metamorphic rocks. 24 Intern Geol Cong, Sec 2: 5–10

Lombardo B, Rolfo F, Compagnoni R (2000) Glaucophane and barroisite eclogites from the upper Kaghan nappe: implications for the metamorphic history of the NW Himalaya. Geological Society London Special Publications 170, London, pp 411–430

Maibam B, Foley S (2008) Abyssal peridotites from the ophiolite belt of northeastern India: implications for the equilibrium conditions and tectonic setting. In: Seminar on Indo-Myanmar Ranges in the Tectonic Framework of the Himalaya and Southeast Asia, Imphal, Abstract, pp 84–85

Mitchell AHG, Mckerrow, WS (1975) Analogous evolution of the Burma orogen and the Scottish Caledonides. Bull Geol Soc Am 86:305–315

Mitra ND and 12 others (eds) (1986) Geology of Nagaland ophiolite. Memoir Geol Surv India 119:113

Mukherjee BK, Sachan HK (2001) Discovery of coesite from Indian Himalaya: a record of ultra-high pressure metamorphism in Indian continental crust. Curr Sci 81(10):1358–1360

Mukhopadhyay M, Dasgupta S (1988) Deep structure and tectonics of the Burmese arc: constraints from earthquake and gravity data. Tectonophysics 149:299–322

Sengupta S, Acharyya SK, Vander Hull HJ, Chattopadhyay B (1989) Geochemistry of volcanic rocks from the Naga Hills ophiolites, northeast India and their inferred tectonic setting. J Geol Soc Lond 146:491–498

Sengupta S, Ray KK, Acharyya SK, de Smith JB (1990) Nature of ophiolite occurrence along eastern margin of Indian plates. Geology 18:439–442

Singh AK (2008) PGE distribution in the ultramafic rocks and chromitites of the Manipur ophiolite complex, Indo-Myanmar orogenic belt, Northeast India. J Geol Soc India 72:649–660

Singh RN, Ghose NC (1981) Geology and stratigraphy of the ophiolite belt of Naga Hills, East of Kiphire. N.E. India. Recent Res Geol 8:359–381

Singh B, Krishnamurthy KV, Razdan ML, Kashkari RL (1989) Geology and petrography of ophiolitic sequence of the Poche area, Markha valley, Ladakh, J & K. In: Ghose NC (ed) Phanerozoic ophiolites of India. Sumna Publications, Patna, pp 57–62

Singh AK, Singh NI, Devala Devi L, Singh RKB (2008) Pillow basalts from the Manipur ophiolite complex, Indo-Myanmar Range, Northeast India. J Geol Soc India 72:168–174

Smith MC, Perfit MR (2007) Petrography and petrogenesis of a mid-ocean ridge lava suite. http://serc.carleton.edu/files/NAGTWorkshops/petrology/teaching-examples/MORB_Petrography_and_petrology_Key_1.1:pdf. Assessed 10 Feb 2009

Srikanth B, Subba Rao MV, Rao BV, Nirmal Charan S, Balaram V, Ejung OC (2004) Geochemical signatures in the basaltic rocks of Naga Hills ophiolite belt: implications for petrogenesis and tectonic environment of emplacement. J Appl Geochem 6:177–189

Srikantia SV (1986) The tectonic design of the Ladakh region, India. In: Ghose NC, Varadarajan S (eds) Ophiolites and Indian plate margin. Sumna Publications, Patna, pp 29–47

Srikantia SV, Razdan ML (1980) Geology of part of Central Ladakh Himalaya with particular reference to Indus Tectonic Zone. J Geol Soc India 21:523–545

Stoneley P (1974) On the origin of ophiolite complexes in the southern Tethys region. Tectonophysics 25:303–322

Treloar PJ (1995) Pressure temperature-time paths and the relationship between collision, deformation and metamorphism in the north-west Himalaya. Geol J 30:77–104

Treloar PJ (1997) Thermal controls on early-Tertiary, short-lived, rapid regional metamorphism in the NW Himalaya. Tectonophysics 180:323–349

Venkataramana P, Bhattacharyya S (1989) Mode of occurrence and origin of chromite, magnetite and Ni-bearing laterite in the Naga Hills ophiolite, N.E. India. In: Ghose NC (ed) Phanerozoic ophiolites of India. Sumna Publications, Patna, pp 213–234

Venkataramana P, Datta AK (1987) Contrasting volcanic suites in Naga Hills and their bearing on the tectonic evolution of the Naga Hills ophiolite belt, N.E. India. J Geol Soc India 30:33–47

Venkataramana P, Datta AK, Acharyya SK (1986) Petrography and petrochemistry. In: Mitra ND & 12 others (eds) Geology of Nagaland ophiolite. Memoir Geol Surv India 119:33–63

Verma R (1985) Gravity field, seismicity and tectonics of the Indian peninsula and the Himalayas. Allied Publications, New Delhi, p 213

Vidyadharan KT, Shukla R, Bhattacharyya S (1986) Blueschists, C-type eclogites and garnet-bearing mafic segregations from Naga Hills ophiolite belt. Geol Surv India Records 114(4):7–13

Virdi NS (1989) Glaucophane metamorphism in the ophiolite belt of the Indus-Tsangpo zone in the Himalaya. In: Ghose NC (ed) Phanerozoic ophiolites of India. Sumna Publications, Patna, pp 73–91

Vohra CP, Haldar D, Ghosh Roy AK (1989) The Andaman–Nicobar ophiolite complex and associated mineral resources – current appraisal. In: Ghose NC (ed) Phanerozoic ophiolites of India. Sumna Publications, Patna, pp 281–315

Watts AB, Talwani M (1974) Gravity anomalies seaward of deep-sea trenches and their tectonic implications. Geophys J R Astronom Soc 36:57–90

Charnockites and Anorthosites

Chapter 14
Age and Origin of the Chilka Anorthosites, Eastern Ghats, India: Implications for Massif Anorthosite Petrogenesis and Break-up of Rodinia

Ramananda Chakrabarti, Asish R. Basu, Pradyot K. Bandyopadhyay, and Haibo Zou

Abstract Massif anorthosites are important Proterozoic continental crustal components with worldwide occurrences along linear belts. Two major questions concerning these massif-anorthosites are whether they have broadly similar ages, and whether their parent magmas are mantle-derived or lower continental crust-derived. The Chilka massif anorthosites, hosted by the Proterozoic Eastern Ghats mobile belt of eastern India, yield 855 ± 31 Ma U–Pb zircon crystallization age that overlaps with the breakup of Rodinia. This age is considerably younger than Grenville-age global massif anorthosites. Trace elements and Nd–Sr–Pb isotopic compositions of the Chilka anorthosites imply derivation from a depleted mantle source contaminated by late-Archean to early-Proterozoic lower continental crust. We suggest that this mantle upwelling to generate the Chilka anorthosites was a result of 'edge-driven convection' given that the Eastern Ghats Belt, hosting the Chilka anorthosites was located 1,000 km away from the margin of the southern Indian cratons in the Neoproterozoic Rodinia. Alternatively, we also consider that the anorthosites' parent magma formed as a result of the regional extensional tectonism at ca. 850 Ma affecting this segment of Rodinia, as recently suggested by some workers.

R. Chakrabarti and A.R. Basu (✉)
Department of Earth and Environmental Sciences, University of Rochester, Rochester, NY 14627, USA
e-mail: rama@eps.harvard.edu; abasu@earth.rochester.edu
Present address: Department of Earth and Planetary Sciences, Harvard University, Cambridge, MA 02138, USA

P.K. Bandyopadhyay
Department of Geology, Presidency College, Kolkata 700073, India
e-mail: rbanerjee2483@gmail.cim

H. Zou
Department of Geology and Geography, Auburn University, Auburn, AL 36849, USA
e-mail: haibo.zou@auburn.edu

J. Ray et al. (eds.), *Topics in Igneous Petrology*, DOI 10.1007/978-90-481-9600-5_14,

14.1 Introduction

Massif-type anorthosites are large batholith-sized igneous complexes mainly Proterozoic in age. They comprise the most voluminous anorthosites on Earth and are almost monomineralic, comprising mostly plagioclase with minor amounts of pyroxene or olivine and Fe–Ti oxides. Plagioclase compositions of massif-anorthosites are typically intermediate (An_{40-60}) in contrast to the more calcic compositions found in the Archean layered anorthosites (An_{80-95}). There is a long-standing debate regarding the magma source and geodynamic setting of massif-anorthosites with no consensus as to whether these magmas primarily originate in the Earth's mantle or crust. Moreover, the mechanism of transport and emplacement of this magma remains poorly understood. Among the other characteristics of massif-type anorthosites are their occurrences in two broad belts when plotted on a predrift continental reconstruction (Herz 1969) spreading over Antarctica, Brazil, Colombia, Mexico, United States, Canada, Greenland, Scandinavia, Poland, former Soviet Union, Africa, Madagascar, India and Australia.

Massif-anorthosite occurrences in India are restricted along the eastern coast hosted mainly by the Eastern Ghats Belt (EGB), a deeply eroded part of an extensive Proterozoic mobile belt. The Eastern Ghat is considered a part of the global Southwest United States-East Antarctica (SWEAT) orogen (Hoffman 1991; Moores 1991). It is found to be juxtaposed with east Antarctica in most paleogeographic reconstructions of Rodinia from middle Proterozoic to the Mesozoic. This correlation is based on the geometric fit of coastlines as well as petrological and large-scale structural similarities. Similar radiometric age estimates of the central and eastern units of the Eastern Ghats belt and the Rayner complex as well as Prydz Bay region of East Antarctica also support the above hypothesis (Mezger and Cosca 1999). Based on this fit, it is thought that the now widely separated massif-type anorthosite-bearing Grenville orogenic belts of Madagascar, eastern India, NE America as well as eastern Antarctica were originally continuous (Moores 1991). Contrastingly, it has been suggested that the EGB and Rayner complex collided during the Grenville period but India, EGB and Rayner complex were juxtaposed to the Australian and Antarctic blocks much later during the Pan-African orogeny (e.g. Powell and Pisarevsky 2002).

In contrast to the extensive geochemical and isotopic data available for the other Proterozoic massif-anorthosites, the eastern Indian anorthosites are not well studied. In this study, we present trace element and Nd–Sr–Pb isotopic data as well as U–Pb zircon ages for the Chilka anorthosite massif (Balugaon unit), which is one of largest anorthosite massifs in India with an estimated area of ~295 km^2 (Sarkar et al. 1981). The goal of this study is to test the time equivalence of the Chilka anorthosite with other anorthosites of similar disposition such as the Grenville anorthosite belt, and to understand the petrogenesis of these rocks. Our precise geochronological results indicate that the Chilka anorthosites are distinctly younger than other global anorthosites and are possibly among the youngest of all Proterozoic massif-anorthosites. The emplacement age of these rocks is contemporaneous with global tectonic events in the Neoproterozoic. The present study expands the geochemical and isotopic database of global massif anorthosites and provides constraints for understanding the genesis of massif-anorthosites in general. Based on this combined trace element and Nd–Sr–Pb isotopic study and in conjunction with other available constraints, we present a model for the genesis of the Chilka anorthosites being derived from the depleted mantle-derived melts with contributions from the lower continental crust. The cause of upper mantle upwelling can be attributed to regional extensional tectonics affecting this part of the Supercontinent Rodinia.

14.2 Geological Setting and Samples

Massif-type anorthosites are a distinctive component of the Proterozoic Eastern Ghats mobile belt of India. They occur mostly to the north of the Mesoproterozoic Godavari rift (Leelanandam 1990) (Fig. 14.1). The Chilka anorthosite massif of the Eastern Ghats is located around the Chilka lake lagoon in the eastern coast of India (Fig. 14.1). The host rocks for these anorthosites comprise a highly deformed and metamorphosed suite of high-grade schists, gneisses and granulites. Common host rock types exposed around this anorthosite massif include khondalites (garnet-bearing sillimanite gneiss), garnetiferous granite gneisses, pyroxene-bearing quartzo-feldspathic gneisses, pyroxene granulites (charnockites), basic granulites and calc-silicate rocks, occurring, respectively, in order of their decreasing abundances (Sarkar et al. 1981). Four major high-grade, metamorphic lithological units can be identified in the Eastern Ghats, all aligned parallel to the Archean Dharwar and Baster cratons located farther west. From west to east, they are the Western Charnockite Zone (WCZ), the Western Khondalite Zone (WKZ), the Charnockite Migmatite Zone (CMZ) and the Eastern Khondalite Zone (EKZ) (Ramakrishnan et al. 1998) (Fig. 14.1), all representing granulite to amphibolite facies meta-igneous and meta-sedimentary rocks.

The granulites and anorthosites around the Chilka region display polyphase deformation structures and three phases of folding episodes have been recognized (Bhattacharya et al. 1994). The oldest tectonothermal event in the Eastern Ghats belt is an episode of granulite facies metamorphism at 1.6–1.4 Ga (Mezger and Cosca 1999) that is recognized in the westernmost unit. This was followed by a high-grade metamorphism that occurred at ca. 960 Ma. This tectonothermal episode, which was the fabric-defining event in most parts of the Eastern Ghats, coincides with the last-phase of the global Grenvillean orogeny at ca. 960 Ma (Mezger and Cosca 1999). Late stage pegmatites indicate that a Pan-African mid-amphibolite facies metamorphism at ~500 Ma affected parts of the Eastern Ghats belt (Mezger and Cosca 1999). The Chilka anorthosites register only the imprint of the last phase of folding and are therefore distinctly younger than the surrounding rocks (Bhattacharya et al. 1994). Recent U–Th–Pb

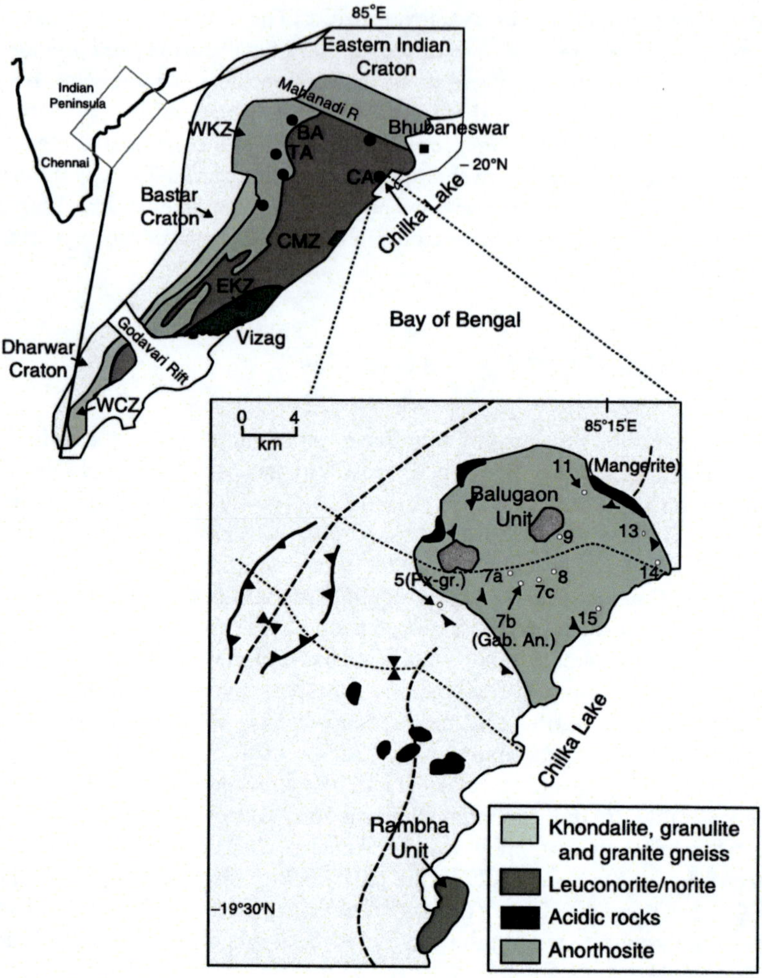

Fig. 14.1 Simplified lithological map (modified after Sarkar et al. 1981) of the Chilka anorthosite complex (*bottom*) showing locations of the samples (*open circles*) analyzed in the present study from the 295 km^2 Balugaon unit. Also shown is a simplified map (*top*) of the Eastern Ghats belt showing the various regional metamorphic and cratonic units: Western Charnockite Zone (WCZ), Western Khondalite Zone (WKZ), Central Migmatite Zone (CMZ), Eastern Khondalite Zone (EKZ), and the Bastar and Dharwar cratons to the west. A number of major anorthosite massifs in the Eastern Ghats occur mostly to the north of the Godavari rift and include the Chilka (CA), Bolangir (BA), and Turkel (TA) massifs, among others (*filled circles*) (Leelanandam 1990)

electron microprobe dating of monazites from synkinematically emplaced leucosomes indicate that the Chilka anorthosites were also affected by granulite facies metamorphism accompanied by a fabric-defining transpressive deformation between 690 and 662 Ma, long after their emplacement (Dobmeier and Simmat 2002).

For the present study we sampled seven anorthosites, one gabbroic anorthosite, one garnet-bearing mangerite and one pyroxene granulite from the Balugaon unit (~295 km^2) of the Chilka anorthosite massif (Fig. 14.1). The anorthosites of this massif typically show a granoblastic texture with nearly equigranular, polygonal, recrystallized plagioclase crystals (Fig. 14.2a). Relict primary plagioclases with euhedral to subhedral grain shapes and straight grain boundaries are also observed in these rocks (Fig. 14.2b). However, the triple junctions of some of these plagioclase grains are annealed to 120° dihedral angles. In addition to plagioclase, varying amounts of clinopyroxene, and subordinate zircons and opaque minerals are also observed in all the anorthosites. The gabbroic anorthosites show mostly plagioclases with up to 25% modal clinopyroxene and comparatively more zircons visible under the petrographic microscope. Although the Chilka anorthosites were affected by granulite facies metamorphism (Dobmeier and Simmat 2002), several zircons with prismatic outlines are still preserved in these rocks (Fig. 14.2c). Some of the cathodoluminescence (CL) images of these zircons (Fig. 14.2c) are c-axis perpendicular and hence show a tetragonal or ditetragonal–dipyramidal outline. Measured high Th/U ratios of these zircons (Table 14.1) at the same spots where U–Pb isotope

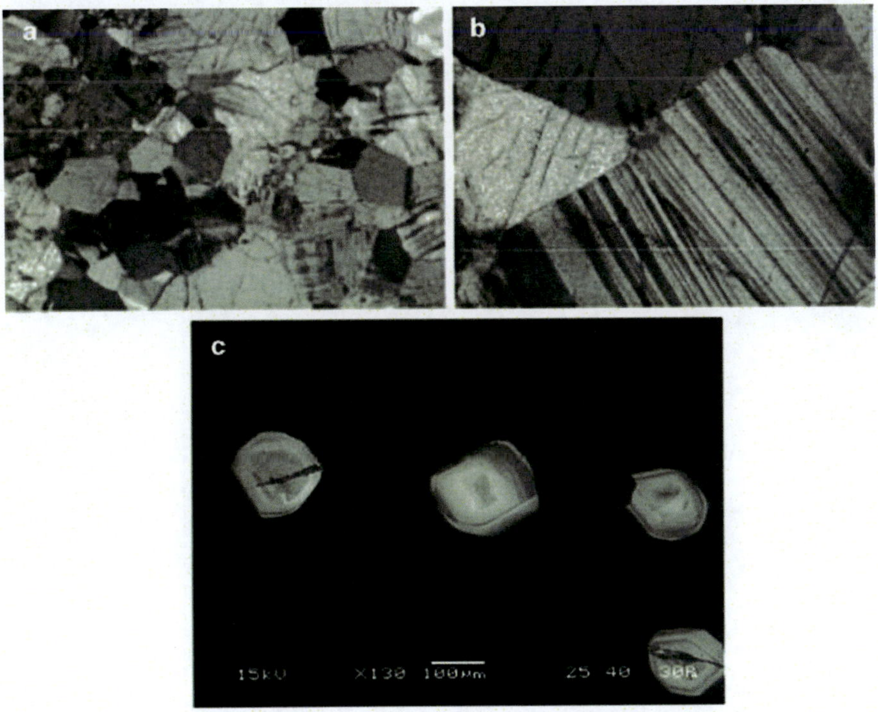

Fig. 14.2 Photomicrographs of the Chilka anorthosites and gabbroic anorthosite showing meta-igneous textures (**a**) with sharp linear plagioclase grain boundaries (**b**). Also shown are CL images of igneous zircons (see text for discussions) commonly found in these rocks showing both sector zoning and oscillatory zoning (**c**)

Table 14.1 U–Pb isotopic ratios and Th/U of 12 zircon grains analyzed with the UCLA Cameca IMS 1270 ion microprobe from the Chilka gabbroic anorthosite sample (EG-7b). The high Th/U ratios (from microprobe analysis) of these zircons attest to their igneous origin. The 2σ errors for the U–Pb ages is 3.6%, which is typical for ionprobe ages in the age range of 800–900 Ma, similar to the zircons of the present study. $^{207}Pb/^{206}Pb$ ages for these zircons are not considered since for zircons less than 1 Ga old, $^{206}Pb/^{238}U$ ages provide better precision and accuracy

Grains	$^{206}Pb^*/^{238}U$	1 s.e.	$^{207}Pb^*/^{235}U$	1 s.e.	$^{207}Pb^*/^{206}Pb^*$	1 s.e.	Th/U	$^{206}Pb/^{238}U$ Age (Ma)	1 s.e.	$^{207}Pb/^{235}U$ Age (Ma)	1 s.e.	$^{207}Pb/^{206}Pb$ Age (Ma)	1 s.e.	Radiogenic ^{206}Pb (%)
g1_s1	0.1369	0.010	1.276	0.117	0.0676	0.004	3.1	827	55	835	52	857	110	99
g2_s1	0.1518	0.007	1.395	0.097	0.0667	0.003	2.1	911	40	887	41	828	89	99
g3_s1	0.1377	0.007	1.309	0.134	0.0690	0.005	2.7	832	39	850	59	898	158	98
g4_s1	0.1271	0.008	1.091	0.130	0.0623	0.006	3.0	771	44	749	63	683	198	98
g5_s1	0.1519	0.011	1.584	0.181	0.0756	0.007	2.3	911	59	964	71	1,086	188	98
g7_s1	0.1556	0.007	1.532	0.109	0.0714	0.004	2.0	932	39	943	44	970	116	99
g8_s1	0.1346	0.010	1.378	0.170	0.0743	0.008	2.8	814	58	880	73	1,049	210	98
g9_s1	0.1274	0.015	1.315	0.245	0.0749	0.010	2.6	773	88	852	108	1,065	275	98
g10_s1	0.1304	0.013	1.119	0.219	0.0622	0.010	3.7	790	71	762	105	682	340	97
g11_s1	0.1444	0.015	1.502	0.347	0.0755	0.014	2.5	869	86	931	141	1,081	366	97
g12_s1	0.1378	0.012	1.425	0.136	0.0750	0.005	2.4	832	66	899	57	1,068	129	99
g13_s1	0.1608	0.021	1.426	0.263	0.0643	0.008	2.3	961	117	900	110	753	259	97

ratios were analyzed, indicate their igneous origin. This feature is discussed in more detail in subsequent sections.

Several of these zircons from a large sample of the gabbroic anorthosite were separated by a combination of heavy liquid, isodynamic magnetic separation and hand-picking methods, and selected for ion microprobe U–Pb age-dating. Both the pyroxene granulite and mangerite samples show granoblastic textures with variable amounts of clino- and orthopyroxenes. The mangerite shows abundant garnet porphyroblasts, but these garnets are rare in the pyroxene granulite.

14.3 Analytical Methods

14.3.1 Zircon U–Pb Dating Method

Several zircon grains were separated from the gabbroic anorthosite sample (EG-7b) of our study. Epoxy grain mounts of hand-selected zircons were polished to expose grain interiors. After ultrasonic cleaning and drying down, samples were coated with about 10 nm layer of Au. U–Pb ages were obtained by the UCLA CAMECA IMS 1270 ion microprobe using a 15-nA $^{16}O^-$ primary beam focused on 25 μm diameter spots. Secondary ions were extracted at 10 kV and the mass spectrometer was tuned to a mass resolution of ~5,000 Å to resolve molecular interferences in the mass range analyzed ($^{94}Zr_2O^+$, $^{204}Pb^+$, $^{206}Pb^+$, $^{207}Pb^+$, $^{208}Pb^+$, $^{232}Th^+$, $^{238}U^+$, $^{238}U^{16}O^+$). The sample analysis surface was flooded with oxygen gas at a pressure of 4×10^{-3} Pa to increase the Pb yields by a factor of 2. The relative sensitivities for Pb and U were determined on a reference zircon AS-3 (Paces and Miller 1993) using a calibration technique similar to Compston et al. (1984). Th/U ratios were obtained by multiplying measured $^{232}Th^+/^{238}U^+$ with the relative sensitivity value determined on reference zircon 91500.

14.3.2 Trace Element and Nd–Sr–Pb Isotopes

Ten rock samples from the Chilka anorthosite complex, collected both from the center and the edges of the Balugaon unit (Fig. 14.1), were analyzed for trace element concentrations and Sr-, Nd-, and Pb-isotopes. Trace element concentrations (Table 14.2) were measured using an Inductively Coupled Plasma Mass Spectrometer (Thermo X-series). 25-mg powdered samples were digested using HF–HNO_3 acid mixtures and diluted to 100 ml of 2% HNO_3 solution with ~10 ppb internal standard of In, Cs, Re, and Bi. BCR-2 and BIR-2 (basalts-USGS) were used as known external standards while AGV-2 (andesite-USGS) and BHVO-2 (basalt-USGS) rock standards were run as unknowns to estimate the error.

Table 14.2 Trace element concentrations and selected element ratios of the Chilka anorthosites (An) (EG- 7a, 7c, 8, 9, 13, 14, 15), the gabbroic anorthosite (Gab-An) (EG-7b), pyroxene granulite (Px-Grlt) (EG-5) and the garnet-quartz-mangerite (Gt-Qtz-Mgt) (EG-11). Analytical uncertainties are less than 5% for all the elements analyzed and less than 2% for the REEs

Sample	EG-5	EG-11	EG-7b	EG-7a	EG-7c	EG-8	EG-9	EG-13	EG-14	EG-15
Rock type	Px-Grlt	Gt-Qtz-Mgt	Gab-An	An	An	An	An	An	An	An
Rb	228	6.4	5.5	1.19	2.06	0.79	0.65	1.59	1.15	1.04
Ba	557	182	76	81	69	59	59	67	63	63
Sr	69	151	393	390	379	329	326	328	359	333
Nb	11.2	33	1.01	1.75	0.091	0.14	0.083	0.38	0.35	0.032
Ta	0.81	2.13	0.077	0.093	0.006	0.008	0.008	0.028	0.023	0.004
La	42	30	7.1	2.5	2.5	1.63	1.88	2.7	1.91	3.2
Ce	74	55	15.9	4.7	4.4	2.9	3.5	4.9	3.5	5.4
Pr	7.9	6.0	1.81	0.53	0.45	0.33	0.35	0.53	0.38	0.52
Nd	29	23	7.2	2.1	1.66	1.27	1.36	2.0	1.44	1.85
Eu	1.25	1.98	0.99	0.93	1.01	0.88	0.85	0.80	0.79	1.21
Sm	5.3	4.7	1.38	0.40	0.24	0.24	0.22	0.37	0.24	0.26
Gd	4.3	5.0	0.99	0.39	0.20	0.23	0.19	0.35	0.22	0.20
Tb	0.60	0.92	0.13	0.067	0.025	0.040	0.031	0.059	0.031	0.028
Dy	3.1	6.0	0.56	0.34	0.12	0.21	0.15	0.35	0.17	0.16
Ho	0.55	1.38	0.10	0.071	0.022	0.049	0.033	0.075	0.032	0.034
Er	1.40	4.5	0.23	0.19	0.053	0.14	0.084	0.23	0.087	0.092
Tm	0.17	0.64	0.026	0.023	0.006	0.020	0.012	0.031	0.010	0.013
Yb	1.20	5.39	0.18	0.16	0.043	0.17	0.10	0.26	0.071	0.10
Lu	0.18	0.92	0.026	0.023	0.005	0.029	0.017	0.041	0.010	0.019

Zr	96	18.3	1.93	1.93	1.64	2.3	1.69	1.51	1.07	1.07
Hf	2.69	0.51	0.056	0.055	0.035	0.060	0.053	0.057	0.038	0.041
Pb	36	7.1	1.66	1.09	1.10	0.74	1.12	1.18	0.90	1.28
Th	4.8	2.40	0.15	1.61	0.14	0.21	0.053	0.14	0.11	0.058
U	0.89	0.18	0.012	0.041	0.016	0.011	0.006	0.017	0.011	0.007
U/Pb	0.025	0.025	0.007	0.037	0.014	0.015	0.005	0.014	0.012	0.006
Th/U	5.3	13.6	13.2	39	9.0	18.6	8.6	8.5	10.6	7.8
Sm/Nd	0.18	0.21	0.19	0.19	0.14	0.19	0.16	0.18	0.17	0.14
Rb/Sr	3.3	0.042	0.014	0.003	0.005	0.002	0.002	0.005	0.003	0.003

Table 14.3 Present day (0) and initial (T) Nd, Sr and Pb isotope data of the Chilka anorthosites and associated rocks, obtained with the Rochester TIMS, and their Rb/Sr, Sm/Nd, U/Pb and Th/Pb ratios estimated from the ICPMS trace element concentration data in Table 14.2

Sample	EG-5	EG-11	EG-7b	EG-7a	EG-7c	EG-8	EG-9	EG-13	EG-14	EG-15
Rock type	Px-Grlt	Gt-Qtz-Mgt	Gab-An	An	An	An	An	An	An	An
$^{147}Sm/^{144}Nd$	0.1155	0.1330	0.1205	0.1216	0.0914	0.1195	0.1023	0.1158	0.1054	0.0889
$^{143}Nd/^{144}Nd(0)^a$	0.511848	0.511878	0.511853	0.511929	0.511801	0.511960	0.511825	0.511874	0.511853	0.511725
$^{143}Nd/^{144}Nd(T)$	0.511201	0.511132	0.511177	0.511247	0.511288	0.511290	0.511251	0.511224	0.511262	0.511227
$\varepsilon Nd(T)^b$	–6.5	–7.9	–7.0	–5.6	–4.8	–4.8	–5.5	–6.1	–5.3	–6.0
$T_{DM}(Ga)^c$	2.02	2.40	2.12	2.02	1.68	1.92	1.81	1.98	1.82	1.74
$^{87}Rb/^{86}Sr$	9.365	0.119	0.039	0.009	0.015	0.007	0.006	0.014	0.009	0.009
$^{87}Sr/^{86}Sr(0)^d$	0.856799	0.719390	0.708468	0.707943	0.708102	0.708192	0.708157	0.708569	0.707992	0.710461
$^{87}Sr/^{86}Sr(T)$	0.742404	0.717938	0.707987	0.707838	0.707915	0.708110	0.708088	0.708401	0.707881	0.710355
$^{238}U/^{204}Pb$	1.58	1.59	0.44	2.44	0.90	0.95	0.34	0.91	0.73	0.36
$^{232}Th/^{204}Pb$	8.71	22.35	5.96	99.54	8.39	18.34	3.02	7.92	8.02	2.89
$^{206}Pb/^{204}Pb(0)^e$	17.852	17.992	17.648	17.951	17.674	17.767	17.496	17.636	17.660	17.627
$^{206}Pb/^{204}Pb(T)$	17.468	17.605	17.542	17.358	17.455	17.536	17.414	17.416	17.482	17.539
$^{207}Pb/^{204}Pb(0)^e$	15.748	15.728	15.698	15.718	15.690	15.688	15.696	15.674	15.683	15.696
$^{207}Pb/^{204}Pb(T)$	15.721	15.701	15.691	15.677	15.674	15.672	15.690	15.659	15.671	15.690
$^{208}Pb/^{204}Pb(0)^e$	38.142	38.946	38.039	41.057	38.222	38.260	37.580	38.148	38.086	37.703
$^{208}Pb/^{204}Pb(T)$	37.598	37.550	37.667	34.843	37.698	37.115	37.391	37.654	37.585	37.523

[a] Measured $^{143}Nd/^{144}Nd$ ratios were normalized to $^{146}Nd/^{144}Nd = 0.7219$. Uncertainties for the $^{143}Nd/^{144}Nd$ ratios in the samples correspond to 2 in the fifth decimal place, representing 2σ of the mean. La Jolla Nd-standard analyzed during the course of this study yielded $^{143}Nd/^{144}Nd = 0.511852 \pm 24$ (n=3)

[b] εNd(T) was calculated using the present day bulk earth (CHUR) value of $^{143}Nd/^{144}Nd = 0.512638$ recalculated at 855 Ma and $^{147}Sm/^{144}Nd = 0.1967$. The εNd(T) values represent the deviation of $^{143}Nd/^{144}Nd(T)$ in parts per 10^4 from the CHUR value at time T = 855 Ma

[c] Depleted mantle model ages (T_{DM}) were calculated using $^{143}Nd/^{144}Nd = 0.513151$ and $^{147}Sm/^{144}Nd = 0.2136$ for the depleted mantle

[d] $^{87}Sr/^{86}Sr$ ratios were normalized to $^{86}Sr/^{88}Sr = 0.1194$. Uncertainties for the measured $^{87}Sr/^{86}Sr$ were less than 4 in the fifth decimal place, corresponding to 2σ of the mean. The NBS-987 Sr standard analyzed during the course of this study yielded $^{87}Sr/^{86}Sr = 0.710245 \pm 23$ (n=4)

[e] Pb data corrected for mass fractionation. Estimated errors are less than ± 0.05%

Nd-, Sr-, and Pb-isotopic ratios were measured with a multi-collector Thermal Ionization Mass Spectrometer (TIMS, VG Sector). 100–200 mg of powder rock samples were dissolved in HF–HNO_3 and HCl acids. Nd- and Sr-isotopes were measured using the procedures established for our laboratory at the University of Rochester (Basu et al. 1991). The La-Jolla Nd standard and the NBS-987 Sr standard were routinely measured during the course of this study and the results of these analyses are reported in Table 14.3. Pb isotopes were measured using the silica-gel technique in our laboratory (Sharma et al. 1992). Filament temperature during Pb-isotope ratio measurements was monitored continuously and raw ratios were calculated as weighted averages of the ratios measured at 1,150°C, 1,200°C and 1,250°C, respectively. The reported Pb-isotopic data are corrected for mass fractionation of 0.12 ± 0.03% per amu based on replicate analyses of the NBS-981 Equal Atom Pb Standard measured in the same fashion. Our laboratory procedural blanks were less than 400 pg for Sr and less than 200 pg for both Nd and Pb. No blank correction was necessary for the isotope ratios measured.

14.4 Results

14.4.1 U–Pb Zircon Ages

Thirteen zircon grains separated from the gabbroic anorthosite of our study yield a mean age of 855 ± 31 Ma as shown in the Concordia diagram of Fig. 14.3. The measured U–Pb isotopic ratios and the Th/U ratios obtained from the cores of these zircon grains are shown in Table 14.1 and the CL images of the spots are also available (Fig. 14.2c). $^{207}Pb/^{206}Pb$ ages for these zircons are not considered since for zircons less than 1 Ga old, $^{206}Pb/^{238}U$ ages provide better precision and accuracy. The 2σ errors for the U–Pb age is 3.6%, which is typical for ion probe ages in the age range of 800–900 Ma, similar to the zircons of the present study. The calibration curve obtained for the standards analyzed in the same session as the Chilka zircons is acceptable. Hence, a better precision could only be achieved if the zircons of the present study had higher U concentrations. The zircons have a high Th/U ratio ranging from 2.1 to 3.7 (Table 14.1) indicating their igneous origin. This is in accordance with the prismatic outlines of the zircon grains and oscillatory zoning as shown in the CL images (Fig. 14.2c). Hence, the age of the zircons determined in this study is considered to reflect the igneous crystallization age of the Chilka anorthosites.

The U–Pb zircon age of the Chilka anorthosite reported in this study is particularly important given the scarcity of zircons in anorthosites. Our age estimate of the Chilka anorthosite is older by at least 32 million years and very likely by

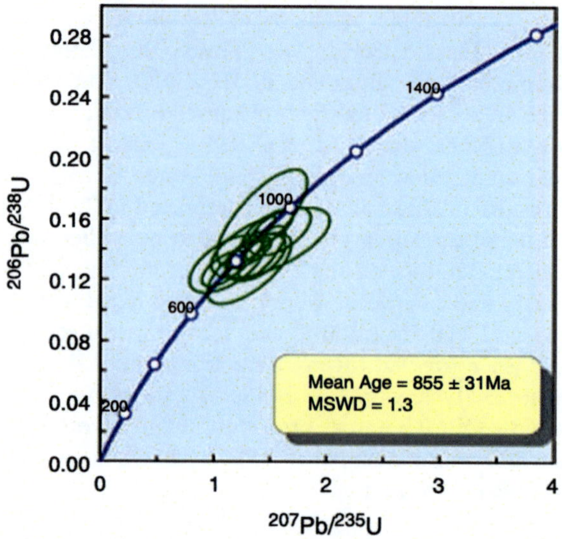

Fig. 14.3 Concordia diagram showing the U-Pb zircon ion micro-probe data for the gabbroic anorthosite (EG-7b) from the Chilka anorthosite complex. Mean age of emplacement of the Chilka anorthosites is 855 ± 31 (2σ) Ma. This age is considerably older than an earlier age estimate of ~792 Ma of this complex by Krause et al. (2001) but younger than the ~983 Ma age estimate by Chatterjee et al. (2008) (see text for discussion). Our geochronological data imply that the Chilka anorthosites are among the youngest Proterozoic massifs in the world

as much as 60 million years than the ~792 Ma age estimated by Krause et al. (2001) based on U–Pb zircon ages of ferrodiorite dikes, which intrude the Chilka anorthosite but are interpreted as almost coeval with these anorthosites. Both of the above-mentioned ages of the Chilka anorthosite are much younger than an earlier whole rock Rb–Sr isochron age of ~1,400 Ma (Sarkar et al. 1981) based on four samples only. This older Rb–Sr 'isochron' (Sarkar et al. 1981) has been recently reinterpreted as a result of mixing between the ascending hot anorthosite pluton and felsic melts generated in its thermal aureole (Krause et al. 2001) in the older country rocks. Thus the nominally concordant 855 Ma-age of the zircons from the gabbroic anorthosite as reported here is considered most likely to be the true crystallization age of the Chilka anorthosite massif. However, a recent U–Pb zircon study (ID-TIMS) suggests that the Chilka anorthosites crystallized ~983 Ma ago (Chatterjee et al. 2008). Although this recent data of Chatterjee et al. (2008) deserves respectful scrutiny, we are puzzled by this age discrepancy of some ca. 130 Ma. The multiple trace elements, Sm–Nd, U–Pb and Rb–Sr isotope systematics that we examined on the same suite of rocks, collected in the field guided by the petrological mapping (Fig. 14.1), and discussed below portray a coherent petrogenetic evolution of the anorthosite suite of rocks and the adjacent country rocks.

14.4.2 Trace Element Geochemistry

Multiple trace element concentrations of the ten whole-rock samples of this study are given in Table 14.2. Chondrite-normalized rare earth element (REE) patterns of the Chilka anorthosites and associated rocks are shown in Fig. 14.4. The anorthosites and the gabbroic anorthosite show an overall uniform light-REE (LREE) enriched pattern ($La/Sm_{(N)}$ = 3.2–7.6) with prominent positive Eu-anomalies (average Eu/Eu* = 10.2), although the gabbroic anorthosite has higher concentrations of LREEs compared to the anorthosites. The pyroxene granulite and the mangerite have much higher REE concentrations than the anorthosites and gabbroic anorthosite. While the mangerite shows a slightly positive Eu-anomaly (Eu/Eu* = 1.25), the pyroxene granulite shows a negative Eu-anomaly (Eu/Eu* = 0.8) similar to average upper continental crustal rocks (Taylor and McLennan 1985). The mangerite also shows enriched, slightly concave upwards HREE patterns ($Gd/Lu_{(N)}$ = 0.67) whereas the anorthosites, gabbroic anorthosite, and the granulite show overall flat to depleted heavy REE (HREE) patterns ($Gd/Lu_{(N)}$ = 0.99–4.75).

Whole rock multiple trace element data for all the above ten samples are shown in Fig. 14.5 in a spider diagram plotted normalized to the NMORB values of these elements (Sun and McDonough 1989) with progressively less incompatible elements arranged from left to the right. Among the most incompatible elements, Rb is characteristically depleted in all the anorthosites. The gabbroic anorthosites as well as the mangerite show smaller Rb depletions (with respect

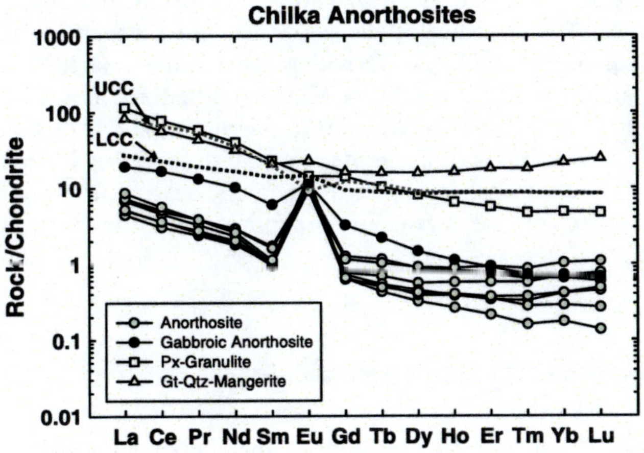

Fig. 14.4 Chondrite-normalized rare earth element patterns of the Chilka anorthosites and associated rocks. Also shown for comparison (*dotted lines*) are the average lower (LCC) and upper continental crustal (UCC) compositions (Taylor and McLennan 1985). The gabbroic anorthosite with LREE pattern similar to those of the anorthosites possibly represents the composition of the parent melt of the anorthosites (see text for discussion). The REE patterns of the pyroxene-granulite and the mangerite are similar to that of the average upper continental crust, while the average LCC pattern is more akin to those of the anorthosites and the gabbroic anorthosite

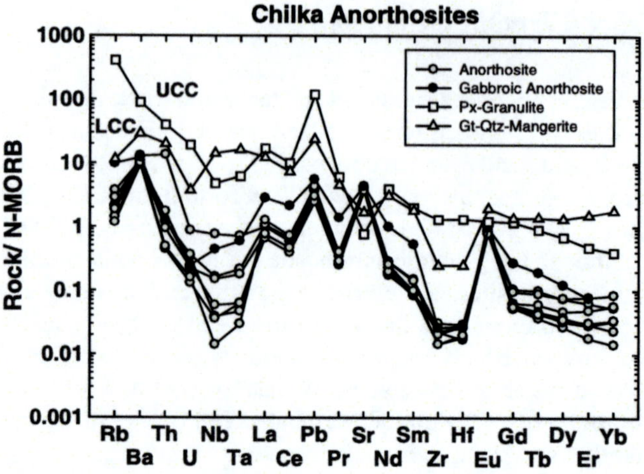

Fig. 14.5 N-MORB-normalized trace element patterns for the Chilka anorthosites and associated rocks. Also shown (*dotted lines*) for comparison are the compositions of the average lower (LCC) and upper continental crust (UCC) (Taylor and McLennan 1985). The anorthosites and the gabbroic anorthosite sample show prominent depletions in Rb, Nb, Ta, Zr, Hf and enrichments in Ba, Pb, Sr and Eu. Note that the apparent depletion of Pr observed for the anorthosites and gabbroic anorthosite is an artifact of data representation, where Pr is plotted between the two enriched elements, Pb and Sr

to Ba) compared to the anorthosites, while the pyroxene granulite is enriched in Rb (Fig. 14.5). The anorthosites also show prominent depletions in Nb, Ta, Zr, Hf, and to some extent in U and Th with respect to the light-REEs, and positive Ba, Sr, Pb and Eu anomalies in this N-MORB-normalized plot. The anorthosites show extremely low U/Pb (0.006–0.037; average = 0.014) (Table 14.2) while the pyroxene granulite and mangerite show slightly higher U/Pb values (average 0.025). The anorthosites and the mangerite show high Th/U values (average 14.4) compared to the pyroxene granulite (Th/U = 5.3) as shown in Table 14.2.

14.4.3 Sr–Nd–Pb Isotope Geochemistry

Nd-, Sr-, and Pb-isotopic compositions of the Chilka anorthosites are given in Table 14.3. The initial isotopic ratios are calculated based on the ~855 Ma U–Pb zircon age of these anorthosites as determined in this study. Initial Sr-isotopic ratios (Fig. 14.6) vary from 0.70784 to 0.71036 for the anorthosites, while those of the pyroxene–granulite and mangerite are more radiogenic at 0.74240 and 0.71794, respectively, and overlap with the compositions of the local crust from the Chilka region (Rickers et al. 2001). The pyroxene–granulite and mangerite

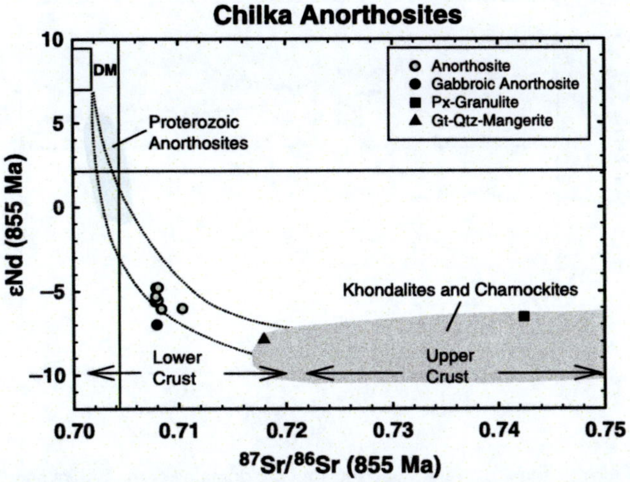

Fig. 14.6 Initial Nd-Sr isotopic compositions of the Chilka anorthosites and associated rocks. Also shown are the domains of the country rocks consisting of khondalites and charnockites from the Chilka region (Rickers et al. 2001), the depleted mantle (DM), lower and upper crust, at the time of emplacement of the anorthosites at ~855 Ma as well as the initial isotopic compositions of other Proterozoic anorthosites from literature (Emslie and Hegner 1993; Scoates and Frost 1996; Demaiffe et al. 1986; Hamilton 1994). The pyroxene granulite and mangerite have compositions similar to those of the country rocks while the anorthosites and the gabbroic anorthosite have less radiogenic Sr. The anorthosites and gabbroic anorthosite fall on a possible mixing trend between the depleted mantle and the continental crust although, all the samples plot closer to the lower crustal end-member, indicating dominant contribution from a lower crustal contaminant to a depleted mantle source (see text for discussion)

samples also have very radiogenic present-day Sr-isotopic ratios along with high $^{87}Rb/^{86}Sr$ ratios (9.37 and 0.12 respectively) as shown in Table 14.3. In contrast, the anorthosites have considerably lower $^{87}Rb/^{86}Sr$ (average 0.013) ratios. Initial ε_{Nd} of the anorthosites and associated rocks range from –7.9 to –4.8 (Fig. 14.6) while the $^{147}Sm/^{144}Nd$ ratios of these rocks are variably low and range from 0.09 to 0.13 (Table 14.3). In the combined Nd–Sr isotopic plot as shown in Fig. 14.6, the Chilka anorthosites lie on an apparent mixing trend between the depleted mantle and the local crust comprising khondalites and charnockites at Chilka (Rickers et al. 2001). Other global Proterozoic massif anorthosites with comparatively less negative to slightly positive initial ε_{Nd} and correspondingly less radiogenic Sr isotopic compositions, e.g. those from the Grenville province (Emslie and Hegner 1993), the Laramie anorthosites (Scoates and Frost 1996) and the Rogaland anorthosites (Demaiffe et al. 1986) also fall on this mixing trend (Fig. 14.6). Anorthositic rocks from the Nain plutonic suite (Hamilton 1994) show less radiogenic Nd and slightly more radiogenic Sr compared to other anorthositic rocks of similar age and overlap with the compositions of the Chilka anorthosites. We also calculated the initial Nd and Sr isotope data at

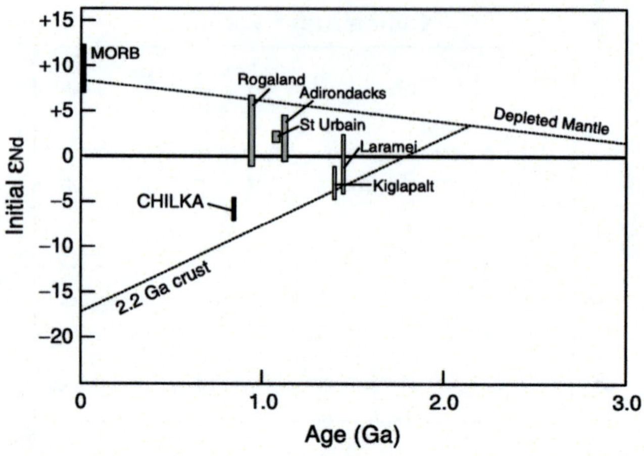

Fig. 14.7 A plot of initial ε_{Nd} versus age showing the domains of the Chilka anorthosite (present study) and a few other global Proterozoic massif anorthosites (Ashwal and Wooden 1983; Demaiffe et al. 1986; Ashwal 1993; Scoates and Frost 1996). The Nd isotopic composition of the Chilka anorthosites fall between the depleted mantle curve and the average local crust beneath the Eastern Ghats, comprising khondalites and charnockites with depleted mantle model ages of ~2.2 Ga (Rickers et al. 2001)

983 Ma, the age of the Chilka anorthosites as suggested by the recent study of Chatterjee et al. (2008). At 983 Ma, the preferred age by Chatterjee et al. (2008), the initial ε_{Nd} values of the Chilka anorthosites and gabbroic anorthosite range from −5.7 to −3.1 that does not alter our conclusions regarding the petrogenesis of these rocks.

Depleted mantle Nd model ages (T_{DM}) for the anorthosites and gabbroic anorthosite range from 1.7 to 2.1 Ga (average 1.9 Ga) while those for the mangerite and pyroxene granulite are 2.4 and 2.0 Ga, respectively (Table 14.3). When plotted on an initial ε_{Nd} versus age diagram, as shown in Fig. 14.7, the initial Nd isotopic composition of the Chilka anorthosites plots between the evolution curves of the depleted mantle and a typical 2.2 Ga old crust similar to the country rocks around Chilka comprising khondalites and charnockites (Rickers et al. 2001). Proterozoic anorthosite massifs like those in Rogaland, Adirondacks, St. Urbain etc. (Ashwal and Wooden 1983; Demaiffe et al. 1986) plot above the chondritic evolution line while others like Kiglapait, Laramie (Ashwal 1993; Ashwal and Wooden 1983; Scoates and Frost 1996) as well as the Chilka anorthosites plot mostly below the chondritic evolution line in the field of negative ε_{Nd} values.

Present day Pb isotopic compositions of the Chilka anorthosites are plotted in Fig. 14.8. In $^{207}Pb/^{204}Pb$ versus $^{206}Pb/^{204}Pb$ and $^{208}Pb/^{204}Pb$ versus $^{206}Pb/^{204}Pb$ plots, all the samples of the present study overlap with the khondalites and charnockites from the Chilka region as reported by Rickers et al. (2001). The initial Pb isotopic ratios at 855 Ma are not much different from the measured values (Table 14.3),

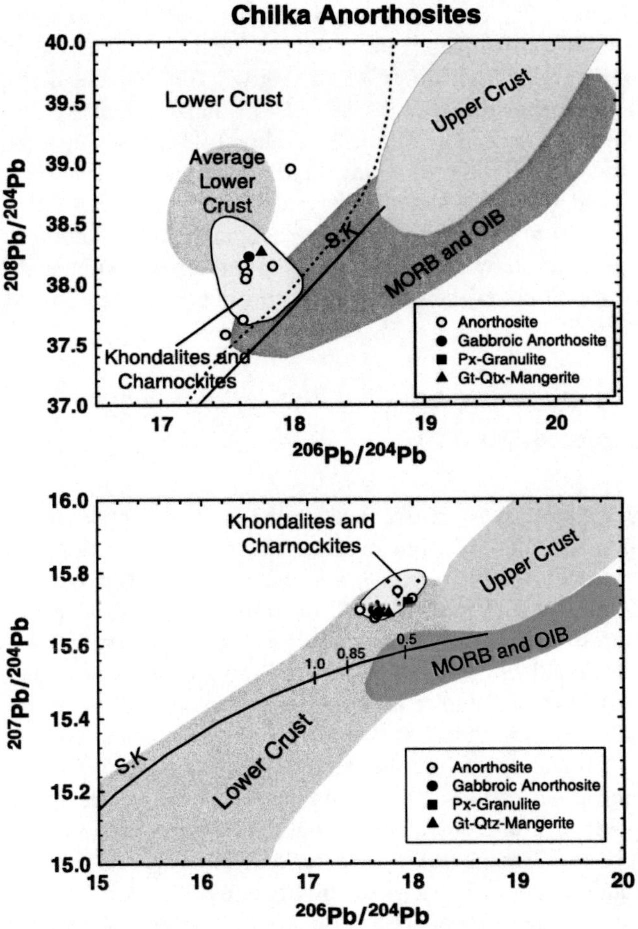

Fig. 14.8 Present day Pb-isotopic composition of the Chilka anorthosites and associated rocks. Also shown for comparison are the domains of the lower (*dashed line*, average composition is shown by the shaded region) and upper crust and mantle-derived rocks (Zartman and Doe 1981) as well as country rocks from the Chilka region comprising khondalites and charnockites (Rickers et al. 2001). All the samples of the present study overlap with the khondalites and charnockites and plot above the Stacey and Kramers (1975) Pb-evolution curves (S-K) in both the diagrams. The tick marks on part of the S-K Pb-evolution curve in the lower diagram represent two-stage model Pb-ages in Ga. The Pb-isotopic composition of the anorthosites and associated rocks are consistent with our interpretation of the geochemical data involving the depleted mantle and the lower crust in their genesis

except for the sample EG-7a, which has a Th concentration of 1.61 ppm that is much higher than the other anorthosites of the present study. The initial $^{206}Pb/^{204}Pb$ of the anorthosites and associated rocks range from 17.36 to 17.61, initial $^{207}Pb/^{204}Pb$ values range from 15.66 to 15.72 while initial $^{208}Pb/^{204}Pb$ ratios range from 37.12 to 37.70. The only exception is the anorthosite sample EG-7a, which has much lower initial $^{208}Pb/^{204}Pb$ of 34.84 (Table 14.3). This sample also has the highest measured $^{208}Pb/^{204}Pb$ of 41.06, high $^{232}Th/^{204}Pb$ and high-μ ($^{238}U/^{204}Pb$). Two other anorthosites EG-9 and EG-15 have relatively lower measured $^{208}Pb/^{204}Pb$ but their initial $^{208}Pb/^{204}Pb$ (Table 14.3) are comparable to the other anorthosites of the present study. The anorthosites have high present-day $^{207}Pb/^{204}Pb$ and $^{208}Pb/^{204}Pb$ and plot above the Stacey and Kramers (1975) evolution curves in Fig. 14.8. The Pb-isotopic composition of the anorthosites and associated rocks (Fig. 14.8) show the involvement of the lower continental crust with the depleted mantle-derived gabbroic anorthositic magma as parental to the anorthosite suite of rocks.

14.5 Discussion

Massif-type anorthosites, mostly found in the mid-Proterozoic are characterized by low Mg-numbers (40–50), high alumina contents (20%) (Ashwal 1993), and there is clear evidence for their intrusion as melts (Valley and O'Neill 1982; Wiebe 1980). The absence of water is notable in these rocks and the depth of emplacement as indicated by oxygen isotopic studies (Bohlen et al. 1985; Valley et al. 1990; Valley and O'Neill 1982) is in the range of 5–13 km, although other estimates from geothermo-barometric studies indicate deeper emplacement (Berg 1977; Berg and Docka 1983; Newton 1985) at the base of the continental crust (~25 km). This discrepancy is clearly due to different equilibration processes involving H_2O, CO_2 and among various cations in different minerals of the anorthosites and also due to varying geological interpretations separating the timing of intrusion and granulite metamorphism. Mafic phases in these rocks are less abundant as a result of plagioclase flotation in the early stages of solidification.

The chondrite-normalized REE pattern of the gabbroic anorthosite sample of the present study with its low overall concentrations and positive Eu-anomaly is similar to the average lower continental crust (LCC) (Rudnick 1995; Rudnick and Fountain 1995; Taylor and McLennan 1985) (Fig. 14.4) except for its strongly HREE-depleted nature compared with the LCC. The Chilka anorthosites have similar REE patterns but lower overall concentrations and more prominent Eu-anomalies compared to the gabbroic anorthosite. The gabbroic anorthosite in our opinion is possibly very close in composition to the parental melt of the anorthositic rocks. This supposition is in concert with the fact that in one of the best-studied massif anorthosite bodies of the world in the Adirondack Mountains of New York, the anorthositic gabbros are also considered to represent the parental melt composition for the anorthosites (e.g. Olson 1992; Olson and Morse 1990; Simmons and Hanson 1978). Anorthosites are formed from Al-rich gabbroic melt by fractional crystallization of plagioclase. The pyroxene granulite of the present study with a slightly negative Eu-anomaly, in contrast to the anorthosites, has REE compositions similar to upper continental crustal rocks (UCC) (Taylor and McLennan 1985) (Fig. 14.4).

The enriched, slightly concave upwards HREE pattern of the mangerite sample, compared to both UCC and LCC, in Fig. 14.4 is due to the presence of abundant garnet in these coarse-grained rocks.

In the N-MORB normalized plot, the anorthosites and the gabbroic anorthosite are characterized by Rb, Th, U, Nb, Ta depletions (Fig. 14.5), positive Eu, Pb, and Sr anomalies as well as low U/Pb ratios (average = 0.014) (Table 14.2) similar to average lower continental crustal rocks (Taylor and McLennan 1985). The low $^{147}Sm/^{144}Nd$ of these rocks (average = 0.108) (Table 14.3) is also similar to those of lower crustal granulites (Ben Othman et al. 1984; Rudnick and Presper 1990). The anorthosites and the mangerite of our study show high Th/U ratios (average 14.4) (Table 14.2), similar to Archean lower crustal granulites (Rudnick and Presper 1990) but much higher than the average depleted MORB mantle (Workman and Hart 2005) or the upper continental crust. The pyroxene granulite however, has much lower Th/U (5.3), as shown in Table 14.2, similar to upper crustal values (Condie 1993). $^{87}Rb/^{86}Sr$ ratios of the anorthosites and the gabbroic anorthosite (average 0.013) (Table 14.3) are however, much lower than lower crustal values and are similar to those of depleted MORB mantle (Workman and Hart 2005). Based on trace element patterns and ratios only, the Chilka anorthosites and gabbroic anorthosite show a compositional similarity with the average lower continental crust. However, to argue that the Chilka anorthosite are derived from melting of the lower crust would require large degree melting of this lower crustal source, which is problematic because of heat source issues as discussed later in this section.

Sr-isotopes, which can easily distinguish between mantle and upper crust-derived rocks, are less sensitive in distinguishing lower crustal rocks from mantle-derived rocks, because of the Rb depletion and mantle-like low Rb/Sr ratios in the lower crust especially under high-pressure conditions (Heier 1973; Spooner and Fairbairn 1970; Weis 1986). Lower crustal melts however, have much lower U/Pb and Sm/Nd ratios compared to the upper mantle-derived melts (Heier 1973; Rudnick and Fountain 1995). On the correlated Nd–Sr isotope diagram (Fig. 14.6), the anorthosites and the gabbroic anorthosite with negative initial ε_{Nd} values fall on a hypothetical mixing trend between the depleted mantle (DM) and the lower continental crust. This relationship is important even though all samples plot closer to the lower crustal end-member, similar to anorthositic rocks from the Nain plutonic suite (Hamilton 1994), indicating dominant contribution from the lower crust and only minor contributions from the depleted mantle. It is possible that there might be samples of Chilka anorthositic rocks, not sampled in our study, that might lie higher up in the trend of the curve in Fig. 14.6 between the anorthosites of the present study and depleted mantle (DM), which have similar initial Nd–Sr isotopic compositions like other global Proterozoic anorthosites from the Adirondacks, the Laramie and Rogaland complexes (Ashwal and Wooden 1983; Demaiffe et al. 1986; Emslie and Hegner 1993; Scoates and Frost 1996). It is noteworthy that the gabbroic anorthosite, which may represent the parent melt of the anorthosites, has an initial ε_{Nd} of −7 at 855 Ma, that is similar to that of an early Proterozoic (~2.2 Ga) age crust consistent with the average age of the lower crust of the Eastern Ghats

(Fig. 14.7). The pyroxene–granulite and mangerite have much more radiogenic Sr-isotopic compositions and most likely indicate contributions from the upper crust (Fig. 14.6). The compositions of the pyroxene granulite and mangerite also overlap with those of the local country rocks from the Chilka region comprising khondalites and charnockites (Rickers et al. 2001).

The initial Nd isotopic composition of the Chilka anorthosites and the gabbroic anorthosite at 855 Ma are akin to those of the lower crust and can be explained by melting of the lower continental crust of the Eastern Ghats region that are predominantly late Archean to early Proterozoic in age (Fig. 14.7). Alternatively, the combined Nd–Sr isotopic composition of these rocks could be explained by assimilation of lower crustal rocks by a depleted mantle-derived magma source. Proterozoic anorthosite massifs such as, for example, in Rogaland and the Adirondacks (Basu and Pettingill 1983), which plot above the chondritic evolution line in Fig. 14.7 were clearly derived from depleted mantle sources (Ashwal 1993), while those like the Chilka anorthosites, which plot below the chondritic evolution line must have formed dominantly by assimilation of ancient lower continental crust by depleted mantle-derived magma source. One might argue that depleted mantle-derived melts provided only the heat required for melting the crust. However, a melt derived from a depleted mantle source with low concentrations of the large ion lithophile elements, such as Rb, U, Th, Pb, Nd, and Sr is likely to be noticeably contaminated by lower continental crustal reservoirs characteristically enriched in these elements. In addition, segregation of monomineralic plagioclase in the anorthosites and gabbroic anorthosites would be more easily imprinted with crustal signatures of these incompatible elements.

The Chilka anorthosites and associated rocks are characterized by low initial $^{206}Pb/^{204}Pb$ ratios (Table 14.3) similar to those from the Rogaland anorthositic complex (Demaiffe et al. 1986; Weis 1986). This suggests their origin in a low U/Pb environment. As shown in Fig. 14.8, the present day Pb isotopic compositions of the Chilka anorthosites overlap with those of the lower continental crust (Zartman and Doe 1981) but some mixing with upper mantle-derived melts that are characteristically low in Pb concentration cannot be ruled out. The Pb-isotopic compositions of the Chilka anorthosites and the associated rocks also overlap with those of the country rocks around Chilka consisting of khondalites and charnockites, (Rickers et al. 2001) that lie characteristically above the Pb evolution curve of Stacey and Kramers (1975) (Fig. 14.8). The high $^{207}Pb/^{204}Pb$ ratios at low $^{206}Pb/^{204}Pb$ also indicate the presence of a late Archean–early Proterozoic crustal component in the source of the Chilka anorthosites, which is in agreement with our interpretations, as mentioned above, of the Nd–Sr isotopic composition of these rocks. High $^{207}Pb/^{204}Pb$ ratios may also develop by evolution in a high U/Pb environment with μ ($^{238}U/^{204}Pb$) values higher than those of the second stage growth curve of Stacey and Kramers (1975). This latter scenario is, however, not viable for the Chilka anorthosites which have low U/Pb ratios.

Depleted mantle Nd model ages (T_{DM}) for the Chilka rocks (Table 14.3), which indicate the approximate time of extraction of the melt from a depleted mantle source (DePaolo 1981), are distinctly older than the 855 Ma crystallization age of these rocks as reported in this study and range from mid- to early Proterozoic, from 1.7 to 2.1 Ga for the anorthosites. The T_{DM} for the mangerite and pyroxene granulite are significantly older at 2.4 and 2.0 Ga, respectively. The mid-Proterozoic model ages of the Chilka anorthosites are younger than those expected for late-Archean to early-Proterozoic lower crustal rocks. It is possible that the comparatively young T_{DM} ages of the anorthositic rocks, which are older than their crystallization ages but younger than the T_{DM} ages of the lower crust of this region, do not reflect the time of extraction of the anorthositic melt from the depleted mantle but indicates the age of the mixed lower crust-depleted mantle source. The relative contribution of the lower crustal assimilant to the depleted upper mantle magma source of the anorthosites can be estimated by the initial Nd-isotopic compositions of these rocks as shown in Fig. 14.7. We argue, based on the relative position of the different global anorthosites in Fig. 14.7, that the contribution of the lower crustal assimilant to the depleted mantle magma source was greater for the Chilka and Kiglapait anorthosites whereas the Rogaland and Adirondack anorthosites were derived from a dominantly depleted mantle source with lesser contributions from the lower crust. Another possibility is that the age of the lower continental crust, the potential contaminant for the gabbroic anorthositic magma, in and around Chilka is much older, possibly late Archean to early Proterozoic. In the case of the other anorthosite massifs as shown in Fig. 14.7, such as for the Adirondacks, Rogaland and St. Urbain, the surrounding country rocks were possibly more juvenile and thus could not produce a significant lowering of the initial Nd-isotopic composition by contaminating the parental mantle-derived magma for the anorthositic gabbro.

There are two proposed models for the formation of dry feldspathic magmas that produce anorthosites. The first one involves fractional crystallization of mantle-derived tholeiitic magma trapped at the crust–mantle boundary due to its high density (Morse 1982; Stolper and Walker 1980; Wiebe 1986). The basaltic affinity of the anorthosite suite, the negative correlation of initial Sr isotopic ratios and anorthite contents and the bimodal nature of plutonism in many terranes have been used as arguments for the mantle origin of these rocks (Menuge 1988; Morse 1982). The second model involves partial melting of a high-Al gabbroic granulitic lower crustal source, which was involved in an earlier episode of granitic melt extraction that lead to the formation of the upper crust (Taylor et al. 1984). Based on the low Mg-numbers of the anorthosites as well as other geophysical evidence indicating little and restricted amounts of ferromagnesian material at lower crustal depth, it was argued that massif anorthosites could not have been derived from "first-stage" mantle-derived melts (Simmons and Hanson 1978). Re–Os isotopic studies from the Rogaland anorthosite province also provide strong evidence for a lower crustal source for these anorthosites (Schiellerup et al. 2000), as originally proposed by Taylor et al. (1984).

Although melting of an Al-enriched lower crust seems to explain most of the geochemical characteristics of massif-anorthosites, the source of heat required to produce this massive scale lower crustal melting remains problematic. Traditionally, two heat sources are invoked for the melting of continental crust: (1) the heat released by the decay of radioactive elements within the crust and (2) the heat released from the underlying mantle and conducted through the crust. Anorthosites are characterized by very low Rb/Sr and U/Pb ratios indicating that their source was depleted in large ion lithophile elements (LILE) including the heat-producing elements K, Th, U. This depletion can be explained by a previous episode of melting of the lower crust and extraction of a granitic upper crust enriched in K, U, Th as well as Rb. Hence, a mantle-heat source is required to generate the anorthositic magmas in the lower crust. The question that now arises is how the lower crustal source can be melted with heat derived from the mantle but without physically involving any significant amount of mantle-derived component in the generation of massif anorthosites. Taylor et al. (1984) speculated that the transfer of mantle heat resulted from the movement of continental crust over a "mantle line plume". A consideration of the paleogeographic location of the Eastern Ghats during the emplacement of the Chilka anorthosites is important in this regard.

The Chilka anorthosites, with U–Pb zircon ages of 855 ± 31 Ma, as reported in the present study, are younger than other global Proterozoic massif-anorthosites, which are 1,100–1,400 Ma old. The 930 Ma old Rogaland massif of Norway (Scharer et al. 1996) however, is younger than the rest of the global Proterozoic massif-anorthosites. While the 1,100–1,400 Ma old anorthosites are linked to the Grenville orogeny, which led to the formation of the Mesoproterozoic supercontinent Rodinia, the timing of emplacement of the Chilka anorthosite massif overlaps with rift-related igneous activity from other parts of the world, e.g. those in the East African Orogen (Stern 1994) as well as magmatism in Madagascar (Kroner 2001), related to the breakup of the Proterozoic supercontinent Rodinia in the Neoproterozoic. A number of 830–820 Ma old granitoid and mafic–ultramafic intrusions from the Yangtze craton in south China are also thought to be related to the breakup of Rodinia (Li et al. 2003). Very recently, 820–880 Ma (U–Pb ages) zircons with igneous cores have been reported from the Uluguru anorthosite massif, hosted by the Eastern granulites, in the Neoproterozoic Mozambique belt in Tanzania (Tenczer et al. 2006). The age of the Uluguru anorthosites are similar to those of the Chilka anorthosites of the present study. Together, these are possibly the youngest ages ever reported for Proterozoic massif-type anorthosites.

The Eastern Ghats mobile belt is juxtaposed with east Antarctica in most paleogeographic reconstructions of Gondwana from the middle Proterozoic to the Mesozoic (Dalziel 1997; Dalziel et al. 2000; Li et al. 2003; Meert and Powell 2001) as shown in Fig. 14.9. This correlation is based on the geometric fit of coastlines as well as petrological and large-scale structural similarities. Similar radiometric age estimates of Eastern Ghats belt and the Rayner complex as well as Prydz Bay region of East Antarctica also support the above hypothesis (Mezger and Cosca 1999). Based on depleted mantle Nd model ages and Sr, Pb-isotopic compositions of whole rocks and leached feldspars, respectively, four crustal domains have been recognized (Fig. 14.1) in the Eastern Ghats belt, each of which might have a different tectonothermal history. Similar T_{DM} ages of the eastern units of the Eastern Ghats belt and the Rayner complex and Prydz Bay regions of East Antarctica support their correlation in Rodinia. However, the

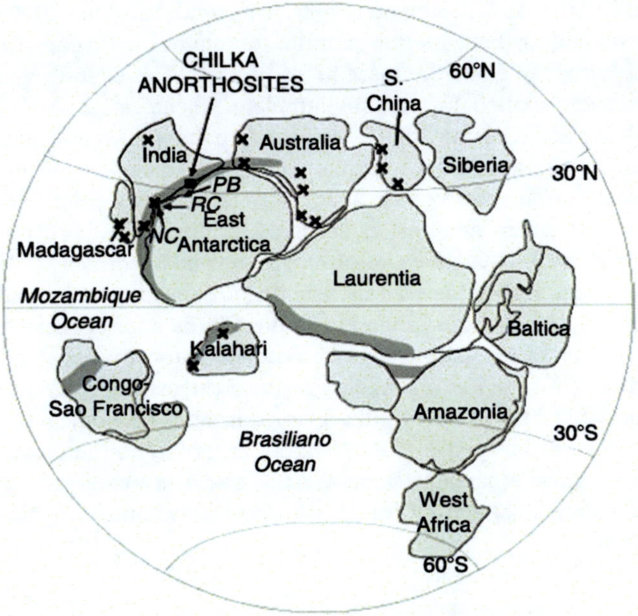

Fig. 14.9 One possible reconstruction of the supercontinent Rodinia in the Neoproterozoic (modified from Dalziel 1997; Dalziel et al. 2000) during the emplacement of the Chilka anorthosites. Also shown is the location of the Chilka anorthosites of the present study, approximate locations of the Napier complex (NC), Rayner complex (RC), Prydz Bay (PB), Grenville-age belts (dark gray bands) as well as 745-830 Ma old volcanics (*crosses*) from around the world related to the breakup of Rodinia (Li et al. 2003). Given that the Eastern Ghats, hosting the Chilka anorthosites, were located ~1000 km away from the margin of the southern Indian cratons in Rodinia ~850 Ma ago, edge-driven convection (King and Anderson 1998; King and Ritsema 2000) could result in mantle upwelling at about ~1000 km away from the margins of cratons below the Eastern Ghats. Chilka anorthosites were derived from this depleted mantle source with contamination from the lower crust. An alternative scenario, based on Rodinia reconstructions by Li et al. (2008), involves rifting and is mentioned in the text

distinctly older ages of the Napier Complex of Antarctica compared to those from the Eastern Ghats precludes its involvement in any orogenic episodes recognized in the Eastern Ghats (Rickers et al. 2001). It must be noted however, that the different Rodinia configurations in the literature do have some drawbacks mainly because of the lack of reliable paleomagnetic data for the different elements of Rodinia (Piper 2000).

In most Rodinia reconstructions at the time of emplacement of the Chilka anorthosites ~850 Ma ago, the Eastern Ghats were located 500–1,000 km away from the margin of the southern Indian cratons (Fig. 14.9). It has been postulated that at the edges of cratons, which are otherwise stable regions of the Earth's lithosphere, changes in lithospheric thickness along with strong lateral viscosity and temperature contrasts induce a small-scale form of convective flow in the mantle beneath the craton margins. This convective flow, known as 'edge-driven convection' (King and Anderson 1998; King and Ritsema 2000), comprises mantle downwelling beneath the margins of cratons and upwelling at about 500–1,000 km away from the margins of cratons. The edge-driven convection model has been applied to explain intraplate volcanism in Africa and South America (King and Ritsema 2000). We postulate that the edge-driven convection model could also be applicable to explain intraplate volcanism during the breakup of Rodinia in the Neoproterozoic. It is possible that the Eastern Ghats, located ~1,000 km from craton margins, as observed in most Rodinia reconstructions of this time (Fig. 14.9), were the site of mantle upwelling driven by 'edge-driven' convection. These mantle-derived melts with subsequent contamination of lower crustal rocks resulted in the formation of the Chilka anorthosites. Alternatively, as shown in the recent paleomagnetic reconstruction by Li et al. (2008) the Rodinia Supercontinent was undergoing regional extensional tectonics ca. 850 Ma. If this is correct, the time of the Chilka anorthosite emplacement and crystallization and lower continental crustal contamination, according to the geochronological, isotopic and trace elemental geochemical study, as presented above, must imply such extensional setting for anorthosite emplacement, crystallization and metamorphism.

14.6 Conclusions

U–Pb zircon ages of the Chilka anorthosites reported in this study indicate that these massif-type anorthosites were emplaced ~855 Ma ago. This age estimate of the Chilka anorthosites together with the 820–880Ma age of the Uluguru anorthosite massif in Tanzania is among the youngest reported for global Proterozoic massif anorthosite occurrences and are much younger than most of the global massif anorthosites, which are of Grenville age.

The Chilka anorthosite massif was possibly emplaced in an extensional setting as the time of emplacement of these rocks coincides with the breakup of the Proterozoic supercontinent Rodinia and associated rifting-related magmatism documented in other parts of the world. This mode of emplacement of the Chilka anorthosite is different from most other Grenville-age massif anorthosites, which are thought to have been emplaced in a compressional regime during the formation of the supercontinent Rodinia.

Our trace element and Nd, Sr, Pb isotopic data indicate that the Chilka anorthosites were formed by melting of a depleted mantle source contaminated by dominantly lower crustal rocks, possibly of late-Archean to early-Proterozoic age. Edge-driven convective flow provides a mechanism for mantle upwelling beneath the Eastern Ghats. Alternatively, regional extensional tectonics of the Rodinia supercontinent at around 850 Ma could also provide a mechanism for mantle upwelling beneath this region.

Our proposed model for the emplacement of the Chilka anorthosites is consistent with the observed geochemical characteristics of these rocks, particularly the low Rb/Sr and U/Pb ratios, depletions in Nb and Ta, positive Eu-anomaly, radiogenic initial Sr isotopic composition, strongly negative initial ε_{Nd} values, and high $^{207}Pb/^{204}Pb$ at low $^{206}Pb/^{204}Pb$. The pyroxene granulite and the mangerite possibly have more complex histories with the upper continental crust also contributing to their formation. T_{DM} ages of the anorthosites possibly reflect the age of this mixed crust–mantle source.

Acknowledgments This study was supported by NSF grants to ARB. We are indebted to late Professor Amitabha Chakrabarti of the Indian Institute of Technology, Kharagpur, India for his encouragement, help and guidance in the field for the sample collection for this study. This paper is dedicated to Professor Mihir K. Bose of Presidency College, Calcutta, who inspired the authors ARB and PKB of this study by his exemplary teaching and enthusiasm in the class room and by his life-long dedication to petrological research in India. The ionprobe facility at UCLA was in part supported by the NSF EAR Instrumentation and Facility Program. The present version of the paper is much improved by the suggestions and reviews by S. A. Morse and two other anonymous reviewers.

References

Ashwal LD (1993) Anorthosites. Minerals and rocks, 21. Springer, New York, 422 pp

Ashwal LD, Wooden JL (1983) Isotopic evidence from the eastern Canadian shield for geochemical discontinuity in the Proterozoic mantle. Nature 306:679–680

Basu AR, Pettingill HS (1983) Origin and age of Adirondack anorthosites re-evaluated with Nd isotopes. Geology 11:514–518

Basu AR, Sharma M, DeCelles PG (1991) Nd, Sr-isotopic provenance and trace element geochemistry of Amazonian foreland basin fluvial sands, Bolivia and Peru: implications for ensialic Andean orogeny. Earth Planet Sci Lett 105:149–169

Ben Othman D, Polve M, Allegre CJ (1984) Nd–Sr isotopic composition of granulites and constraints on the evolution of the lower continental crust. Nature 307:510–515

Berg JH (1977) Regional geobarometry in contact aureoles of anorthositic Nain complex, Labrador. J Petrol 18:399–430

Berg JH, Docka JA (1983) Geothermometry in the Kiglapait contact aureole, Labrador. Am J Sc 283:414–434

Bhattacharya S, Sen SK, Acharyya A (1994) The structural setting of the Chilka lake granulite-migmatite-anorthosite suite with emphasis on the time relation of charnockites. Precam Res 66:393–409

Bohlen SR, Valley JW, Essene EJ (1985) Metamorphism in the Adirondacks. I. Petrology, pressure and temperature. J Petrol 26:971–992

Chatterjee N, Crowley JL, Mukherjee A, Das S (2008) Geochronology of the 983-Ma Chilka Lake Anorthosite, Eastern Ghats Belt, India: implications for pre-Gondwana tectonics. J Geol 116:105–118

Compston W, William IS, Meyer C (1984) U-Pb geochronology of zircons from lunar breccia 73217 using a sensitive high mass-resolution ion microprobe. J Geophys Res Suppl 89:B525–B534

Condie KC (1993) Chemical composition and evolution of the upper continental crust: contrasting results from surface samples and shales. Chem Geol 104:1–37

Dalziel IWD (1997) Neoproterozoic–Paleozoic geography and tectonics: Review, hypothesis, environmental speculation. GSA Bull 109:16–42

Dalziel IWD, Mosher S, Gahagan LM (2000) Laurentia–Kalahari collision and the assembly of Rodinia. J Geol 108:499–513

Demaiffe D, Weis D, Michot J, Duchesne JC (1986) Isotopic constraints on the genesis of the Rogaland Anorthositic Suite (Southwest Norway). Chem Geol 57:167–179

DePaolo DJ (1981) Neodymium isotopes in the Colorado Front Range and crust-mantle evolution in the Proterozoic. Nature 291:193–196
Dobmeier C, Simmat R (2002) Post-Grenvillean transpression in the Chilka Lake area, Eastern Ghats Belt – implications for the geological evolution of peninsular India. Precam Res 113:243–268
Emslie RF, Hegner E (1993) Reconnaissance isotopic geochemistry of anorthosite-mangerite-charnockite-granite (AMCG) complexes, Grenville Province, Canada. Chem Geol 106:279–298
Hamilton MA (1994) Review of isotopic data for the Nain plutonic suite. In: Berg JH (ed) Anorthosite, granitoid and related rocks of the Nain plutonic suite. IGCP-290 & 315 Field excursion guidebook, pp 15–19
Heier KS (1973) Geochemistry of granulite facies rocks and problems of their origin. Phil Trans R Soc A 230:429
Herz N (1969) Anorthosite belts, continental drift, and the anorthosite event. Science 164:944–946
Hoffman PF (1991) Did the breakout of Laurentia turn Gondwanaland inside-out? Science 252:1409–1412
King SD, Anderson DL (1998) Edge-driven convection. Earth Planet Sci Lett 160:289–296
King SD, Ritsema J (2000) African hotspot volcanism: small-scale convection in the upper mantle beneath cratons. Science 290:1137–1140
Krause O, Dobmeier C, Raith MM, Mezger K (2001) Age of emplacement of massif-type anorthosites in the Eastern Ghats belt, India: constraints from U-Pb zircon dating and structural studies. Precam Res 109:25–38
Kroner A (2001) The Mozambique belt of East Africa and Madagascar: significance of zircon and Nd model ages for Rodinia and Gondwana supercontinent formation and dispersal. S Afr J Geol 104:151–166
Leelanandam C (1990) The anorthosite complexes and Proterozoic mobile belt of Peninsular India: a review. In: Naqvi SM (ed) Precambrian continental crust and its economic resources. Developments in Precambrian geology. Elsevier, Amsterdam, The Netherlands, pp 409–435
Li ZX, Li XH, Kinny PD et al (2003) Geochronology of Neoproterozoic syn-rift magmatism in the Yangtze Craton, South China and correlations with other continents: evidence for a mantle superplume that broke up Rodinia. Precam Res 122:85–109
Li ZX, Bogdanova SV, Collins AS et al (2008) Assembly, configuration, and break-up history of Rodinia: a synthesis. Precam Res 160:179–210
Meert JG, Powell CM (2001) Editorial – assembly and break-up of Rodinia: introduction to the special volume. Precam Res 110:1–8
Menuge JF (1988) The petrogenesis of massif anorthosites: a Nd and Sr isotopic investigation of the Proterozoic of Rogaland/Vest-Agder, SW Norway. Contrib Mineral Petrol 98:363–373
Mezger K, Cosca MA (1999) The thermal history of the Eastern Ghats Belt (India) as revealed by U-Pb and 40Ar/39Ar dating of metamorphic and magmatic minerals: implications for the SWEAT correlation. Precam Res 94:251–271
Moores EM (1991) Southwest U.S.-East Antarctic (SWEAT) connection: a hypothesis. Geology 19:425–428
Morse SA (1982) A partisan review of Proterozoic anorthosites. Am Miner 67:1087–1100
Newton RC (1985) Temperature, pressure and metamorphic fluid regimes in the amphibolite facies to granulite facies transition zones. In: Tobi AC, Touret JLR (eds) The deep Proterozoic crust in the north Atlantic provinces. D. Reidel Publishing Company, Dordrecht, The Netherlands, pp 75–104
Olson KE (1992) The petrology and geochemistry of mafic igneous rocks in the anorthosite-bearing Adirondack Highlands, New York. J Petrol 33:471–502
Olson KE, Morse SA (1990) Regional Al–Fe mafic magmas associated with anorthosite-bearing terranes. Nature 344:760–762
Paces JB, Miller JD (1993) Precise U-Pb ages of Duluth Complex and related mafic intrusions, northeastern Minnesota; geochronological insights to physical, petrogenetic, paleomagnetic,

and tectonomagmatic processes associated with the 1.1 Ga Midcontinent Rift System. J Geophys Res 98:13997–14013
Piper JDA (2000) The Neoproterozoic supercontinent: Rodinia or Paleopangea? Earth Planet Sci Lett 176:131–146
Powell CM, Pisarevsky SA (2002) Late Neoproterozoic assembly of East Gondwana. Geology 30:3–6
Ramakrishnan M, Nanda JK, Augustine PF (1998) Geological evolution of the Proterozoic Eastern Ghats Mobile Belt. Geol Surv India Spec Publ 44:1–21
Rickers K, Mezger K, Raith MM (2001) Evolution of the continental crust in the Proterozoic Eastern Ghats belt, India and new constrains for Rodinia reconstruction: implications from Sm-Nd, Rb-Sr and Pb-Pb isotopes. Precam Res 112:183–210
Rudnick RL (1995) Making continental crust. Nature 378:571–578
Rudnick RL, Fountain DM (1995) Nature and composition of the continental crust: a lower crustal perspective. Rev Geophys 33:267–309
Rudnick RL, Presper T (1990) Geochemistry of intermediate to high-pressure granulites. In: Vielzeufand D, Vidal P (eds) Granulites and crustal evolution. Kluwer, Amsterdam, The Netherlands, pp 523–550
Sarkar A, Bhanumathi L, Balasubrahmanyan MN (1981) Petrology, geochemistry and geochronology of the Chilka Lake igneous complex, Orissa state, India. Lithos 14:93–111
Scharer U, Wilmart E, Duchesne JC (1996) The short duration and anorogenic character of anorthosite magmatism: U-Pb dating of the Rogaland complex, Norway. Earth Planet Sci Lett 139:335–350
Schiellerup H, Lambert RD, Prestvik T et al (2000) Re-Os isotopic evidence for a lower crustal origin of massif-type anorthosites. Nature 405:781–784
Scoates J, Frost CD (1996) A strontium and neodymium isotopic investigation of Laramie anorthosites, Wyoming, USA: implications for magma chamber processes and the evolution of magma conduits in Proterozoic anorthosites. Geochim Cosmochim Acta 60:95–107
Sharma M, Basu AR, Nesterenko GV (1992) Temporal Sr, Nd and Pb isotopic variations in the Siberian flood basalts: Implications for the plume-source characteristics. Earth Planet Sci Lett 113:365–381
Simmons EC, Hanson GN (1978) Geochemistry and origin of massif-type anorthosites. Contrib Miner Petrol 66:119–135
Spooner CM, Fairbairn HW (1970) 87Sr/86Sr initial ratios in pyroxene granulite terranes. J Geophys Res 75:6706–6713
Stacey JS, Kramers JD (1975) Approximation of terrestrial lead isotope evolution by a two-stage model. Earth Planet Sci Lett 26:207–221
Stern RJ (1994) Arc assembly and continental collision in the Neoproterozoic east African orogen. Ann Rev Earth Planet Sci 22:319–351
Stolper E, Walker D (1980) Melt density and the average composition of basalt. Contrib Miner Petrol 74:7–12
Sun S-S, McDonough WF (1989) Chemical and isotopic systematics of oceanic basalts: implications for mantle composition and processes. Magmatism in the Ocean Basins. Geol Soc Spec Publ 42:313–345
Taylor SR, McLennan SM (1985) The continental crust: its composition and evolution. Blackwell Scientific Publications, Oxford, 312 pp
Taylor SR, Campbell IH, McCulloch MT, McLennan SM (1984) A lower crustal origin for massif-type anorthosites. Nature 311:372–374
Tenczer V, Hauzenberger CA, Fritz H et al (2006) Anorthosites in the Eastern Granulites of Tanzania – New SIMS zircon U–Pb age data, petrography and geochemistry. Precam Res 148:85–114
Valley JW, O'Neill JR (1982) Oxygen isotope evidence for shallow emplacement of Adirondack anorthosite. Nature 300:497
Valley JW, Bohlen SR, Essene EJ, Lamb W (1990) Metamorphism in the Adirondacks: II The role of fluids. J Petrol 31:555–596

Weis D (1986) Genetic implications of Pb isotopic geochemistry in the Rogaland anorthositic complex (Southwest Norway). Chem Geol 57:181–199

Wiebe RA (1980) Anorthositic magmas and the origin of Proterozoic anorthosite massifs. Nature 286:564

Wiebe RA (1986) Lower crustal cumulate nodules in Proterozoic dikes of the Nain Complex: evidence for the origin of Proterozoic anorthosites. J Petrol 27:1253–1275

Workman RK, Hart SR (2005) Major and trace element composition of the depleted MORB mantle (DMM). Earth Planet Sci Lett 231:53–72

Zartman RE, Doe BR (1981) Plumbotectonics – the model. Tectonophysics 75:135–162

Chapter 15
Geochemical and Geochronological Data from Charnockites and Anorthosites from India's Kodaikanal–Palani Massif, Southern Granulite Terrain, India

Elizabeth J. Catlos, Kaan Sayit, Poovalingam Sivasubramanian, and Chandra S. Dubey

Abstract The Kodaikanal–Palani Massif is an important component of India's Southern Granulite Terrain; understanding the tectonic history of its rocks lends considerable insight into its role within South India. The massif is located south of the Palghat Cauvery Shear Zone (PCSZ). Compilations of available geochronologic and geochemical information from charnockites north and south of the PCSZ show these rocks largely differ in age, with northern samples recording Archaean crystallization events, whereas those to the south yielding Cambro-Ordovician and Neoproterozoic ages. The Kodaikanal–Palani charnockitic rocks contain monazite grains that fall within the Cambro-Ordovician timescale. The Oddanchatram anorthosite, located along the northern boundary of the Kodaikanal–Palani Massif, contains zircon grains that record mid-Neoproterozoic to Cambro-Ordovician crystallization ages. This anorthosite differs in texture and composition depending on location, that may be the result of its multi-stage metamorphic and/or intrusion history. Charnockitic rocks north and south of the PCSZ also differ geochemically. For example, north of the PCSZ, these rocks become more calcic with increasing SiO_2 contents, whereas those to the south become alkali-calcic. Southern charnockitic

E.J. Catlos (✉)
Department of Geological Sciences, Jackson School of Geosciences, University of Texas at Austin, University Station, C1100, Austin, TX 78712, USA
and
Department of Geological Engineering, Middle East Technical University, Ankara, Turkey
e-mail: ejcatlos@gmail.com

K. Sayit
Department of Geological Sciences, San Diego State University, San Diego, CA 92182-1020, USA
e-mail: ksayit@hotmail.com

P. Sivasubramanian
Marine Geochemistry Research Lab, VO Chidambaram College, SPIC Research Center, Tuticorin 628008, India
e-mail: spsiva@hotmail.com

C.S. Dubey
Department of Geology, University of Delhi, Delhi 110007, India
e-mail: csdubey@gmail.com

J. Ray et al. (eds.), *Topics in Igneous Petrology*, DOI 10.1007/978-90-481-9600-5_15,

rocks tend to have higher K_2O/TiO_2, Zr/SiO_2, Rb/Sr, Ba and Rb contents, but lower Sr/Ba ratios. Using available geochemical data, we find more charnockitic rocks south of the PCSZ record zircon saturation temperatures between 800°C and 900°C than those to the north. Although samples of charnockitic rocks within the Kodaikanal–Palani Massif yield similar monazite ages, the rocks differ in their whole rock geochemistry and zircon and monazite saturation temperatures depending on location. The geochemical data from these rocks suggest that charnockitic rocks within the Kodaikanal–Palani Massif possibly experienced different mechanisms of generation and/or metamorphic histories.

15.1 Introduction

India's Southern Granulite Terrain (Fig. 15.1) is largely composed of charnockite massifs with scattered anorthosite bodies and carbonatite plugs. Charnockites have been thought to be comparatively rare on Earth (Frost and Frost 2008) whereas carbonatites have been termed volumetrically insignificant (e.g., Bell 2001). Intermediate plagioclase massif-type anorthosites appear restricted in Earth's history and are

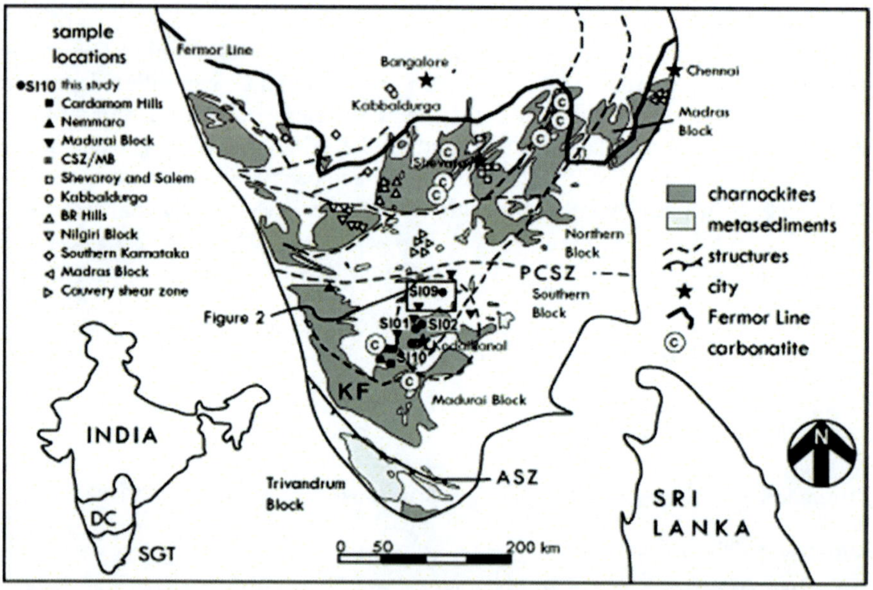

Fig. 15.1 Simplified geologic map of South India after Ghosh et al. (2004) and Fermor (1936). Map of India shows the location of the Southern Granulite Terrain (SGT) and Dharwar Craton (DC). FL Fermor Line, ASZ Anchankovil Shear Zone, PCSZ Palghat Cauvery Shear Zone, KF Kambam fault, CSZ/MB Cauvery Shear Zone/Madurai Block (see Tomson et al. 2006). *Inset* on map shows the location of Fig. 15.2. *Dashed circle* outlines the approximate extent of the Kodaikanal–Palani Massif

likely to have been formed only during the Proterozoic (e.g., Drüppel et al. 2007). The geochemical nature of these rock types is highly variable (e.g., Woolley and Kempe 1989; Frost and Frost 2008), yet numerous papers have proliferated over the years seeking to identify element variations of these rocks specifically in South India (e.g., Holland 1900; Howie and Subramanian 1957; Janardhan et al. 1982; Allen et al. 1985; Bernard-Griffiths et al. 1987; Narayana and Nijagunappa 1987; Staehle et al. 1987; Chacko et al. 1992; Raith et al. 1997; 1999; Bhattacharya and Sen 2000; Rajesh et al. 2000; Kumar 2004; Rajesh and Santosh 2004; Suresha and Srikantappa 2005; Tomson et al. 2006; Rajesh 2007; Catlos et al. 2008). Their presence is linked to the lower crust, thus they can provide insight into the mechanisms operating to generate the Earth's lithosphere. Newton (1992) states that the origin of charnockitic massifs and the mechanisms of incompatible element depletion are major unsolved problems of petrology which must bear fundamentally on the origin of continents.

India's Southern Granulite Terrain is located south of an orthopyroxene isograd identified in 1936 as the Fermor Line (Fig. 15.1; Fermor 1936). The terrain is cut into "northern" and "southern" blocks by the Palghat Cauvery Shear Zone (PCSZ) (e.g., Drury and Holt 1980; Drury et al. 1984; Harris et al. 1994; Yoshida et al. 1998; Jain et al. 2003; Santosh et al. 2004; John et al. 2005; Mishra et al. 2006). The east–west-trending PCSZ is a wide (~70 km) zone marked by E–W trending shear zones (Bhavani, Mettur and Palghat-Cauvery) that extend ~400 km in length (Prasad et al. 2007). It has been described as a right-lateral strike-slip fault, collision zone and/or terrane boundary (Drury et al. 1984; Harris et al. 1994; Ghosh et al. 2004; Cenki and Kriegsman 2005; Chetty and Bhaskar Rao 2006; Shimpo et al. 2006; Rao and Prasad 2006). The PCSZ appears to mark the northern limit of Pan African activity in South India (Mohan and Jayananda, 1999), but relics of Archean petrogenesis may be present in the Palghat–Cauvery area (Rao et al. 2003). North of the PCSZ, charnockite rocks typically yield ages ranging from the Archaean to Paleoproterozoic, whereas southern block charnockitic rocks have a mid-Neoproterozoic to Devonian history and/or overprint (see Table 15.1 and references therein).

Charnockites have recently been technically redefined (Frost and Frost 2008) as Opx-(or Fay-) bearing granitic rocks clearly of igneous origin or that are present as orthogneiss within a granulite terrane. South Indian charnockites have long been described by the latter part of the definition (e.g., Rama Rao 1945; Pichamuthu 1960; Newton et al. 1980; Janardhan et al. 1982; Newton 1989, 1992; Pichamuthu 1990; Santosh and Omori 2008). These specific rocks are thought to have lost their primary igneous identity due to carbonic metamorphism, becoming depleted in Rb and other high field strength elements (HFSE), and heavy rare earth elements (HREE) (Newton 1989). A key unknown and a matter of debate in South Indian tectonics is the source and scale of the CO_2 metamorphism (see Frost and Frost 1987; Santosh et al. 1991; Bhattacharya and Sen 2000; Frost and Frost 2008). The CO_2 may be a coeval, but subsidiary component of a granulite-forming fluid dominated by brines of igneous origin (Hansen et al. 1995).

Here we report ages and geochemical data from charnockite and anorthosite rocks located south of the PCSZ (in the Madurai Block) near the towns of Kodaikanal and Oddanchatram (Figs. 15.1 and 15.2). The rocks represent key locations both within the core and outer topographic expression of the Kodaikanal–Palani (or Palni)

Table 15.1 Ages, saturation temperatures, and references for Southern Granulite Terrain charnockites

Charnockite body	Age (Ma)[a]	Zr temp[b]	Mnz temp[b]	Ref[c]
Cardamom Hill Massif				
various locations	588 ±6 Ma	782 ± 61°C	753 ± 76°C	C1992, R2007, M1996
Anaimalai Hill Massif				
Nemmara	Cambro-Ordovician	844 ± 43°C	827 ± 55°C	K2004
Kodaikanal–Palani Massif				
Kodaikanal	461 ± 26 Ma	706 ± 14°C	664 ± 6°C	This study
Oddachatram	409 ± 21 Ma	837 ± 6°C	798 ± 1°C	This study
Between Kodaikanal and Palani	618 ± 21–786 ± 84 Ma	686 ± 93°C	746 ± 99°C	T2006, C2008
Palani	~550 Ma	808 ± 17°C	–	R1997, J1995, B1995
Dingidul	–	814 ± 51°C	–	S2005
Munnar	Cambro-Ordovician	~900–800°C	–	R2000
Nilgri Block				
various locations	2,460 ± 81 Ma; 2,506 ± 70 Ma; ~2,535 Ma; 2.6 Ga	756 ± 51°C	725 ± 56°C	N1987, J1994, R1999, J1982, S1988
Shevaroy Hills				
Salem	Archaean	737 ± 39°C	721 ± 43°C	N1987, A1985
Dharmapuri	3.3–2.5 Ga		~680°C	A1985
Southern Karnakata				
Kabbaldurga	2,528 ± 5 Ma	796 ± 20°C	779 ± 43°C	A1985, B2000, F1991, J1982, S1987
Yelachipalaiyam	–	~790°C	–	J1982
Hullahalli	–	715 ± 25°C	–	J1982
Swarnavati	–	~770°C	–	J1982
Anechowkur	–	762 ± 31°C	–	J1982
Biligirirangan Hills	3.4; 2.51 Ga	751 ± 29°C	719 ± 42°C	N1987, J1994, J1982, A1985
Madras Block				
various locations	2,555 ± 140 Ma	656 ± 112°C	601 ± 183°C	B1987, H1990, H1957, H1955, S1970, W1978, W1980

[a] See references for details about the age data

[b] If possible, zircon saturation temperatures were calculated using Watson and Harrison (1983) and monazite saturation temperatures were calculated using Montel (1993)

[c] References: A1985, Allen et al. (1985); B1987, Bernard-Griffiths et al. (1987); B1995, Bartlett et al. (1995); B2000, Bhattacharya and Sen (2000); C1992, Chacko et al. 1992; C2008, Catlos et al. (2008); F1991, Friend and Nutman (1991); H1955, Howie (1955); H1957, Howie and Subramanian (1957); H1990, Holland (1900); J1982, Janardhan et al. (1982); J1994, Janardhan et al. (1994); J1995, Jayananda et al. (1995); K2004, Kumar (2004); M1996, Miller et al. (1996); N1987, Narayana and Nijagunappa (1987); R1997, Raith et al. (1997); R1999, Raith et al. (1999); R2000, Rajesh et al. (2000); R2007, Rajesh (2007); S1970, Spooner and Fairbairn (1970); S1987, Staehle et al. (1987); S1988, Srikantappa (1988); S2005, Suresha and Srikantappa (2005); T2006, Tomson et al. (2006); W1978, Weaver (1978); W1980, Weaver (1980)

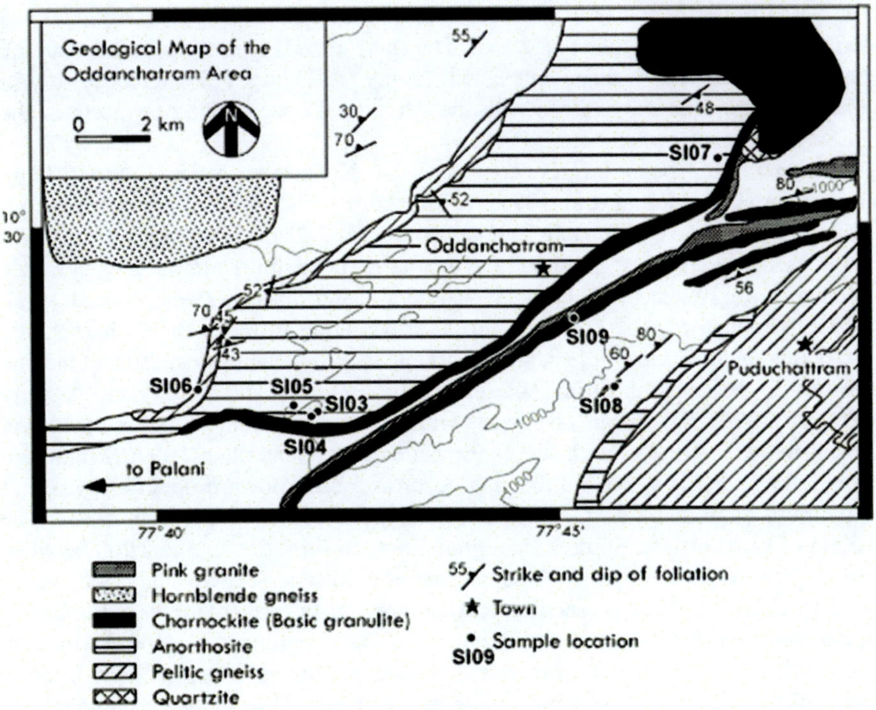

Fig. 15.2 Geological map of the Oddanchatram anorthosite massif (After Sivasubramanian 1993; Satyanarayana et al. 2003)

Massif (e.g., Prakash 1999; Sajeev et al. 2006). The goal is to compare the geochemical and geochronological results of these rocks to each other, and to those collected from nearby assemblages, for better understanding the evolution of the massif. We also compare our results to those reported from other charnockite and anorthosite bodies in the Southern Granulite Terrain.

15.2 Geological Background of the Kodaikanal–Palani Massif

The Kodaikanal–Palani Massif (Fig. 15.1) is the largest highland block of granulites exposed in South India (Drury et al. 1984; Brown and Raith 1996; Raith et al. 1997). The massif is largely composed of charnockitic rocks, mafic granulites interbedded with metapelite, calc-silicate lithologies, and anorthosite complexes, rocks characteristic of extreme crustal temperatures and pressures. The eastern part of the massif is cut by prominent lineament termed the Kambam fault (Balakrishnan et al. 1985; Anto et al. 1997; 1999) or the Karur–Kambam (Kambum or Kamban)–Painavu–Truchur Shear Zone (KKPTSZ) (Ghosh et al. 2004; Tomson et al. 2006; Singh et al. 2006). The fault roughly divides charnockitic rocks and metapelites in the west from orthogneisses and calcsilicates to the east (Fig. 15.1) (see Cenki and Kriegsman 2005; Tomson et al. 2006). The shear zone is speculated to be a terrane boundary with the v-shape attributed to late refolding (Ghosh et al. 2004). The Kambam fault may merge with the Attur fault zone located far north of the PCSZ (Grady 1971; Anto et al. 1999).

Charnockitic rocks largely comprise the Kodaikanal–Palani Massif (e.g., Janardhan 1999; Mohan et al. 1996a, b; Tadokoro et al. 2007), and commonly contain orthopyroxene + sillimanite + K-feldspar + Na-feldspar ± sapphirine ± cordierite ± garnet ± spinel ± phlogopite ± apatite ± monazite ± zircon ± calcite (see Mohan et al. 1996a, b; Prakash 1999; Prakash and Shastry 1999; Prakash et al. 2006; Tadokoro et al. 2007). The presence of garnet in Kodaikanal charnockitic rocks (and charnockites in general) is variable (e.g., Pichamuthu 1970) and may be related to bulk composition, barometric conditions, and/or oxygen fugacity (Prakash 1999). Sapphirine also exists within metasedimentary lithologies that are found to be associated with the charnockites (e.g., Janardhan 1999; Sriramguru et al. 2002), and is a mineral that can occur as a breakdown product of pyrope at conditions as extreme as 17–25 kbar and 1,000–1,600°C (Deer et al. 1993). The origin of sapphirine in South Indian granulites can form due to a reaction between orthopyroxene + sillimanite → sapphirine + cordierite (Brown and Raith 1996; Mohan et al. 1996b) or a reaction between previously formed garnet + sillimanite inclusions → sapphirine + quartz (Braun et al. 2007). Braun et al. (2007) speculate that sapphirine can also be the result of melt-producing (e.g., orthopyroxene + cordierite + sillimanite → sapphirine + quartz + liquid) or melt-consuming reactions (e.g., garnet + liquid → orthopyroxene + sapphirine). Minerals within the charnockitic rocks commonly show reaction textures suggestive of decompression, including sillimanite replacing kyanite, and garnets partially replaced by symplecitic orthopyroxene–plagioclase (e.g., Brown and Raith 1996; Mohan et al. 1996a, b; Anto 1998).

Charnockitic rocks in the Kambam Valley record decompression from 9.2–9.7 kbar to 6.5–7.3 kbar (~800–950°C; Anto 1998). Charnockitic rocks within the massif itself have recorded peak conditions as high as ~12 kbar and 900–1,000°C and may have followed a decompressional pressure–temperature (P–T) path in a clockwise direction towards ~5 kbar and 800°C (Brown and Raith 1996; Raith et al. 1997; Sriramguru et al. 2002; Sajeev et al. 2006). Kodaikanal charnockitic rocks commonly record P–T conditions that range from 750–1,000°C and 4.8–9.5 kbar (Harris 1981; Janardhan 1999; Mohan et al. 1996b; Prakash et al. 2006; Sajeev et al. 2006; Tadokoro et al. 2007). Calculated thermobarometric conditions and proposed P–T paths have led suggestions that the Kodaikanal rocks were buried (to ~25 km) during collision and experienced rapid uplift and extension (Mohan et al. 1996b). Uplift may have occurred in two phases separated by a period of cooling (Brown and Raith 1996; Raith et al. 1997; Tadokoro et al. 2007).

The timing of the tectonic events in the Kodaikanal–Palani Massif have largely centered around the Archaean and Cambro-Ordovician (Table 15.1). Archaean ages have been reported for the Kodaikanal region (see review by Rao et al. 2003), and the rocks may have a 2.6–2.5 Ga protolith (e.g., Mohan and Jayananda 1999; Mohan et al. 1996b). Mid-Neoproterozoic to Cambrian monazite inclusions in garnet are present within pelitic enclaves within Kodaikanal charnockitic rocks (Table 15.1; Catlos et al. 2008). These monazite grains often contain older cores and younger rims, and may represent crystallization during Mid-Neoproterozoic collision followed by Cambrian extension. However, some consider a ~550 Ma as the timing of peak metamorphism in the region (e.g., Mohan and Jayananda 1999; Sajeev et al. 2006).

The Oddanchatram area is located at the northern edge of the Kodaikanal–Palani Massif (Figs. 15.1 and 15.2). This region is characterized by one of the largest exposures of massive anorthosite bodies in South India (Janardhan and Wiebe 1985), and has been suggested to be part of a larger suture between east and west Gondwana (Janardhan 1999; Janardhan and Srikarni 2001). Anorthosites are plutonic rocks composed chiefly or primarily of plagioclase and are commonly found adjacent to charnockites. The Oddanchatram anorthosite has been reported to consist of almost 90% plagioclase (Janardhan and Wiebe 1985) although the composition and textures of the plagioclase can vary. Janardhan and Wiebe (1985) indicate that plagioclase from the region's anorthosites fall into two major categories: primary An60 and secondary An94 that occurs as a breakdown product of garnet. Garnet-bearing assemblages located in close proximity to the Oddanchatram yield peak P–T conditions of ~920°C and ~10.6 kbar (Janardhan and Wiebe 1985; Janardhan 1999). The highest P–T conditions are recorded by anorthosite garnet-bearing assemblages that lack symplectite rims (~920°C and ~11 kbar), whereas rocks with garnets in contact with anorthite and hypersthene yield lower grade conditions (~775°C and 5.7–7.3 kbar; Wiebe and Janardhan 1988).

The Oddanchatram anorthosite cuts surrounding charnockites, high-grade gneisses and migmatities and quartzites. Rao et al. (2003) reports the intrusion to have taken place at 560 ± 12 Ma, although the anorthosite preserves remnant magnetization during 1,100–1,000 Ma and may be related to the Eastern Ghat Orogeny (Satyanarayana et al. 2003).

15.3 Methods

Prior to the geochronological analyses, rocks were petrographically analyzed using an optical microscope and a JEOL733 electron microprobe. We used the electron microprobe to not only locate and image the monazite and zircon grains chosen for geochronology, but also to identify and characterize rock textures. X-ray element maps of some rocks were made using an accelerating potential of 20 kV and a primary current of 150–200 nA; we compositionally analyzed plagioclase in the Oddanchatram anorthosite using conditions of 15 kV and 15–20 nA. Maximum count times are 20 s and raw data was reduced using the ZAF correction.

Monazite and zircon grains in South Indian samples were dated in rock thin section using the CAMECA ims1270 ion microprobe (Tables 15.2 and 15.3) (Harrison et al. 1995; Schneider et al. 1999). Dating monazite in thin section is ideal for

Table 15.2 Geochronological data from Kodaikanal–Palani charnockites

Grain_spot[a]	$^{208}Pb/^{232}Th$ Age (Ma, ±1σ)	ThO_2^+/Th^+ (±1σ)[b]	$^{208}Pb^*$ (%, ±1σ)[c]	$^{208}Pb^*/^{232}Th^+$ (±1σ, × 10^{-2})[d]	$^{207}Pb/^{206}Pb$ Age (Ma, ±1σ)[e]
Charnockite SI02 (N10°27′33″ E77°33′13″)[f]					
m2_s1	606.1 (38.8)	2.761 (0.011)	99.9 (0.01)	3.044 (0.198)	517.0 (26.6)
m7_s2	535.7 (41.9)	2.459 (0.021)	99.9 (0.01)	2.686 (0.213)	512.5 (24.7)
m7_s1	492.3 (83.8)	2.628 (0.015)	100 (0.08)	2.465 (0.425)	530.8 (21.5)
m4_s1	449.9 (22.3)	3.710 (0.035)	70.9 (1.40)	2.251 (0.113)	–
Charnockite SI09 (N10°28′41″ E77°45′16″)					
m2_s1	409.1 (21.7)	3.178 (0.021)	99.9 (0.02)	2.045 (0.109)	–
Charnockite SI10 (N10°13′60″ E77°28′32″)					
m1_s1	456.2 (25.9)	3.361 (0.069)	99.2 (0.4)	2.283 (0.131)	–
m2_s1	465.5 (25.1)	3.777 (0.067)	89.2 (1.8)	(0.127)	–

[a] The nomenclature is the grain #_ ion microprobe spot #

[b] Measured in the sample. Ideally, the unknown ThO_2^+/Th^+ ratio lies within the range defined by the standard (from 2.837 ± 0.031 to 3.960 ± 0.021)

[c] Percent radiogenically derived ^{208}Pb

[d] The reported Pb*/Th (±1σ) ratio is includes the radiogenically derived $^{208}Pb^*$ calculated using a common Pb ratio $^{208}Pb/^{204}Pb$ of 39.5 ± 0.1 (Stacey and Kramers 1975)

[e] In some cases we were able to calculate a meaningful $^{207}Pb/^{206}Pb$ age. This age is independent of any calibration errors and was only possible when the U content of the anlayzed monazite grain was enough to produce detectable ^{207}Pb

[f] Sample number and GPS location. See Figs. 15.1 and 15.2 for sample locations

Table 15.3 U–Pb ion microprobe zircon ages from the Oddanchatram anorthosite

Grain_spot[a]	$^{238}U–^{206}Pb$ Age (±1σ)	$^{235}U–^{207}Pb$ Age (±1σ)	$^{207}Pb–^{206}Pb$[e] Age (±1σ)	%$^{206}Pb^*$[b]
SI04 (N10°27′35″ E77°41′59″)[c]				
z1_s1	828 (18)	892 (64)	1,055 (201)	78.1 (0.5)
z3_s1[d]	309 (13)	385 (180)	871(1,114)	90.4 (1.9)
z4_s1[d]	710 (16)	723 (20)	764 (62)	97.8 (0.2)
z2_s2	505 (22)	528 (71)	627 (240)	95.0 (0.9)
z2_s1	603 (21)	574 (35)	458 (105)	98.8 (0.3)

[a] The nomenclature is the grain #_ ion microprobe spot #. See Fig. 15.5 for concordia diagram and BSE images of the dated grains

[b] Percent radiogenically derived ^{206}Pb

[c] Sample number and GPS location. See Fig. 15.2 for sample locations

[d] inclusion in garnet

[e] In some cases we were able to calculate a meaningful $^{207}Pb/^{206}Pb$ age. This age is independent of any calibration errors and was only possible when the U content of the anlayzed monazite grain was enough to produce detectable ^{207}Pb

petrologic investigations because the chemistry and the textural relationships of mineral grains are preserved and analyses of small grains (~10 μm) and zones within larger grains are feasible. The ion microprobe method requires no rock crushing, providing information usually lost during conventional geochronologic processes.

The method also removes the need to assume that all Pb in the mineral analyzed is of radiogenic origin.

Each grain chosen for geochronology was identified using qualitative energy dispersive spectrometer measurements and imaged in high contrast using backscattered electrons (BSE) (e.g., Scherrer et al. 2000; Catlos et al. 2002). The grains were then cut out of thin section using a high-precision saw and mounted in epoxy with a block of ion microprobe age standards (monazite_554; Harrison et al. 1995) and analyzed using the UCLA CAMECA ims 1270 ion microprobe. For zircon analysis, the age standards (zircon_AS3; Schneider et al. 1999) were located on a separate epoxy block.

During ion microprobe analysis, an oxygen ion beam of ~30 μm diameter focuses on the grain age standard and sputters isotopes, oxides, and ions of U, Th, and Pb, which are filtered through an energy analyzer and mass spectrometer. The ratios of $^{208}Pb^+/Th^+$ versus ThO_2^+/Th^+ form a calibration curve for monazite, with the slope and intercept of the curve controlled by the age of the standard grain. Forty-eight analytical spots on monazite 554 produced a calibration line of: $ThO_2^+/Th^+ = 0.095(^{208}Pb^+/Th^+) + 0.973 \pm 0.116$, which reproduced the age of the standard to 45.1 ± 2.3 Ma (±1σ). For the zircon analyses, a calibration with a fixed slope was used: $UO^+/U^+ = 1.000(^{206}Pb^+/U^+) + 2.990 \pm 0.031$ and reproduced the age of the standard to 1,094 ± 31 Ma (±1σ). Reported ages are corrected for common Pb using ^{204}Pb and an assumed $^{208}Pb/^{204}Pb = 38.6$ (Stacey and Kramers 1975).

Compared to U–Pb isotope dilution analysis, ion microprobe dating has the disadvantage of lower precision, which can range from ±1% to ±10%, depending on each day's calibration and specific mineral compositions. Less precision may also result from common Pb contamination and/or poor reproduction of the calibration curve. In some cases, we were able to obtain $^{206}Pb/^{207}Pb$ ages from the monazite and zircon grains (Tables 15.2 and 15.3), which are not calibration dependent. These ages could be obtained when the ^{207} Pb contents of the analyzed grains were at levels sufficient enough to be detected.

To facilitate in the interpretation of the geochronological results, all monazite grains were imaged using secondary electron (SE) after dating to determine where the ion microprobe spot was located. The oxygen beam in some cases was larger than the grain itself. To address this issue, aperture windows were used to constrict the measured ions sputtered from monazite grain to a ~20 micron^2 region in the center of the beam. Aperture windows are an essential part of in situ analyses, as they decrease the amount of non-radiogenic Pb from the surrounding minerals and matrix.

Whole rock fusion inductively coupled plasma (ICP) and ICP/mass spectrometry (MS) analyses were obtained from chips of charnockitic rocks SI09 and SI10 and anorthosites SI04B and SI05 (Figs. 15.1 and 15.2) from Activation Laboratories (Tables 15.4 and 15.5). As these types of rocks were coarse-grained and appeared compositionally heterogeneous, we obtained three analyses from each sampling locality. These data were used to identify trends in the geochemistry of the rocks and compare them to those previously reported, as well as to obtain monazite and zircon saturation temperatures (Watson and Harrison 1983; Montel 1993).

Table 15.4 Major and trace element data from South Indian charnockites[a]

Element[b]	SI09			SI10		
SiO_2	70.88	71.56	70.20	59.26	59.32	58.95
Al_2O_3	14.58	14.64	15.17	15.98	15.83	15.78
Fe_2O_3[c]	3.18	3.17	3.31	7.66	7.79	7.83
MnO	0.06	0.06	0.07	0.11	0.11	0.11
MgO	0.20	0.20	0.20	3.80	3.89	3.87
CaO	1.44	1.43	1.47	6.17	6.25	6.20
Na_2O	3.58	3.57	3.73	4.51	4.53	4.54
K_2O	5.15	5.14	5.18	1.69	1.83	1.81
TiO_2	0.18	0.18	0.19	0.69	0.69	0.68
P_2O_5	0.06	0.06	0.07	0.17	0.16	0.17
LOI[d]	0.21	0.24	0.32	0.44	0.44	0.44
Total	99.5	100.2	99.9	100.5	100.9	100.4
Sc	2	2	3	19	20	20
Be	2	2	2	3	3	3
V	<5	<5	<5	128	131	134
Cr	<20	<20	<20	50	30	40
Co	2	2	2	21	13	18
Ni	<20	<20	<20	50	30	40
Cu	<10	<10	<10	60	40	50
Zn	50	50	40	140	50	80
Ga	22	23	20	20	13	18
Ge	1	1	1	2	<1	<1
Rb	160	159	145	24	19	25
Sr	171	172	182	742	749	741
Y	33	33	39	18	19	17
Zr	279	277	310	111	170	131
Nb	7	7	7	9	8	9
Sn	2	1	2	6	3	2
Sb	<0.5	<0.5	0.7	6	1	<0.5
Ba	1,432	1,437	1,532	690	682	669
La	47.2	47.3	45.5	33.8	32.4	37.1
Ce	84.5	83.3	83	62	58.1	67.6
Pr	8.83	8.6	8.41	6.96	6.51	7.66
Nd	31.3	31.2	30.4	25.7	24.4	28.9
Sm	6.4	6.3	6.2	4.9	4.9	5.7
Eu	1.5	1.46	1.4	1.43	1.38	1.55
Gd	6.5	6.6	6.4	4.3	4.3	4.1
Tb	1.2	1.2	1.1	0.6	0.6	0.7
Dy	6.9	6.7	6.5	3.4	3.5	3.7
Ho	1.3	1.3	1.3	0.6	0.6	0.7
Er	3.9	3.9	3.9	1.8	1.9	2
Tm	0.59	0.58	0.59	0.27	0.29	0.3
Yb	3.9	3.8	3.8	1.9	1.9	2
Lu	0.59	0.58	0.58	0.27	0.27	0.29
Hf	7.6	7.6	8.0	2.8	4.2	3.5

(continued)

Table 15.4 (continued)

Element[b]	SI09			SI10		
Ta	<0.1	<0.1	<0.1	0.2	0.2	0.2
Tl	0.7	0.7	0.7	0.2	0.1	0.1
Pb	46	38	45	40	39	34
Bi	<0.4	<0.4	<0.4	0.5	<0.4	<0.4
Th	7.8	7.7	8.0	0.2	0.2	0.2
U	0.6	0.6	0.6	<0.1	<0.1	<0.1
T(°C)zircon[e]	833	833	843	694	722	702
T(°C)mon[e]	798	799	798	668	656	666

[a]All oxides as well as Sc, Be, V, Sr, Y, Zr, and Ba were obtained using whole rock fusion ICP methods. The other elements were obtained using whole rock fusion ICP/MS. Elements below detection limits include As (<5 ppm), Mo (<2 ppm), Ag, Cs (<0.5 ppm), In (<0.2 ppm), and W (<1 ppm)

[b]Oxides reported as %, whereas trace elements reported at the ppm level

[c]Fe_2O_3 = Total reported as Fe_2O_3

[d]LOI = Loss of Ignition

[e]Zircon saturation temperatures were calculated using Watson and Harrison (1983) and monazite saturation temperatures were calculated using Montel (1993)

15.4 Sample Descriptions and Petrography

15.4.1 Charnockitic Rocks

Charnockite samples were collected from the Kodaikanal–Palani Massif located in the Madurai block of the India's Southern Granulite Terrain. Sample SI10 is located within the core of the massif, ~1 km northwest of the town of Kodaikanal (Fig. 15.1). Oddanchatram sample SI09 is found at the northeast exposure of the massif, in close association with basic granulites, pink granites, and anorthosites (Fig. 15.2). In between these rocks along the Kodaikanal–Palani road, we collected sample SI02 (Fig. 15.1). The rocks contain orthopyroxene + plagioclase + K-feldspar + magnetite + chlorite + apatite + biotite + monazite + zircon + quartz ± garnet ± sapphirine ± calcite. We found both calcite and sapphirine in sample SI10 (Fig. 15.3). Sample SI10 also contains flame perthite, thin (5–30 μm) anastomosing lamellae of albite in K-feldspar (Fig. 15.3) (e.g., Pryer and Robin 1996; Vernon 1999). Many of the lamellae appear to be connected to thicker (50–100 μm) veins of albite located at the grain boundaries of K-feldspar; the albitization suggests the perthite is retrogressive. Flame perthite can be the result of saussuritization of calcic plagioclase and hydration of K-feldspar during replacement (Pryer and Robin 1995, 1996). Although it can form under greenschist facies conditions, flame perthite can also develop at high pressures where fluid activity is low (e.g., Altenberger and Wilhelm 2000). Feldspar thermometry of charnockitic rocks from the Palani Hills with similar textures yield temperatures >900°C at 9 kbar (Prakash et al. 2006). The orientation of the flames is crystallographically controlled and can be linked to stress imposed during metamorphism (Pryer and Robin 1996).

Table 15.5 Major and trace element data from the Oddanchatram anorthosite[a]

Element[b]	SI04B			SI05		
SiO_2	54.33	54.47	54.27	54.08	54.46	54.37
Al_2O_3	28.40	28.27	28.28	28.23	28.14	28.19
Fe_2O_3[c]	0.53	0.54	0.56	0.79	0.81	0.8
MnO	0.003	0.003	0.003	0.008	0.010	0.010
MgO	0.33	0.33	0.34	0.24	0.24	0.24
CaO	11.23	11.38	11.25	11.04	11.20	11.12
Na_2O	4.49	4.48	4.51	4.52	4.53	4.56
K_2O	0.65	0.72	0.75	0.64	0.70	0.73
TiO_2	0.061	0.061	0.059	0.023	0.026	0.026
P_2O_5	0.03	0.04	0.04	0.02	0.03	0.02
LOI[d]	0.67	0.67	0.67	0.79	0.79	0.79
Total	100.8	101.0	100.7	100.4	100.9	100.9
Sc	<1	<1	<1	1	1	2
Be	<1	<1	<1	<1	1	1
V	6	9	7	<5	<5	6
Co	1	<1	1	<1	<1	<1
Ga	21	15	17	19	14	17
Ge	<1	<1	<1	1	<1	<1
Rb	2	<2	2	<2	<2	<2
Sr	490	505	499	446	456	453
Nb	2	<1	<1	1	<1	<1
Sn	8	<1	<1	2	<1	<1
Sb	2.9	1.5	0.7	5.9	<0.5	0.6
Ba	285	287	285	287	292	289
La	4.0	3.7	4.2	4.0	3.9	4.4
Ce	5.8	5.2	6.1	5.2	5.1	5.7
Pr	0.54	0.49	0.57	0.43	0.40	0.48
Nd	1.8	1.6	2.1	1.3	1.2	1.5
Sm	0.3	0.3	0.4	0.2	0.2	0.2
Eu	0.52	0.50	0.56	0.52	0.54	0.60
Gd	0.2	0.2	0.2	0.2	0.2	0.2
Dy	<0.1	<0.1	<0.1	0.3	0.2	0.3
Er	<0.1	<0.1	<0.1	0.2	0.2	0.2
Yb	<0.1	<0.1	<0.1	0.2	0.2	0.2
W	<1	<1	<1	<1	1	<1
Pb	35	38	33	49	31	40
ΣREE	13.2	12.0	14.1	12.6	12.1	13.8
Ba/Sr	0.582	0.568	0.571	0.643	0.640	0.638

[a]All oxides as well as Sc, Be, V, Sr, Y, Zr, and Ba were obtained using whole rock fusion ICP methods. The other elements were obtained using whole rock fusion ICP/MS. Elements below detection limits include Zn (<30 ppm), Cr, Ni (<20 ppm), Cu (<10 ppm), As (<5 ppm), Zr (<4 ppm), Y, Mo (<2 ppm), Ag, Cs (<0.5 ppm), Bi (<0.4 ppm) In, Hf (<0.2 ppm), Tb, Ho, Ta, Tl, Th, U (<0.1 ppm), Tm (<0.05 ppm), and Lu (<0.04 ppm)

[b]Oxides reported as %, whereas trace elements reported at the ppm level

[c]Fe_2O_3 = Total reported as Fe_2O_3

[d]LOI = Loss of Ignition

The presence of calcite in the Kodaikanal charnockite (Fig. 15.3) may be taken as a petrographic evidence of CO_2 fluids, which have been invoked as a driving force for metamorphism of these rocks (e.g., Newton et al. 1980; Janardhan et al. 1982). However, the mineral could also form during alteration and/or retrogression. Calcite in the rock is found in contact with quartz, albite and K-feldspar, and as lobes filling cracks.

Further evidence for fluid interaction and/or retrogression is observed in sample SI02, where monazite exists both as individual grains 10–50 μm in diameter and as ~1–10 μm-thick bands in contact with subhedral apatite (Fig. 15.3). Similar apatite-monazite textures observed elsewhere are interpreted as products of metasomatism (e.g., Harlov and Förster 2002, 2003; Catlos et al. 2008). Monazite crystallizes using the REE and P locally available at the apatite grain boundary. Sample SI02 also shows feldspar exsolution textures (Fig. 15.3d), typical of charnockitic rocks from this area (Prakash et al. 2006).

We petrographically examined rocks near the charnockite sampling localities to identify any textural evidence for fluid-mediated reactions or decompression textures.

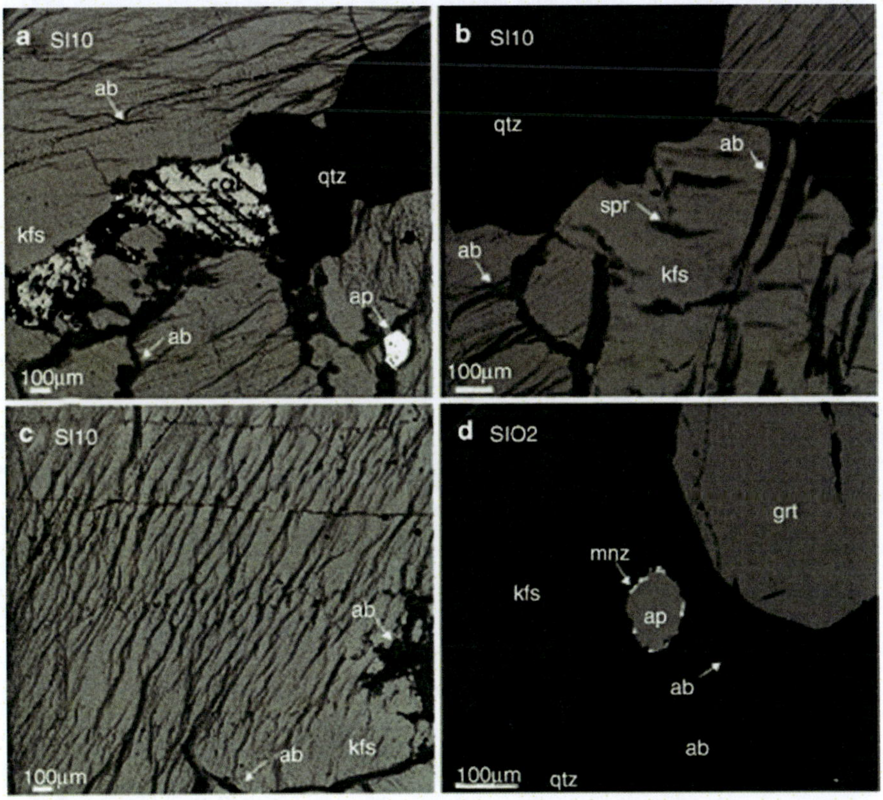

Fig. 15.3 Backscattered electron (BSE) images of charnockitic rocks SI10 (**a–c**) and SI02 (**d**). Mineral abbreviations after Kretz (1983). See Fig. 15.1 for sample locations

For example, sample SI01A located ~1 km south of SI02 (Fig. 15.1; Catlos et al. 2008) is a high-grade metamorphic gneiss found as an enclave within the surrounding charnockite. The rock is composed of garnet + magnetite + plagioclase + K-feldspar + biotite + monazite + zircon + quartz. Monazite grains from this sample yields ages ranging from 786 ± 94 to 618 ± 21 Ma (Catlos et al. 2008). Myrmekite, a symplectite of quartz and plagioclase adjacent to K-feldspar, is present in the rock (Fig. 15.4). This type of texture can form from different mechanisms, including exsolution or replacement (Passchier and Trouw 1998; Menegon et al. 2006; Yuguchi and Nishiyama 2008). X-ray Na and K maps of the texture are useful in identifying how the minerals are distributed within the myrmekite and surrounding grains (Fig. 15.4b). Myrmekite textures are commonly seen as a breakdown product of K-feldspar during retrograde metamorphism and metasomatism (e.g., Yuguchi and Nishiyama 2008).

Samples SI01B and SI03 are charnockitic rocks that contain orthopyroxene + clinopyroxene + magnetite + plagioclase + K-feldspar + apatite + quartz (see Figs. 15.1 and 15.2 for locations). These rocks show plagioclase-orthopyroxene symplectites that are likely pseduomorphs after garnet (Fig. 15.4) (see also Prakash et al. 2006; Prakash and Mohan 2007). These types of symplectites are consistent with isothermal decompression (e.g., Choudhuri and Silva 2000; Prakash et al. 2006) and are commonly observed in Kodaikanal charnockites (e.g., Harris 1981; Mohan et al. 1996a, b; Anto 1998; Prakash and Mohan 2007). X-ray element maps of the plagioclase-orthopyroxene symplectites were used to identify the distribution of minerals within the texture (Figs. 15.4). We commonly found grains of magnetite that increased in size (from 20 to >300 μm) from core to rim of the textures, but could identify no relict garnet.

15.4.2 Anorthosites

Samples SI04, SI04B, SI05, and SI07 were collected from the Oddanchatram anorthosite (Fig. 15.2) (Janardhan and Wiebe 1985; Wiebe and Janardhan 1988; Satyanarayana et al. 2003). Here we report plagioclase compositions from each sample (Fig. 15.5a; Table 15.6), whole rock data from SI04B and SI05 (Table 15.5) and zircon ages from SI04 (Fig. 15.5b; Table 15.3).

Anorthosites from this body differ in texture and composition depending on location (Janardhan and Wiebe 1985). Sample SI05 contains plagioclase + quartz + magnetite + calcite + orthopyroxene. Calcite is present within plagioclase-orthopyroxene symplectites in this sample (Figs. 15.4g–h). The mineral appears in cracks and as blebs within the texture, suggesting the presence of fluids. SI04, an anorthosite located adjacent to SI05 (Fig. 15.2; Tables 15.3 and 15.5) contains garnets of up to 8 mm in diameter, plagioclase, K-feldspar appearing in exsolution textures, quartz, muscovite, and abundant, large (50–>100 micron-diameter) compositionally-zoned zircons (Fig. 15.5). We labelled the compositionally analyzed sample of SI04 as

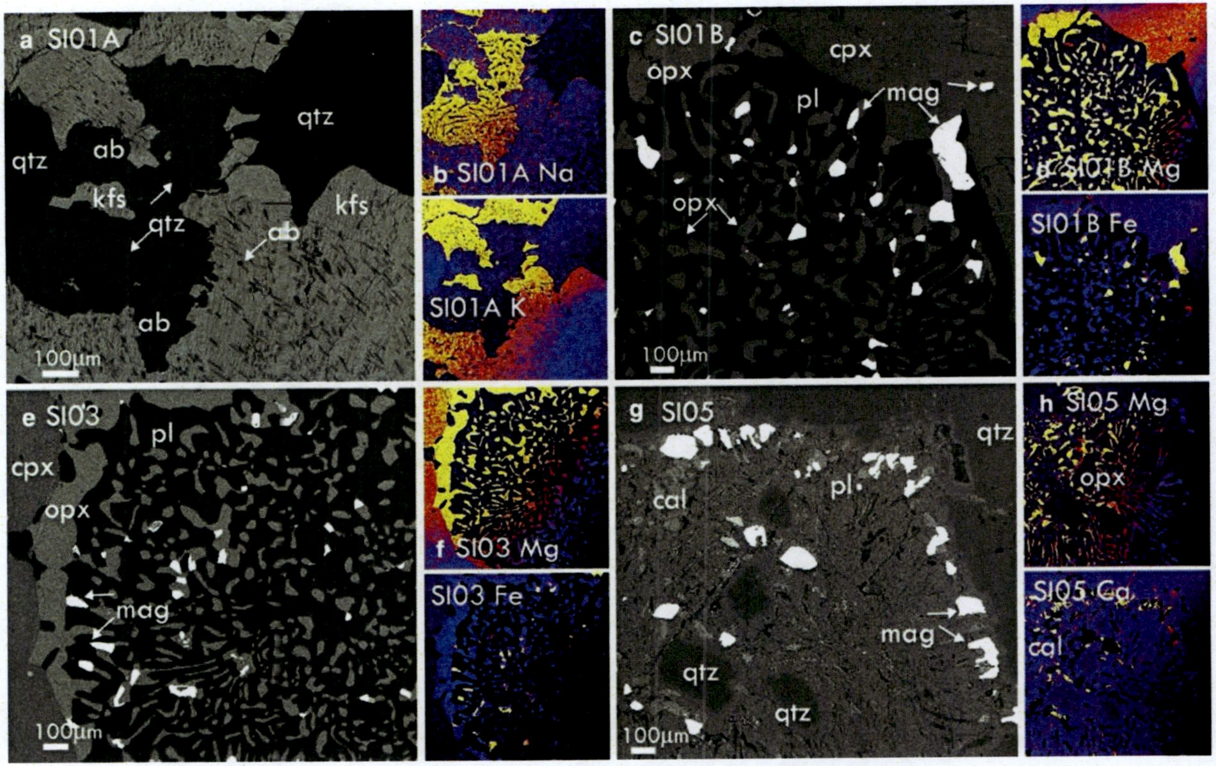

Fig. 15.4 BSE images and X-ray element maps of rocks from the Kodaikanal–Palani Massif. (**a** and **b**) Myrmekite texture and X-ray Na and K maps from sample SI01A, a high-grade garnet-bearing gneiss located within the Kodaikanal charnockitic rocks. (**c–f**) Symplectite textures and X-ray Mg and Fe maps of charnockite samples SI01B and SI03. (**g** and **h**) Symplectite textures and X-ray Mg and Ca maps of Oddanchatram anorthosite SI05. Mineral abbreviations after Kretz (1983). See Figs. 15.1 and 15.2 for sample locations

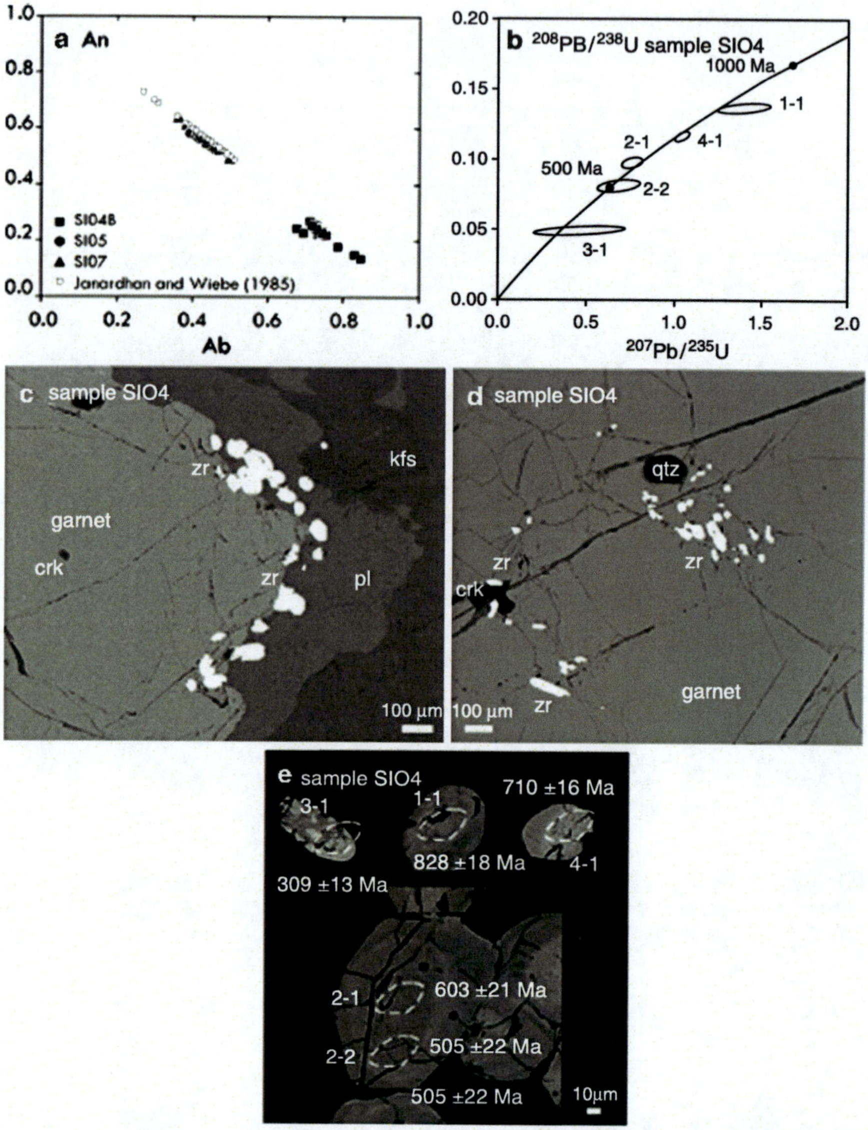

Fig. 15.5 **(a)** Plagioclase compositions from Oddanchatram anorthosite samples SI04B, SI05, and SI07 (Fig. 15.2) and Janardhan and Wiebe (1985). **(b)** Concordia diagram of dated grains in sample SI04. Nomenclature is grain number-spot number. **(c and d)** BSE images of zircons in sample SI04. Abbreviations: qtz, quartz; zr, zircon; pl, plagioclase; crk, crack. **(e)** High resolution images of the dated zircon grains in sample SI04 with $^{238}U/^{206}Pb$ ages indicated, $\pm 1\sigma$)

Table 15.6 Average plagioclase compositions from the Oddanchatram anorthosite

	SI04 (n = 44)	SI05 (n = 93)	SI07 (n = 49)
SiO_2	64.1 (1.3)	55.3 (0.6)	55.7 (0.6)
Al_2O_3	22.2 (1.0)	27.5 (0.5)	27.4 (0.7)
Na_2O	7.5 (1.6)	4.6 (0.2)	4.7 (0.3)
CaO	4.22 (1.04)	10.8 (0.17)	10.5 (0.45)
K_2O	1.08 (2.72)	0.36 (0.03)	0.25 (0.05)
FeO	0.04 (0.04)	0.19 (0.04)	0.20 (0.05)
Total	99.1 (0.8)	98.8 (0.8)	98.7 (0.5)
Si	11.4 (0.2)	10.1 (0.1)	10.1 (0.1)
Al	4.66 (0.19)	5.91 (0.08)	5.87 (0.13)
Na	2.56 (0.54)	1.61 (0.07)	1.66 (0.11)
Ca	0.80 (0.20)	2.10 (0.04)	2.04 (0.09)
K	0.25 (0.64)	0.08 (0.01)	0.06 (0.01)
Fe	0.01 (0.01)	0.03 (0.01)	0.03 (0.01)
Total	19.7 (0.1)	19.8 (0.1)	19.8 (0.1)
Ab	0.71 (0.15)	0.42 (0.01)	0.44 (0.02)
An	0.22 (0.05)	0.55 (0.01)	0.54 (0.02)
Or	0.07 (0.20)	0.022 (0.002)	0.016 (0.03)

Normalized to 32 oxygens

SI04B because it contains <4 ppm Zr (Table 15.5), which conflicts with petrographic observations of its mineralogy. Sample SI07 is a coarse-grained anorthosite which is composed dominantly of plagioclase with olivine, quartz, biotite, pyrite, and calcite.

Janardhan and Wiebe (1985) indicate that the Oddanchatram anorthosite is characterized by bimodal plagioclase compositions of primary An60 and secondary plagioclase of An94 due to garnet breakdown. Plagioclase in samples SI05 and SI07 are An55 (±0.01; n = 93 analyses) and An54 (±0.02; n = 49 analyses), respectively, whereas those in sample SI04 are only An23 (±0.03; n = 44 analyses) (Table 15.6). Samples SI05 and SI07 fall within the primary plagioclase distributions reported by Janardhan and Wiebe (1985) (Fig. 15.5a). Note that sample SI04 differs from the other anorthosite samples, as it contains garnet and zircon.

15.5 Results

15.5.1 Geochemistry

Numerous researchers have reported geochemical analyses of Indian charnockites in an effort to decipher their emplacement history, origin, and/or relationships to surrounding rocks (e.g., Table 15.1 references; reviews by Rajesh and Santosh 2004; Rajesh 2007, 2008). Here we report major and trace element chemistry from samples SI09 and SI10 taken from the Kodaikanal charnockitic rocks (Table 15.4) and compare the results to other charnockite localities in the Southern Granulite Terrain, including those to the south (Cardamom Hills, Nemmara, other Madurai block charnockites) and north (Shevaroy, Kabbaldurga, Biligirirangan Hills, Nilgiri, Southern Karnataka, Madras Block, and Cauvery Shear Zone) of the PCSZ (see Fig. 15.1 for locations). Our approach treats South Indian charnockites as suites of differentiated igneous rocks rather than purely metamorphic assemblages. Although the diagrams we employ to understand the geochemistry of these rocks are specifically for igneous assemblages, many have been used to understand the petrogenesis of metamorphic charnockites in South India and elsewhere (Kilpatrick and Ellis 1992; Rajesh et al. 2000; Tomson et al. 2006; Rajesh 2007; 2008).

15.5.1.1 Major Elements

Southern Granulite Terrain charnockites or charnockitic rocks range from 45 to 81 wt% SiO_2 (Figs. 15.6 and 15.7). The lowest (<50 wt%) and highest (~80 wt%) SiO_2 contents are from the Madras block (Fig. 15.1) (Holland 1900; Howie and Subramanian 1957; Bernard-Griffiths et al. 1987) considered the type-locality (e.g., Holland 1900; Sen and Ray 1971; Weaver 1980; Frost and Frost 2008). Charnockites with high SiO_2 contents (~75 wt%) have been reported elsewhere (Kilpatrick and Ellis 1992; Duchesne and Wilmart 1997; Bucher and Frost 2006). The lowest SiO_2 samples are from what are classically termed basic or ultrabasic charnockites (e.g., Holland 1900; Bernard-Griffiths et al. 1987) or basic or mafic granulites (Howie and Subramanian 1957; Sen and Ray 1971; Weaver 1980). In a total alkali-silica (TAS) diagram (Middlemost 1994), Southern Granulite Terrain charnockites are observed to be of subalkaline to midalkaline compositions. Although they span a wide range, we see that most of the samples fall in the field of granite-granodiorite, whereas the rest largely plot in the diorite-gabbro field (Fig. 15.6a).

The definition of charnockite has been a subject of contention since their discovery (Naidu 1963; Frost and Frost 2008). The low and extremely high SiO_2 samples may have at the time of their publication been considered charnockites, but today be considered as other types of rocks. As stated by Naidu (1963) "there is nothing like an igneous, consanguineous series of rocks called the charnockite series which have differentiated from the ultrabasic to the acidic end." In any case, we include all analyses of rocks defined as charnockites in our data compilation to ascertain their geochemical relationships.

Fig. 15.6 (continued) (**e**) TiO_2 (wt%) versus SiO_2 (wt%) variation diagram. Isotherms show the Fe–Ti oxide saturation temperatures at 7.5 kbar (Green and Pearson 1986). Cardamom Hills analyses fit the equation TiO_2 (wt%) = $3.07 \times 10^9 \pm 2.21 \times 10^{10}$ *exp[$-SiO_2$(wt%)/(2.50 ± 0.84)] + 0.554 ± 0.118 with r^2 of 0.708. (F) P_2O_5 (wt%) versus SiO_2 (wt%) variation diagram. Isotherms in show the apatite saturation temperatures at 7.5 kbar (Harrison and Watson 1984), after Rajesh (2007). Cardamom Hills analyses fit the equation P_2O_5 (wt%) = $6.695 \times 10^3 \pm 1.643 \times 10^4$*exp[$-SiO_2$ (wt%)/(5.83 ± 1.58)] + 0.182 ± 0.055 with r^2 of 0.914

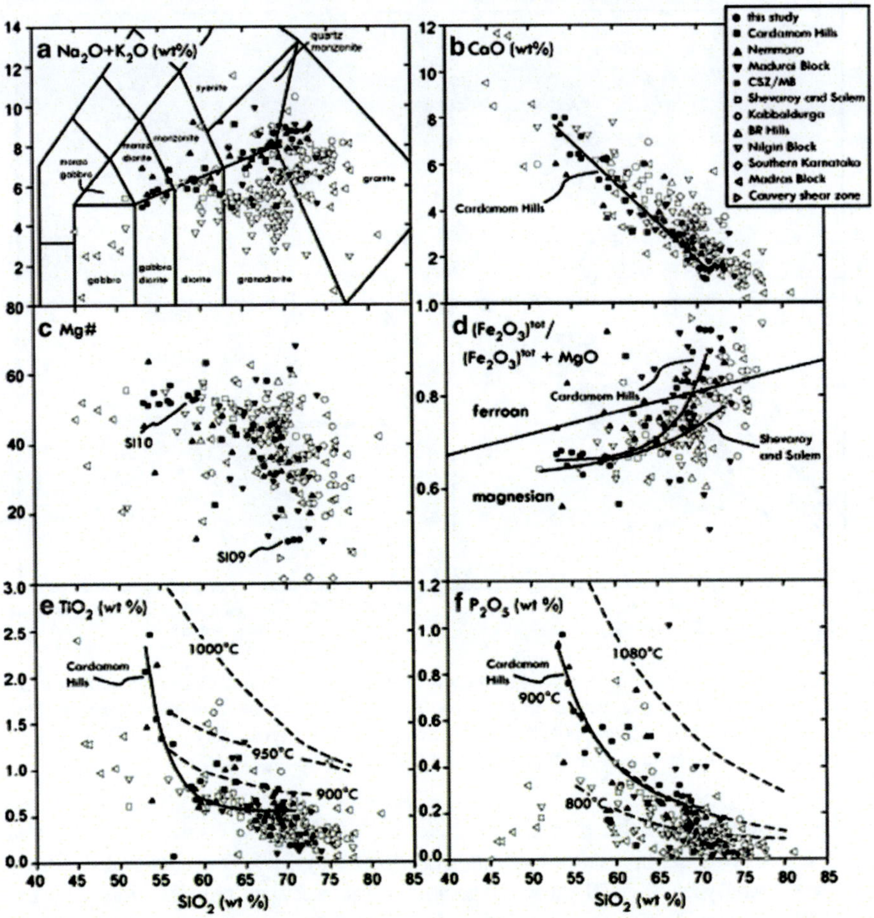

Fig. 15.6 (**a**) Total alkali-silica (TAS) diagram for Southern Granulite Terrain charnockitic rocks. Fields after Middlemost (1994). *Open symbols* are samples north of the PCSZ, whereas *filled symbols* are from south of the PCSZ. Data from this study, Cardamom Hills (Chacko et al. 1992; Rajesh 2007), Nemmara (Kumar and Sukumaran 2003; Kumar 2004), Madurai Block (Raith et al. 1997; Rajesh et al. 2000; Suresha and Srikantappa, 2005; Tomson et al. 2006), Cauvery Shear Zone/ Madurai Block (CSZ/MB; Tomson et al. 2006), Shevaroy (Allen et al. 1985; Narayana and Nijagunappa 1987), Kabbaldurga (Janardhan et al. 1982; Staehle et al. 1987; Bhattacharya and Sen 2000), Biligirirangan (BR) Hills (Janardhan et al. 1982; Narayana and Nijagunappa 1987; Janardhan et al. 1994), Nilgiri Block (Janardhan et al. 1982; Narayana and Nijagunappa 1987; Srikantappa 1988; Janardhan et al. 1994; Raith et al. 1999), Southern Karnataka (Janardhan et al. 1982), Madras Block (Holland 1900; Howie and Subramanian 1957; Bernard-Griffiths et al. 1987), Cauvery Shear Zone (CSZ; Tomson et al. 2006). See Fig. 15.1 for approximate sample locations. (**b**) CaO (wt%) versus SiO_2 (wt%) variation diagram. The regression fits the Cardamom Hills analyses only, and yields CaO (wt%) = 25.98±0.95 – 0.345 ± 0.015*SiO_2 (wt%) with r^2 of 0.968. (**c**) Mg number versus SiO_2 (wt%). (**d**) $(Fe_2O_3)^{tot}/[(Fe_2O_3)^{tot} + MgO]$ versus SiO_2 (wt%). Cardamom Hills analyses fit the equation MgO wt% = $1.67 \times 10^{-9} \pm 1.24 \times 10^{8}$ * exp[($Fe_2O_3^{tot}/Fe_2O_3^{tot}$ + MgO) /(3.79 ± 1.52)] + 0.658 ± 0.020 with a r^2 of 0.634. Shevaroy and Salem analyses fit the equation MgO wt% = $7.60 \times 10^{-5} \pm 3.27 \times 10^{-4}$ * exp[($Fe_2O_3^{tot}/Fe_2O_3^{tot}$ + MgO)/(9.60 ± 5.19)] + 0.623 ± 0.041 with a r^2 of 0.697.

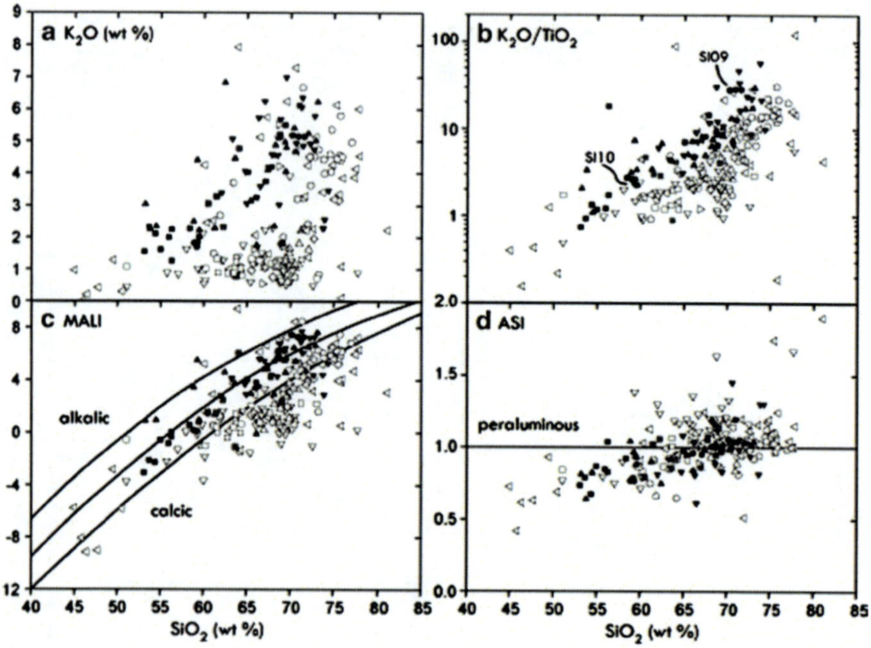

Fig. 15.7 (**a**) K_2O versus SiO_2 (wt%) variation diagram for Southern Granulite Terrain charnockitic rocks. (**b**) K_2O/TiO_2 versus SiO_2 (wt%). (**c**) Modified alkaline lime index (MALI, $Na_2O + K_2O - CaO$) versus SiO_2 (wt%). (**d**) Aluminum saturation index (ASI; Al/Ca – 1.67P + Na + K) versus SiO_2 (wt%). Symbols and data sources are the same as in Fig. 15.6

The samples from the Kodaikanal–Palani Massif (Table 15.4; Figs. 15.1 and 15.2) reflect two distinct compositions as reflected by SiO_2 contents that differ by ~12 wt%. Analyses of SI10 show the rock is of intermediate composition, whereas the SI09 is felsic. With the exception of K_2O and SiO_2, major element oxide compositions are observed to be higher in sample SI10 than in SI09. The Mg numbers of SI10 average 52.8, whereas that of SI09 is considerably lower (12.3), implying a more evolved nature (Fig. 15.6c). The Mg numbers from sample SI09 are some of the lowest reported for Southern Granulite Terrain rocks.

In general, charnockitic rocks from South India become more iron rich with increasing SiO_2 contents (Fig. 15.6d). Charnockitic rocks from the Cardamom and Shevaroy Hills and near the town of Salem exponentially increase in $Fe_2O_3^{tot}/(Fe_2O_3^{tot} + MgO)$ with increasing SiO_2. Their trends suggest that charnockites from the south of the PCSZ become more Fe-rich at lower SiO_2 values than those north of the structure. Our analyses from the Kodaikanal–Palani Massif lie along the trend defined by the Cardamom Hills (Fig. 15.6d).

We observed that the Southern Granulite Terrain charnockitic rocks have low TiO_2 contents, mostly confined in an interval below 1.0 wt% (average 0.6 ± 0.3 wt%). Figure 15.6e and f shows the Fe–Ti (Green and Pearson 1986) and apatite saturation temperatures (Harrison and Watson 1984) for the Southern Granulite

Terrain charnockitic rocks at 7.5 kbar. Most Southern Granulite Terrain charnockitic rocks experienced crystallization of these minerals at temperatures below 900°C. Charnockitic rocks from Cardamom show a systematic decrease in temperature with increasing SiO_2 contents (Fig. 15.6e), which is expected from a fractionating charnockite intrusion (e.g., Grantham et al. 2001).

In general, charnockitic rocks south of the PCSZ increase in K_2O with increasing SiO_2, whereas those to the north increase or remain at levels below 2 wt% (Fig. 15.7a). Depletions in K_2O have been suggested to reflect increasing levels of metamorphism and fluid-rock interactions (e.g., Kilpatrick and Ellis 1992). Charnockitic rocks north of the PCSZ tend to have lower K_2O/TiO_2 values for a given SiO_2 content than those south of the structure (Fig. 15.7b). These rocks can range significantly in K_2O/TiO_2. For example, sample SI10 averages 2.6 ± 0.1 K_2O/TiO_2, whereas SI09 has 28.1 ± 0.2 K_2O/TiO_2 (Table 15.4).

Southern Granulite Terrain charnockitic rocks increase in modified alkaline lime index (MALI, $Na_2O + K_2O$ –CaO) and aluminium saturation index (ASI; molar Al/Ca-1.67P + Na + K) with increasing SiO_2 contents (Fig. 15.7c and d). We find that charnockitic rocks north of the PCSZ become more calcic with increasing SiO_2 contents, whereas those to the south become alkali-calcic (Fig. 15.7c). Sample SI10, taken from the core of the Kodaikanal–Palani Massif (Fig. 15.1) has a lower ASI and is less iron rich and is more calcic than sample SI09 (Figs. 15.6 and 15.7). Sample SI09 is marginally peraluminous (ASI = 1.05 ± 0.01), whereas SI10 is metaluminous (ASI < 1.0).

Geochemical analyses of the Oddanchatram anorthosite samples SI04 and SI05 are reported in Table 15.5. These samples are of high Al_2O_3 and CaO contents as expected from their mineralogy which is dominated by plagioclase. Analyses reported here and those from Janardhan and Wiebe (1985) show that the anorthosite body increases in ASI, MALI and becomes more ferroan with increasing SiO_2.

15.5.1.2 Trace Elements

Trace element patterns reveal that the Kodaikanal–Palani Massif charnockitic rocks are highly depleted in terms of Nb, Ta, Ti and P (Fig. 15.8). The more felsic SI09 samples display high Th abundances with distinct negative Nb anomaly ($[Th/Nb]_{PM}$ = 9.4 ± 0.2). However the intermediate group samples SI10 are considerably depleted in terms of Th but with nearly the same Nb contents as the former, and thus characterized by considerably low $(Th/Nb)_{PM}$ ratios (0.19 ± 0.01). This depletion in Th can be attributed to removal of this element in volatile-rich fluids during high-grade metamorphism and it is a common feature observed in the charnockitic-granulitic suites worldwide (e.g. Zhao et al. 1997; Larin et al. 2006; Mikhalsky et al. 2006; Tomson et al. 2006). Although the felsic samples appear to be more enriched in incompatible elements, Sr, P and Ti are observed to be more depleted in this group. Some accessory phases like zircon, apatite as well as Fe–Ti oxides are known to partition strongly some HFSE and REE elements (e.g. Nagasawa 1970; Fujimaki 1986; LeMarchand et al. 1987). The negative Ti and P anomalies can be interpreted as the early crystallization of Fe–Ti oxides (ilmenite and apatite, respectively). Depleted Sr contents may be indicative of plagioclase fractionation, or alternatively it could have resulted from the presence of plagioclase as a residual phase in the source region. The remarkable Nb depletion in both groups, on the other hand, may have been inherited from the source that is Nb-depleted itself (i.e. subduction-modified source). The Kodaikanal–Palani Massif charnockitic rocks also differ in their behaviour with Rb, with the element being more depleted in the intermediate group samples.

The chondrite-normalized rare earth element variation plots show both groups of Kodaikanal–Palani Massif charnockitic rocks are enriched in terms of LREE and display similar patterns [SI09 $(La/Sm)_N = 4.78 \pm 0.06$; SI10 $(La/Sm)_N = 4.31 \pm 0.13$] (Fig. 15.8b). However, the HREE patterns differ between the two groups. The intermediate SI10 samples display moderate fractionation in HREE with lower abundances (ΣHREE = 13.4 ± 0.3 ppm), whereas the felsic SI09 samples shows rather unfractionated patterns with noticeably higher abundances (ΣHREE = 24.6 ± 0.4 ppm). The significant depletion of HREE relative to LREE observed in the intermediate samples [$(La/Yb)_N = 12.77 \pm 0.54$] may suggest the presence of residual garnet in the source region of these charnockites since HREE can be strongly fractionated into garnet (e.g., Johnson 1998). However, it is also possible that the source itself may have possessed high La/Yb signatures. Fractionation of zircon may also have lead to HREE depletion observed in intermediate SI10 group, since zircon is known to strongly partition HREE over LREE (e.g. Nagasawa 1970). However, the slight depletion in Zr in these samples suggests only a minor importance of zircon as a fractionating phase. The Zr enrichment in the felsic SI09 samples, on the other hand, may imply accumulation of zircon during their magmatic history. This is supported by relatively higher concentrations of HREE with flat patterns compared to the intermediate group. The SI09 group displays moderate negative Eu anomalies (Eu/Eu* = 0.69; where $Eu/Eu^* = Eu_N/[(Sm_N)(Gd_N)]^{0.5}$) (Fig. 15.8b), which is absent in the analyses of the SI10 charnockitic rocks (Eu/Eu* = 0.95), suggesting early crystallization of feldspar for the former. Alternatively, this could be attributed to the retention of Eu by plagioclase in the source region.

We compile the available trace element data from Southern Granulite Terrain charnockites in Figs. 15.8, 15.9, 15.10 and 15.11. Many of these rocks show similar trace element behaviour as the Kodaikanal–Palani massif charnockites, including depletions in Rb, Nb, Ta, P, and Ti. With the exception of analyses of samples near Chennai and Kabbaldurga, charnockitic rocks north of the PCSZ are more depleted in Ba and Rb than those south of the PCSZ (Fig. 15.11a). The variable Rb depletion is reflected by high K/Rb ratios (average K/Rb = 590). We find that charnockites from the Southern Granulite Terrain tend to show increase in Rb/Sr and decrease in Sr/Ba ratios with increasing SiO_2 contents (Fig. 15.11c and d). This may indicate importance of plagioclase fractionation, since Sr is typically concentrated in plagioclase in contrast to Ba that would be removed by K-feldspar. The slight decreases or near constant variations in Sr/Ba ratios in some of the suites, however, suggests K-feldspar fractionation has also taken place in addition to

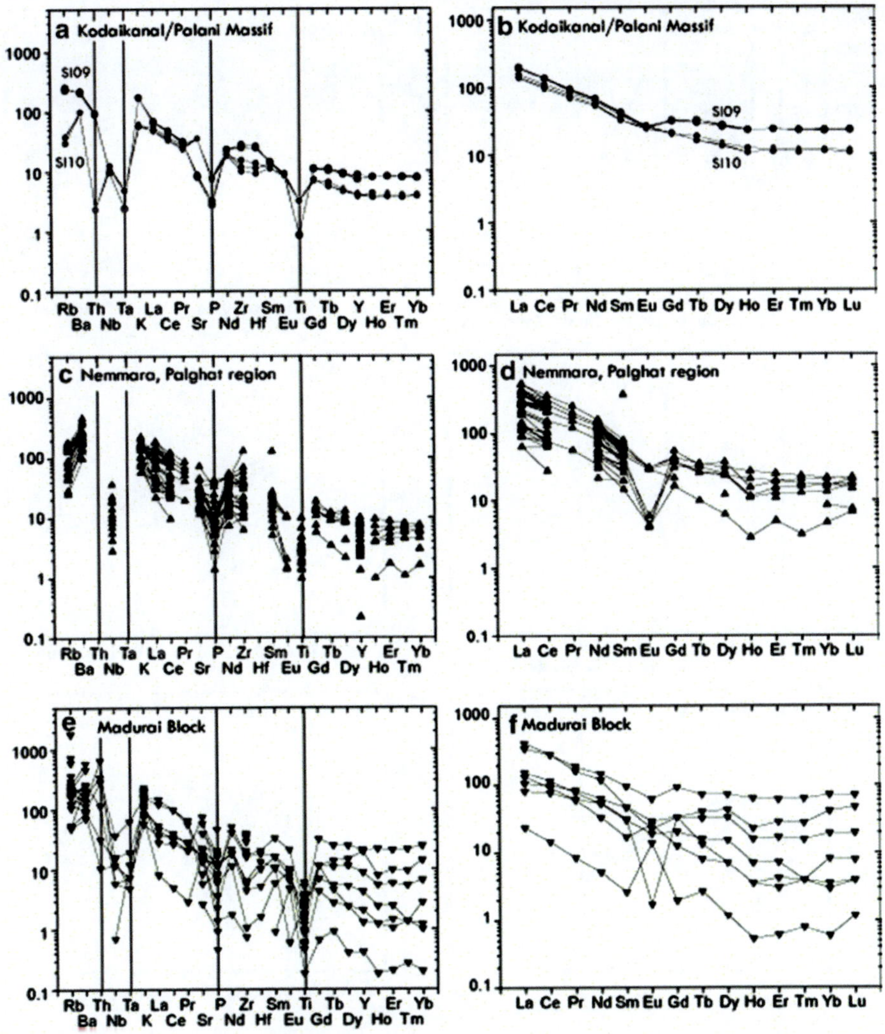

Fig. 15.8 Spider and REE diagrams for charnockite samples south of the PCSZ. (**a** and **b**) Kodaikanal–Palani massif (**c** and **d**) Nemmara in the Palghat region and (**e** and **f**) Madurai Block. Data sources the same as in Fig. 15.6. See Fig. 15.1 for locations. REE elements normalized to C1 chondrite; elements in the spider diagram normalized to primitive mantle (Sun and McDonough 1989)

plagioclase. In general, those south of the PCSZ have higher Rb/Sr and Zr, but lower Sr/Ba ratios at a given SiO_2 content compared to those north of the PCSZ. REE patterns from the rocks vary, but in general, Southern Granulite Terrain charnockitic rocks are LREE enriched and can be depleted in HREE (Figs. 15.8, 15.9 and 15.10). The degree of flatness and shape of the REE patterns from these rocks differ, even within the same charnockite massif (e.g., Nemmara region or from the

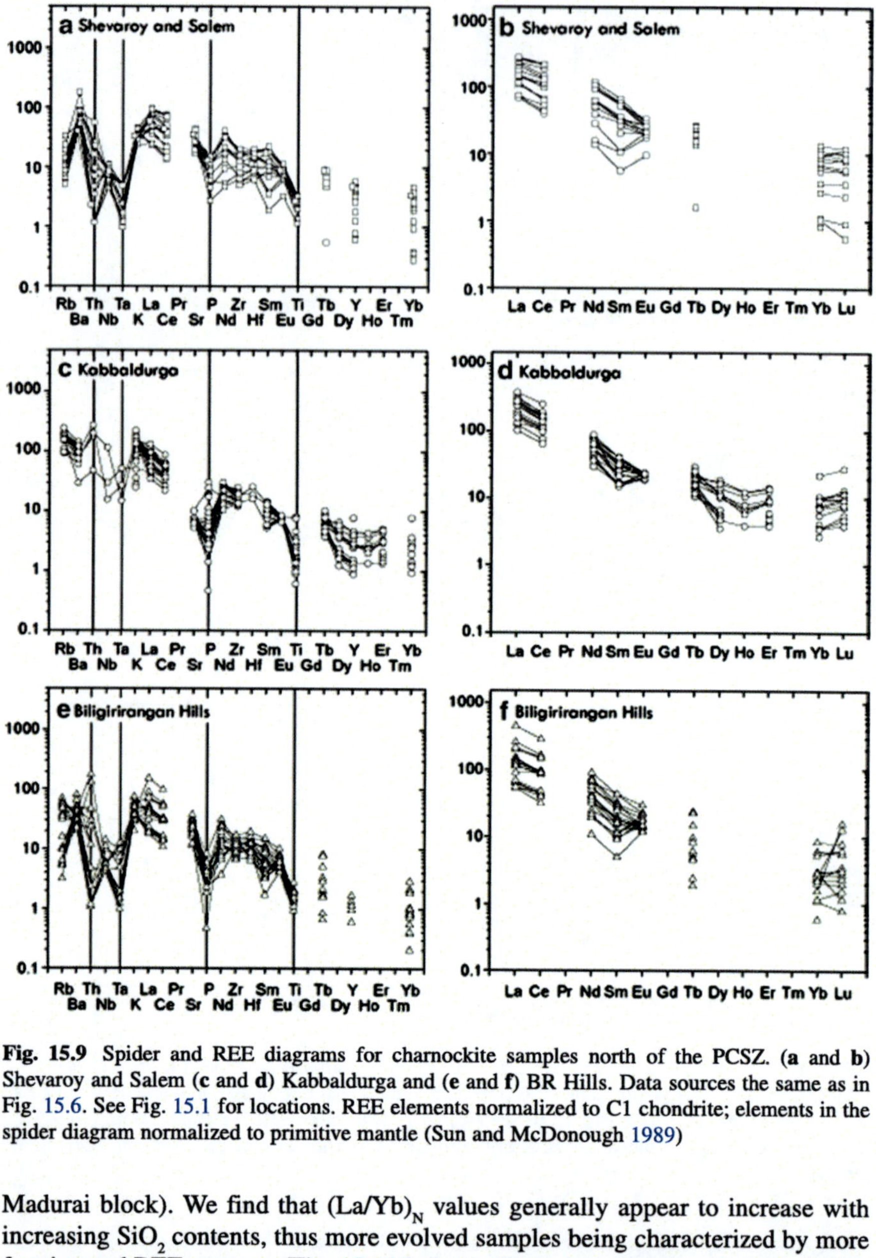

Fig. 15.9 Spider and REE diagrams for charnockite samples north of the PCSZ. (**a** and **b**) Shevaroy and Salem (**c** and **d**) Kabbaldurga and (**e** and **f**) BR Hills. Data sources the same as in Fig. 15.6. See Fig. 15.1 for locations. REE elements normalized to C1 chondrite; elements in the spider diagram normalized to primitive mantle (Sun and McDonough 1989)

Madurai block). We find that $(La/Yb)_N$ values generally appear to increase with increasing SiO_2 contents, thus more evolved samples being characterized by more fractionated REE patterns (Fig. 15.11f). Lack of data prevented us from exploring correlations between sample evolution and Eu anomalies or REE abundances. However, the presence of variable Eu enrichment/depletion suggests the importance of plagioclase as a fractionating/accumulating phase and/or as a residual

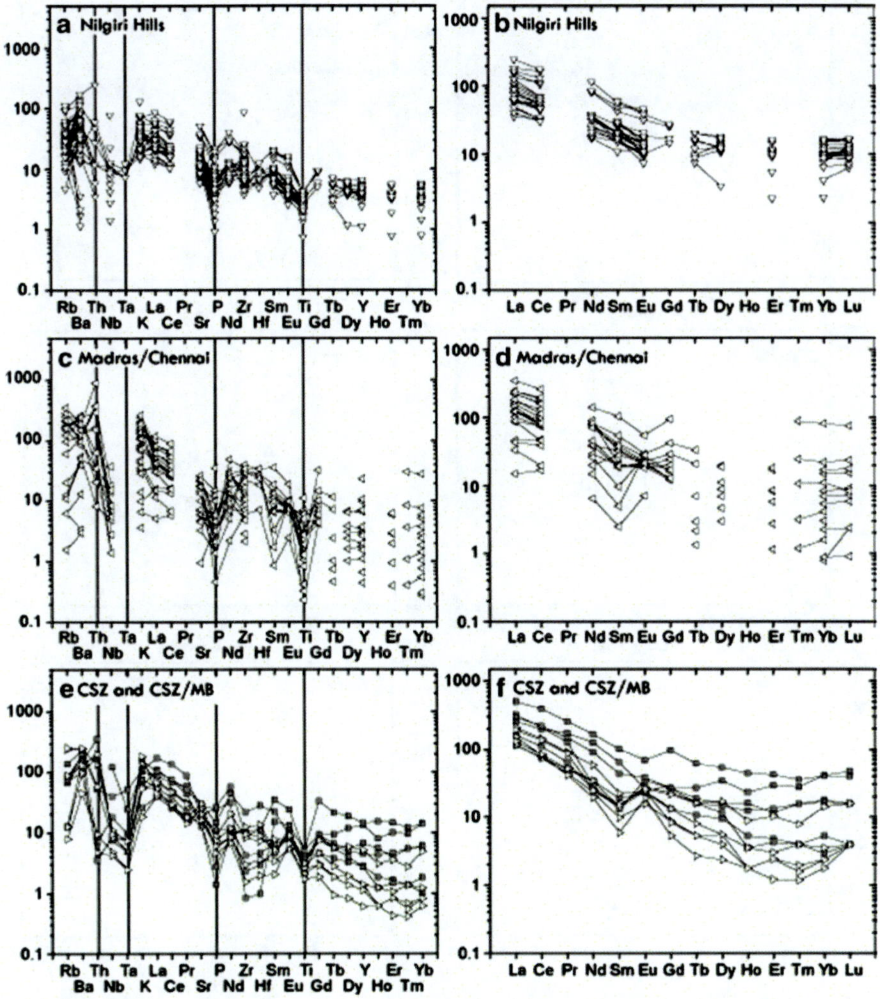

Fig. 15.10 Spider and REE diagrams for charnockite samples. (**a** and **b**) Nilgiri Block analyses (**c** and **d**) Madras Block and (**e** and **f**) CSZ and CSZ/MB analyses. Data sources the same as in Fig. 15.6. See Fig. 15.1 for locations. REE elements normalized to C1 chondrite; elements in the spider diagram normalized to primitive mantle (Sun and McDonough 1989)

phase in the source region of these charnockitic rocks. The common Nb depletion, however, seems to be linked to the source region and/or source processes. These Nb-depleted characteristics are also displayed by the discrimination diagram of Rb versus (Yb + Nb) (Fig. 15.11b) (Pearce et al. 1984) where most of the South Indian charnockites fall in the field of volcanic-arc granites.

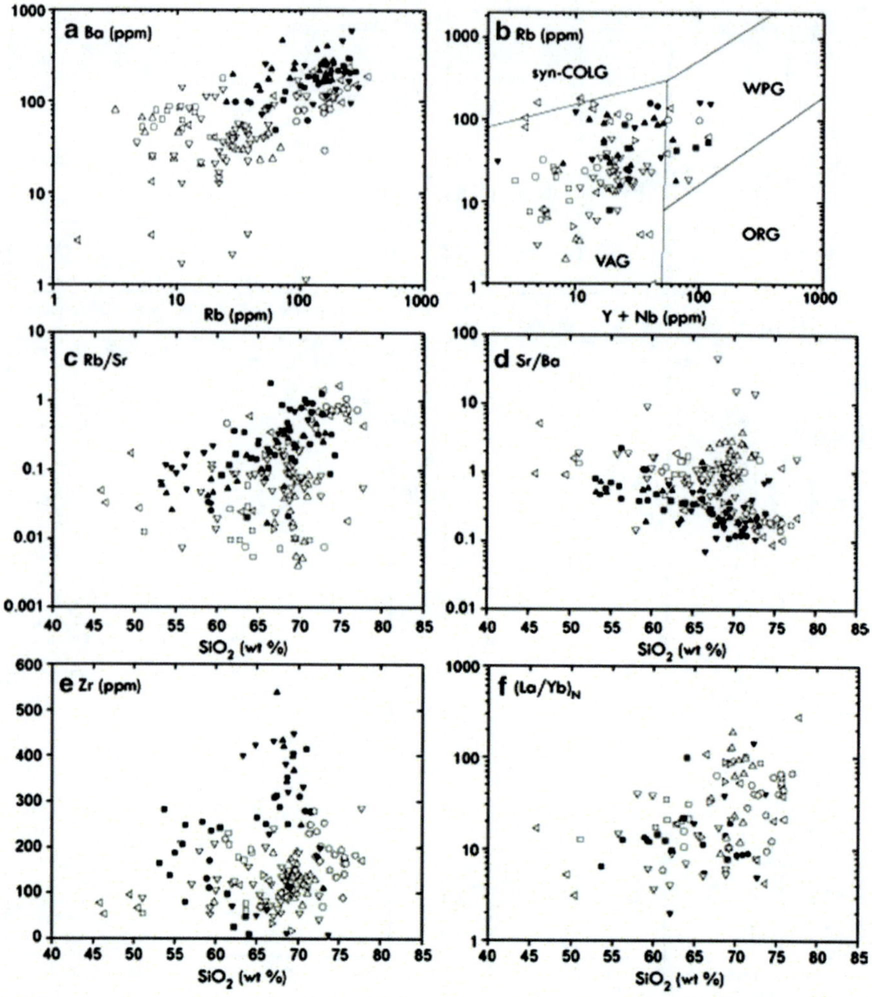

Fig. 15.11 **(a)** Ba (ppm) versus Rb (ppm), **(b)** Rb (ppm) versus Y + Nb (ppm) discrimination diagram (Pearce et al. 1984), **(c)** Rb/Sr versus SiO_2 (wt%), **(d)** Sr/Ba versus SiO_2 (wt%), **(e)** Zr (ppm) versus SiO_2 (wt%) and **(f)** $(La/Yb)_N$ versus SiO_2 (wt%) for Southern Granulite Terrain charnockites. Data sources the same as in Fig. 15.6. See Fig. 15.1 for locations. La and Y are normalized to C1 chondrite (Sun and McDonough 1989)

The Oddanchatram anorthosite samples SI04 and SI05 are enriched in terms of Ba, K, and Sr and depleted in Nb and Rb. Both sample groups display highly fractionated LREE pattern [$(La/Sm)_N$ = 6.8–14.2] and pronounced positive Eu anomalies (average Eu/Eu* = 7.4), however their MREE and HREE concentrations appear to be considerably low. The Sr/Ba contents of the Oddanchatram anorthosite samples differ between the two samples, likely due to plagioclase crystallization.

15.5.2 Charnockite Saturation Temperatures

Using available trace element data, we calculated the zircon and monazite saturation temperatures for the charnockitic rocks (Watson and Harrison 1983; Montel 1993). Although this approach is generally reserved for H_2O-saturated granitic rocks, these thermometers have been applied to analyses of charnockites (Grantham et al. 2001; Rajesh 2008). Charnockitic rocks north and south of the PCSZ show similar saturation temperatures within uncertainty (753 ± 56°C and 785 ± 80°C, for zircon; 726 ± 81°C and 770 ± 70°C for monazite) (Fig. 15.12). However, we find more charnockitic rocks south of the PCSZ record 800–900°C zircon saturation temperatures than those to the north. This result may be due to available data, or could be indicative of a difference in the tectono-magmatic history of these rocks. The histograms suggest that the charnockite generation event south of the PCSZ, which may have occurred during the Cambro-Ordovician, may have been hotter than those experienced by charnockitic rocks in the north. Sample SI09 has higher (by 100–150°C) zircon and monazite saturation temperatures than sample SI10 (Table 15.4), consistent with geochemical trends that suggest the rocks differed in their petrologic history.

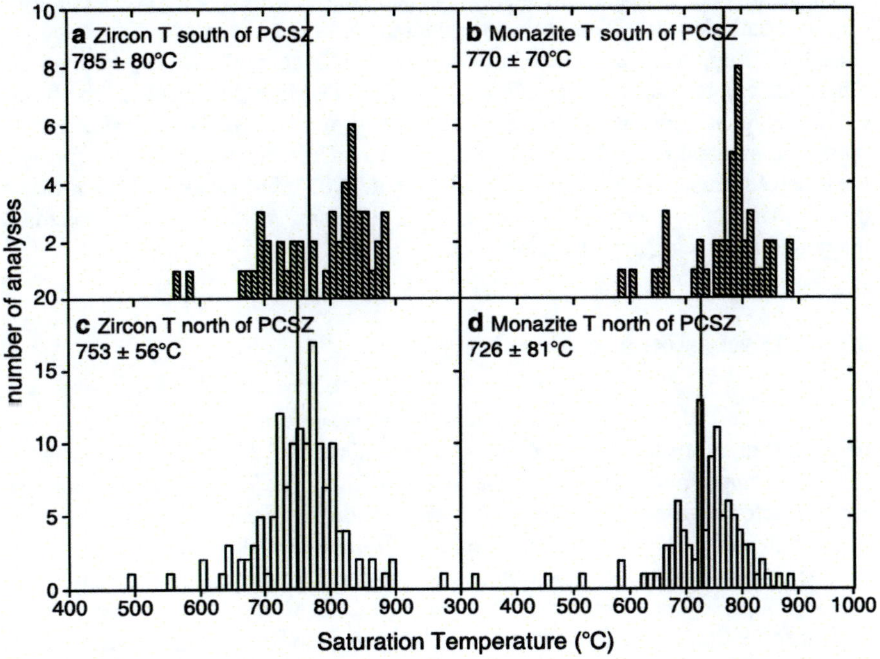

Fig. 15.12 Zircon (Watson and Harrison 1983) and monazite (Montel 1993) saturation temperatures for charnockitic rocks south of the PCSZ (**a** and **b**) and north of the PCSZ charnockitic rocks (**c** and **d**). Calculated using data in Table 15.4 and available trace element data; see references listed in Fig. 15.6 for sources. *Vertical lines* on the figures indicate the average temperatures

15.5.3 Geochronology

Monazite grains from the Kodaikanal–Palani Massif yield Cambro-Ordovician Th–Pb ages (Table 15.2), consistent with geochronological results from the Southern Granulite Terrain that show the region experienced extreme crustal metamorphism during this time (Table 15.1). The age is similar to results obtained from Southern Granulite Terrain charnockites south of the PCSZ and adds to the growing database of Cambro-Ordovician dates from rocks of this region.

Here we also report $^{238}U/^{206}Pb$ zircon ages from the Oddanchatram anorthosite (sample SI04), which range from 309 ± 13 Ma to 828 ± 18 Ma (Table 15.3; Fig. 15.5). The youngest age overlaps a metamict region within the zircon and has no geological significance, whereas the oldest age is found within a zircon core and likely times crystallization of the grain within the anorthosite. Zircon is strikingly abundant in sample SI04B and is found both as inclusions in garnet and in clusters at the rims of garnet. Zircon inclusions in garnet appear to be in contact with cracks throughout the grains, suggesting that they could have precipitated in the rock via a fluid phase that infiltrated the garnet host. Alternatively, the cracks may have developed during retrogression of the garnet during exhumation. Most zircons in the anorthosite show distinct differences in zoning in BSE from core to rim. We could identify no clear pattern shared by each zircon; for example cores could be bright or dark depending on grain. All obtained ages are in contact with concordia (Fig. 15.5), suggesting each date has not been significantly affected by Pb loss and represents crystallization. The oldest age is 828 ± 18 Ma lies within the time frame of the polyphase and prolonged Eastern Ghats orogeny (e.g. Satyanarayana et al. 2003; Simmat and Raith 2008), while the Cambro-Ordovician ages could represent the time of intrusion of the body (e.g., Rao et al. 2003). The range of dates indicate that this anorthosite body has undergone a polyphase and complicated tectono-magmatic and metamorphic history.

15.6 Discussion and Conclusions

Here we report geochemical and geochronological information from charnockitic rocks and anorthosites collected from the Kodaikanal–Palani Massif. The massif forms an important component of the Southern Granulite Terrain, and understanding the tectonic history of this massif lends considerable insight into its role within South India as well as the formation of these important rock types.

Charnockitic rocks from the Southern Granulite Terrain can be divided into two major groups separated by the E–W trending PCSZ. Compilations of geochronologic information from these rocks (Table 15.1, see also Rao et al. 2003) show these rocks largely differ in age, with northern samples recording Archaean crystallization events, whereas those to the south yielding Cambro-Ordovician and more recently, Neoproterozoic ages (Catlos et al. 2008). The Kodaikanal charnockitic rocks dated here contain monazite grains that fall within the Cambro-Ordovician timescale, similar to charnockitic rocks in the Cardamom and Anaimalai Hill massifs, and others dated elsewhere south of the PCSZ (Table 15.1).

The Oddanchatram anorthosite, however, located in along the northern boundary of the Kodaikanal–Palani Massif contains zircon grains that record mid-Neoproterozoic to Cambro-Ordovician crystallization ages (Fig. 15.5). Late-Mesoproterozoic ages have also been reported elsewhere from this body (Satyanarayana et al. 2003). This anorthosite differs in texture and composition depending on location (see Janardhan and Wiebe 1985), likely the result of its multi-stage metamorphic and/or intrusion history.

Compilations of geochemical data for Southern Granulite Terrain charnockitic rocks illustrate what has been reported for these rock types: they can cover nearly the whole range of granite chemistry (Frost and Frost 2008). However, some differences between charnockitic rocks north and south of the PCSZ can be observed. North of the PCSZ, charnockitic rocks become more calcic with increasing SiO_2 contents, whereas those to the south become alkali-calcic (Fig. 15.7c). Southern charnockitic rocks tend to have higher $Na_2O + K_2O$, K_2O, K_2O/TiO_2 for a given SiO_2 content than those north of the PCSZ. In addition, they show higher Rb/Sr ratios, Ba, Rb and Zr contents. However, the southern charnockitic rocks demonstrate lower CaO and lower Sr/Ba at a given SiO_2 content compared to those north of the PCSZ.

In terms of the Kodaikanal–Palani Massif, sample SI10 fits some geochemical criteria (low SiO_2, high TiO_2, low K_2O/Na_2O, magnesian Fe-number, calc-alkalic MALI, low Rb/Sr, high Sr/Y; see Rajesh 2007) for an intermediate charnockite, whereas SI09 can be considered a silicic charnockite. The trace element data from these rocks may suggest their generation from a source region devoid of Nb that may have been previously fluxed by subduction-related events, though we do not entirely preclude the possibility of a mantle-derived component. It seems unlikely, however, that the felsic Kodaikanal–Palani samples have been produced by fractionation of the intermediate ones. Because, although the major oxides variations are indicative of fractionation, the distinct HREE patterns with similar LREE variations rule out fractionation process for the derivation of felsic group from the intermediate ones. Therefore, the similarity in LREE but different HREE variations can be attributed to their derivation from different sources. Indeed, the Sr depletion in the felsic group as well as negative Eu anomalies could be rather suggestive of a shallow, probably lower crustal (plagioclase-facies) source. The Sr enrichment and HREE depleted chemistry observed in the intermediate group, on the other hand, may be related to derivation from deeper levels where garnet is a stable residual phase.

Using available geochemical data, we find a larger number of samples recording between 800 and 900°C zircon saturation temperatures for charnockitic rocks south of the PCSZ (Fig. 15.12). The observation may reflect differences in how the charnockitic rocks were generated. The charnockite generation event to the south of the PCSZ during the Cambro-Ordovician may have been at a higher temperature than that which occurred to the north during the Archaean. Sample SI09 records higher (by 100–150°C) zircon and monazite saturation temperatures than sample SI10 (Table 15.4). The result is also consistent with the geochemical data that suggests that charnockitic rocks within the Kodaikanal–Palani Massif experienced different mechanisms of generation and/or metamorphic histories. Further exploration of this hypothesis requires isotopic data.

Acknowledgments This paper is dedicated to Dr. A. S. Janardhan, who passed away in November 2004. He was an essential member of the field expedition and provided important insights to the authors of this paper into the geology of the Southern Granulite Terrain. We thank the UCLA National Ion Microprobe facility. Rock samples were analyzed for their bulk and trace element chemistry by Activation Laboratories. We thank Ms. Kelli Wakefield (Tarelton State University) for generating the plagioclase compositions as part of a subcontract from NSF0138942, New Frontiers-Research Experience for Undergraduates in the Space and Planetary Sciences. We thank the Oklahoma State University Vice President for Research Office for funding.

References

Allen P, Condie KC, Narayana BL (1985) The geochemistry of prograde and retrograde charnockite-gneiss reactions in southern India. Geochim Cosmochim Acta 49:323–336

Altenberger U, Wilhelm S (2000) Ductile deformation of K-feldspar in dry eclogite facies shear zones in the Bergen Arcs, Norway. Tectonophysics 320:107–121

Anto KF (1998) Fluid inclusion studies granulites from Kambam Valley, Tamil Nadu, South India, in relation to their metamorphic evolution. Indian Mineral 32:27–42

Anto KF, Janardhan AS, Sivasubramanian P (1997) A new sapphirine occurrence from Kambam valley, Tamil Nadu and its possible relation to Pan-African tectonothermal event. Curr Sci 73:792–796

Anto KF, Janardhan AS, Basavalingu B (1999) Metamorphic history of calc-silicate lithologies from the Kambam Valley, Tamil Nadu and its bearing on the evolution of the Southern Granulite Terrain. J Geol Soc India 53:27–37

Balakrishnan P, Bhattacharya S, Anilkumar (1985) Carbonatite body near Kambammetu, Tamil Nadu. J Geol Soc India 26:418–421

Bartlett JM, Harris NBW, Hawkesworth CJ, Santosh M (1995) New isotope constraints on the crustal evolution of South India and Pan-African granulite metamorphism. Mem Geol Soc India 34:391–397

Bell K (2001) Carbonatites: relationships to mantle-plume activity. Geol Soc Am Spec Pap 352:267–290

Bernard-Griffiths J, Jahn B-M, Sen SK (1987) Sm-Nd isotopes and REE geochemistry of Madras granulites, India: an introductory statement. Precambrian Res 37:343–355

Bhattacharya S, Sen SK (2000) New insights into the origin of Kabbaldurga charnockites, Karnataka, South India. Gondwana Res 3:489–506

Braun I, Cenki-Tok B, Paquette J-L, Tiepolo (2007) Petrology and U-Th-Pb geochronology of the sapphirine-quartz bearing metapelites from Rajapalayam, Madurai Block, Southern India: Evidence for polyphase Neoproterozoic high-grade metamorphism. Chem Geol 241:129–147

Brown M, Raith M (1996) First evidence of ultrahigh-temperature decompression from the granulite province of southern India. J Geol Soc London 153:819–822

Bucher K, Frost BR (2006) Fluid transfer in high-grade metamorphic terrains intruded by anorogenic granites: the Thor Range, Antartica. J Petrol 47:567–593

Catlos EJ, Gilley LD, Harrison TM (2002) Interpretation of monazite ages obtained via in situ analysis. Chem Geol 188:193–215

Catlos EJ, Dubey CS, Sivasubramanian P (2008) Monazite ages from carbonatites and high-grade assemblages along the Kambam Fault Southern Granulite Terrain, South India. Am Miner 93:1230–1244

Cenki B, Kriegsman LM (2005) Tectonics of the Neoproterozoic Southern Granulite Terrain, South India. Precambrian Res 138:37–56

Chacko T, Kumar GRR, Meen JK, Rogers JJW (1992) Geochemistry of high-grade suprarustal rocks from the Kerala Khondalite Belt and adjacent massif charnockites, South India. Precambrian Res 55:469–489

Chetty TRK, Bhaskar Rao YJ (2006) The Cauvery Shear Zone, Southern Granulite Terrain, India: a crustal-scale flower structure. Gondwana Res 10:77–85

Choudhuri A, Silva D (2000) A clinopyroxene-orthopyroxene-plagioclase symplectite formed by garnet breakdown in granulite facies, Guaxupe, Minas Gerais, Brazil. Gondwana Res 3:445–452

Deer WA, Howie RA, Zussman J (1993) An Introduction to the rock-forming minerals, 2nd edn. Longman Scientific and Technical Press, Harlow, Essex, 696 pp

Drüppel K, Littmann S, Romer RL, Okrusch M (2007) Petrology and isotope geochemistry of the Mesoproterozoic anorthosite and related rocks of the Kunene Intrusive Complex, NW Namibia. Precambrian Res 156:1–31

Drury SA, Holt RW (1980) The tectonic framework of the South Indian Craton: a reconnaissance involving LANDSAT imagery. Tectonophysics 65:T1–T15

Drury SA, Harris NBW, Holt RW, Reeves-Smith GJ, Wrightman RT (1984) Precambrian tectonics and crustal evolution in South India. J Geol 92:3–20

Duchesne J-C, Wilmart E (1997) Igneous charnockites and related rocks from the Bjerkreim-Sokndal layered intrusion (Southwest Norway); a jotunite (hypersthene monzodiorite)-derived A-type granitoid suite. J Petrol 38:337–369

Fermor LL (1936) An attempt at the correlation of the ancient schistose formations of peninsular India (part 1). Indian Geological Survey, Memoir 70, Calcutta, India

Friend CRL, Nutman AP (1991) SHRIMP U–Pb Geochronology of the Closepet granite and Peninsular gneiss, Karnataka, South India. J Geol Soc India 38:357–368

Frost BR, Frost CD (1987) CO_2 melts and granulite metamorphism. Nature 327:503–506

Frost BR, Frost CD (2008) On charnockites. Gondwana Res 13:30–44

Fujimaki H (1986) Partition coefficients of Hf, Zr and REE between zircon, apatite and liquid. Contrib Mineral Petrol 94:42–45

Ghosh JG, de Wit MJ, Zartman RE (2004) Age and tectonic evolution of Neoproterozoic ductile shear zones in the Southern Granulite Terrain of India, with implications for Gondwana studies. Tectonics. doi:10.1029/2002TC 001444

Grady JC (1971) Deep main faults in South India. J Geol Soc India 12:56–62

Grantham GH, Eglington BM, Thomas RJ, Mendonidis P (2001) The nature of the Grenville-age charnockitic A-type magmatism from the Natal, Namaqua and Maud Belts of southern Africa and western Dronning Maud Land, Antarctica. Mem Natl Polar Res Spec Issue 55:59–86

Green TH, Pearson NJ (1986) Ti-rich accessory phase saturation in hydrous mafic-felsic compositions at high P, T. Chem Geol 54:185–201

Hansen EC, Newton RC, Janardhan AS, Lindenberg S (1995) Differentiation of late Archean crust in the eastern Dharwar Craton, Krishnagiri-Salem area, South India. J Geol 103:629–651

Harlov DE, Förster H-J (2002) High-Grade fluid metasomatism on both a local and a regional scale: the Seward Peninsula, Alaska, and the Val Strona di Omegna, Ivrea–Verbano Zone Northern Italy. Part II: Phosphate Mineral Chemistry. J Petrol 43:801–824

Harlov DE, Förster H-J (2003) Fluid-induced nucleation of (Y + REE)-phosphate minerals within apatite: nature and experiment Part II. Fluorapatite. Am Miner 88:1209–1299

Harris NBW (1981) The application of spinel-bearing metapelites to P/T determinations: an example from South India. Contrib Mineral Petrol 76:229–233

Harris NBW, Santosh M, Taylor PN (1994) Crustal evolution in South India: constraints from Nd isotopes. J Geol 102:139–150

Harrison TM, Watson EB (1984) The behavior of apatite during crustal anatexis: equilibrium and kinetic considerations. Geochim Cosmochim Acta 48:1464–1477

Harrison TM, McKeegan KD, Le Fort P (1995) Detection of inherited monazite in the Manaslu leucogranite by $^{208}Pb/^{232}Th$ ion microprobe dating: crystallization age and tectonic implications. Earth Planet Sci Lett 133:271–282

Holland TH (1900) The charnockite series, a group of Archaean hypersthenic rocks in peninsular India. Mem Geol Soc India 17:7–27

Howie RA (1955) The geochemistry of the charnockite series of Madras, India. Trans R Soc Edin 82:725–769

Howie RA, Subramanian AP (1957) The paragenesis of garnet in charnockite, enderbite, and related granulites. Min Mag 31:565–586

Jain AK, Singh S, Manickavasagam RM (2003) Intracontinental shear zones in the Southern Granulite Terrain; their kinematics and evolution. Mem Geol Soc India 50:225–253

Janardhan AS (1999) Southern Granulite Terrain, South of the Palghat-Cauvery Shear Zone: implications for India-Madagascar Connection. Gondwana Res 2:463–469

Janardhan AS, Srikarni C (2001) Pan-African Granulite Facies Assemblages in Kodaikanal-Anaimalai Ranges, Tamil Nadu and Mt. Abu-Balaram Areas of Gujarat, India: Madagascar-India Connection in Eastern Gondwana Assembly. Gondwana Res 4:643–644

Janardhan AS, Wiebe RA (1985) Petrology and geochemistry of the Oddanchatram anorthosite and associated basic granulites, Tamil Nadu, South India. J Geol Soc India 26:163–176

Janardhan AS, Newton RC, Hansen EC (1982) The transformation of amphibolite facies gneiss to charnockite in southern Karnataka and northern Tamil Nadu, India. Contrib Mineral Petrol 79:130–149

Janardhan AS, Jayananda M, Shankara MA (1994) Formation and tectonic evolution of granulites from the Biligriri Ranan and Niligiri Hills S. India: geochemical and isotopic constraints. J Geol Soc India 44:27–40

Jayananda M, Janardhan AS, Sivasubramanian P, Peucat JJ (1995) Geochronology and isotopic constraints on granulite formation in the Kodaikanal area, Southern India. Mem Geol Soc India 34:373–390

John MM, Balakrishnan JS, Bhadra BK (2005) Contrasting metamorphism across Cauvery Shear Zone, south India. J Earth SystSci 114:143–158

Johnson KTM (1998) Experimental determination of partition coefficients for rare earth and high-field-strength elements between clinopyroxene, garnet, and basaltic melt at high pressures. Contrib Mineral Petrol 133:60–68

Kilpatrick JA, Ellis DJ (1992) C-type magmas: igneous charnockites and their extrusive equivalents. Trans R Soc Edinb: Earth Sci 83:155–164

Kretz R (1983) Symbols for rock-forming minerals. Am Miner 68:277–279

Kumar GRR (2004) Mechanism of arrested charnockite formation at Nemmara, Palghat region, southern India. Lithos 75:331–358

Kumar GRR, Sukumaran S (2003) Petrology and geochemistry of gneiss, charnockite and charno-enderbite of Palghat Region, Southern India. Mem Geol Soc India 50:409–434

Larin AM, Kotov AB, Sal'nikova EB, Glebovitskii VA, Sukhanov MK, Yakovleva SZ, Kovach VP, Berezhnaya NG, Velikoslavinskii SD, Tolkachev MD (2006) The Kalar Complex, Aldan-Stanovoi Shield, an ancient anorthosite-mangerite-charnockite-granite association: geochronologic, geochemical, and isotopic-geochemical characteristics. Petrology 14:2–20

LeMarchand F, Villemant B, Calas G (1987) Trace element distribution coefficients in alkaline series. Geochim Cosmchim Acta 51:1071–1081

Menegon L, Pennacchioni G, Stünitz H (2006) Nucleation and growth of myrmekite during ductile shear deformation in metagranites. J Metamorph Geol 24:553–568

Middlemost EAK (1994) Naming materials in the magma/igneous rock system. Earth Sci Rev 37:215–224

Mikhalsky EV, Sheraton JW, Hahne K (2006) Charnockite composition in relation to the tectonic evolution of East Antarctica. Gondwana Res 9:379–397

Miller JS, Santosh M, Pressley RA, Clemens AS, Rogers JJW (1996) A Pan-African thermal event in southern India. J Southeast Asian Earth Sci 14:127–136

Mishra DC, Kumar VV, Rajasekhar RP (2006) Analysis of airborne magnetic and gravity anomalies of peninsular shield, India integrated with seismic and magnetotelluric results and gravity anomalies of Madagascar, Sri Lanka, and Antarctica. Gondwana Res 10:6–17

Mohan A, Jayananda M (1999) Metamorphism and isotopic evolution of granulites of southern India; reference to Neoproterozoic crustal evolution. In: Roy AB (ed) Neoproterozoic crustal evolution and Indian-Gondwana linkage, Gondwana Res 2:251–262
Mohan A, Prakash D, Motoyoshiti Y (1996a) Decompressional P-T history in sapphirine-bearing granulites from Kodaikanal, southern India. J Southeast Asian Earth Sci 14:231–243
Mohan A, Prakash D, Sachan KH (1996b) Fluid inclusions in charnockites from Kodaikanal massif (South India): P-T record and implications for crustal uplift history. Mineral Petrol 57:167–184
Montel J-M (1993) A model for monazite/melt equilibrium and application to the generation of granitic magmas. Chemical Geol 110:127–146
Nagasawa H (1970) Rare earth concentrations in zircons and apatites and their host dacites and granites. Earth Planet Sci Lett 9:359–364
Naidu PRJ (1963) Hypersthene-bearing rocks of the Madras State, India. Mem Geol Soc India 17:290–299
Narayana BL, Nijagunappa R (1987) Rare-earth element geochemistry of charnockites from highland areas of South India. Geophys Res Bull 25:96–113
Newton RC (1989) Charnockitic alteration: evidence for CO2 infiltration in granulite facies metamorphism. J Metamorphic Geol 10:383–400
Newton RC (1992) An overview of charnockite. Precambrian Res 55:399–405
Newton RC, Smith JV, Windley BF (1980) Carbonic metamorphism, granulites, and crustal growth. Nature 288:45–50
Passchier CW, Trouw RAJ (1998) Microtectonics. Spinger, New York
Pearce KA, Harris NBW, Tindle AG (1984) Trace element discrimination diagrams for tectonic interpretation of granitic rocks. J Petrol 25:956–983
Pichamuthu CS (1960) Charnockite in the making. Nature 188:135–136
Pichamuthu CS (1970) On the occurrence of garnet in charnockite. J Geol Soc India 11:273–275
Pichamuthu CS (1990) The charnockites of South India. Mem Geol Soc India 17:133–135
Prakash D (1999) Petrology of the basic granulites from Kodaikanal, South India. Gondwana Res 2:95–104
Prakash D, Mohan A (2007) Colour-coded compositional mapping of orthopyroxene-plagioclase symplectites in mafic granulites from Panrimalai, south India. J Geol Soc India 69:285–290
Prakash D, Shastry A (1999) Geochemical studies on the sapphirine-granulites from Kodaikanal, South India. Gondwana Res 2:603–604
Prakash D, Arima M, Mohan A (2006) Ultrahigh-temperature metamorphism in the Palni Hills, South India: insights from feldspar thermometry and phase equilibria. Int Geol Rev 48:619–638
Prasad RB, Kesava RG, Mall DM, Koteswara RP, Raju S, Reddy MS, Rao GSP, Sridher V, Prasad ASSSRS (2007) Tectonic implications of seismic reflectivity pattern observed over the Precambrian Southern Granulite Terrain, India. Precambrian Res 153:1–10
Pryor LL, Robin P YF (1995) Retrograde metamorphic reactions in deforming granites and the origin of flame perthite. J Metamorph Geol 14:645–658
Pryer LL, Robin P-YF (1996) Differential stress control on the growth and orientation of flame perthite: a palaeostress-direction indicator. J Struct Geol 18:1151–1166
Raith M, Karmakar S, Brown M (1997) Ultra-high temperature metamorphism and multistage decompressional evolution of sapphirine granulites from the Palni Hills, southern India. J Metamorph Geol 15:379–399
Raith M, Srikantappa C, Buhl D, Koehler H (1999) The Nilgiri enderbites, South India: nature and age constraints on protolith formation, high-grade metamorphism and cooling history. Precambrian Res 98:129–150
Rajesh HM (2007) The petrogenetic characterization of intermediate and silicic charnockites in high-grade terrains: a case study from southern India. Contrib Mineral Petrol 154:591–606
Rajesh HM (2008) Petrogenesis of two granites from the Nilgiri and Madurai blocks, southwestern India: implications for charnockite – calc-alkaline granite and charnockite – alkali (A-type) granite link in high-grade terrains. Precambrian Res 162:180–197

Rajesh HM, Santosh M (2004) Charnockitic magmatism in southern India. Proc Indian Acad Sci (Earth Planet Sci) 113:565–585
Rajesh HM, Santosh M, Yoshida M (2000) Petrogenesis of a aluminous A-type granite from Munnar, southwestern India. J Geosci (Osaka City University) 43:203–225
Rama Rao B (1945) The charnockitic rocks of Mysore. Bull Mysore Geolog Dept 18:159
Rao VV, Prasad BR (2006) Structure and evolution of the Cauvery Shear Zone system, Sothern Granulite Terrain, India: evidence from deep seismic and other geophysical studies. Gondwana Res 10:29–40
Rao BYJ, Janardhan AS, Kumar TV, Narayana BL, Dayal AM, Taylor PN, Chetty TRK (2003) Sm-Nd model ages and Rb-Sr isotopic systematics of charnockites and gneisses across the Cauvery Shear Zone, Southern India: implications for the Archaean-Neoproterozoic terrane boundary in the Southern Granulite Terrain. Mem Geol Soc India 50:297–317
Sajeev K, Santosh M, Kim HS (2006) Partial melting and P–T evolution of the Kodaikanal Metapelite Belt, southern India. Lithos 92:465–483
Santosh M, Omori S (2008) CO_2 flushing; a plate tectonic perspective. Gondwana Res 13:86–102
Santosh M, Jayananda M, Mahabaleswar B (1991) Fluid evolution in the Closepet Granite: a magmatic source for CO2 in charnockite formation at Kabbaldurga? J Geol Soc India 38:55–65
Santosh M, Tsunogae T, Koshimoto S (2004) First report of sapphirine-bearing rocks from the Palghat-Cauvery shear zone system, southern India. Gondwana Res 7:620–626
Satyanarayana KVV, Arora BR, Janardhan AS (2003) Rock magnetism and palaeomagnetism of the Oddanchatram anorthosite, Tamil Nadu, South India. Geophys J Int 155:1081–1092
Scherrer NC, Engi M, Gnos E, Jakob V, Liechti A (2000) Monazite analysis; from sample preparation to microprobe age dating and REE quantification. Schweiz Mineral Petrogr Mitt 80:93–105
Schneider DA, Edwards MA, Kidd WSF, Zeitler PK, Coath CD (1999) Early Miocene anatexis identified in the western syntaxis, Pakistan Himalaya. Earth Planet Sci Lett 167:121–129
Sen SK, Ray S (1971) Hornblende-pyroxene granulites versus pyroxene granulites: a study from the type charnockite area. N Jb Miner Abh 115:291–314
Shimpo M, Tsunogae T, Santosh M (2006) First report of garnet-corundum rocks from southern India: implications for prograde high-pressure (eclogite-facies?) metamorphism. Earth Planet Sci Lett 242:111–129
Simmat R, Raith MM (2008) U–Th–Pb monazite geochronometry of the Eastern Ghats Belt, India: timing and spatial disposition of poly-metamorphism. Precambrian Res 162:16–39
Singh AP, Kumar N, Singh B (2006) Nature of the crust along Kuppam–Palani geotransect (South India) from Gravity studies: implications for Precambrian continental collision and delamination. Gondwana Res 10:41–47
Sivasubramanian P (1993) Geology and metamorphic history of parts of Kodaikanal ranges, Ph.D. thesis, University of Mysore, India
Spooner CM, Fairbairn HW (1970) Strontium 87/strontium 86 initial ratios in pyroxene granulite terranes. J Geophys Res 32:6706–6713
Srikantappa C (1988) High pressure charnockites of the Nilgiri Hills, southern India. Mem Geol Soc India 25:95–110
Sriramguru K, Janardhan AS, Basava S, Basavalingu B (2002) Prismatine and sapphirine bearing assemblages from Rajapalaiyam area, Tamil Nadu: origin and metamorphic history. J Geol Soc India 59:103–110
Stacey JS, Kramers JD (1975) Approximation of terrestrial lead isotope evolution by a two-stage model. Earth Planet Sci Lett 26:207–221
Staehle HJ, Raith M, Hoernes S, Delfs A (1987) Element mobility during incipient granulite formation at Kabbaldurga, Southern India. J Petrol 28:803–834
Sun S-s, McDonough WF (1989) Chemical and isotopic systematic of ocean basalts: implications for mantle composition and processes. Geol Soc Lond Spec Publ 42:313–345
Suresha KJ, Srikantappa C (2005) Igneous charno-enderbites and charnockites (C-type magmas) around Dindigul, Tamil Nadu. Geol Soc India 44:27–40

Tadokoro H, Tsunogae T, Santosh M, Yoshimura Y (2007) First report of the spinel + quartz assemblage from Kodaikanal in the Madurai Block, southern India; implications for ultrahigh-temperature metamorphism. Int Geol Rev 49:1050–1068

Tomson YJ, Bhaskar Rao T, Vijaya Kumar J, Mallikharjuna R (2006) Charnockite genesis across the Archaean–Proterozoic terrane boundary in the South Indian Granulite Terrain: constraints from major–trace element geochemistry and Sr–Nd isotopic systematics. Gondwana Res 10:115–127

Vernon RH (1999) Flame perthite in metapeltic gneisses at Cooma, SE Australia. Am Miner 84:1760–1765

Watson EB, Harrison TM (1983) Zircon saturation revisited: temperature and composition effects in a variety of crustal magma types. Earth Planet Sci Lett 64:295–304

Weaver BL (1980) Rare-earth geochemistry of Madras granulites. Contrib Mineral Petrol 71:271–279

Weaver BL, Tarney J, Windley BF, Sugavanam EB, Rao VV (1978) Madras granulites: geochemistry and P-T conditions of crystallisation. In: Archaean geochemistry; Proceedings of the first international symposium on Archaean geochemistry; The origin and evolution of Archaean continental crust, pp 177–204

Wiebe RA, Janardhan AS (1988) Metamorphism of the Oddanchatram anorthosite, Tamil Nadu, South India. J Geol Soc India 31:163–165

Woolley AR, Kempe DRC (1989) Carbonatites: Nomenclature, average chemical compositions, and element distribution. In: Bell K (ed) Carbonatites: genesis and evolution. Unwin Hyman, London, pp 1–14

Yoshida M, Bindu RS, Santosh M, Shirahata H (1998) Electron microprobe monazite ages from the Trivandrum Granulite Belt of South India. J Afr Earth Sci 27:216–217

Yuguchi T, Nishiyama T (2008) The mechanism of myrmekite formation deduced from steady-diffusion modeling based on petrography: case study of the Okueyama granitic body, Kyushu, Japan. Lithos 106:237–260

Zhao J-X, Ellis DJ, Kilpatrick JA, McCulloch MT (1997) Geochemical and Sr-Nd isotopic study of charnockites and related rocks in the northern Prince Charles Mountains, East Antarctica: implications for charnockite petrogenesis and proterozoic crustal evolution. Precambrian Res 81:37–66

Mineralogy, Mineralization and Earth Dynamics

Chapter 16
Kimberlites, Supercontinents and Deep Earth Dynamics: Mid-Proterozoic India in Rodinia

Stephen E. Haggerty

Abstract Diamonds and the mineral inclusions in diamonds demonstrate unequivocally that the kimberlites and lamproites (kimberlite clan rocks-KCRs) that transport these exotic assemblages to Earth's surface from subcratonic lithospheres (~200 km), the transition zone (410–660 km), and the lower mantle (>660 km), must have a thermal source to induce the eruptive process that is deeper than the deepest samples. The diamonds are old (2.8–3.2 Ga) but the KCRs are young. A synchronously global diamond eruptive event occurred in the Mesozoic at ~100 Ma, with eruptions of huge volumes of basalt in large igneous provinces (LIPS), fragmentation and disruption of Pangaea, superchron behavior of Earth's geomagnetic field, and superplumes from the D″ core–mantle boundary layer at 2,900 km. There was another significant diamond eruptive episode in the mid-Proterozoic at ~1.1 Ga in the supercontinent of Rodinia. This event is examined in the context of the Mesozoic paradigm of LIPS, superplumes and superchrons. The conclusion is positive, and is particularly fitting, given the nature of this dedicated volume, and the fact that the largest number (>100) of known diamondiferous KCRs are in India.

16.1 Introduction

Diamonds are natural antiques (2.8–3.2 Ga), that are erupted to the surface in volcanoes of opportunity at specific times in Earth's history; exclusively in cratons and from depths in excess of >200 km, these extraordinary deep Earth conduits, and their view through windows of diamond, provide an unprecedented record and understanding of crust, mantle and core dynamics. In a series of innovative contributions on the "Geological consequences of superplumes", Larson (1991a, b) showed an empirical relationship between and among the supercontinent of Pangaea, large igneous provinces (LIPS), anaerobic oil and gas production, and a Mid-Cretaceous superchron of

S.E. Haggerty (✉)
Department of Earth and Environmental Sciences, Florida International University, Miami, FL 33155, USA
e:mail: haggerty@fiu.edu

J. Ray et al. (eds.), *Topics in Igneous Petrology*, DOI 10.1007/978-90-481-9600-5_16,

Earth's magnetic field. These studies, along with a new dynamic model of kimberlites and diamond genesis (Haggerty 1986), the recognition of ultra deep (>300 km) diamonds (Moore and Gurney 1985), transition zone xenoliths in kimberlites (Haggerty and Sautter 1990), and diamonds from the lower mantle >660 km (Scott-Smith et al. 1984), led to a revised and integrated model for the globally synchronous eruption of diamondiferous kimberlites at 80–120 Ma (Haggerty 1994). Driven by superplumes, fingerprinted paleomagnetically by the 40 million year long normal superchron event, it was proposed that the Mesozoic breakup of Pangaea was additionally coupled to kimberlites and their mineral trove of diamonds from lithospheric and sublithospheric depths; the diamonds were dynamically erupted, and the core-triggering mechanism was critically recorded by events in the Earth's magnetic field (Haggerty 1994, 1999a). Having established that diamonds are geological and geophysical expressions of deep Earth processes for the global Mesozoic event, the next and obvious question is: do these correlations apply to other time periods? The other major, global kimberlite event is mid-Proterozoic, at ~1,100 Ma and with the largest concentration of eruptions being in India, it is appropriately suitable for detailed discussion in this dedicated volume to Professor Mihir K. Bose.

Diamonds were first discovered in India some 4 millennia ago. Thousands of miners were reported along the banks of the Krishna River, SE Andhra Pradesh, and it was here that the historically famous diamonds (Great Mogal – 900 ct; Nissam – 440 ct: Regent – 410 ct; Orloff – 195 ct; Darya-I-Noor – 186 ct; Shah – 95 ct), as well as many stones with a rich history (Hope – 112 ct; Koh-I-Nur – 793 ct), were recovered. The primary source of these exotic gems, however, has never been found, possibly because the host rocks are neither archetypal kimberlite nor classic olivine lamproite (Haggerty and Birkett 2004); hence, the term, kimberlite clan rock (KCR), is employed. About 50 diatremes, in two major clusters, are recognized in the Dharwar Craton alone in south India; and more than 50 are scattered in cratons to the N and NE of the subcontinent. Majhgawan, in the Bundelkhand Craton, is the only operating diamond mine in India.

16.2 The Geology and Genesis of Diamond

It is generally accepted, through successful and futile exploration programs over the past 5 decades, that diamondiferous kimberlites and petrochemically related lamproites are restricted to old (>2 Ga), cold and stable cratons (Levinson et al. 1992 and refs. therein). Based on mineral inclusions in diamonds (e.g. garnet, pyroxenes, sulfides), and the age determination (most are in the range 2.8–3.2 Ga) of these inclusions by a variety of isotopic systems (Rb/Sr, Nd/Sm, U/Pb, Re/Os), we may conclude that nucleation and growth of diamond was concurrent with the development of subcratonic lithospheres (Boyd et al. 1985; Haggerty 1999a; Stachel and Harris 2008), in which melt extraction to form embryonic, continental crust was a fundamental process; and from stable isotopic compositions, the carbon in diamond is primordial (Deines 1980; Haggerty 1999a). The extreme alternative views are that cratons grew by lateral accretion (De Wit et al. 1992), that subduction played a significant role in generating kimberlites and producing diamonds (Ringwood et al. 1993), and that carbon, albeit Archean, is biologically mediated (Tappert et al. 2009; Albach et al. 2009). Theories for the crystallization of diamond are less extreme with a majority view that converges on the reduction of carbon dioxide [$CO_2 + 2H_2 = C + 2H_2O$], the dissociation of methane [$CH_4 = C + 2H_2$], and the catalytic reaction of sulfides [e.g. $2FeS + CO_2 = 2FeO + S + C$], or nitrogen [$2N_2 + 3CH_4 = 3C + 4NH_3$], from mantle melts or supercritical metasomatic fluids (Haggerty 1999b).

16.3 Lithospheric and Sublithospheric Diamonds

A schematic view of a typical cratonic cross-section is shown in Fig. 16.1, representing ancient crust with marginal fold belts at the surface, to upper mantle lithospheric depths of ~250 km. The keel-shaped lithosphere is dominated by depleted harzburgite and dunite with enclaves of eclogite and a morphological variety of diamonds; a zone of intense metasomatism is present at 80–100 km (Haggerty 1986 and refs. therein). Isotherms parallel the contour of the subcratonic lithosphere, and the crossing diamond–graphite stability curve is accurately depicted based on experimental P–T determinations (Bundy et al. 1996). The lithosphere has a lower density, lower oxygen fugacity, and lower temperature than the surrounding asthenosphere which is composed of spinel lherzolite at shallow depths and garnet lherzolite at greater horizons; the lithosphere has trapped pods of eclogite and is underplated by magmatic eclogite (i.e. high pressure basalt). Dissolved gasses in melts are dominated by C, O, H, N, and S, from which diamonds nucleate and grow over millions of years (Haggerty 1999a, b and refs. therein). Four typical sampling profiles are shown in Fig. 16.1 representing: K1 as a non-diamond bearing kimberlite, carbonatite, or alkali basalt vent; K2 would sample lithospheric and asthenospheric diamonds; K3 is a typical lithospheric diamondiferous kimberlite; and L1 is a geochemically enriched diamond-bearing lamproite with a complex plumbing system that transgresses and assimilates the zone of intense metasomatism at 80–100 km and is intruded at the edge of a craton, as in Argyle, NW Australia (Jaques et al. 1984).

The lower panel of Fig. 16.1 represents the contact of the upper mantle (410 km) with the transition zone (TZ 410–660 km); and the TZ with the lower mantle down to the D″ layer at the core–mantle boundary (2,900 km). Rising plumes from the metastable D″ layer are assumed to flatten at and in the thermal and mechanical boundary layers of the TZ, because of density contrasts, may spurn mini-plumes, and may become convoluted by convection above 410 km (Courtillot et al. 2003). The mineralogy of these deeper horizons changes dramatically as a function of P–T conditions: orthorhombic olivine [$(FeMg)_2SiO_4$], the major mineral in the lithosphere, becomes structurally more compact (wadsleyite), and then cubic with the spinel structure (ringwoodite); this is the TZ. Associated pyroxene also becomes cubic, dissolves in garnet and becomes majorite. Below the TZ and into the lower mantle, ringwoodite ultimately dissociates into (FeMg)O [FeO – wüstite; MgO – periclase] + $(FeMg)SiO_3$ [i.e. pyroxene composition with a cubic ($CaTiO_3$ – perovskite sensu stricto) structure)].

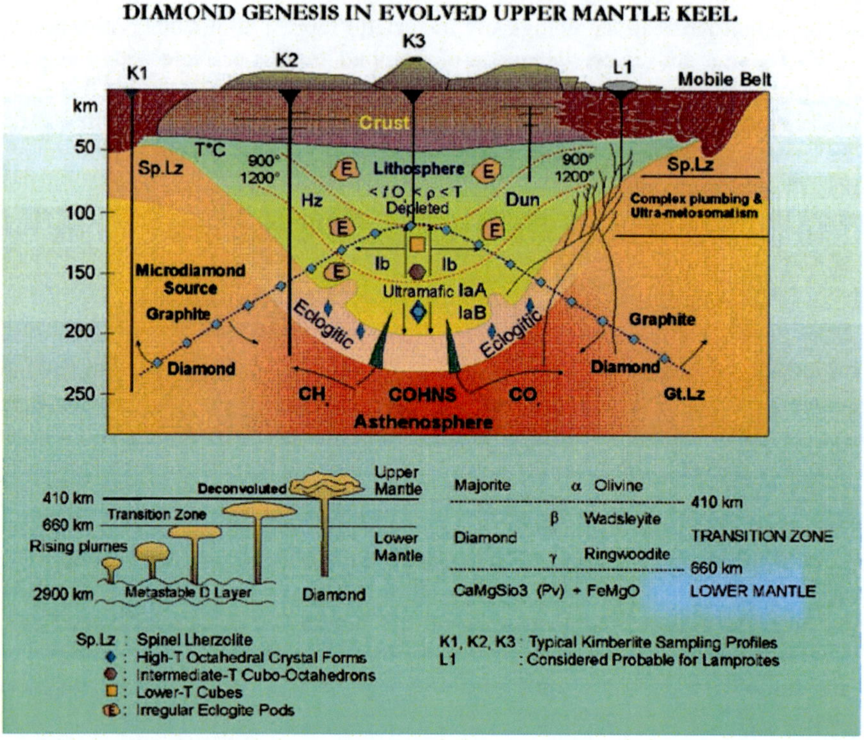

Fig. 16.1 A schematic cross-section of an Archean cratonic keel, from crust to the asthenosphere. K1–K3 are typical conduits for kimberlites; L1 is characteristic of lamproites. Note the isotherms in the lithosphere and the form of the diamond–graphite stability curve. *E* is eclogite; *Dun* is dunite; *Hz* is harzburgite; and *Sp.Lz* is spinel lherzolite. An explanation and significance of components in the lower panel are discussed in the text and given more fully in Haggerty (1986) (Modified from Haggerty 1999a)

It is from xenoliths and diamond inclusions (e.g. Nixon 1987) that the above distributions of minerals, phase assemblages and rocks have been determined, and on which the mineralogy and petrology of the lithosphere, asthenosphere, and the lower mantle is based. The materials profile is supported by seismology (e.g. Anderson 1989) and high P–T experimental data (e.g. Manghnani and Yagi 1998).

16.4 Spatial, Temporal and Geodynamic Sampling of Diamonds

We know from mineral inclusions in diamonds, and from corresponding suites of xenoliths in kimberlites that diamond populations fall into distinct petrological classes (Fei et al. 1999): *peridotitic* (Fo-rich olivine, orthopyroxene, Mg–Cr-rich garnet, chromite); *eclogitic* (Ca–Fe-rich garnet, clinopyroxene, coesite, zircon, moissanite-SiC, metallic Fe); *lherzolitic* (more Fa-rich olivine, orthopyroxene, clinopyroxene); and *sulfidic* (pentlandite, pyrite, pyrrhotite, mono-sulfide solid solution, chalcopyrite; high-Ni sulfides in diamonds have peridotitic affinities, and low-Ni sulfides are typical of eclogitic diamonds). These four classes are from the upper mantle and are in contrast to the lower mantle suites that contain majorite, majoritic-garnet, magnesio-wüstite; ferro-periclase, and perovskite-structured pyroxene. Upper mantle diamonds are globally ubiquitous; asthenospheric and TZ diamonds are much rarer (Stachel 2001); and lower mantle diamonds have so far been recognized only in single localities in Brazil, Australia, Guinea, South Africa and Canada (Tappert et al. 2009 and refs. therein).

Although the lower mantle diamonds are volumetrically small, a factor possibly related to sampling statistics, these super deep diamonds are important markers, and must be taken into account, if any transport model is to be viable and the mechanism of sampling sound for diamonds from depths >660 km. Kimberlite intrusions are CO_2 – charged and cryogenically explosive at crustal levels but require a substantial heat source to initiate the intrusion process (Sparks et al. 2006; Wilson and Head 2007). Mantle melting is required, but there are a limited number of known ways (extension, subduction and plumes) in which this can occur. Whatever mode of melting is favored there are constraints to be met, the most severe of which are: (i) diamondiferous eruptions are globally synchronous; (ii) sampling is from extreme depths (>600 km); (iii) eruptions are restricted to super-continental cratons (non-crustal injections result in ocean island basalts); (iv) kimberlite eruptions are related to LIPS; (v) kimberlite eruptions in the Mesozoic correlate with superchron behavior of the Earth's magnetic field; and (vi) the driving force must be sufficient to aid injection through the TZ and the crust at supersonic velocities to preserve crystalline carbon as diamond.

Among the possible thermal alternatives of decompression, subduction, and plumes that best satisfy the above constraints is heat liberated from the D″ core-mantle boundary layer at 2,900 km. This is not steady state heat loss from the core but dynamically focused thermal emissions in superplumes. It is superplumes, therefore, that were strongly favored, particularly for the Mesozoic diamondiferous kimberlite event (Haggerty 1994); and it is the superplume hypothesis that will be tested in the sections that follow for the mid-Proterozoic, 1.1 Ga diamond event.

16.5 Paleozoic to Mesozoic Kimberlite–Diamond Events

As a prelude to the central discussion it is appropriate to present the diverse correlations that have thus far been determined for Paleozoic to Mesozoic kimberlite events, and then to use this correlative template in a similar manner for diamond intrusions at 1.1 Ga.

A time (present – 500 Ma) – depth (surface – 2,900 km) profile is illustrated in Fig. 16.2. At the base in shaded blocks are superchron events (N & R); mushroom-like

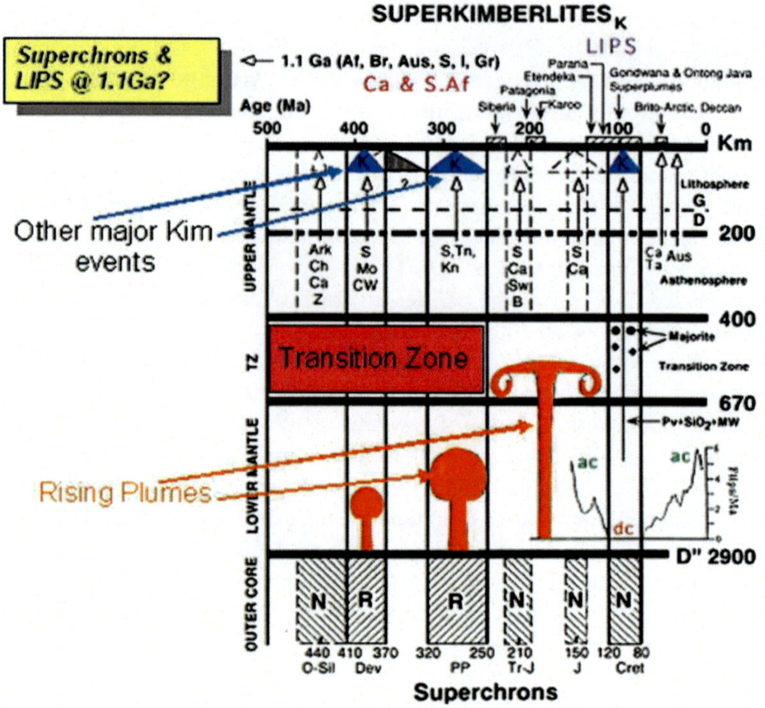

Fig. 16.2 An orthogonal cross-section of age (0–500 Ma) versus depth (surface to the outer core), with horizons at 2,900 km (D″); the transition zone (410–660 km); and the asthenosphere–lithosphere boundary (200 km). The hachured N and R time-slots are superchrons and the ac–dc curve refers to reversals per Ma (see text). Columns are times of kimberlite eruptions: Archangelsk (Ark), China (Ch), Canada (Ca), Zimbabwe (Z), Siberia (S), Missouri (Mo), Colorado-Wyoming (CW), Tennessee (Tn), Kentucky (Kn), Swaziland (Sw), Botswana (B), Australia (Aus). Note that other events in Africa (Af), Brazil (Br), Greenland (Gr), and India (I) are off scale at 1.1 Ga. LIPS are listed in the upper panel (Modified from and more fully discussed in Haggerty 1994)

plumes rise from the D″ layer into the TZ which is dominated by majorite, wadsleyite and ringwoodite (Fig. 16.1); the lithosphere is marked at 200 km with a series of intrusive kimberlites penetrating the surface; the most significant global-wide, intrusive diamond event is at ~100 Ma with well over a 2,000 bodies in all of the major kimberlite fields; this event corresponds with the long (40 Ma) Cretaceous N-superchron. Other major kimberlite events, both with R-superchron correlations, are at ~300 Ma (Siberia, Tennessee, Kentucky), and ~400 Ma (Siberia, Missouri, Colorado–Wyoming) in the Pennsylvanian – Permian (the 70 Ma Kiaman event), and the Devonian, respectively. Lesser events (i.e. fewer intrusions over smaller areas), but still with excellent superchron correlations exist in the Jurassic (Normal – Siberia, Canada), Triassic–Jurassic (Normal – Siberia, Canada, Swaziland, Botswana), and the Ordovician–Silurian (Normal – Archangelsk, China, Canada, and Zimbabwe). LIP events are listed in the upper-most panel: the Tertiary Brito-Arctic,

Ethiopian and Deccan Traps are related to very young (50–80 Ma) kimberlites in Canada, Tanzania, and Australia; the widespread Mesozoic diamondiferous intrusions are preceded by the Karoo, Etendeka, Parana, Patagonia, and Ferrar basalts; and the Triassic – Jurassic kimberlite intrusions in Siberia, Canada, Swaziland and Botswana, are related to the volumetrically enormous Siberian Traps at ~250 Ma.

Not indicated, but implicit in the data outlined in Fig. 16.2, is the fact that the greatest number of kimberlite intrusions, from the Neo-Paleozoic to the Present, erupted in association with LIPS and Gondwana rifting at ~180 Ma. There are no clear correlations between and among rifting, drifting, and kimberlite intrusive episodes for the other five diamond clusters shown in Fig. 16.2; this would imply that supercontinent fragmentation is *not* a fundamental prerequisite to diamondiferous transport from Earth's deep interior. But it does not preclude the possibility that intrusions took place during intermittent rifting, in the overall process of supercontinent assembly, as concluded below for KCR intrusions in Rodinia. Intrusions during overall superchron events, however, are indisputable (Fig. 16.2). A first order conclusion, therefore, for primary diamond deposits intruded since 500 Ma, is that the heat source is the outer core, and the thermal driving force that facilitated sampling from the lower and upper mantles were in superplume conduits. It was also this heat loss that changed the dynamics of outer core convection, and as argued previously (Haggerty 1994), and discussed below, led to superchron behavior of the Earth's geomagnetic field.

16.6 The Mid-Proterozoic (1.1 Ga) Diamond–Kimberlite Event

The Premier kimberlite in South Africa was the first to be recognized as having an age distinct from the many Mesozoic kimberlites that are spread far and wide across the Kaapvaal Craton (Allsopp et al. 1967). Once dated, it became the standard for this early event, but it was not until the discovery of the diamond-bearing Argyle lamproite in NW Australia (Jaques et al. 1984), that the mid-Proterozoic event was considered to be a possible indicator for global diamond volcanism. That inference was finally substatiated when kimberlites in India were dated at 1.1 Ga (Kumar et al. 1993). It was evident, as more dates became available, that the mid-Proterozoic diamond intrusive event is, indeed, global with examples now from Africa, Australia, and India to Brazil, Siberia, and Greenland (Haggerty 1994; Heaman et al. 2003; Kumar et al. 2007).

Based on geologic and paleomagnetic data, the mid-Proterozoic time period of 1,100–1,000 Ma is considered critical to the consolidation of the supercontinent Rodinia (Dalziel 1997), followed by breakup between 825 and 740 Ma (Li et al. 2008). It is important to appreciate, however, that accretion and dispersal may overlap in mini-cycles (Silver and Behn 2008), so that the most extensive and coherent continental mass in any supercontinent cannot be defined by a single point in time (Rogers and Santoch 2003); moreover, as noted by Torsvik et al. (2008) "……*amalgamation, continental makeup, and fragmentation is hampered by the fact at any given time, the latitudes for only a few continents are known.*"; in this case the reference is to Rodinia. At first glance, therefore, it would seem that the 1.1 Ga diamond intrusive episode into Rodinia is in stark contrast to the supercontinent association of Gondwana and its global Mesozoic diamond event; the former being related to supercontinent amalgamation, and the latter to supercontinental rifting and drifting.

Closer examination of magmatic events in Rodinia, however, although complex in distribution (i.e. whether blocks were in contact or dispersed during the same datable event), and compounded by interpretations of thermal sources (e.g. continental insulation versus plumes), show that there are, indeed, common factors between the disparate events at ~1.1 Ga and ~100 Ma. The first of these uncertainties will require substantially more paleomagnetic measurements, but the second, which is really more important and highly controversial (Foulger et al. 2005), can be addressed and critically evaluated because diamond events are multiply constrained by geology and geophysics. Geologically in the broadest context that includes exotic, deep Earth minerals (majorite, silicate–perovskite, magnesiowüstite, ferro-periclase, coesite, and of course diamond), cratons, lithospheric xenoliths, including rocks from the TZ, and highly explosive volcanism of ultramafic compositions. Geophysically, the appeal is to D″ superplumes with collateral effects in the outer molten core, and records held in superchrons of Earth's geomagnetic field.

A variant on the simple continental-insulation model (Anderson 1994), is top-down tectonics (Bebout et al. 1996; Anderson 2001). In the general model, subducted plates flatten out in the TZ, accumulate, and subsequently avalanche into the lower mantle (Tackley et al. 1993); whether such catastrophically – driven plates ever reach the core–mantle boundary (Maruyama et al. 2007), is speculative. Supercontinent subduction, as for Rodinia, is circumferential as illustrated in Fig. 16.3 (e.g. Torsvik 2003; Li et al. 2008). For this scenario, the geometry invoked for draw-down subduction is conical, and the question is: can this subducted conical plate generate a plume in response to its impact on D″? There are at least two simple tests to evaluate the process, one petrochemical, and the other geophysical. If the avalanche hypothesis is correct, these plumes should contain evidence of *substantial amounts* of crustal protoliths and arc-related entities; that only small volumes are inferred (e.g. Hergt et al. 1991), suggests the process unlikely. Secondly: Is the process sufficient to severely disrupt thermal convection in the outer core to induce a geomagnetic superchron; it's unknown, but intuitively has a low probability.

The other important question is: can subducted oceanic crust produce a plume and release sufficient heat from the core–mantle boundary to generate LIPS? It seems unlikely because from the very extensive global data base, LIPS are defined as being aerially extensive 10,000–100,000 km^2, volumetrically large 3–5 km^3, and with eruptions over short (1–3 Ma) periods of time (Mahoney and Coffin 1997; Ernst and Buchan 2001). And the fact that LIPS are associated with supercontinent disruption makes subduction an even less likely possibility.

As outlined in Fig. 16.2, the evidence for multiple correlations between diamond eruptives, LIPS, Gondwana and superchrons, particularly in the Mesozoic at ~100 Ma,

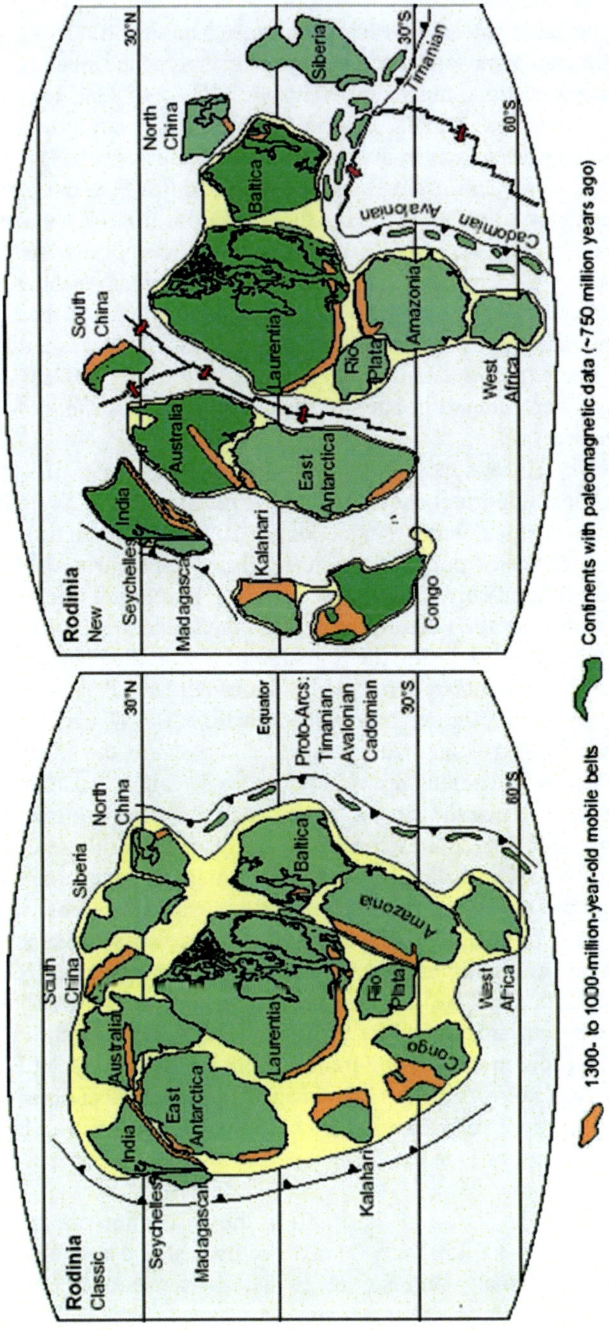

Fig. 16.3 Two images, one *classic*, the other *new* of reconstructed Rodinia; the former is for the period 1,300–1,000 Ma, and the latter, with paleomagnetic data, is at ~750 Ma (Torsvik 2003). The implications of circular subduction are discussed in the text

is robust. The question raised throughout is: can a similar set of correlations be established for the mid-Proterozoic diamond episode? A significant inhibitor is the difficulty of obtaining reliable paleomagnetic measurements on very old rocks, as well as the uncertainties in isotopic age dating. These were overcome in three major studies: (1) Some 28 Mesoproterozoic, magmatic events (1.20 Ga–1.07 Ga), for kimberlites, lamproites, and carbonatites in India, Australia, Africa and North America are presented in a compelling discussion by Kumar et al. (2007) for a global 1.1 Ga link to alkali and ultrapotassic volcanism. (2) The global record for LIPS, as outlined by Ernst et al. (2008), consists of 16 major Provinces that date from 1.0 to 1.2 Ga, the largest of which are the Keweenawan (1,117–1,085 Ma) in Canada, and the Umkondo (1,112–1,106 Ma) in Southern Africa and Antarctica. (3) In a far reaching collaborative program, undertaken and reported on by Hanson et al. (2004), both excellent paleomagnetic and radiometric data were obtained on coeval LIPS from the Kalahari (Umkondo) and Laurentian (Keweenawan) Cratons in Rodinia. Collectively, these studies are a major step forward in attempting to answer the 1.1 Ga globally-active diamond intrusive events.

The results from Hanson et al. (2004), summarized in Fig. 16.4, are self explanatory. The most important conclusions to be reached from this study are: (1) the 1.1 Ga diamond events coincide with a long (~20 Ma) N-superchron; (2) LIPS were intruded in two major pulses, one during the N-superchron, and the other during an earlier R-superchron. Ernst et al. (2008) interpret these events as episodic-fragmentation in the overall cycle of supercontinent assembly, and both are related to superplumes. In India, the intrusion of kimberlite clan rocks (KCRs) are *locally* controlled by intersecting major faults (Babu 1998); on a *regional* scale, diamond-bearing KCRs are in a well defined, NE–SW trending corridor, ~200 km wide and 1,000 km long (Chetty 2000); and *globally*, the mid-Proterozoic KCRs of India lie on a smooth arc that includes Brazil, the Central African Republic, and Argyle (Australia) in the reconstructed supercontinent of Rodinia (Haggerty 1999b). Crustal tectonics controlled the geometry of these intrusions in tensional settings, consistent with deep source volcanism during rifting. An important adjunct to appreciate is that each continental block had its own subcratonic lithosphere, and its own inventory of ancient diamonds, that were synchronously erupted over a short time span induced by a common thermal source.

From the above discussion, we may confidently conclude that the mid-Proterozoic diamond intrusive events correlate with LIPS and geomagnetic superchrons, and that the ultimate thermal source and driving force is from the D″ core–mantle boundary layer at 2,900 km. It is the last of these broad conclusions that are outlined schematically in Fig. 16.5 to more fully capture the core–superchron relationship and the intimate role played by plumes in removing heat and altering the convection dynamics of Earth's outer liquid core.

Aspects relevant to the figure of the Earth in this discussion are the solid and liquid cores, and the 200–300 km thick D″ layer composed of a light element (H, C, K, S, O) enriched Si–Fe matt (Gurnis et al. 1998). The sequence of events leading to superplume eruptions are given in actions 1–6, labeled to the right of Fig. 16.5 (Griffiths and Campbell 1991; Larson and Olson 1991). (1) D″ disrupts on reaching

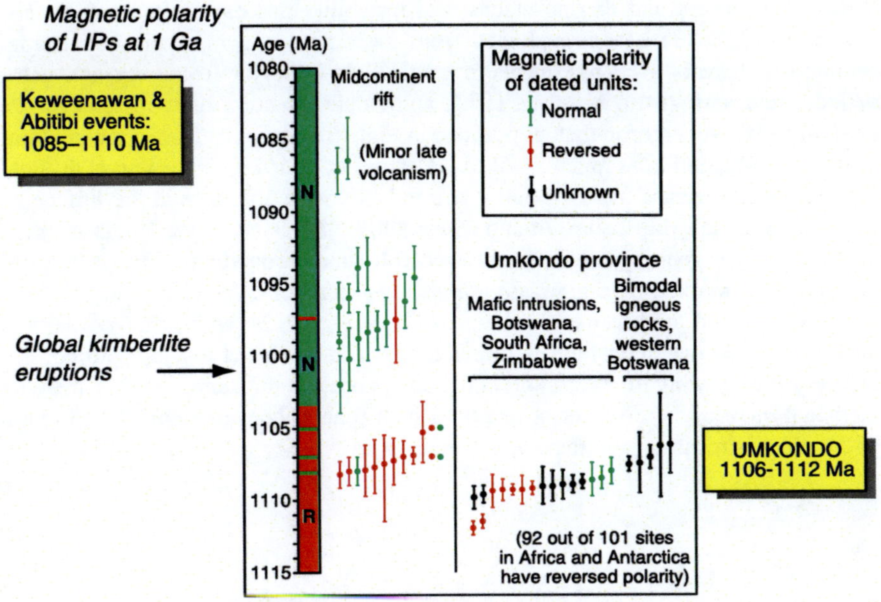

Fig. 16.4 Magneto-stratigraphy for N and R events for two very large LIPS, one in Canada, and the other in South Africa and Antarctica for the period 1,080–1,115 (Hanson et al. 2004). The global 1.1 Ga diamond kimberlite event is global

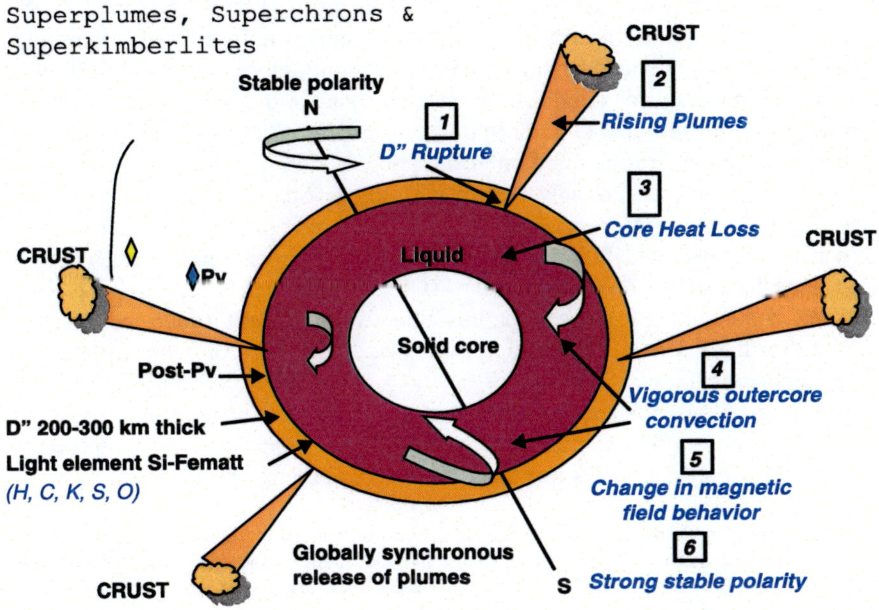

Fig. 16.5 Superplumes, superchrons and superkimberlites centered on the core and D″ zone with events (1–6) as discussed in the text

a critical thickness and thermo-mechanical instability ensues; (2) a rising superplume develops; (3) heat is siphoned from the outer core; (4) rapid convection results; (5) there is a change in geomagnetic field behavior from constant, non-periodic reversals; (6) to a strong, stable and single direction of the geomagnetic field, namely a superchron that, as outlined in Fig. 16.2, may last from a few million years to 40 Ma, and to as much as 70 Ma (Harland et al. 1989). The interval of time between plume release from the outer core and explosive eruptions at the surface is of the order of 20 Ma (Courtillot and Olson 2007); this is about the length of time required for the geomagnetic field to reach full superchron status. The onset of a superchron is marked by a gradually *diminishing* number of reversals per Ma (see the ac–dc inset in the Mesozoic column at D″ of Fig. 16.2); with no reversals, the direction of the magnetic field is immobilized, and a superchron is achieved; on recovery (i.e. slowing of convection in the outer core and thickening of D″), there is a gradual *increase* in the reversal rate until such time as conventional N-R flipping of the poles is resumed and the *status quo* retained.

16.7 Conclusions

The globally synchronous mid-Proterozoic diamondiferous intrusive event was closely associated with LIPS and superchrons, and is most reasonably related to superplume activity at the D″ core–mantle boundary. Although the episode appears to have arisen during the amalgamation of Rodina, the overall distribution of KCRs, and the tectonic control in particular of KCRs in India, strongly favors intermittent episodes of rifting induced by superplumes. The correlations are strikingly similar to those documented for diamond intrusive episodes in the Mesozoic and for the interval of intrusions back to ~500 Ma in the Cambrian. The trigger to plume generation is uncertain but the more interesting observation is that LIPS and diamondiferous intrusions are possibly related not only in time (Fig. 16.2), but also in space insofar as the very region on the CMB (at the margins of low shear velocity provinces, (Torsvik et al. 2006; Burke et al. 2008) that serve to generate LIPS, may also be the heat source from which proto-KCRs are initiated (T. Torsvik, June 2009, personal communication). If correct, this attractive hypothesis would more readily account for the Rodinia relationship in space and time for KCRs and LIPS in the mid-Proterozoic.

We may, at this point, conclude that most, if not all, diamond volcanoes are driven by deep Earth superplumes that are brilliantly embossed with superchron diamond signatures, as befits this dedication to Professor Mihir K. Bose.

Acknowledgments Thanks to Gautam Sen for the invitation and opportunity to contribute to this volume; it is, indeed, an honor to acknowledge the contributions and scholarship of Professor Mihir K. Bose. Thanks are also extended to Professor Jyotisankar Ray, two reviewers, and to Trond Torvik for instructive comments. Funding from the NSF over many years, with support from the Fulbright Foundation, the Government of India, and industry, provided unprecedented opportunities to explore the original land of diamonds: Namaste India!

References

Albach S, Shirey SB, Stachel T, Creighton S, Muehlenbachs K, Harris JW (2009) Diamond formation episodes at the southern margin of the Kaapvaal Craton: Re-Os systematics of sulfide inclusions from the Jagersfontein Mine. Contrib Mineral Petrol 157:525–540

Allsopp HL, Burger AJ, Van Zyl C (1967) A minimum age for the Premier kimberlite pipe yielded by biotite Rb-Sr measurements, and related galena isotopic data. Earth Planet Sci Lett 3:161–166

Anderson D (1989) Theory of the Earth. Blackwell, Boston

Anderson D (1994) Superplumes or supercontinents? Geology 22:39–42

Anderson D (2001) Top-down tectonics. Science 293:2016–2018

Babu TM (1998) Diamonds in India. Geological Society of India, Bangalore, 331p

Bebout GE, Scholl DW, Kirby SH, Platt JP (1996) Subduction top to bottom. American Geophysical Union, Monograph 96, Washington, DC

Boyd FR, Gurney JJ, Richardson SH (1985) Evidence for a thick 150–200 km Archean lithosphere from diamond inclusion thermobarometry. Nature 315:387–389

Bundy FP, Bassett WA, Weathers MS, Hemley RJ, Mao HK, Goncharov AF (1996) The pressure–temperature phase and transformation diagram for carbon: updated through 1994. Carbon 34:141–153

Burke K, Steinberger B, Torsvik TH, Smethurst MA (2008) Plume generation zones at the margins of large low velocity provinces on the core-mantle boundary. Earth Planet Sci Lett 265:49–60

Chetty TRK (2000) Superplume and the Indian diamond corridor. In: Raval U (ed) Plate tectonics abstract vol. National Geophysical Research Institute, Hyderabad, pp 6–8

Courtillot V, Olson P (2007) Mantle plumes link magnetic superchrons to Phanerozoic mass depletion events. Earth Planet Sci Lett 260:495–504

Courtillot V, Davaille A, Besse J, Stock JM (2003) Three distinct types of hot spots in the Earth's mantle. Earth Planet Sci Lett 205:295–308

Dalziel IWD (1997) Neoproterozoic-Paleozoic geography and tectonics: Review, hypothesis, environmental speculation. Geol Soc Am Bull 109:16–42

De Wit MJ et al (1992) Formation of an Archean crust. Nature 357:553–562

Deines P (1980) The carbon isotopic composition of diamonds: relationship to shape, color, occurrence, and vapor composition. Geochem Cosmochim Acta 44:943–961

Ernst RE, Buchan KL (2001) Mantle plumes: their identification through time, Geological Society of America Special Paper. GSA, Boulder

Ernst RE, Wingate MTD, Buchan KL, Li ZX (2008) Global record of 1600–700 Ma large igneous provinces (LIPS): Implications for the reconstruction of the proposed Nuna (Columbia) and Rodinia supercontinents. Precam Res 160:159–178

Fei Y, Bertka M, Mysen BO (1999) Mantle petrology: field observations and high pressure experimentation. A tribute to FR Boyd, The Geochemical Society Special Publications 6. The Geochemical Society University, Houston

Foulger GR, Natland JH, Presnall DC, Anderson DL (2005) Plates, plumes and paradigms, Geological Society of America Special Paper 388. GSA, Boulder

Griffiths RW, Campbell IH (1991) Interaction of mantle plume heads with the Earth's surface and onset of small scale convection. J Geophys Res 96:18295–18310

Gurnis M, Wysession ME, Knittle E, Buffett BA (1998) The core–mantle boundary, vol 28, American Geophysical Union, Geodynamic series. AGU, Washington, DC

Haggerty SE (1986) Diamond genesis in a multiply-constrained model. Nature 320:34–38

Haggerty SE (1994) Superkimberlites: a geodynamic diamond window to the Earth's core. Earth Planet Sci Lett 122:57–69

Haggerty SE (1999a) A diamond trilogy: superplumes, supercontinents and supernovae. Science 285:851–860

Haggerty SE (1999b) Diamond formation and kimberlite-clan magmatism in cratonic settings. In: Fei Y, Bertka CM, Mysen BO (eds) Mantle petrology: field observations and high pressure experimentation – a tribute to FR Boyd, The Geochemical Society, Special Publications 6. The Geochemical Society University, Houston

Haggerty SE, Birkett T (2004) Geological setting and chemistry of kimberlite clan rocks in the Dharwar Craton, India. Lithos 76:535–549

Haggerty SE, Sautter V (1990) Ultra-deep (>300 km), ultramafic, upper mantle xenoliths. Science 248:993–996

Hanson RE et al (2004) Coeval large-scale magmatism in the Kalahari and Laurentian Cratons during Rodinia assembly. Science 304:1126–1129

Harland BW, Armstrong RL, Cox AV, Lorraine EG, Smith A, Smith DG (1989) A geologic time scale. Cambridge University Press, Cambridge, UK

Heaman LM, Kjarsgaard RA, Creaser RA (2003) The timing of kimberlite magmatism in North America: Implications for global kimberlite genesis and diamond exploration. Lithos 71:153–184

Hergt JM, Peate DW, Hawkesworth CJ (1991) The petrogenesis of Mesezoic Gondwana low-Ti flood basalts. Earth Planet Sci Lett 105:134–148

Jaques AL et al (1984) The diamond-bearing ultra-potassic (lamproitic) rocks of the West Kimberley region, Western Australia. In: Kornprobst J (ed) Kimberlites and related rocks. Dev Petrol 11A:225–254. Elsevier, Amsterdam

Kumar A, Padmakumara VM, Dayal AM, Murthy DSN, Gopalan K (1993) Rb-Sr ages of Proterozoic kimberlites of India: evidence for contempoareous emplacement. Precam Res 62:227–232

Kumar A, Heaman LM, Manikyamba C (2007) Mesoproterozoic kimberlites in India: a possible link to 1.1 Ga global magmatism. Precam Res 154:192–204

Larson RL (1991a) Latest pulse of Earth: evidence for a Mid-Cretaceous superplume. Geology 19:963–966

Larson RL (1991b) Geological consequences of superplumes. Geology 19:963–966

Larson RL, Olson P (1991) Mantle plumes control magnetic reversal frequency. Earth Planet Sci Lett 107:437–447

Levinson AA, Gurney JJ, Kirkley MB (1992) Diamond sources and production: past, present and future. Gems Gemol Winter:234–253

Li ZX et al (2008) Assembly, configuration, and break-up history of Rodinia: a synthesis. Precam Res 160:179–210

Mahoney JJ, Coffin MF (1997) Large igneous provinces. Continental, oceanic, and planetary flood volcanism, vol 100, Geophysical monograph. American Geophysical Union, Washington, DC

Manghnani MH, Yagi T (1998) Properties of earth and planetary materials at high pressure and temperature, vol 101, Geophysical monograph. American Geophysical Union, Washington, DC

Maruyama S, Santoch M, Zhao D (2007) Superplume, supercontinent, and post-perovskite: Mantle dynamics and anti-plate tectonics on the core-mantle boundary. Precamb Res 11:7–37

Moore RO, Gurney JJ (1985) Pyroxene solid solution in garnet included in diamonds. Nature 318:553–555

Nixon PH (1987) Mantle xenoliths. Wiley, New York

Ringwood AE, Kesson SE, Hibberson W, Ware N (1993) Origin of kimberlites and related magmas. Earth Planet Sci Lett 113:521–538

Rogers JJW, Santoch M (2003) Supercontinents in Earth history. Gondwana Res 6:357–368

Scott-Smith BH, Danchin RV, Harris Jw, Stracke, KJ (1984) Kimberlites from Ororooroo, South Australia. In: Kornbrobst J (ed) Kimberlites and related rocks. Dev Petrol 11A:121–142. Elsevier, Amsterdam

Silver PG, Behn MD (2008) Intermittent plate tectonics? Science 319:85–88

Sparks RSJ et al (2006) Dynamical constraints on kimberlite volcanism. J Volc Geotherm Res 155:18–48

Stachel T (2001) Diamonds from the asthenosphere and the transition zone. Eur J Miner 13:883–892

Stachel T, Harris JW (2008) The origin of cratonic diamonds – constraints from mineral inclusions. Ore Geol Rev 34:5–32

Tackley PJ, Stevenson DJ, Glatzmaer GA, Schubert G (1993) Effects of an endothermic phase transition at 670 km in a spherical model of convection in the Earth's mantle. Nature 361:699–704

Tappert R, Foden J, Stachel T, Muehlenbachs K, Tappert M, Wills K (2009) Deep mantle diamonds from South Australia: a record of Pacific subduction at the Gondwanan margin. Geology 37:43–46

Torsvik TH (2003) The Rodinia jigsaw puzzle. Science 300:1379–1381

Torsvik TH, Smethurst MA, Burke K, Steinburger B (2006) Large igneous provinces generated from the margins of the large low-velocity provinces in the deep mantle. Geophys J Int 167:1447–1460

Torsvik TH, Steinburger B, Cocks LRM, Burke K (2008) Longitude: linking Earths ancient surface to its deep interior. Earth Planet Sci Lett 276:273–282

Wilson L, Head JW (2007) An integrated model of kimberlite ascent and eruption. Nature 447:53–57

Chapter 17
Petrological Evolution and Emplacement of Siwana and Jalor Ring Complexes of Malani Igneous Suite, Northwestern Peninsular India

G. Vallinayagam and N. Kochhar

Abstract The Siwana and Jalor Ring Complexes of Malani Igneous suite, northwestern Peninsular India consist predominantly of acid magmatic rocks with minor amount of basic rocks. The Siwana Ring Complex (SRC) has mainly acid volcanic and acid volcano-plutonic associations in northern and southern flank respectively. Acidic dyke rocks trending NE–SW directions cut these rocks. The Jalor Ring Complex (JRC) has mainly acid plutonic rocks and cut by numerous dolerite dyke rocks. The granites in Siwana are hypersolvus, alkali amphibole–alkali pyroxene-bearing, peralkaline (mildly peraluminous) in character whereas the granites of Jalor are subsolvus, mica-bearing, peraluminous (mildly peralkline) in character. The granites show a high abundance of aluminum, calcium, magnesium in Jalor and sodium, potassium in Siwana areas. The elements of LIL, HFS, REE groups in the granites of Siwana show enriched pattern as compared to Jalor. Both complexes show low oxygen and Pb radiogenic isotope values. Petrogeochemical studies define them as A-type, anorogenic, HHP magmatism in the trans-Aravalli region and were formed by low degree partial melting of parent stock of acid crust rock due to underplating of mantle material via plume activity.

17.1 Introduction

The A-type magmatism plays a significant role in the crust–mantle interaction vis-a-vis genesis of several types of mineral deposits viz rare metal, rare earth, radioactive minerals are associated with them (Kochhar 1992, 2000, 2008a, b;

G. Vallinayagam (✉)
Department of Geology, Kurukshetra University, Kurukshetra 136119, India
e-mail: gvallinayagam@rediffmail.com

N. Kochhar
Department of Geology, Panjab University, Chandigarh 160014, India
e-mail: nareshkochhar2003@yahoo.com

J. Ray et al. (eds.), *Topics in Igneous Petrology*, DOI 10.1007/978-90-481-9600-5_17,

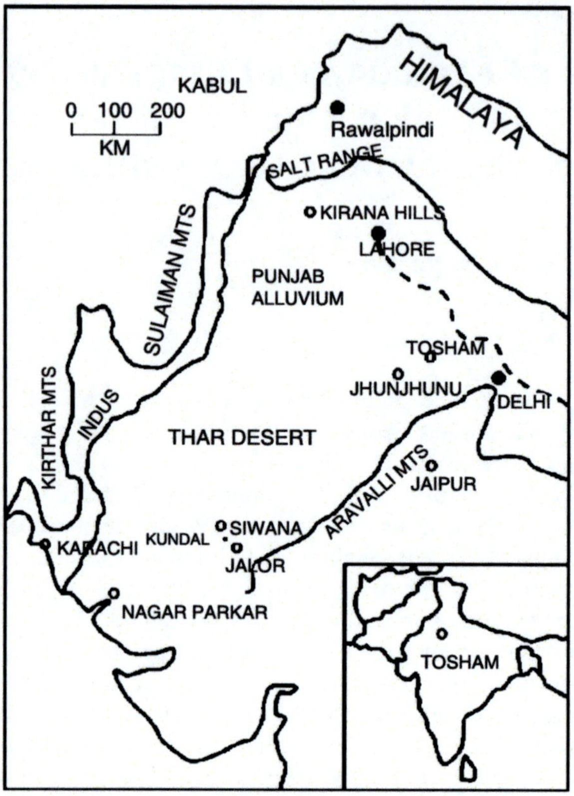

Fig. 17.1 Occurrence of the rocks of Malani Igneous Suite, Indian Subcontinent (Modified after Kochhar 2004)

Vallinayagam 2006). The Neoproterozoic Malani Igneous Suite (MIS) is the largest anorogenic, A-type acid magmatism of this kind in the Indian Shield (Kochhar 2000). It occurs in the trans-Aravalli mountain region (Fig. 17.1) and occupies an area of approximately 55,000 km^2. The MIS consists mainly of acid volcano-plutonic rocks with a minor amount of basic counterparts in the form of ring structure, ring complex and ring dyke (Kochhar 1984, 2004; Eby and Kochhar 1990; Bhushan and Chandrasekaran 2002). The Ring Complex at Siwana is the world class representative of this nature (Vallinayagam 1988). This paper attempts to bring a holistic account of geological setting, petrography and geochemistry on MIS to evaluate its petrogenetic evolution, mode of emplacement and tectonic implication in the northwestern Peninsular India.

17.2 Geological Setting

Based on the field studies in these complexes (Figs. 17.2 and 17.3), the various rocks exposed are grouped into three phases which are as follows:

Siwana Ring Complex (SRC) (Vallinayagam and Kochhar 1998)

3. Dyke phase: Trachyandesite, trachydacite, rhyolite and microgranites
2. Intrusive phase: Granite and gabbro with marginal facies variations

(other than dyke phase)
1. Extrusive phase: Basalt, trachyte, rhyolite and welded tuff

Jalor Ring Complex (JRC) (Kochhar and Dhar 1993)
3. Dyke phase: Rhyolite, olivine dolerite and microgranite
2. Intrusive phase: Olivine gabbro, biotite hornblende granite, biotite granites and alkali granites

(other than dyke phase)
1. Extrusive phase: Basalt, Rhyolite and welded tuff

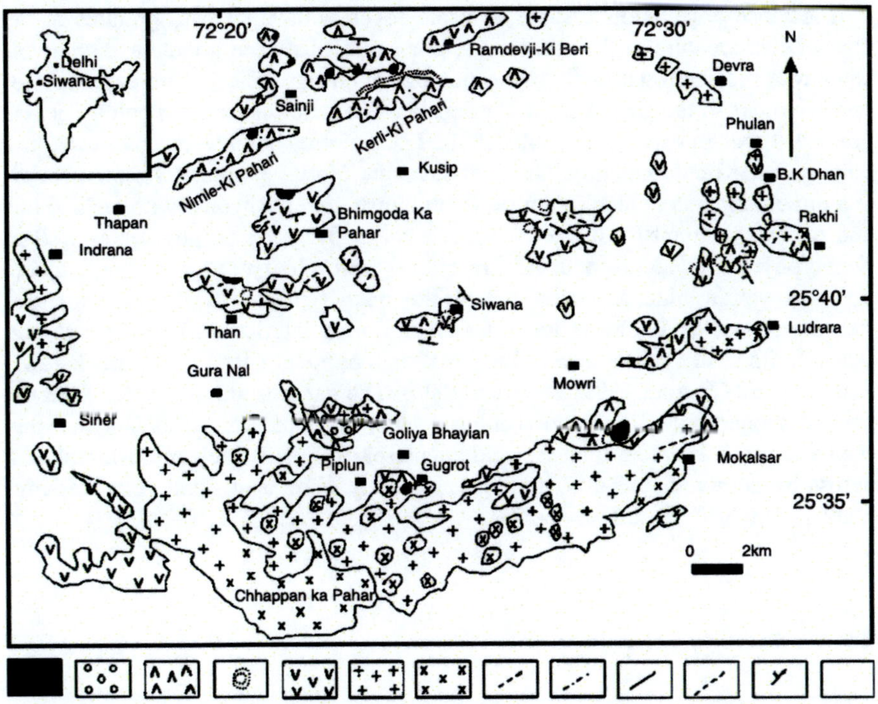

Fig. 17.2 Geological map of Siwana Ring Complex. 1. Basalt, 2. Gabbro, 3. Trachyte, 4. Ash bed, 5. Rhyolite, 6. Pink Granite, 7. Grey granite, 8. Dacite dyke, 9. Trachyandesite dyke, 10. Rhyolite dyke, 11. Microgranite dyke, 12. Dip, 13. Alluvium and blown sand (After Vallinayagam 2004)

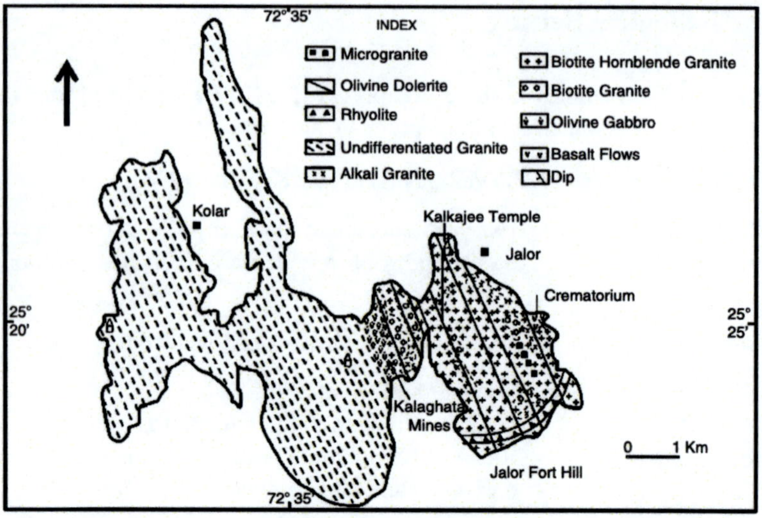

Fig. 17.3 Geological map of Jalor Ring Complex (After Kochhar and Dhar 1993)

The close association of acid volcanic rocks with high level granites is the characteristic feature of the complexes. The contact between granite and acid volcanic rocks is sharp without any morphological change. There is no exposures of country rocks in the area. The SRC granites are pink, bluish grey in colour, coarse grained, hard, massive and compact. In JRC, biotite–hornblende granites (pink colour) occur predominantly. Other granites are biotite granites (white coloured) and minor amounts of alkali granites (with bluish tinge). Rhyolite with welded tuff and trachyte account for the acid volcanic rocks. They are porphyritic as well as nonporphyritic in nature and vary in colour from cherry red, brick red, green, brownish yellow, blue, grey etc. They display many typical acid volcanic structures like flow, vesicles and amygdules. Basalt flows are interleaved with several acid volcanic flows at various levels. Thus MIS is a bimodal volcanism (Pareek 1981; Bhushan and Chittora 1999; Bhushan 2000). The gabbros are associated with the acid components. The occurrence of basic xenoliths in the acid rocks indicates that the basic rocks are older than the acid counterparts. Finally they are invariably cut across by numerous NE–SW trending acid/basic dykes and stand very predominantly due to differential weathering.

17.3 Petrography

Under the microscope, granites display hypidiomorphic and granophyric textures. SRC granites are hypersolvus (one feldspar type) and consist of alkali feldspar (modal composition 49–64%), quartz (28–34%), arfvedsonite (9–18%), riebeckite and aegirine (3–7%) as essential minerals. The accessory minerals present are apatite, sphene, zircon, haematite and aenigmatite. JRC granites are predominantly subsolvus – two feldspar type and consist of quartz (22–35%), perthite (8–19%), orthoclase (27–43%), albite (5–14%), biotite (8–15%), muscovite (3–9%), hornblende (1–6%) as constituent minerals. The accessory minerals are magnetite, zircon, fluorite, apatite and tourmaline. The rhyolites consist of alkali feldspar, high quartz, arfvedsonite, riebeckite, aegirine as essential minerals. The groundmass is composed of microlite of alkali feldspar, quartz, arfvedsonite and riebeckite. The basic components show ophitic and subophitic textures and consist of labradorite, albite, augite and olivine as essential minerals. The accessory minerals are magnetite, haematite, rutile, chlorite and sphene. The dyke rocks also consist of analogous petromineralogical characters to the extrusive and intrusive phases.

17.4 Mineral Chemistry

In SRC the mineral chemical characters and substitutional schemes portray that the amphiboles of granites evolved from richterite to arfvedsonite (magmatic subsolidus trend). In trachyte they evolve from arfvedsonite to riebeckite (oxidizing) and in rhyolite from richterite to riebeckite through arfvedsonite (magmatic subsolidus to oxidizing) (Vallinayagam 1997; Baskar and Kochhar 2006; Giret et al. 1980; Strong and Taylor 1984; Czamanske and Dillet 1988). The pyroxene evolves from hedenbergite to aegirine through aegirine augite (acmite–hedenbergite trend) in granites whereas in rhyolite they are represented as aegirine (acmite trend) (Stephenson 1972; Bonin and Giret 1985; Morimot 1989). In basic rocks, mineral chemical studies show the presence of arfvedsonite and aegirine in them. The alkali granites of JRC have a ferro-hornblende and ferro-edenitic composition of amphibole (Dhar and Kochhar 1997). The chemistry of biotite from JRC granites shows an iron enrichment trend (FeO*/MgO ratio: 6.72) and indicates the dominance of Mg, Fe and 2Al, $3Fe^{2+}$ substitution (Dhar et al. 2002).

17.5 Geochemistry

The geochemistry (major, trace including rare earth elements) of these complexes has been provided in Eby and Kochhar (1990), Kochhar and Dhar (1993), Vallinayagam and Kochhar (1998), Bhushan and Chittora (1999) Kochhar (2000) and Bhushan (2000). The important geochemical signatures pertaining to the theme of the paper are as follows:

The granites of SRC are high in SiO_2 (68.99–73.87%), alkali content (7.54–11.35%), total iron (6.48–9.81%), TiO_2 (1.5%), Zr (983–4870 ppm), Hf (23–132 ppm), Ta (9–31 ppm), Th (16–61 ppm), U (3–17 ppm), Rb 97–421 ppm), REE (627–5382 ppm) and low in Al_2O_3 (7.74–12.98%), MgO (0.10%), CaO (1%), P_2O_5

(0.1%), MnO(0.2%), Ba (33–194 ppm), Cs (1–11), Sc(1–4), Cr (1–2), Co. The rhyolites have higher SiO_2 (68.80–75.06%), Al_2O_3 (9.51–12.32%), MgO (0.4%), CaO (0.42%), TiO_2 (0.2–0.9%), Sc(2–1 ppm), Co (1–3)and low alkalis (4.68–9.65%), total iron (4.25–8.64%), P_2O_5 (0.1%), MnO (0.1%), Zr (780–3980 ppm), Hf (19–107), Ta (2–16), Th (6–60), U (1–14), Rb (60–207), Cs (1–2), REE (280–1050) as compared to the granites. The dyke rocks show enriched LIL, HFSE, REE content as compared to the acid volcano-plutonic rocks, but their major element geochemical characters are similar to them. The basic rocks are high in titanium (1.5–4%), iron (6.11–13.46%), calcium (7.72–12.11%), aluminum (13.42–17.94%), transition elements and low in HFSE, REE. The volcano-plutonic rocks of JRC show the reverse order of geochemical element concentrations except for silica. The agpaitic index is 1–1.5 and acmite appears in SRC granites. In the granites of JRC, the agpaitic index ranges between 0.73–0.97 and corundum appears in the norm. Presence of normative corundum and acmite in the few samples of SRC and JRC acid rocks attests the mildly peraluminous and peralkaline in SRC and JRC complexes respectively. In geochemical variation diagrams they plot in A-type (Fig. 17.4), Anorogenic and Within Plate Granite fields of Whalen et al. (1987), Batchelor and Bowden (1985), and Pearce et al (1984), respectively.

The granites of SRC show a high total REE content and relatively flat chondrite normalized pattern with distinct negative Eu anomaly (Fig. 17.5a). The rhyolite and trachytes also show similar REE contents and pattern of granites with less marked Eu anomaly. The dyke rocks show highest REE contents. Low REE abundance without any Eu anomaly is present in their basic components. The granites of JRC show low REE contents (962 ppm) with moderate LREE and negative Eu anomaly (Fig. 17.5b).

Dhar et al. (1996) provided a Rb/Sr ages of 723 ± 6 Ma and 725 ± 41 Ma for the granites of SRC and JRC respectively. Pb (Fig. 17.6) and Nd isotope data show

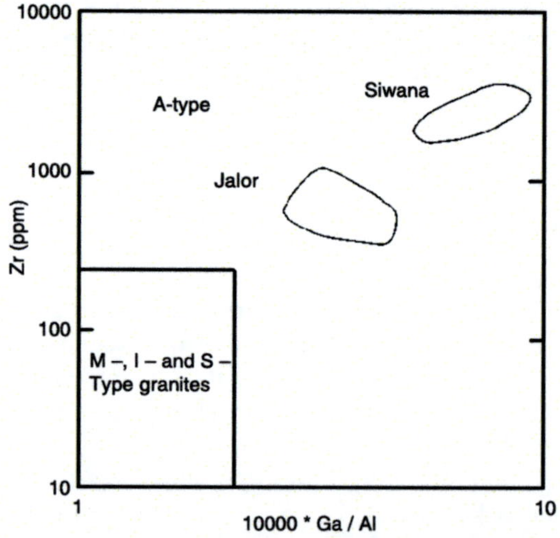

Fig. 17.4 10000*Ga/Al–Zr diagram (Whalen et al. 1987)

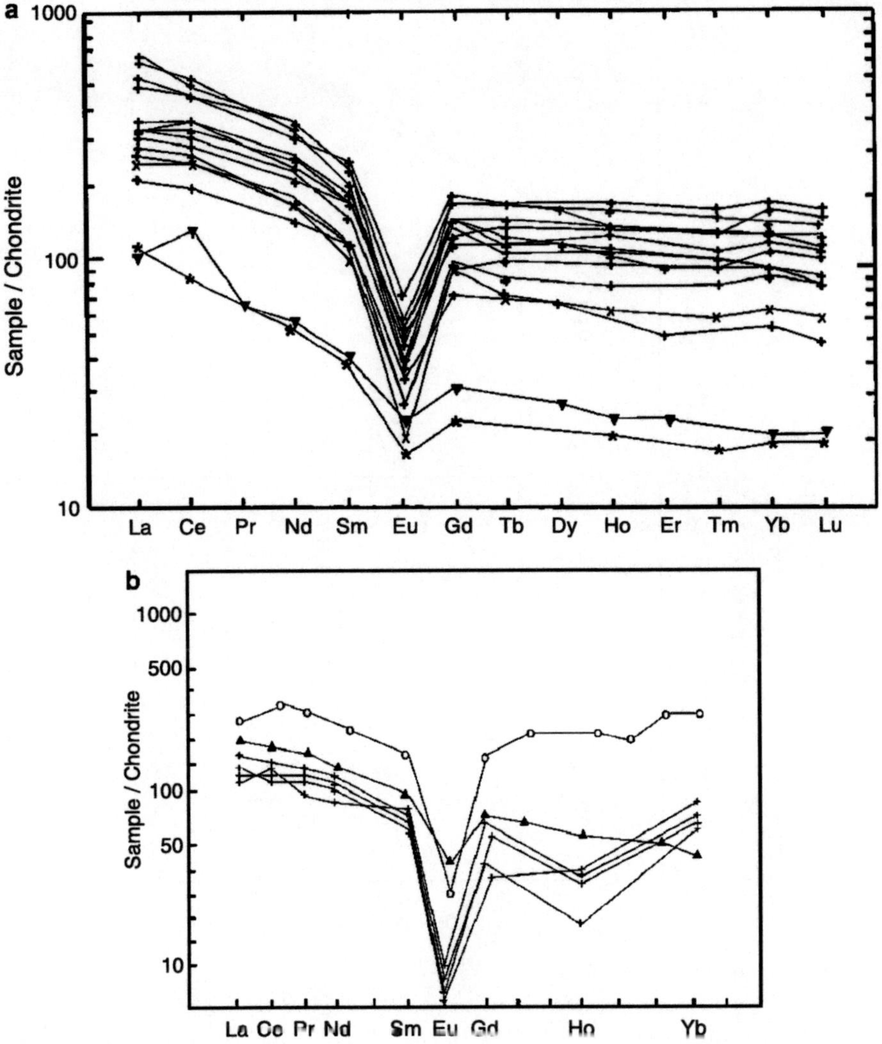

Fig. 17.5 REE diagrams (**a**) REE plot for granites of Siwana; (**b**) REE plot for Jalor subsolvus granite with A-type granite of SE Australia (Δ) and Nigerian Younger granites (O)

these complexes are derived from mantle magma with variable degree of crustal contamination which is attested by the wide scatter of the data. The low (+1 to 1.8‰) radiogenic oxygen isotope are encountered by them for these complexes as compared to the most terrestrial rock (+5 to +11‰). The low value of ^{18}O is suggestive of interaction with low ^{18}O rift related meteoric/hydrothermal systems generated by these granites. Based on the radio elements (U, Th, K) concentration for SRC and JRC, the heat production values of 5.90 and 2.80 uMw-3 are obtained.

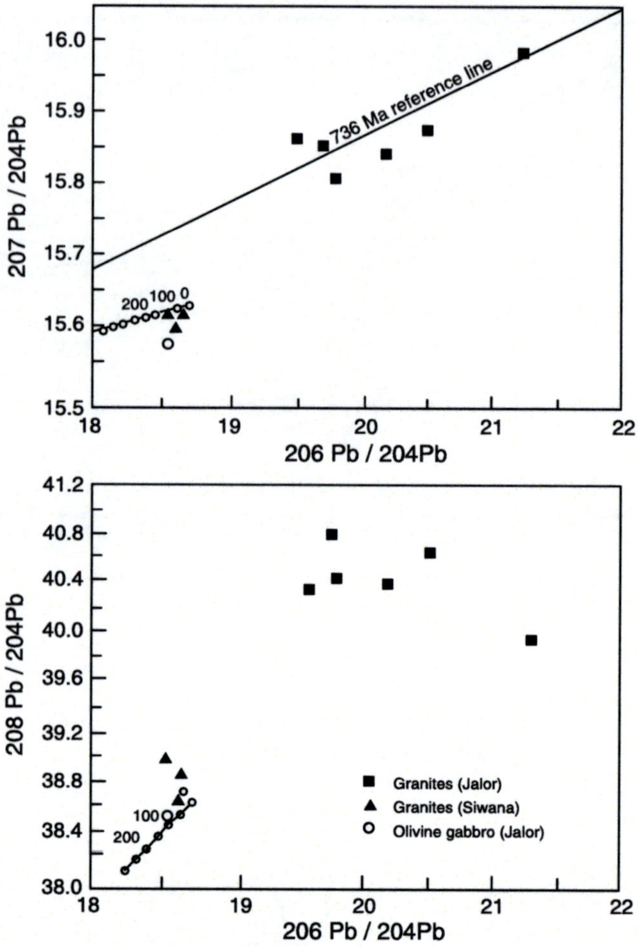

Fig. 17.6 Pb Isotope data for these complexes

Thus they are identified as high heat production (HHP) type. The A-type, anorogenic, HHP granites have the potential for rare metal and rare earth mineralization. The granites of SRC correspond to niobium–yttrium–fluorine association and JRC correspond to lithium–cesium–tantalum association (Kochhar 1992, 2000; Vallinayagam 2006).

17.6 Magmatic Evolution and Emplacement

Subvolcanic ring structures/dykes together with their petromineralogical and geochemical data manifest them for shallow crust residency A-type magmatism. This in turn reflects their emplacement in a non-compressive tectonic regime where the crust tends to be thin and magmatic advection of heat can approach the Earth's surface. Pitcher (1997) also advocated the role of hot plumes, crust extension and rifting in the evolution of silicic, subvolcanic magma chamber. Douce Patina (1997) experimentally worked out that profuse crystallization of plagioclase + orthopyroxene takes place during low pressure ($P \leq 4$ kbar) incongruent dehydration melting of hornblende and biotite bearing assemblages and it is the key factor in the generation of A-type melt with its typical geochemical characteristics. Feldspar fractionation plays an important role in the evolution of A-type granites. Plot of Eu/Eu* versus Ba and Eu/Eu* ratio are important indicators for feldspar fractionations (Eby 1990). Eu/Eu* ratios less than 1 can be due to feldspar fractionation, residual feldspar in the source region and/or source region with a negative anomaly. An Eu/Eu* ratio greater than 1 is usually ascribed to the presence of cumulus feldspar. Plot of Ba versus Eu/Eu* (Fig. 17.7) shows that the SRC granites are largely controlled by alkali feldspar fractionation which is in general agreement with its peralkaline characters. The enrichment of Rb and the marked depletion of Eu also support the alkali feldspar fractionation. For the JRC granites, the linear trend is indicative of both plagioclase and alkali feldspar fractionation. Low contents of CaO, MgO, Al_2O_3, Cr, Co, Sc in SRC suggest a felsic source. Hence the SRC magma is derived by a partial melting of a tonalitic to granodioritic rocks of lower crust. It is also possible that, if SRC magmatic activity is related to mantle plume (Kochhar 1984, 2008a, b), then fluid moving upwards from this plume could metasomatise lower crust material thus providing chemically anomalous region of enriched HFS, LIL elements in which the melt is generated. However Dhar et al (1996) based on Sr, Pb and Nd isotope studies have suggested a mantle derived primary magma for SRC rocks. The field and petrographical data together with petrogenetic modelling studies of Vallinayagam (2004) infer a low degree partial melting of a crustal source similar to rhyolite tuff sample # R1 for the origin of rhyolite of Siwana. Harris and Marriner (1980) have suggested that high level emplacement associated with volcanic activity of volatile

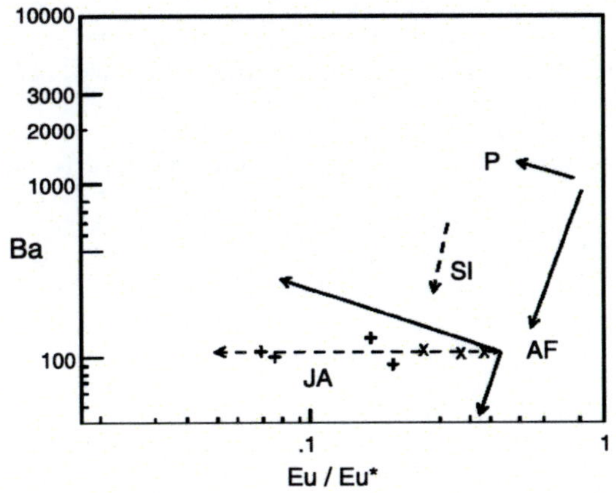

Fig. 17.7 Eu/Eu* versus Ba plot for Siwana and Jalor granites; P and AF are Raleigh fractionation trends calculated for removal of alkali feldspar (AF) and plagioclase feldspar (P)

rich peralkaline magma could result in a pressure drop in the crust underlying the granite. The pressure drop would allow partial melting to occur at depth, a condition likely to satisfy since the available volatiles have lost to the peralkaline magma. Probably similar mechanism operated to generate JRC magmatism. Curie (1989) also advocated that mildly peralkaline siliceous ash flow sheets associated with non peralkaline rocks occur worldwide. Thus above field, petromineralogical and geochemical studies suggest that the peralkaline and peraluminous magma are cogenetic and derived from a parent stock of peralkaline and peraluminous rhyolite (Kochhar and Dhar 1993). These complexes mark a period of A-type anorogenic magmatism related to hotspot activity in the northwestern Peninsular India during Neoproterozoic period and cratonisation. In view of the close association of acid and basic volcano-plutonic rocks in these complexes, it is advocated that the polymagmatic chamber (Chapman 1966) model can be evoked for their emplacement. According to this model two magmas, basaltic and granitic in composition, could have formed by selective melting of crustal materials by the hot basaltic magma under plating during the ring fracturing initiated by plume activity. According to Bonin (2007) the A-type granite is likely to come from mantle derived transitional to alkaline mafic to intermediate magmas. Rare felsic minerals found in the meteorites and lunar record yield dominantly A-type features. As such A-type granites are not typical of earth and were produced in a planetary environment differing from those prevailing on Earth. More work is needed to understand the enigma of A-type granites.

Acknowledgments The authors are thankful to Professor Jyotisankar Ray, Kolkata for the invitation to contribute this paper and to their parent Universities for research facilities. G. Vallinayagam expresses gratitude to DST, New Delhi for FIST grant facilities. Useful comments of the referees are thankfully acknowledged.

References

Baskar R, Kochhar N (2006) Alkali amphiboles and alkali pyroxenes from Siwana granite and the associated acid volcanics, Rajasthan. J Geol Soc India 66:427–437

Batchelor RA, Bowden P (1985) Petrogenetic interpretation of granitiod rock series using multicationic parameters. Chem Geol 48:43–55

Bhushan SK (2000) Malani rhyolites – a review. Gondwana Res 3:65–77

Bhushan SK, Chandrasekaran V (2002) Geology and geochemistry of the magmatic rocks of the Malani igneous suite and Tertiary Volcanic Province of the western Rajasthan. Mem Geol Surv India 126:1–29

Bhushan SK, Chittora VK (1999) Late Proterozoic bimodal volcanic assemblage of Siwana subsidence structure, western Rajasthan, India. J Geol Soc India 53:433–452

Bonin B (2007) A-type granites and related rocks: evolution of a concept, problems and prospects. Lithos 97:1–29

Bonin B, Giret A (1985) Clinopyroxene compositional trends in oversaturated and undersaturated alkaline ring complexes. J Afr Earth Sci 3:175–183

Chapman CA (1966) Paucity of mafic ring dykes: evidence for floored polymagmatic chambers. Am J Sci 264:66–77

Curie KL (1989) New ideas on an old problem: the peralkaline rocks. Mem Geol Soc India 15:117–136

Czamanske GK, Dillet B (1988) Alkali amphibole, tetrasilicic mica and sodic pyroxene in peralkaline rocks, Questa Caldera, New Mexico. Am J Sci 288-A:358–392

Dhar S, Kochhar N (1997) Mineral chemistry of amphibole from the Jalor ring complex, Rajasthan. Indian Miner 31:24–30

Dhar S, Frei R, Kramer JD, Nagler TF, Kochhar N (1996) Sr, Pb, and Nd isotope studies and their bearing on the petrogenesis of the Jalor and Siwana complexes, Rajasthan, India. J Geol Soc India 48:151–160

Dhar S, Kochhar N, Sharma R, Gupta LN (2002) Mineral chemistry of biotites from Jalor, Tusham and Jhunjhunu ring complexes, Malani igneous suite, India. J Geol Soc India 60:567–571

Douce Patina AE (1997) Generation of meta-aluminous A-type granites by low pressure melting of calc-alkaline granitoids. Geology 25:743–756

Eby GN (1990) The A-type granitiods: a review of their occurrence and chemical characteristics and speculation on their petrogenesis. Lithos 26:115–134

Eby GN, Kochhar N (1990) Geochemistry and petrogenesis of the Malani igneous suite, north Peninsular India. J Geol Soc India 36(2):109–130

Giret A, Bonin B, Leger J (1980) Amphibole compositional trends I oversaturated and undersaturated alkaline plutonic ring complexes. Can Miner 18:481–495

Harris NBW, Marriner GF (1980) Geochemistry and petrogenesis of a peralkaline granite complex from Midian Mountains, Saudi Arabia. Lithos 13:325–337

Kochhar N (1984) Malani igneous suite: hot spot magmatism and cratonization of the northern part of the Indian Shield. J Geol Soc India 25:155–161

Kochhar N (1992) Mineralisation associated with A-type Malani magmatism, northwest Peninsular India. In: Sarkar SC (ed) Metallogeny related to tectonics of the Proterozoic mobile belt. Oxford – IBH, New Delhi, pp 209–224

Kochhar N (2000) Attributes and significance of the A-type Malani magmatism, north western Peninsular India. In: Deb M (ed) Crustal evolution and metallogeny in the northwestern Indian Shield. Narosa Publishing House, New Delhi, pp 158–188

Kochhar N (2004) Geological evolution of the Trans-Aravalli Block (TAB) of the NW Indian Shield: constraints from the Malani Igneous Suite (MIS) and its Seychelles connection during Late Proterozoic. Spl Publ Geol Surv India 84:247–264

Kochhar N (2008a) A-type Malani Magmatism, NW Peninsular India. In: Singhvi AK, Bhattacharya A, Guha S (eds) Glimpses of geoscience research in India The Indian reports to IUGS 2004–2008, INSA Platinum Jubilee (1935–2010) pp 176–181

Kochhar N (2008b) Geochemical signatures of the A-type Malani magmatism, NW Peninsular India and the assembly of the Late Proterozoic Malani super continent. Mem Geol Surv India 73:21–36

Kochhar N, Dhar S (1993) The association of hypersolvus–subsolvus granites: a study of Malani igneous suite, India. J Geol Soc India 42:449–467

Morimot N (1989) Nomenclature of pyroxenes. Can Mineral 27:143–156

Pareek HS (1981) Petrochemistry and petrogenesis of the Malani Igneous Suite, India. Bull Geol Soc Am 92:206–273

Pearce JA, Harris NBW, Tindel AG (1984) Trace element discrimination diagrams for the tectonic interpretation of granitic rocks. J Petrol 25:956–983

Pitcher WS (1997) The nature and origin of granites. Chapman & Hall, London, pp 258–279, Chapter 15

Stephenson D (1972) Alkali clinopyroxenes from nepheline syenites of the south Qoroq centre, South Greenland. Lithos 5:187–201

Strong DF, Taylor RP (1984) Magmatic – subsolidus and oxidation trends in composition of amphiboles from silica-saturated peralkaline igneous rocks. Tschermaks Miner Petrol Mitt 32:211–222

Vallinayagam G (1988). Geology and geochemistry of alkali granites and the associated acid volcanics around Mokalsar, Distt. Barmer, W. Rajasthan, India and their bearing on the rift tectonics. Ph.D. thesis, Panjab University, Chandigarh, pp 1–113

Vallinayagam G (1997) Mineral chemistry of Siwana ring complex, W. Rajasthan, India. Indian Mineral 31:37–47

Vallinayagam G (2004) Peralkaline – peraluminous A-type rhyolites, Siwana ring complex, Northwestern India: petrogenetic modelling and tectonic implications. J Geol Soc India 64:336–344

Vallinayagam G (2006) Geochemistry of rare metals and rare earth ore deposits: a study of Malani igneous suite Trans-Aravalli Block of NW Peninsular India, Indian. Indian J Geochem 21(2):439–446

Vallinayagam G, Kochhar N (1998) Geochemical characterisation and petrogenesis of A-type granites and the associated acid volcanics of Siwana ring complex, northern Peninsular India. In: Paliwal BS (ed) The Indian precambrian. Scientific Publishers, Jodhpur, pp 460–481

Whalen JB, Currie KL, Chappel BW (1987) A-type granites: geochemical characteristics, discrimination and petrogenesis. Contrib Mineral Petrol 96:407–419

Chapter 18
Occurrence and Origin of Scapolite in the Neoproterozoic Lufilian–Zambezi Belt, Zambia: Evidence/Role of Brine-Rich Fluid Infiltration During Regional Metamorphism

Crispin Katongo, Friedrich Koller, Theodoros Ntaflos, Christian Koeberl, and Francis Tembo

Abstract Scapolite is an important mineral in the metamorphism of calc-silicates, marbles, amphibolites, and metagabbros in the Lufilian–Zambezi belt of Zambia. Both field and petrographic studies on granite gneisses, amphibolites, and metagabbros from the Munali hills area, in the southern part of the Lufilian–Zambezi belt indicated that scapolitization was due to metasomatism. The scapolite occurs as a pervasive replacement of plagioclase in the Munali hills granite gneiss, amphibolites, and metagabbros, and is associated with mineral assemblages that are indicative of amphibolite-facies metamorphism. Results of mineral analyses show that all the scapolites have calcian–marialite compositions, which range from 27 to 47 Me % and X_{Cl} contents of 0.37–0.50 a.p.f.u. The anorthite equivalent (31–46 EqAn %) of the scapolites overlaps with that of coexisting plagioclase (21–48 X_{An} %). The composition of scapolite is similar to that of the Copperbelt region of Zambia, where there is documented evidence of evaporite horizons. Moderate to high NaCl salinities, which range from 0.2–0.5 mol, and high contents of Cl in scapolite indicate that metamorphism in the belt was accompanied by metasomatic introduction of NaCl-rich fluids, which were derived from evaporite horizons that existed in the metasedimentary succession in the belt. This study shows that evaporites, from which NaCl-rich fluids were derived, were widespread in the Lufilian–Zambezi belt and played an important role in the metamorphic history of the belt.

C. Katongo†, F. Koller, T. Ntaflos, and C. Koeberl (✉)
Department of Lithospheric Research, University of Vienna, Althanstrasse 14, A-1090 Vienna, Austria
e-mail: friedrich.koller@univie.ac.at; theodoros.ntaflos@univie.ac.at; christian.koeberl@univie.ac.at

F. Tembo
School of Mines, Department of Geology, University of Zambia, P.O. Box 32379, Lusaka, Zambia
e-mail: ftembo@mines.unza.zm

†Deceased

J. Ray et al. (eds.), *Topics in Igneous Petrology*, DOI 10.1007/978-90-481-9600-5_18,

18.1 Introduction

The fluid evolution in metamorphic terrains may be recorded by fluid inclusions and mineral chemical equilibria. Several studies have shown that scapolite is a mineral indicator of chlorine (Cl) concentration in coexisting fluid (e.g., Orville 1975; Ellis 1978; Vanko and Bishop 1982; Oterdoom and Wenk 1983; Mora and Valley 1989; Moecher and Essene 1991; Harley et al. 1994). Unlike other chlorine-bearing minerals, such as those in the mica and amphibole groups, scapolite contains little OH; as such, its Cl content at a given pressure and temperature conditions reflects the NaCl activity in the fluid, independent of the f_{H2O}, thereby making scapolite an ideal indicator of Cl composition in the fluid (Ellis 1978). Scapolite-group minerals are generally considered as the salt-bearing analogues of (Na, Ca)-feldspars, which are stable under conditions of high volatile activity. Compositions within the scapolite group are represented generally as 3(Ab, An, Ksp).($CaCO_3$, NaCl, $CaSO_4$), where Ab, An, and Ksp refer to the components $NaAlSi_3O_8$, $CaAl_2Si_2O_8$, and $KAlSi_3O_8$, respectively (Rebbert and Rice 1997). The end-member formulae were originally derived by analogue with plagioclase and may be written as marialite (Ma) = $3NaAlSi_3O_8$. NaCl (or Ab_3. NaCl) and meionite (Me) = $3CaAl_2Si_2O_8$. $CaCO_3$ (An_3 . $CaCO_3$) (Evans et al. 1969; Teertstra and Sherriff 1997). Beside Silvialite (Si) = $3CaAl_2Si_2O_8$. $CaSO_4$ (An_3 . $CaCO_4$) Marialite and meionite are the only two currently accepted species (Bayliss 1987).

Scapolite is a common metamorphic mineral in the Lufilian–Zambezi belt. It is widespread in most marbles and calc-silicate lithologies in southern Zambia (Munyanyiwa 1985) and is also widespread in carbonate and calc-silicate rocks in central Zambia (Simpson et al. 1965; Drysdall and Stillman 1966; Katongo 1999), but rare in granitic gneisses and pelitic schists elsewhere in the belt. Small metagabbroic bodies intrusive into the supracrustal sequence in central and southern Zambia have abundant scapolite (Simpson and Stillman 1963; Vrána et al. 1975; John et al. 2003). Although rare in most granitic rocks, scapolite has been reported in granite gneisses in southern Zambia (Wilson et al. 1993; Hanson et al. 1994). In the Copperbelt region, scapolite is abundant in the Katangan metasediments, and metagabbros (Mendelsohn 1961; Cosi et al. 1992; Tembo et al. 1999).

In spite of the prevalence of scapolite in the Neoproterozoic Lufilian–Zambezi belt, good scapolite compositional data that could be used independently in regional evaluation of scapolitization are limited. Simpson et al. (1965) determined the composition of scapolite as "dipyre" ($Na_3Ca_{14}Si_8O_{24}Cl$) from physical properties such as specific gravity, density, refractive index etc., whereas Drysdall and Stillman (1966) found the composition of scapolite as sodic-meionite (Me_{55}) by wet-chemical analyses. Munyanyiwa (1990) reported Cl-free scapolite (Me_{60-65}) in calc-silicates from southern Zambia, but did not present mineral data or statement of how this mineral composition was determined. Similarly Cosi et al. (1992), Hanson et al. (1994), and Wilson et al. (1993) did not provide compositional data of scapolites from their study areas in the belt. The methods used in determining compositions of scapolite in some of these studies are grossly inaccurate and cast doubts on the reliability of the claimed compositions of scapolite. The only previous reliable scapolite compositional data (calcic marialite-Me_{25-38}) in the belt are those that were obtained by electron microprobe analysis (Tembo 1994).

The origin of scapolite in most metamorphic terrains is generally attributed to evaporite precursors. In the Lufilian–Zambezi belt, in spite of limited scapolite compositional constraints, NaCl-rich fluids are generally considered by most authors to have been derived from evaporite beds that were present in metasedimentary successions.

In this paper, we present the first scapolite compositional data from the Munali hills granite gneiss, amphibolites, and metagabbros from the Munali hills area, in the southern part of the Lufilian–Zambezi belt. We also attempt to evaluate the source and composition of fluids that were associated with metamorphism in the belt. We have integrated compositional data of scapolite from the Copperbelt region in the northern part of the belt (Tembo 1994) with our data to assess the regional extent of the source of fluids, which caused extensive scapolitization in the Lufilian–Zambezi belt.

18.2 Regional Geological Framework of the Lufilian–Zambezi Belt

Until recently, the Lufilian arc and the Zambezi belt (Fig. 18.1) were believed to form discrete orogenic belts separated by the Mwembeshi dislocation zone (MDZ), on the basis of presumed marked differences in metamorphic grade, structural vergence, and orogenic histories (De Swardt and Drysdall 1964; Unrug 1983; Coward and Daly 1984; Hanson et al. 1993). The MDZ is a prominent crustal-scale ENE trending shear zone across central Zambia, which is believed to extend to the northeast into Malawi and to the southwest into Namibia (De Swardt and Drysdall 1964; Coward and Daly 1984; Daly 1986; Porada 1989; Kampunzu and Cailteux 1999). Isotopic age data of key lithotectonic units, coupled with re-interpretations of the geology in the belts in Zambia and Zimbabwe, have shown that the two belts are coeval and form part of a network of Neoproterozoic-early Palaeozoic orogenic belts in central-southern Africa (e.g., Hanson et al. 1993; Dirks and Sithole 1999; Kampunzu and Cailteux 1999; Vinyu et al. 1999; Porada and Berhorst 2000; De Waele et al. 2009). The transcontinental network of belts separates the Congo and Kalahari cratons (Coward and Daly 1984; Unrug 1983; Johnson et al. 2007; De Waele et al. 2009) and formed during the reconstruction of Rodinia (Unrug 1996; Weil et al. 1998). The disposition and composition of metasedimentary rocks in the belt are regarded to indicate deposition in a rift basin (Wilson et al. 1993; Hanson et al. 1994; Porada and Berhorst 2000).

A detailed review and synthesis of the evolutionary history for much of the Lufilian–Zambezi belt was provided by Porada and Berhorst (2000). Key et al. (2001) presented the most recent comprehensive interpretation of tectono-thermal evolution of the Lufilian arc based on their work in NW Zambia. An overview of

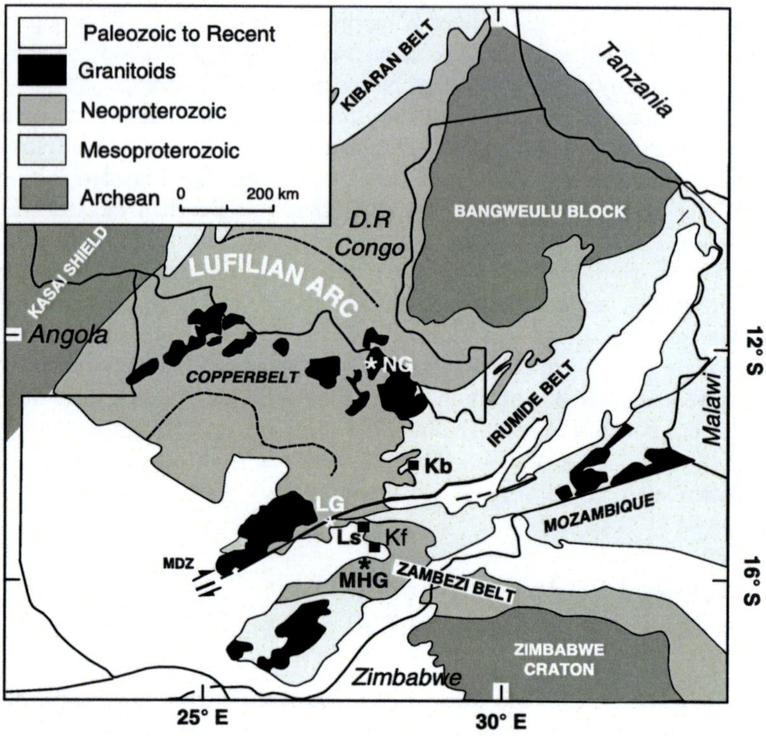

Fig. 18.1 Simplified geological map of Zambia and parts of neighbouring countries. Symbols; *: sample locations; MHG: Munali hills granite gneiss and metagabbro and amphibolitic dykes; NG: Nchanga granite; LG: Lusaka granite. Reference towns indicated by ■; Kb: Kabwe; Ls: Lusaka and Kf: Kafue (After Porada and Berhorst 2000)

the structure and geology of the Zambezi belt was given by Hanson et al. (1994). Here, we summarize previous work on the Lufilian arc and Zambezi belt to provide a regional geological framework for this study.

The Lufilian arc is an arcuate, northward convex-shaped geological structure covering eastern Angola, the southern Democratic Republic of Congo (DRC), northwestern Zambia, and is well known for its vast world-class copper–cobalt deposits (Fig. 18.1). In the Copperbelt region, which straddles northwestern Zambia and southern DRC, the stratigraphy is well known due to exploration and mining activities. The succession comprises basement and metasedimentary rocks of the Neoproterozoic Katanga Supergroup. In the Zambian Copperbelt region, the basement is composed of Lufubu schists and gneisses (Mendelsohn 1961) that are intruded by Eburnian (ca. 2,200–1,800 Ma) granites (Cahen et al. 1984). Key et al. (2001) dated basement rocks of Archaean age (ca. 2.6 Ga) in the western arm of the Lufilian arc. Overlying the Lufubu schists and gneisses, is a sequence of schists and quartzites of the Muva Supergroup, which are in turn intruded by early Neoproterozoic granites (e.g., Nchanga granite). The revised lithostratigraphy of the Katanga Supergroup (Porada and Berhorst 2000) consists, in stratigraphic succession, of arenaceous Roan Group, carbonaceous Mwashia Group with metavolcanic interbeds, and conglomeratic Kundelungu Group. It has long been observed by many authors that there are several horizons of evaporites in the Katangan succession (e.g., Mendelsohn 1961; Binda 1994; Cailteux et al. 1994). The former existence of evaporites in the Katangan metasediments is indicated by sabkha facies, crystals and pseudomorphs of gypsum and anhydrite, stratigraphic gaps underlain by collapse breccias, chloride inclusions in ores, and saline springs (Porada and Berhorst 2000; Jackson et al. 2003). Evaporites have played an important role in deformation (Cailteux and Kampunzu 1995) and formation of megabreccias (olistostromes)(Wendorff 2000) in the Katanga succession. In the Damara belt, the southwestern extension of the Lufilian arc, former evaporite deposits are characterized by high sodium contents, abundant albite pseudomorphs after evaporite minerals and concordant collapse breccia (Porada and Berr 1988).

Metamorphism in the Lufilian arc increases gradually from prehnite–pumpellyite facies in the northern boundary of the Lufilian arc, up to amphibolite facies in the southern part of the Copperbelt region (Kampunzu and Cailteux 1999), but higher grades of up to eclogite facies have been locally recorded (Vrána et al. 1975; Cosi et al. 1992). From studies of mineral associations and phengite compositions in the ore-shale, in the Copperbelt region of Zambia, Moine et al. (1986) estimated the metamorphic conditions of 420–550°C and 2–6 kbar in the Copperbelt region. Deformation in the Lufilian arc mainly involved thin-skinned tectonics with northerly directed thrusting (Coward and Daly 1984; Daly 1986; Kampunzu and Cailteux 1999; Porada and Berhorst 2000; Key et al. 2001).

The east-west trending Zambezi belt stretches from central Zambia into northern Zimbabwe, where it merges with the north–south trending Mozambique belt (Fig. 18.1). In Zambia, the Zambezi belt consists of wide zones of remobilized Mesoproterozoic to Neoproterozoic crystalline basement, unconformably overlain by late Neoproterozoic supracrustal rocks (Hanson et al. 1988; Wilson et al. 1993). The Basement is composed of the Mpande gneiss, a megacrystic, K-feldspar, biotite augen gneiss and the Ngoma gneiss, a protomylonitic, K-feldspar gneiss. The two gneissic units form an extensive terrain in the central parts of the belt. The Munali hills granite gneiss is a small lensoidal body that is intrusive into the Mpande gneiss, and adjacent metasedimentary rocks (Smith 1963) Katongo et al. (2004) report an U–Pb zircon age of 1,090.1 ±1.3 Ma. The supracrustal sequence starts with a thick pile of rift-related bimodal volcanics, which are overlain by a thick sequence of psammites and pelites that are in turn succeeded by an extensive unit of marbles and calc-silicates rocks (De Swardt and Drysdall 1964; Wilson et al. 1993; Hanson et al. 1994). The supracrustal sequence in the Zambezi belt in Zambia has been correlated based on structural continuity and broad lithological similarities with the Makuti and Rushinga Groups in Zimbabwe (Broderick 1981; Barton et al. 1991). Evaporite horizons have not really been observed in the Zambezi belt, but former evaporites are indicated by relict chicken-wire structures, reflecting replacement of original anhydrite nodules by calcite in marbles at Nampundwe Mine, in central-southern Zambia (Burnard et al. 1993). Metamorphism in the Zambezi belt was primarily amphibolite facies (Barton et al. 1991; Hanson et al. 1994; Hargrove et al. 2003), but locally there are occurrences of tectonically exhumed high-pressure rocks comprising eclogites and talc-kyanite whiteschists in Zambia and northwest Zimbabwe (Vrána et al. 1975; Dirks and Sithole 1999; Johnson and Oliver 2000). Diopside-plagioclase assemblages are uniformly developed in the calc-silicate rocks and carbonate formations in southern Zambia. Garnet-biotite thermometry and plagioclase-hornblende thermobarometry on amphibolite layers in the Carbonate Formations indicate peak metamorphic conditions of 560–600°C and 5–6 kbars (Munyanyiwa and Hanson 1988; Munyanyiwa 1990). These P-T conditions are consistent with widespread occurrence of kyanite in the supracrustal sequence south of Kafue. Temperatures between 565°C and 580°C and pressures ranging from 6 to 8 kbars were indicated by the garnet–hornblende thermobarometry on garnet amphibolites in the Chisamba area north of Lusaka (Katongo 1999). Occurrences of several small bodies of eclogite (Vrána et al. 1975; Vrána 1978), and talc-kyanite-whiteschists (Vrána and Barr 1972; Cosi et al. 1992; John et al. 2003) indicate that local medium to high pressure were attained during metamorphism in most parts of the Zambezi belt. Scapolite metamorphism also affected metagabbros, and eclogites in the southeastern extension of Zambezi belt in northern Zimbabwe (Dirks and Sithole 1999; Munyanyiwa et al. 1997). In contrast, to the Lufilian arc, deformation in the Zambezi belt is characterised by thick-skinned tectonics, involving both supracrustal and basement rocks (Coward and Daly 1984).

18.3 Field Relations and Petrography of Scapolite Bearing Rocks

In this study, scapolite was identified in the Munali hills granite gneiss, amphibolites, and metagabbros in the Munali hills area (Fig. 18.1). Scapolite occurs as pervasive replacement of plagioclase in all the studied lithologies, as shown in microphotos (Fig. 18.2) and field occurrences (Fig. 18.3).

18.3.1 Munali Hills Granite Gneiss

The Munali hills granite gneiss forms a small part of the Munali hills, a WNW trending elongate ridge, underlain mainly by granitic gneiss of the Mpande Formation (Smith 1963; Mallick 1966). The southern boundary of the granite gneiss is marked by an alternating sequence of kyanite schist, quartzite, calc-silicates, and marbles. The Munali hills granite gneiss is a porphyritic K-feldspar-rich granite gneiss composed of a range of fine- to coarse-grained varieties, varying from leucocratic to melanocratic types and displaying variable degrees of deformation. The granite gneiss is composed of pink microcline porphyroclasts set in fine-medium grained felsic groundmass. Isolated, dark, medium-grained irregularly

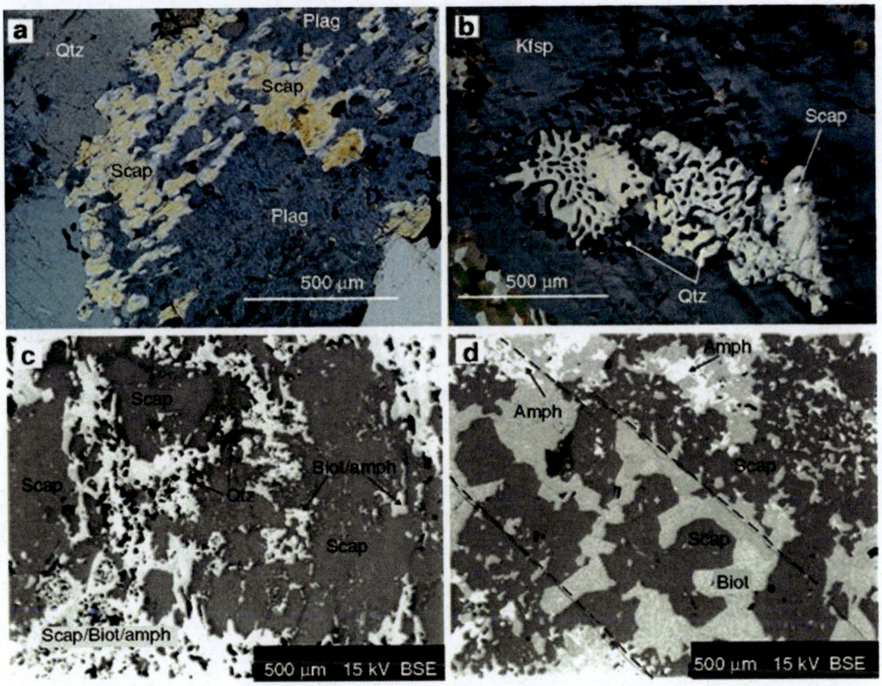

Fig. 18.2 (**a**) Photomicrograph showing characteristic partial replacement of plagioclase by scapolite in sample MHG2. (**b**) Photomicrograph showing of coronal intergrowth of scapolite and quartz in sample MHG9. Note: photograph was taken with quartz in partial extinction to enhance contrast against scapolite. (**c**) Back-scattered electron (BSE) image of coarse porphyroblastic scapolite with inclusions of quartz, biotite, and amphibole. (**d**) NW trending veinlet in central part of BSE image of relatively coarse-grained intergrowth of scapolite and biotite. Scap: scapolite; Biot: Biotite; Amp: amphibole; qtz: quartz; Plag: plagioclase: Kfsp: K-feldspar

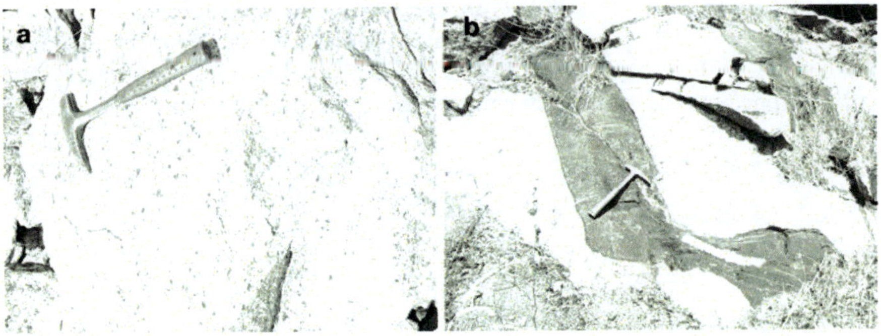

Fig. 18.3 Field pictures of the scapolite-bearing rocks in the Kafue area. (**a**) Leukocratic, medium- to fine-grained variety of the Munali Hill granite, dark spots are biotite-rich areas. (**b**) Deformed mafic dyke (now amphibolite) intruding into a medium-grained Munali Hill granite

shaped, biotite-rich enclaves are common. Deformation is generally weak, and exhibited by a crude foliation defined by aligned biotite warping around slightly elongated porphyroclasts of feldspar and quartz aggregates. The gneisses are composed of microcline, quartz, plagioclase/scapolite, biotite and calcite ± muscovite. The microcline crystals are generally poikiloblastic, hosting smaller grains of other mineral constituents, non-perthitic and slightly kaolinized. Accessory minerals include sphene, epidote, apatite, zircon, and ±tourmaline.

The granite displays varying degrees of alteration, ranging from chloritization of biotite, sericitization and ubiquitous scapolitization of plagioclase. Scapolite constitutes up to 30 modal % and invariably intergrown with plagioclase. The replacement of plagioclase by scapolite is mainly along grain margins, cleavage planes and less commonly in the cores (Fig. 18.2a). In some grains, scapolite hosts fine inclusions of twinned plagioclase, sericite, epidote, and carbonate, probably indicating replacement after slight retrogression of plagioclase. In a few porphyroblasts of scapolite, calcite occurs along cleavage or fractures planes. A few grains of scapolite display symplectite-like intergrowths with quartz (Fig. 18.2b). Undulose extinction in quartz, kink bands in biotite and deformation twins in plagioclase indicate low temperature deformation.

18.3.2 Amphibolites

The Munali hills granite gneiss is intruded by several steep, WNW trending amphibolitic dykes, measuring up to 5 m in width. At outcrop scale, the amphibolitic dykes exhibit a strong shear fabric, suggesting emplacement in extensional shear fractures. The amphibolites are dark green, and fine-grained, with white plagioclase phenocrysts and scapolite porphyroblasts, and are composed mainly of blue green amphibole and subordinate biotite, plagioclase, scapolite and minor calcite. Plagioclase is partly retrogressed to sericite. The scapolite is up to 40 modal % and partially or completely replaces plagioclase. Scapolite in the amphibolite displays various textures ranging from large porphyroblasts to fine grains and coarse-grained veinlets (Fig. 18.2c, d). The veinlets are composed of coarse-grained scapolite and biotite in stable equilibrium. Most scapolite crystals host inclusions of plagioclase, calcite, and epidote. Accessory minerals include epidote, sphene, carbonates, and opaque minerals. The opaque minerals are chalcopyrite and pyrite.

18.3.3 Metagabbros

18.3.3.1 Munali Hills Area

The metagabbro is one of the varieties of a metagabbroic composite body that is associated with nickel sulphide mineralization. The sulphide mineralization consists mainly of pyrrhotite with lesser amounts of pyrite, pentlandite, chalcopyrite, and accessory magnetite. The metagabbro is dark green, medium- to coarse-grained and consists mainly of amphiboles, augite, plagioclase, and biotite, which have undergone various degrees of alteration. Plagioclase (20–30 modal %) is invariably partly replaced by scapolite. Other secondary minerals include chlorite, tremolite, epidote, sericite and zoisite.

18.3.3.2 Copperbelt Region

Detailed descriptions of metagabbros in the Copperbelt region are provided by Tembo et al. (1999). The metagabbros are composed of calcic amphiboles, sodic plagioclase, and variable amounts of scapolite and have preserved primary igneous textures. The metagabbros are invariably affected by hydrothermal alteration. As in other scapolite bearing rocks in the Munali hills area, scapolite in these metagabbros occurs as a pervasive replacement of plagioclase. Textural relationships among the minerals in the metagabbros indicate that scapolitization post-dated amphibolitization (Tembo 1994).

18.4 Analytical Procedures

Mineral analyses on scapolite, and associated minerals such as plagioclase, amphiboles, and biotite from a few selected samples, were conducted on a four-spectrometer CAMECA SX100 microprobe at the University of Vienna, Department of Geological Sciences. Counts were obtained simultaneously from four spectrometers, using a 15 keV accelerating voltage, a 1–2 μm beam diameter, and a beam current of 20 nA. Natural and synthetic standards were used for calibrations and the PAP correction (Pouchou and Pichoir 1991) was applied to the data.

Mineral compositional data of scapolite by Tembo (1994) from the Copperbelt region were obtained on a scanning electron microscope (SEM)-Cambridge Stereoscan 250 MK3-equipped with an energy dispersive (ED) analytical system, at the University of Göttingen, Germany. Operating conditions for the quantitative analyses were 15 keV accelerating potential, 20 nA probe current and 100 s counting time. Natural and synthetic standards were used to calibrate the analyses. Matrix effects on elements and their oxides weight percentages were corrected for and calculated using the Cambridge online program MACRO ZAF. Representative results of mineral analyses are presented in Tables 18.1–18.5.

18.5 Results

18.5.1 Scapolite

Scapolite compositions are commonly reported in terms of equivalent anorthite content ($EqAn = 100 * (Al - 3)/3$), where Al is calculated on the basis of tetrahedral 16 cations (Evans et al. 1969; Teertstra and Sherriff 1997), mole fractions of chlorine ($X_{Cl} = Cl$

Table 18.1 Representative mineral compositions of scapolite from the Munali hills area

	Munali hills granite gneisses												
	MHG2					MHG9					MHG10		
Sample	1r	2c	4r	5c	6r	2c	4c	5c	6c	8c	5r	6r	7r
SiO_2	53.40	53.10	53.50	53.60	54.10	52.90	53.00	53.40	52.50	52.80	57.10	56.70	55.00
TiO_2	0.03	<0.02	<0.02	<0.02	0.02	<0.02	<0.02	0.03	0.04	<0.02	<0.02	0.02	0.02
Al_2O_3	25.00	24.80	25.30	25.10	25.50	25.20	25.40	24.90	25.30	25.20	23.70	23.50	24.40
FeO	0.12	0.12	0.12	0.12	0.18	0.12	0.06	0.10	0.08	0.36	0.05	<0.02	0.04
MnO	0.05	0.02	<0.02	0.03	0.03	0.03	0.05	0.02	0.02	<0.02	<0.02	<0.02	<0.02
MgO	<0.02	<0.02	<0.02	<0.02	0.02	<0.02	<0.02	<0.02	<0.02	<0.02	<0.02	<0.02	<0.02
CaO	10.10	9.90	10.00	9.80	9.50	10.50	10.80	10.20	11.20	10.70	5.90	6.60	8.10
Na_2O	7.50	8.10	8.00	8.40	7.00	7.80	7.40	7.80	7.50	7.50	9.70	9.90	9.10
K_2O	0.78	0.84	0.71	0.71	0.64	0.69	0.70	0.77	0.84	0.73	0.92	0.85	0.81
Cl	1.75	1.86	1.79	1.95	1.76	1.73	1.73	1.84	1.61	1.78	2.60	2.50	2.38
Total	98.73	98.75	99.44	99.71	98.75	98.98	99.14	99.06	99.09	99.08	99.97	99.45	99.85
Si	7.735	7.745	7.705	7.733	7.711	7.685	7.672	7.739	7.648	7.679	8.057	8.060	7.888
Al	4.265	4.255	4.295	4.267	4.289	4.315	4.328	4.261	4.352	4.321	3.943	3.940	4.112
Ti	0.006	0.003	0.001	<0.001	0.003	<0.001	<0.001	0.006	0.008	0.001	<0.001	0.004	0.005
Fe	0.015	0.014	0.014	0.015	0.022	0.014	0.007	0.012	0.010	0.044	0.006	0.004	0.004
Mn	0.006	0.002	0.001	0.003	0.003	0.004	0.006	0.002	0.002	<0.001	<0.001	0.000	<0.001
Mg	<0.001	<0.001	<0.001	<0.001	0.004	0.001	<0.001	<0.001	<0.001	<0.001	<0.001	0.000	<0.001
Ca	1.567	1.554	1.543	1.510	1.452	1.626	1.675	1.578	1.742	1.670	0.893	1.002	1.243
Na	2.112	2.282	2.235	2.359	1.930	2.182	2.073	2.190	2.116	2.107	2.654	2.560	2.530
K	0.144	0.156	0.130	0.131	0.116	0.127	0.130	0.142	0.157	0.136	0.165	0.154	0.149
Cl	0.422	0.450	0.429	0.468	0.416	0.415	0.415	0.444	0.390	0.430	0.609	0.590	0.566
Me %	42.6	40.5	40.8	39.0	42.9	42.7	44.7	41.9	45.2	44.2	25.2	28.1	33.0
EqAn	0.422	0.418	0.432	0.422	0.430	0.438	0.443	0.420	0.451	0.440	0.314	0.313	0.371

	Amphibolites											Metagabbro		
	MHG5a			MHG5c								NDG		
Sample	1c	2c	4c	1c	2c	3c	4c	5c	6c	7c	8c	2c	3r	4r
SiO_2	52.90	52.90	52.70	52.96	53.40	52.90	53.00	53.20	53.00	52.90	53.10	51.70	54.70	55.00
TiO_2	<0.02	<0.02	<0.02	0.07	0.04	0.04	<0.02	<0.02	<0.02	0.02	<0.02	<0.02	<0.02	<0.02
Al_2O_3	25.41	25.30	25.30	25.18	25.50	25.10	25.00	24.90	25.00	25.00	24.60	25.90	24.20	24.10
FeO	0.06	0.26	0.26	0.25	0.50	0.38	0.29	0.15	0.21	0.15	0.09	0.14	0.19	0.11
MnO	<0.02	0.04	<0.02	<0.02	<0.02	<0.02	<0.02	0.02	<0.02	0.03	<0.02	0.04	<0.02	<0.02
MgO	<0.02	<0.02	<0.02	<0.02	<0.02	<0.02	<0.02	<0.02	<0.02	<0.02	<0.02	<0.02	<0.02	<0.02
CaO	11.10	10.80	10.80	10.45	10.20	10.60	10.50	10.10	10.60	10.40	10.10	11.30	8.20	8.00
Na_2O	7.40	7.30	7.80	7.79	7.50	7.80	8.00	8.10	7.80	7.70	8.10	7.60	9.30	9.60
K_2O	0.56	0.59	0.72	0.68	0.78	0.76	0.64	0.70	0.65	0.81	0.78	0.40	0.50	0.44
Cl	1.71	1.70	1.71	1.79	1.93	1.82	1.83	1.94	1.79	1.92	1.90	1.69	2.30	2.27
Total	99.16	98.91	99.30	99.18	99.85	99.41	99.28	99.11	99.06	98.93	98.68	98.77	99.40	99.52
Si	7.665	7.670	7.665	7.690	7.684	7.695	7.709	7.729	7.714	7.704	7.764	7.548	7.886	7.914
Al	4.335	4.330	4.335	4.310	4.316	4.305	4.291	4.271	4.286	4.296	4.236	4.452	4.114	4.086
Ti	<0.001	0.003	<0.001	0.014	0.008	0.008	0.003	<0.001	0.002	<0.001	0.003	0.001	0.002	<0.001
Fe	0.007	0.032	0.032	0.030	0.061	0.046	0.035	0.018	0.025	0.025	0.011	0.018	0.023	0.013
Mn	0.001	0.005	<0.001	<0.001	<0.001	0.001	<0.001	0.003	<0.001	0.001	<0.001	0.005	<0.001	<0.001
Mg	0.002	0.003	0.002	0.003	<0.001	<0.001	0.003	<0.001	<0.001	<0.001	<0.001	<0.001	<0.001	<0.001
Ca	1.716	1.671	1.686	1.626	1.578	1.655	1.629	1.573	1.663	1.610	1.582	1.770	1.270	1.236
Na	2.074	2.051	2.184	2.194	2.100	2.201	2.267	2.279	2.212	2.263	2.295	2.156	2.606	2.665
K	0.104	0.109	0.134	0.126	0.144	0.141	0.118	0.129	0.120	0.122	0.146	0.075	0.092	0.081
Cl	0.411	0.410	0.413	0.431	0.461	0.440	0.442	0.469	0.433	0.457	0.463	0.409	0.550	0.543
Me %	45.3	44.9	43.6	42.6	42.9	42.9	41.8	40.8	42.9	41.6	40.8	45.1	32.8	31.7
EqAn	0.361	0.361	0.361	0.359	0.360	0.359	0.358	0.356	0.357	0.358	0.353	0.484	0.371	0.362

Oxides in wt%, calculation based on 32 oxygens, Me % = 100 * (Ca/(Ca + Na), EqAn = (Al – 3)/3

Table 18.2 Representative mineral composition of scapolite from metagabbros in the Copperbelt region, Zambia (After Tembo 1994)

	Metagabbros																
Sample	FTK64				FTK44					FTK47				NN44			
	1c	1b	2a	3c	1b	2r	3c	3r	5r	1	2	4	5	1c	2r	3c	5r
SiO_2	55.69	55.18	55.06	55.81	54.00	54.58	53.45	54.12	53.87	54.16	54.34	54.28	54.61	55.02	55.06	54.87	55.60
Al_2O_3	22.59	22.72	22.99	22.62	23.45	23.23	23.56	22.95	23.14	23.22	23.20	24.02	23.63	22.81	22.94	23.19	22.79
CaO	7.42	8.42	7.73	7.60	9.47	8.93	9.75	8.95	9.20	9.00	8.83	8.94	9.10	8.74	9.04	9.09	8.12
Na_2O	9.50	8.85	9.32	9.90	8.68	8.50	8.56	8.91	8.50	8.40	8.30	8.57	8.44	9.17	8.65	8.69	9.09
K_2O	0.47	0.34	0.45	0.39	0.32	0.20	0.14	0.27	0.30	0.89	0.81	1.06	0.97	0.36	0.42	0.34	0.45
Cl	3.16	3.04	3.21	3.24	2.55	2.76	2.46	2.61	2.69	3.01	2.81	3.12	3.05	2.72	2.71	2.59	2.84
CO_2	0.47	0.52	0.46	0.47	<0.01	<0.01	<0.01	<0.01	<0.01	0.32	0.61	<0.01	0.20	0.17	0.19	0.20	0.09
SO_4	0.83	0.94	0.78	0.75	1.81	1.58	1.90	1.71	1.61	0.81	0.67	1.17	0.97	1.39	1.39	1.52	1.37
Total	99.26	99.31	99.27	100.03	99.69	99.15	99.25	98.92	98.69	98.31	98.35	99.27	99.30	98.36	98.39	98.37	98.98
Si	8.12	8.08	8.04	8.12	7.94	7.99	7.90	8.00	7.97	7.97	7.98	7.89	7.95	8.06	8.05	8.01	8.09
Al	3.89	3.92	3.96	3.88	4.06	4.01	4.10	4.00	4.03	4.03	4.02	4.11	4.05	3.94	3.95	3.99	3.91
Ca	1.16	1.32	1.21	1.19	1.49	1.40	1.54	1.42	1.46	1.42	1.39	1.39	1.42	1.37	1.42	1.42	1.27
Na	2.69	2.51	2.64	2.79	2.47	2.41	2.45	2.55	2.44	2.40	2.39	2.42	2.38	2.61	2.45	2.46	2.56
K	0.08	0.06	0.08	0.07	0.06	0.04	0.03	0.05	0.06	0.17	0.15	0.20	0.18	0.07	0.08	0.06	0.09
Cl	0.78	0.76	0.79	0.80	0.64	0.69	0.62	0.65	0.67	0.75	0.70	0.77	0.75	0.67	0.67	0.64	0.70
CO_2	0.05	0.06	0.05	0.05	<0.01	<0.01	<0.01	<0.01	<0.01	0.09	0.17	<0.01	0.06	0.05	0.05	0.06	0.03
SO_4	0.17	0.19	0.16	0.15	0.36	0.32	0.38	0.35	0.33	0.16	0.13	0.23	0.19	0.28	0.28	0.30	0.27
Me %	30.1	34.5	31.4	29.9	37.6	36.7	38.6	35.8	37.4	37.2	36.8	36.5	37.4	34.4	36.7	36.6	33.2
EqAn	0.30	0.31	0.32	0.29	0.35	0.34	0.37	0.33	0.34	0.34	0.34	0.37	0.35	0.31	0.32	0.33	0.30

Oxides in wt%, calculation based on 32 oxygens, Me % = 100 * (Ca/(Ca + Na), EqAn = (Al – 3)/3

Table 18.3 Representative mineral compositions of plagioclase from the Munali hills area

	Munali hills granite gneiss										Amphibolite		Lusaka granite		Nchanga granite	
Sample	MHG2					MHG9					MHG5a		LG2		NG4	
	1	3	4	5	6	1	2	4	5	6	5	7	1	2	2	3
SiO_2	62.00	61.10	62.00	62.40	60.70	58.30	60.20	58.80	58.20	58.10	51.80	58.20	65.10	65.10	67.40	63.90
TiO_2	<0.02	<0.02	<0.02	<0.02	0.01	0.01	0.02	0.02	0.01	<0.02	0.10	0.04	0.01	<0.02	<0.02	<0.02
Al_2O_3	25.00	25.20	24.60	25.00	25.80	26.90	25.30	26.60	26.90	26.90	30.80	26.60	21.90	22.20	21.20	22.90
FeO	0.07	0.04	0.02	0.34	0.17	0.06	0.30	0.43	0.11	0.12	0.37	0.25	0.03	0.04	0.03	0.02
MnO	<0.02	<0.02	<0.02	<0.02	<0.02	<0.02	<0.02	<0.02	<0.02	<0.02	0.02	<0.02	0.02	<0.02	<0.02	0.02
MgO	<0.02	<0.02	<0.02	<0.02	<0.02	<0.02	0.02	<0.02	<0.02	<0.02	0.11	<0.02	<0.02	<0.02	<0.02	<0.02
CaO	5.30	5.50	5.00	5.10	6.50	7.60	5.80	7.40	7.80	8.00	12.60	8.20	2.80	3.10	0.76	3.40
Na_2O	8.60	8.50	9.00	9.10	8.30	7.40	8.50	7.50	7.40	7.40	4.30	6.90	10.20	10.00	10.50	9.60
K_2O	0.07	0.10	0.17	0.09	0.07	0.20	0.11	0.13	0.06	0.11	0.17	0.05	0.16	0.18	0.08	0.08
Total	101.04	100.46	100.80	102.04	101.55	100.48	100.25	100.89	100.43	100.63	100.27	100.24	100.22	100.62	99.97	99.92
Si	10.890	10.800	10.922	10.880	10.661	10.390	10.704	10.434	10.371	10.362	9.393	10.388	11.453	11.412	11.761	11.282
Al	5.163	5.250	5.104	5.128	5.333	5.647	5.292	5.560	5.647	5.633	6.570	5.600	4.534	4.582	4.355	4.751
Ti	<0.001	<0.001	<0.001	<0.001	0.002	0.001	0.002	0.002	0.001	<0.001	0.013	0.005	0.002	<0.001	<0.001	<0.001
Fe	0.011	0.005	0.003	0.049	0.025	0.008	0.044	0.063	0.017	0.018	0.057	0.037	0.005	0.005	0.004	0.003
Mn	<0.001	0.001	0.001	<0.001	<0.001	<0.001	0.001	0.002	<0.001	0.003	0.003	<0.001	0.002	<0.001	<0.001	0.003
Mg	<0.001	0.002	<0.001	0.001	<0.001	0.002	0.005	0.004	<0.001	0.001	0.029	0.001	<0.001	<0.001	<0.001	<0.001
Ca	0.989	1.035	0.938	0.947	1.229	1.442	1.114	1.410	1.488	1.517	2.449	1.576	0.527	0.574	0.143	0.637
Na	2.912	2.921	3.083	3.075	2.809	2.542	2.933	2.583	2.549	2.553	1.526	2.374	3.475	3.407	3.558	3.298
K	0.016	0.023	0.039	0.019	0.016	0.045	0.026	0.028	0.014	0.024	0.040	0.012	0.037	0.041	0.019	0.018
Cations	19.983	20.038	20.084	20.097	20.075	20.073	20.120	20.083	20.082	20.106	20.080	19.993	20.032	20.018	19.842	19.992
X_{An}	25.2	26	23.1	23.4	30.3	35.8	27.4	35.1	36.7	37.1	61	39.8	13	14.3	3.8	16.1

Oxides in wt%, calculation based on 32 oxygens, $X_{An} = 100 * (Ca/(Ca + Na + K)$

Table 18.4 Representative mineral compositions of amphiboles in amphibolites from the Munali hills area

	Amphiboles													
Sample	MHG5a					MHG5c								
	1	2	4	5	6	1	2	3	4	5	6	7	8	9
SiO_2	38.28	38.3	38.58	38.35	38.49	38.79	38.73	38.49	39.07	38.85	38.67	38.86	38.78	38.24
TiO_2	0.64	0.56	0.73	0.63	0.67	0.59	0.45	0.42	0.43	0.55	0.42	0.59	0.47	0.56
Al_2O_3	15.27	14.7	15.15	14.47	14.82	14.19	13.92	14.03	13.83	13.86	13.97	13.69	13.94	14.28
FeO	22.77	23.38	23.46	22.84	23.15	21.94	22.44	22.45	22.18	21.86	22.53	22	21.96	22.32
MnO	0.27	0.29	0.29	0.28	0.28	0.21	0.21	0.18	0.2	0.19	0.16	0.21	0.2	0.18
MgO	5.04	5.04	5.33	5.31	5.27	6.86	6.73	6.67	7.09	6.91	6.48	7	6.87	6.57
CaO	11.31	11.33	11.23	10.98	11.46	11.74	11.7	11.63	11.53	11.61	11.77	11.73	11.76	11.65
Na_2O	1.23	1.2	1.28	1.2	1.36	1.1	1.12	1.12	1.06	1.08	1	1.07	1.08	1.1
K_2O	1.49	1.49	1.59	1.55	1.51	2.27	2.03	2.06	1.97	2.02	1.97	2.08	1.94	2.24
Cl	1.01	0.99	0.99	1	0.99	1.43	1.39	1.4	1.29	1.33	1.44	1.36	1.27	1.52
Total	97.31	97.28	98.63	96.61	98	99.12	98.72	98.45	98.65	98.26	98.41	98.59	98.27	98.66
Si	5.980	5.995	5.946	6.031	5.984	5.968	5.978	5.958	6.000	6.006	5.993	5.995	5.989	5.925
Al^{IV}	2.020	2.005	2.054	1.969	2.016	2.032	2.022	2.042	2.000	1.994	2.007	2.005	2.011	2.075
Al^{VI}	0.789	0.706	0.696	0.710	0.696	0.539	0.507	0.516	0.501	0.528	0.536	0.483	0.525	0.532
Fe^{3+}	0.525	0.613	0.650	0.597	0.553	0.654	0.746	0.762	0.810	0.700	0.733	0.719	0.727	0.711
Ti	0.075	0.065	0.085	0.074	0.079	0.068	0.052	0.049	0.050	0.064	0.049	0.069	0.055	0.065
Mg	1.175	1.176	1.225	1.245	1.221	1.573	1.547	1.539	1.623	1.593	1.496	1.610	1.582	1.518
Fe^{2+}	2.451	2.448	2.375	2.407	2.456	2.169	2.151	2.144	2.039	2.125	2.186	2.120	2.111	2.181
Mn	0.036	0.038	0.038	0.038	0.019	0.028	0.028	0.024	0.026	0.024	0.022	0.028	0.026	0.024
Ca	1.893	1.901	1.854	1.850	1.908	1.935	1.935	1.929	0.013	1.923	1.955	1.940	1.946	1.934
Na	0.372	0.365	0.382	0.365	0.408	0.328	0.335	0.335	0.314	0.322	0.301	0.320	0.324	0.330
K	0.297	0.297	0.312	0.310	0.300	0.445	0.399	0.407	0.385	0.398	0.388	0.409	0.383	0.442
Cations	15.612	15.608	15.616	15.595	15.659	15.738	15.699	15.704	15.645	15.679	15.660	15.698	15.678	15.737
Cl	0.268	0.264	0.259	0.266	0.261	0.372	0.364	0.368	0.335	0.349	0.379	0.357	0.333	0.399
X_{Mg}	0.324	0.325	0.340	0.341	0.332	0.424	0.418	0.418	0.443	0.428	0.406	0.432	0.428	0.410

Oxides in wt%, calculation based on 23 oxygens (average of 15-NK and 13-CNK); $X_{Mg} = Mg/(Mg + Fe^{2+})$

Table 18.5 Representative mineral compositions of biotite from the Munali hill area

	Munali hills granite gneiss								Amphibolites					Lusaka granite		Metagabbro	
Sample	MHG2			MHG9			MHG10		MHG5a			MHG5c		LG2		NDG	
	1	3	4	1		3	1	2	1	2	3	1	2	1	2	1	2
SiO_2	34.20	33.70	34.00	34.60	34.60	34.50	34.60	34.40	35.30	35.40	35.50	36.80	35.50	35.70	35.50	39.70	39.20
TiO_2	2.60	2.60	2.46	2.35	3.40	3.40	2.60	2.50	2.54	2.60	2.60	1.72	2.40	1.68	1.73	0.54	0.50
Al_2O_3	16.40	16.40	16.40	17.30	16.50	16.50	16.00	15.90	16.20	16.00	16.10	15.60	16.60	15.30	15.10	12.50	12.50
FeO	29.00	30.50	29.00	25.60	25.90	26.00	28.60	28.90	23.10	22.70	23.10	18.60	22.60	27.60	28.20	22.50	23.30
MnO	0.31	0.40	0.32	0.04	0.10	0.08	0.05	0.05	0.15	0.14	0.16	0.10	0.18	0.41	0.45	0.13	0.12
MgO	4.50	3.40	4.50	7.00	6.20	6.10	5.20	5.30	9.40	9.50	9.70	13.30	9.60	5.80	5.60	7.40	6.90
CaO	0.04	<0.02	0.05	<0.02	<0.02	<0.02	0.02	<0.02	0.04	0.04	0.06	0.02	0.04	0.02	<0.02	11.50	11.40
Na_2O	0.04	0.03	0.03	0.04	0.05	0.04	0.02	0.04	0.08	0.06	0.07	0.07	0.06	0.04	0.06	1.67	1.71
K_2O	9.20	9.10	8.90	9.20	9.30	9.30	8.90	9.10	9.10	9.00	8.60	9.50	9.10	9.10	9.00	1.31	1.40
Cl	0.45	0.59	0.50	0.44	0.43	0.47	0.79	0.82	0.53	0.53	0.53	0.64	0.52	0.64	0.68	1.38	1.61
Total	96.74	96.73	96.16	96.57	96.49	96.39	96.78	97.01	96.44	95.97	96.42	96.35	96.60	96.29	96.32	98.63	98.64
Si	5.438	5.409	5.436	5.397	5.417	5.413	5.483	5.471	5.453	5.474	5.467	5.569	5.450	5.652	5.636	5.938	5.911
Al(IV)	2.562	2.591	2.564	2.603	2.583	2.587	2.517	2.529	2.547	2.526	2.533	2.431	2.550	2.348	2.364	2.062	2.089
Al(VI)	0.505	0.505	0.525	0.571	0.463	0.463	0.461	0.439	0.401	0.397	0.394	0.352	0.449	0.512	0.466	0.146	0.123
Ti	0.313	0.316	0.296	0.276	0.397	0.396	0.310	0.294	0.295	0.302	0.296	0.196	0.277	0.200	0.207	0.061	0.057
Fe^{2+}	3.849	4.103	3.871	3.339	3.398	3.420	3.792	3.839	2.986	2.940	2.976	2.360	2.908	3.659	3.749	2.807	2.942
Mn	0.042	0.055	0.043	0.005	0.014	0.011	0.007	0.006	0.020	0.018	0.021	0.013	0.023	0.055	0.060	0.016	0.015
Mg	1.059	0.809	1.069	1.617	1.451	1.435	1.239	1.245	2.160	2.193	2.215	2.993	2.207	1.359	1.334	1.650	1.554
Ca	0.006	0.002	0.009	0.001	0.002	0.001	0.003	<0.001	0.007	0.007	0.010	0.003	0.006	0.003	<0.001	1.848	1.849
Na	0.011	0.008	0.010	0.012	0.015	0.013	0.006	0.012	0.023	0.018	0.022	0.021	0.017	0.014	0.017	0.483	0.499
K	1.862	1.857	1.804	1.841	1.849	1.854	1.798	1.834	1.785	1.783	1.689	1.826	1.779	1.837	1.829	0.249	0.269
Cations	15.647	15.655	15.627	15.662	15.589	15.593	15.616	15.669	15.677	15.658	15.623	15.764	15.666	15.639	15.662	15.260	15.308
Cl	0.120	0.161	0.137	0.118	0.115	0.130	0.214	0.222	0.139	0.138	0.138	0.163	0.133	0.172	0.183	0.350	0.412
X_{Mg}	0.216	0.165	0.216	0.326	0.299	0.296	0.246	0.245	0.420	0.427	0.427	0.559	0.432	0.271	0.262	0.370	0.346

Oxides in wt%, calculation based on 22 oxygens, $X_{Mg} = Mg/(Mg + Fe^{2+})$

atoms per formula unit (a.p.f.u)), CO_2 (X_{CO2} = 1 − Cl − S a.p.f.u), and meionite composition (Me % = Ca/(Ca + Na + K)). Scapolite from the Munali hills area has a range of compositions from 31–46 EqAn % and 0.37 to 0.50 X_{Cl}. The scapolites are sodium-rich with 27–47 Me % compositions that fall within calcian marialite scapolite compositional group (15< %Me < 50, Fig. 18.4a; Rebbert and Rice 1997). The highest X_{Cl} contents are recorded in the Munali hills granite gneiss sample (MHG10). The compositions of scapolite plot close mid-way between the ideal end-members marialite and meionite, indicating compositions represented by solid solutions between these two end-members and a possible silvialite component $Ca_4[Al_6Si_6O_{24}]SO_4$ (Fig. 18.4b).

Mineral compositions of scapolite in metagabbros from the Copperbelt region (Tembo 1994) have a restricted range between 29 and 38 EqAn % comparable to the compositions of scapolite from the Munali hills area. Compared to those of the Munali hills area, the scapolites from the Copperbelt region have higher X_{Cl} values ranging from 0.61 to 0.84 and data points lie on and above the tie line between the marialite-meionite ideal end-member compositions.

18.5.2 *Plagioclase*

Plagioclase coexists with scapolite in all samples. The mineral formula was calculated on the basis of 32 oxygens. The plagioclase has a wide compositional range from 13 to 61 anorthite content with lower contents (An_{23-37}–oligoclase-andesine) in granite gneisses and higher contents (An_{36-61} –andesine-labradorite) mainly in the amphibolite samples (Fig. 18.4b, Table 18.3). The anorthite contents of plagioclase overlap with coexisting scapolite (Fig. 18.4b). Plagioclase from the non-scapolite bearing granites in the belt, e.g., Lusaka and Nchanga granites, are more sodium-rich with a compositional range of An_{4-16}.

18.5.3 *Amphiboles*

The amphiboles are Cl-bearing ferroan paragasite and magnesian hastingsite. Chlorine contents in pargasite are rather uniform at about 0.26 X_{Cl}, whereas that in hastingsite are slightly higher, ranging from 0.34 to 0.40 X_{Cl}. Similarly, X_{Mg} values range from 0.32 to 0.34 and 0.41 to 0.42 in pargasite and hastingsite, respectively.

18.5.4 *Biotite*

Biotite mineral formulae were calculated on the basis of 22 oxygens. Biotite was classified into two groups on the basis of Al(IV) a.p.f.u contents and X_{Fe} (Fig. 18.5a). Most biotites have high annite components, except some from one amphibolite sample (MHG5a), which are rich in the phlogopite component. On the X_{Cl}-X_{Mg} diagram (Fig. 18.5b), the biotite (annite-rich) from the metagabbros have the

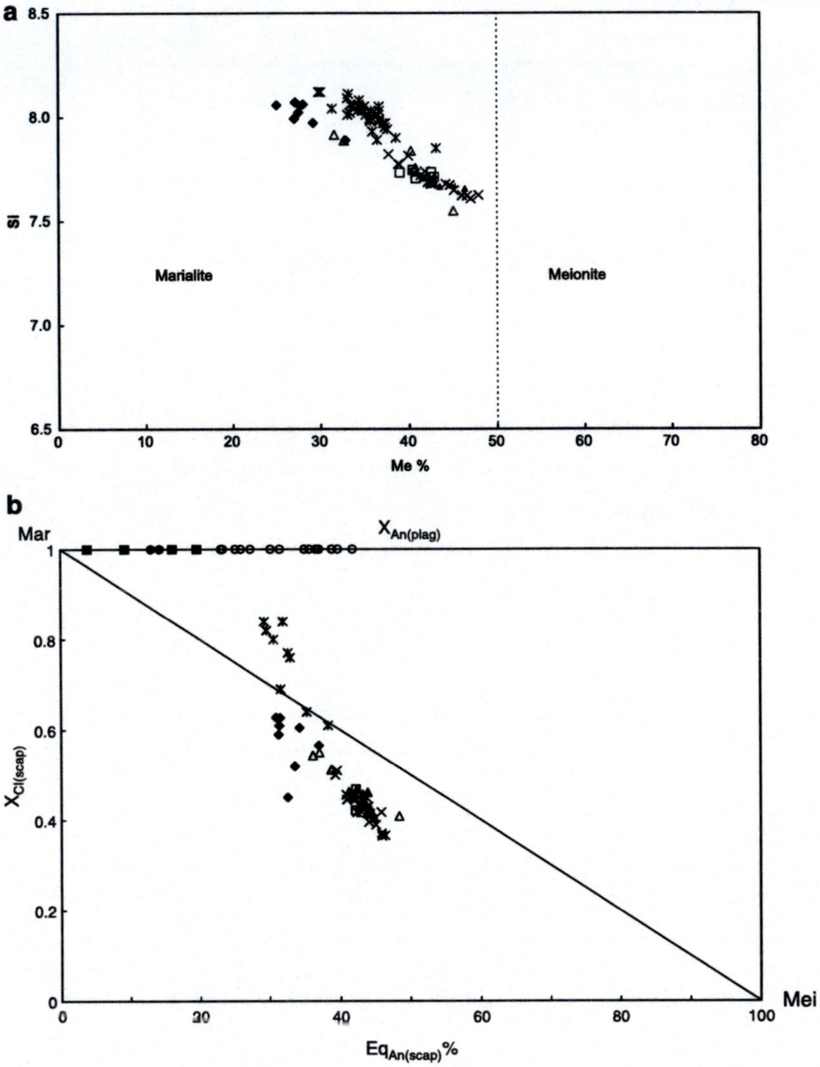

Fig. 18.4 **(a)** Classification of scapolites from the Lufilian–Zambezi belt on the basis of meionite component: Me% = 100 * (Ca/Ca + Na + K). All the scapolite data points plot in the calcian marialite group (After Rebbert and Rice 1997). **(b)** Plot of X_{Cl} versus $EqAn_{(scap)}$ and $X_{An(plag)}$. Scapolite data points from the Munali hills area plot between the three ideal end member tie lines, whereas some data points from the Copperbelt region lie above the ideal compositions. Symbols: ♦MHG10; □ MHG2; +- MHG5a; ▲ MHG5c; ×- MHG9; Δ- NDG; CB- metagabbros; ○ Plag coexisting with scapolite; ● LG-plag; ■- NG-plag

highest Cl contents ranging from 0.6 to 0.9 wt%, whereas biotites from amphibolites have the highest X_{Mg} values. Biotites from non-scapolite bearing rocks such as Lusaka granite also have compositions and Cl contents similar to those coexisting in scapolite-bearing rocks.

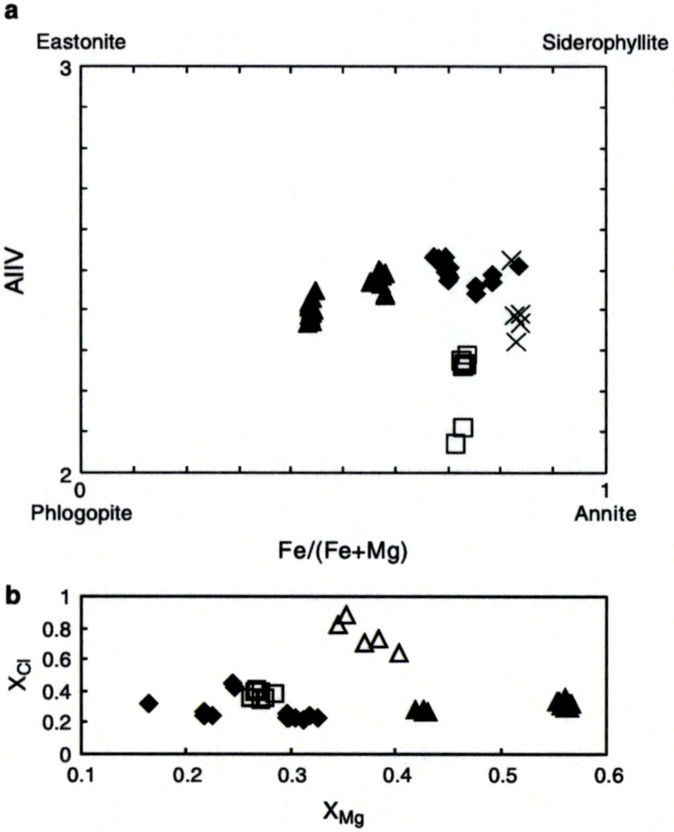

Fig. 18.5 (**a**) Classification of biotite from scapolite and non-scapolite bearing rocks from the Lufilian–Zambezi belt. Most biotites plot in the annite–siderophyllite field, except some from the amphibolitic dyke sample MHG5a. (**b**) Plot of X_{Cl} versus X_{Mg} of biotites from scapolite and non-scapolite bearing rocks from the Lufilian–Zambezi belt. Metagabbros have biotites with the highest Cl contents, whereas some from the amphibolitic dykes have high Mg (the phlogopite-rich biotites in (**a**)). Symbols: ♦ Munali hills granite gneiss; ▲ Amphibolite; Δ Metagabbro; □ Lusaka granite

18.6 Discussion

18.6.1 Scapolite and Fluid Compositions

The NaCl content of a fluid can be estimated from scapolite coexisting with calcite according to the reaction (Ellis 1978):

$$(NaCl)_{scp} + (CaCO_3)_{cc} = (CaCO_3)_{scp} + (NaCl)_{fluid}$$

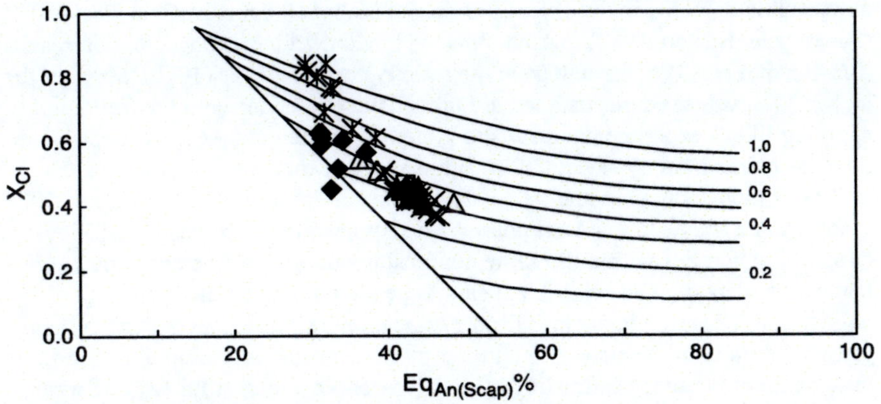

Fig. 18.6 Plot of $X_{Cl,Scap}$-EqAn of scapolites from the Lufilian–Zambezi belt, superimposed on the NaCl salinity curves of Ellis (1978). Data points of scapolite from the Munali hills area indicate salinities ranging from 0.2–0.5 mol, whereas those from the Copperbelt region are unrealistically high. Same symbols as in Fig. 18.4

Ellis (1978) investigated this fluid exchange equilibrium at 750°C and 4 kbar, consistent with metamorphic conditions in the granulite facies. Mora and Valley (1989) outlined the limitations underlying Ellis's expression that are mainly because of thermodynamic mixing properties and inappropriate P–T conditions done by Ellis (1978). In the absence of thermodynamic data different from that of Ellis (1978), we used X_{NaCl} curves of Ellis (1978) to estimate, as a first approximation, the NaCl salinity of coexisting fluids in the belt. The X_{NaCl} in scapolite from the Munali hills area range from 0.2 to 0.5 mol, whereas those from the Copperbelt region (Tembo 1994) are higher and vary from 0.5 to 1 (Fig. 18.6). Salinities in this range are known from areas where evaporite sequences occur (Rich 1979; Roeder 1984; Behr and Horn 1982).

Fluid inclusion studies from carbonate hosted Kabwe Pb–Zn deposit in central Zambia indicate H_2O–NaCl–$CaCl_2$ and H_2O–$MgCl_2$–$CaCl_2$ fluid compositions; salinities ranging from 11 to 31 wt% NaCl equivalent (EQ) and homogenisation temperatures varying from 257°C to 305°C (Kamona 1993). In the Chisamba area, north of Lusaka, H_2O-NaCl fluid inclusions in quartz from vein quartz yielded fluid salinities ranging from 34 to 44 wt% NaCl EQ, and temperatures between 250°C and 350°C (Katongo 1999). These independent fluid inclusion studies indicate relatively high NaCl salinities at conditions consistent with greenschist facies metamorphism, but the host rocks are not scapolite bearing.

Scapolites from both the Munali hills area and Copperbelt region have strikingly similar calcian-marialitic compositions (23–48 Me%), but this may well just be a coincidence, as the two areas are several hundreds of kilometres apart. However, similar compositions were also found with the crude analytical techniques mentioned earlier. However, Munyanyiwa (1990) reported scapolite of mizzonite composition from southern Zambia without providing mineral compositional data.

Experiments by Vanko and Bishop (1982) showed that marialitic scapolite could be formed in fluids with little or no calcite. A minimum salinity for the production of marialitic scapolite in the absence of calcite is about 0.5 mol NaCl at 700–750°C (Vanko and Bishop 1982), which falls within the salinity range of our results (0.2–0.5 mol NaCl), estimated from the salinity curves of Ellis (1978). Some of the higher X_{NaCl} values in our data set are unrealistic and underscore the limitation of applying Ellis's experimental results to systems of different thermodynamic conditions, as pointed-out by several other authors (e.g., Mora and Valley 1989).

The composition of scapolite in the Munali hills area is independent of the lithology and coexisting mineral phases. The composition of coexisting plagioclase ($An_{23–48}$), which lies within the same compositional range as the anorthite equivalent of the scapolite ($EqAn_{29–48}$), appears to be an important factor in favoring scapolitization, because plagioclase from non-scapolite bearing Lusaka and Nchanga granites is more albite-rich ($An_{5–20}$) than that coexisting with scapolite. Biotite is most commonly used in estimating halogen contents of fluids (Munoz and Swenson 1981) but we found no direct correlation between scapolite and biotite in the Munali hills granite gneiss, especially that Cl-bearing biotite also occurs in non-scapolite bearing Lusaka and Nchanga granites. However, phlogopite-rich biotite, which occurs in equilibrium with scapolite in veinlets in some amphibolites, was evidently formed under similar metamorphic conditions with scapolite. Similarly, the metamorphic Cl-bearing amphiboles also occur in equilibrium with scapolite and probably formed during the same metamorphic event.

18.6.2 Mineral Assemblages and Metamorphism in the Lufilian–Zambezi Belt

In the Lufilian–Zambezi belt, scapolites occur in mineral assemblages that are indicative of amphibolite facies metamorphism. In southern Zambia, amphibolite facies metamorphism is reflected by the scapolite–tremolite–diopside assemblage in calc-silicates, and scapolite–garnet–hornblende assemblage in garnet amphibolites (Simpson et al. 1965; Munyanyiwa 1990; Katongo 1999). However, in the Copperbelt region, Tembo (1994) reported greenschist facies assemblages comprising scapolite–chlorite–epidote–sericite in metagabbros. However, elsewhere around the world, scapolites occur in association with mineral assemblages that are stable at metamorphic conditions ranging from amphibolite to granulite facies (e.g., Vanko and Bishop 1982; Markl and Piazolo 1998; Faryad 2002). In our sample suite, scapolite formed mainly at the expense of plagioclase, especially in the Munali hills granite gneiss, however the scapolite–plagioclase assemblage is stable over a wide range of metamorphic conditions. Breakdown of plagioclase to sericite is indicative of greenschist facies metamorphism, which post-dated scapolitization. In amphibolites, however, scapolite is in equilibrium with both Cl-bearing biotite and amphiboles, suggesting that the assemblage formed during amphibolite facies metamorphism.

The symplectite-like intergrowth of scapolite with quartz (Fig. 18.2b) observed in the Munali hills granite gneiss is not common in most scapolite bearing rocks. Because scapolite invariably replaced plagioclase in most rocks, this texture could have formed in two ways. Myrmekite-like intergrowths between scapolite and quartz adjacent to K-feldspar are common in most samples, therefore the first and most likely possibility is that scapolite replaced plagioclase, which had already formed myrmekite intergrowth with quartz. The second possibility is that scapolite replaced calcium-rich plagioclase. Since scapolite has less SiO_2 in the crystal lattice than calcium-rich plagioclase, the excess SiO_2 was exsolved after replacement and formed worm-like blebs as inclusions in scapolite, resulting in the symplectite-like intergrowth. A replacement of K-feldspar by scapolite was not found in the investigated samples.

18.6.3 Origin of Scapolite and NaCl-Rich Fluids

The distribution of scapolite in the field is often the best evidence for the origin of this mineral. Several modes of field occurrence that indicate the possible origin of scapolite are proposed. Scapolite distributed as fine-scale interbedded units of scapolite-bearing beds in meta-evaporites suggests strongly for an in-situ salt precursor (e.g., Hietanen 1967; Mora and Valley 1989; Markl and Piazolo 1998). In southern Zambia, most scapolite-bearing marble and calc-silicate rocks probably represent in-situ former evaporitic sequences because metamorphism in most cases was essentially isochemical (Munyanyiwa 1990; Munyanyiwa and Hanson 1988). Any other mode of occurrence is attributed to metasomatism (e.g., Vanko and Bishop 1982). Metasomatic introductions of NaCl fluids are associated with scapolite occurrences in meta-igneous rocks, which otherwise do not contain scapolite. In the Copperbelt region, evidence of former evaporites provides unambiguous precursor for the external source of NaCl-rich fluids for the origin of the scapolite. In the Munali hills and surrounding areas, there is no hard evidence indicating the occurrence of former evaporites in the area. However, Cl-rich scapolite in the Munali hills area, coupled high NaCl salinities of coexisting fluids determined in this study, are indicative of an external source of NaCl-rich fluids from former evaporite horizons in the supracrustal rocks that correlate well with the Katangan metasediments in the Copperbelt region. Thus, the most credible source for NaCl fluids in the Lufilian–Zambezi belt, as previously suggested by many authors, is the dissolution of evaporite horizons in metasedimentary sequences during regional amphibolite facies metamorphism that channeled through fault and shear zones in the area. These investigations show that the study area is of great importance to understand the relation between evaporites and scapolites.

18.7 Summary and Conclusions

We conducted fieldwork in the Munali hills area, in the southern part of the Lufilian–Zambezi belt, and performed petrographic and mineral analyses of scapolite, amphibole, biotite and plagioclase from the Munali hills granite gneiss, amphibolites and metagabbros, in an effort to evaluate the nature of fluids, their source and attending metamorphic conditions. Results of this study show that:

1. The scapolites from the Munali hills area and from the Copperbelt region have strikingly similar Ca-rich marialite compositions (28–46 Me%), indicating that the mineral formed from hydrothermal fluids from precursor rocks of similar composition.
2. The scapolite in the Munali hills area occurs as pervasive replacement of plagioclase (21–48 X_{An}%) by metasomatic processes and occurs in stable equilibrium with mineral assemblages that are indicative of amphibolite facies metamorphism. The abundance of scapolite throughout the belt suggests that the grade of metamorphism in most parts of the belt was at least amphibolite facies, at which scapolite is more stable than plagioclase in the presence of NaCl-rich fluids.
3. The source of NaCl-rich fluids for the formation of scapolite in the Copperbelt region is evidently from evaporite horizons in the Katangan metasedimentary succession. In the southern part of the belt, including the Munali hills area, there is no direct evidence of the former existence of the evaporite beds. The prevalence of Cl-rich scapolite in the Munali hills area and high NaCl salinities of coexisting fluids, indicate former extensive evaporite successions in the supracrustal rocks that correlate well with the Katangan metasediments in the Copperbelt region.

Acknowledgements We thank the Austrian Academic Exchange Service (ÖAD) for a Ph.D. stipend to C. Katongo. Analytical expenses were supported by the Austrian science Foundation FWF. The co-authors would like to dedicate this paper to the memory of Crispin Katongo, who passed away after a short illness on July 18, 2004, at the age of 38, shortly after completing his PhD at the University of Vienna. We are grateful to three anonymous reviewers, and Prof. Jyotisankar Ray, whose extensive and constructive comments helped to improve this manuscript.

References

Barton CM, Carney JN, Crow MJ, Dunkley PN, Simango S (1991) The geology of the country around Rushinga and Nyamapanda. Bull Geol Surv Zimbabwe 92:220

Bayliss P (1987) Mineral nomenclature: scapolite. Mineralog Mag 41:176

Behr HJ, Horn EE (1982) Fluid inclusion systems in metaplaya deposits and their relationship to mineralization and tectonics. Chem Geol 37:173–189

Binda PL (1994) Stratigraphy of the Zambian Copperbelt orebodies. J Afr Earth Sci 19:251–264

Broderick TJ (1981) The Zambezi metamorphic belt in Zimbabwe. In: Hunter DR (ed) Precambrian of the southern hemisphere. Elsevier, Amsterdam, The Netherlands, pp 739–743

Burnard P, Sweeney MA, Vaughan DJ, Spiro B, Thirwall MF (1993) Sulfur and lead isotope constraints on the genesis of a southern Zambian massive sulphide deposit. Econ Geol 88:418–436

Cahen L, Snelling NL, Delhal J, Vail JR, Bonhomme M, Ledent D (1984) Geochronology and evolution of Africa. Clarendon, Oxford, p 512

Cailteux J, Kampunzu AB (1995) The Katangan tectonic breccias in the Shaba Province (Zaire) and their genetic significance. In: Wendorff M, Tack L (eds) Late Proterozoic belts in Central and Southern Africa. Annales Sciences Géologigues Musée Royal de I´Afrique Centrale 101, Tervuren, pp 49–62

Cailteux J, Binda PL, Katekesha WM, Kampunzu AB, Intiomale MM, Kapenda D, Kaunda C, Ngongo K, Tshiauka T, Wendorff M (1994) Lithostratigraphical correlation of the neoproterozoic Roan Supergroup from Shaba (Zaire) and Zambia, in central African copper-cobalt metallogenic province. J Afr Earth Sci 19:265–278

Cosi M, De Bonis A, Gosso G, Hunziker J, Martinotti G, Moratto S, Robber JP, Ruhlman F (1992) Late Proterozoic thrust tectonics, high pressure metamorphism and uranium mineralisation in the Domes Area, Lufilian Arc, northwestern Zambia. Precambrian Res 58:215–240

Coward MP, Daly MC (1984) Crustal lineaments and shear zones in Africa: their relationship to plate movements. Precambrian Res 24:27–45

Daly MC (1986) Crustal shear zones and thrust belts: their geometry and continuity in central Africa. Philos Trans R Soc Lond A317:111–128

De Swardt AML, Drysdall AR (1964) Precambrian geology and structure in central northern Rhodesia. Mem Geol Surv N Rhodesia 2:82

De Waele B, Fitzsimons ICW, Wingate MTD, Tembo F, Mapani B, Belousova EA (2009) The geochronological framework of the Irumide belt: a prolonged crustal history along the margin of the Bangweulu craton. Am J Sci 309:132–187

Dirks PHGM, Sithole TA (1999) Eclogites in the Makuti gneisses of Zimbabwe: implications for the tectonic evolution of the Zambezi belt in southern Africa. J Metamorph Geol 17:593–612

Drysdall AR, Stillman CJ (1966) Scapolite from the Katanga carbonite rocks of the Lusaka district. Records Geol Soc Zambia 10:20–24

Ellis DE (1978) Stability and phase equilibria of chloride-and carbonate bearing scapolites at 750 °C and 4000 bar. Geochim Cosmochim Acta 42:1271–1281

Evans BW, Shaw DM, Haughton DR (1969) Scapolite stoichiometry. Contrib Mineralog Petrol 24:293–305

Faryad SW (2002) Metamorphic conditions and fluid compositions of scapolite bearing rocks from the Lapis Lazuli deposit at Sare Sang, Afghanistan. J Petrol 13:725–747

Hanson RE, Wilson TJ, Brueckner HK, Onstott TC, Wardlaw MS, Johns CC, Hardcastle KC (1988) Reconnaissance geochronology, tectono-thermal evolution and regional significance of the Middle Proterozoic Choma-Kalomo block, southern Zambia. Precambrian Res 42:39–61

Hanson RE, Wardlaw MS, Wilson TJ, Mwale G (1993) U-Pb zircon ages from the Hook granite massif and Mwembeshi dislocation: constraints on Pan-African deformation and transcurrent shearing in central Zambia. Precambrian Res 63:189–209

Hanson RE, Wilson TJ, Munyanyiwa H (1994) Geologic evolution of the Neoproterozoic Zambezi orogenic belt in Zambia. J Afr Earth Sci 18:135–150

Hargrove US, Hanson RE, Martin MW, Blenkinsop TG, Bowring SA, Walker N, Munyanyiwa H (2003) Tectonic evolution of the Zambezi belt: geochronological, structural and petrographical constraints from northern Zimbabwe. Precambrian Res 123:159–186

Harley SL, Fitzsimons ICW, Buick IS (1994) Reactions and textures in wollastonite-scapolite granulites and their significance for pressure-temperature-fluid histories of high-grade terranes. Precambrian Res 66:309–323

Hietanen AH (1967) Scapolite in the Belt series in the St Joe-Clear water Region, Idaho. Geol Soc Am Spec Pap 86:56

Jackson MAP, Warin ON, Woad GM, Hudec MR (2003) Neoproterozoic allochthonous salt tectonic during the Lufilian orogeny in the Katangan Copperbelt, central Africa. Geol Soc Am Bull 115:314–330

John T, Schenk V, Haase K, Scherer E, Tembo F (2003) Evidence of a Neoproterozoic ocean in south-central Africa from mid-ocean-ridge-type geochemical signatures and pressure-temperature estimates of Zambian eclogites. Geology 31:243–246

Johnson SP, Oliver GJH (2000) Mesoproterozoic oceanic subduction, Island-arc formation and the initiation of back-arc spreading in the Kibaran belt of central southern Africa: evidence from the Ophiolite Terrane, Chewore Inliers, northern Zimbabwe. Precambrian Res 103:125–146

Johnson SP, Fitzsimons ICW, Wingate MTD, Tembo F, Mapani B, Belousova EA (2007) The geochronological framework of the Irumide belt: a prolonged crustal history along the margin of the Bangweulu craton. Am J Sci 309:132–187

Kamona F (1993) The carbonate-hosted Kabwe Pb-Zn deposit, central Zambia. Ph.D thesis, Technical University of Archean, 207 pp

Kampunzu AB, Cailteux J (1999) Tectonic evolution of the Lufilian arc central Africa Copperbelt during Neoproterozoic Pan-African orogenesis. Gondwana Res 2:135–150

Katongo C (1999) Structural and petrographic fabrics in the Chisamba and Kabwe areas, central Zambia: implications for the Late Proterozoic Mwembeshi Dislocation Zone. M.Sc. thesis (unpublished), University of Zambia, 209 pp

Katongo C, Koller F, Kloetzli U, Koeberl C, Tembo F, De Waele B (2004) Petrography, geochemistry and geochronology of key granitoid rocks in the Neoproterozoic-Paleozoic Fufilian-Zambesi belt, Zambia: implications for tectonic setting and regional correlation. J Afr Earth Sci 40:219–244

Key RM, Liyungu AK, Njamu FM, Somwe V, Banda J, Mosley PN, Armstrong RA (2001) The western arm of the Lufilian arc in NW Zambia and its potential for copper mineralisation. J Afr Earth Sci 33:503–528

Mallick DIJ (1966) The stratigraphy and structure of the Mpande Dome, southern Zambia. Trans Geol Soc SA 69:211–230

Markl G, Piazolo S (1998) Halogen-bearing minerals in syenites and high-grade marbles of Dronning Maud Land, Anrtartica: monitors of fluid compositional changes during late-magmatic fluid-rock interaction processes. Contrib Mineralog Petrol 132:246–268

Mendelsohn F (1961) The geology of the Northern Rhodesia copper belt. McDonald, London, p 523

Moecher DP, Essene EJ (1991) Calculations of CO_2 activities using scapolite equilibria: constraints on the presence and composition of fluids phase during high grade metamorphism. Contrib Mineralog Petrol 108:219–240

Moine B, Guilloux L, Audeoud D (1986) Major element geochemistry of host rocks in some sediment-hosted copper deposits. In: Friedrich GH (ed) Geology and metallogeny of copper deposits. Springer, Berlin, pp 443–460

Mora CI, Valley JW (1989) Halogen-rich scapolite and biotite: implications for metamorphic fluid-rock interaction. Am Mineralog 74:721–737

Munoz JL, Swenson A (1981) Chloride-hydroxyl exchange in biotite and estimation of relative HCL/HF activities in hydrothermal fluids. Econ Geol 76:2212–2221

Munyanyiwa H (1985) The geochemistry and metamorphism of calc-silicate rocks, marbles and amphibolites in a portion of the Zambezi belt, southern Zambia. M.Sc. thesis (unpublished) University of Zambia, 193 pp

Munyanyiwa H (1990) Mineral assemblages in calc-silicates and marbles in Zambezi mobile belt: their implications on mineral-forming reactions during metamorphism. J Afr Earth Sci 10:693–700

Munyanyiwa H, Hanson RE (1988) Geochemistry of marbles and calc-silicate rocks in the Pan-African Zambezi belt, Zambia. Precambrian Res 38:177–200

Munyanyiwa H, Hanson RE, Blenkinsop TG, Treloar PJ (1997) Geochemistry of amphibolites and quartzofeldspathic gneisses in the Pan-African Zambezi belt, northwest of Zimbabwe: evidence for bimodal magmatism in a continental rift setting. Precambrian Res 81:179–196

Orville PM (1975) Stability of scapolite in the system Ab-An-NaCl-$CaCO_3$ at 4 kbar and 750°C. Geochim Cosmochim Acta 39:1091–1105

Oterdoom WH, Wenk HR (1983) Ordering and composition of scapolite; field observations and structural interpretations. Contrib Mineralog Petrol 83:330–341

Porada H (1989) Pan-African rifting and orogenesis in southern to equatorial Africa and eastern Brazil. Precambrian Res 44:103–136

Porada H, Berr HJ (1988) Setting and sedimentary facies of late Proterozoic alkali lake (playa) deposits in the southern Damara belt of Namibia. Sed Geol 58:171–194

Porada H, Berhorst V (2000) Towards a new understanding of the Neoproterozoic-Early Palaeozoic Lufilian and northern Zambezi belts in Zambia and the Democratic Republic of Congo. J Afr Earth Sci 30:727–771

Pouchou J, Pichoir F (1991) Quantitative analysis of homogeneous or stratified microvolumes applying the model "PAP". In: Heinrich KFJ, Newbury DE (eds) Electron probe quantitation. Plenum, New York, pp 31–75

Rebbert CR, Rice JM (1997) Scapolite-plagioclase exchange: Cl-CO_3 scapolite solution chemistry and implications for peristerite plagioclase. Geochim Cosmochim Acta 61:555–567

Rich RA (1979) Fluid inclusion evidence of Silurian evaporites in southern Vermont. Geol Soc Am Bull 90:1628–1643

Roeder E (1984) Low-medium grade metamorphic environments. Mineralog Soc Am Rev Mineral 12:413–472

Simpson JG, Stillman CJ (1963) Metamorphism and reactions phenomena in gabbros of the Mwembeshi and Lusaka areas. Records Geol Surv Zambia 9:10–14

Simpson JG, Drysdall AR, Lambert HH (1965) The geology and groundwater resources of the Lusaka area: explanation of degree sheet 1528, NW quarter. Rep Geol Surv N Rhodesia 16:59

Smith AG (1963) The geology of the country around Mazabuka and Kafue: explanation of degree sheets 1527, SE quarter and 1528 SW quarter. Rep Geol Surv N Rhodesia 2:32

Teertstra DK, Sherriff BL (1997) Substitution mechanisms, composition trends and the end-member formulae of scapolite. Chem Geol 136:233–260

Tembo F (1994) The geology, geochemistry and tectonic significance of metagabbroic rocks in the Lufilian Arc of Zambia. Ph.D thesis, University of Göttingen, Cuvillier Verlag, Göttingen, 118 pp

Tembo F, Kampunzu AB, Porada H (1999) Tholeiitic magmatism associated with continental rifting in the Lufilian belt of Zambia. J Afr Earth Sci 28:403–425

Unrug R (1983) The Lufilian arc: a microplate in the Pan-African collision zone of the Congo and Kalahari cratons. Precambrian Res 21:181–196

Unrug R (1996) The assembly of Gondwanaland. Episodes 19:11–20

Vanko DA, Bishop FC (1982) Occurrence and origin of marialitic scapolite in Humboldt Lapolith, N.W Nevada. Contrib Mineralog Petrol 81:277–289

Vinyu ML, Hanson RE, Martin MW, Bowring SA, Jelsma HA, Krol MA, Dirks PHGM (1999) U-Pb and $^{40}Ar/^{39}Ar$ geochronological constraints on tectonic evolution of the eastern part of the Zambezi orogenic belt, northeast Zimbabwe. Precambrian Res 98:67–82

Vrána S (1978) Metamorphic patterns in Zambia and their bearing on problems of Zambian tectonic history. Precambrian Res 5:127–130

Vrána S, Barr MWC (1972) Talc-Kyanite-quartz schists and other high-pressure assemblages from Zambia. Mineralog Mag 38:837–846

Vrána S, Prasad R, Fediuková E (1975) Metamorphic kyanite eclogites in the Lufilian arc of Zambia. Contrib Mineralog Petrol 51:1–22

Wendorff M (2000) Genetic aspects of the Katangan megabreccias: Neoproterozoic of central Africa. J Afr Earth Sci 30:703–715

Weil AB, Van der Voo R, MacNiocall C, Meert JG (1998) The Proterozoic supercontinent Rodinia: paleomagnetically derived reconstructions for 1100–800 Ma. Earth Planet Sci Lett 154:13–24

Wilson T, Hanson RE, Wardlaw MS (1993) Late proterozoic evolution of the Zambezi belt, Zambia: implications for the regional Pan-African and shear displacements in Gondwana. In: Findlay RH, Unrug R, Banks MR, Veevers JJ (eds) Gondwana Eight: assembly, evolution and dispersal. Balkema, Rotterdam, The Netherlands, pp 69–82

Index

J. Ray et al. (eds.), *Topics in Igneous Petrology*, DOI 10.1007/978-90-481-9600-5,

N